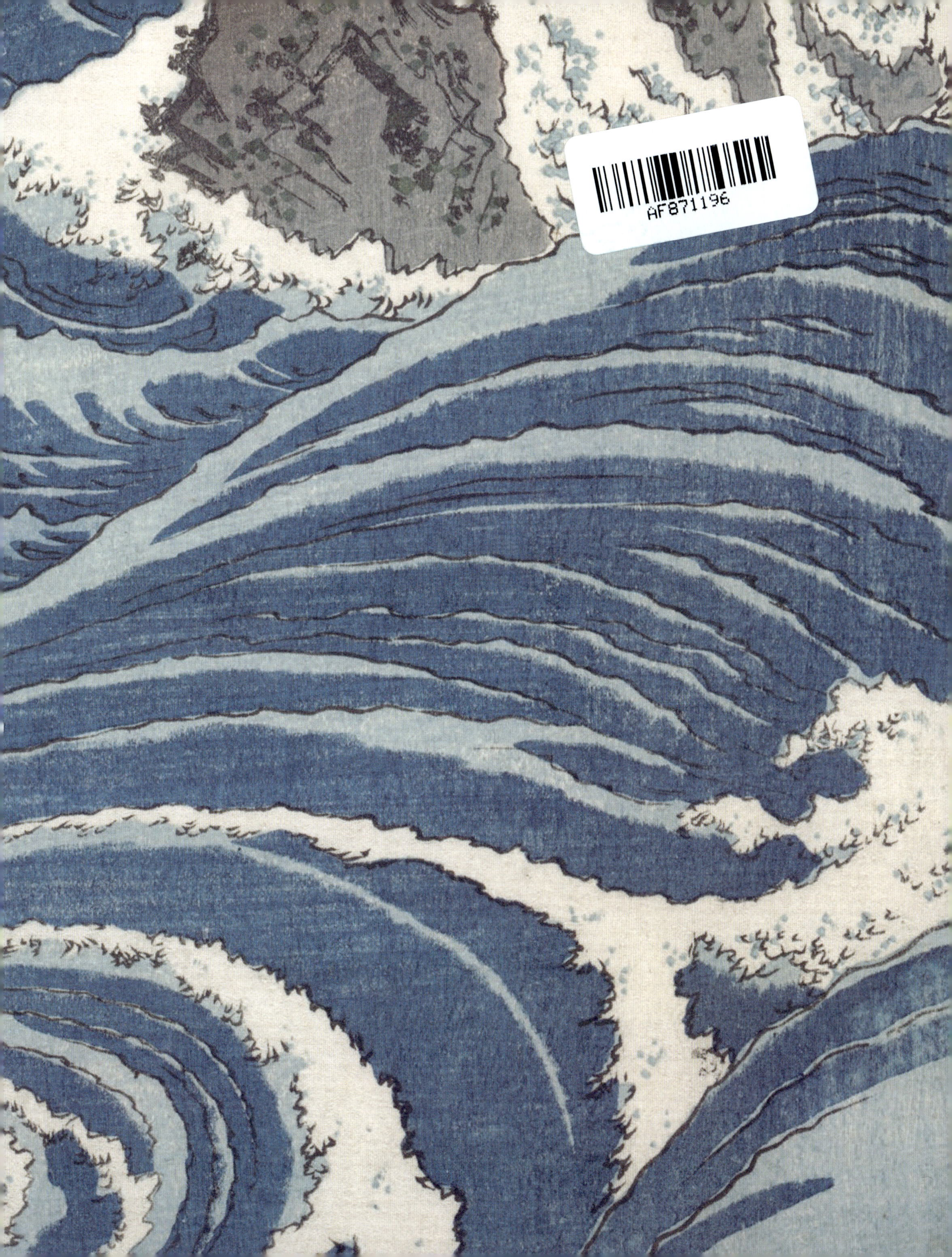

ARK

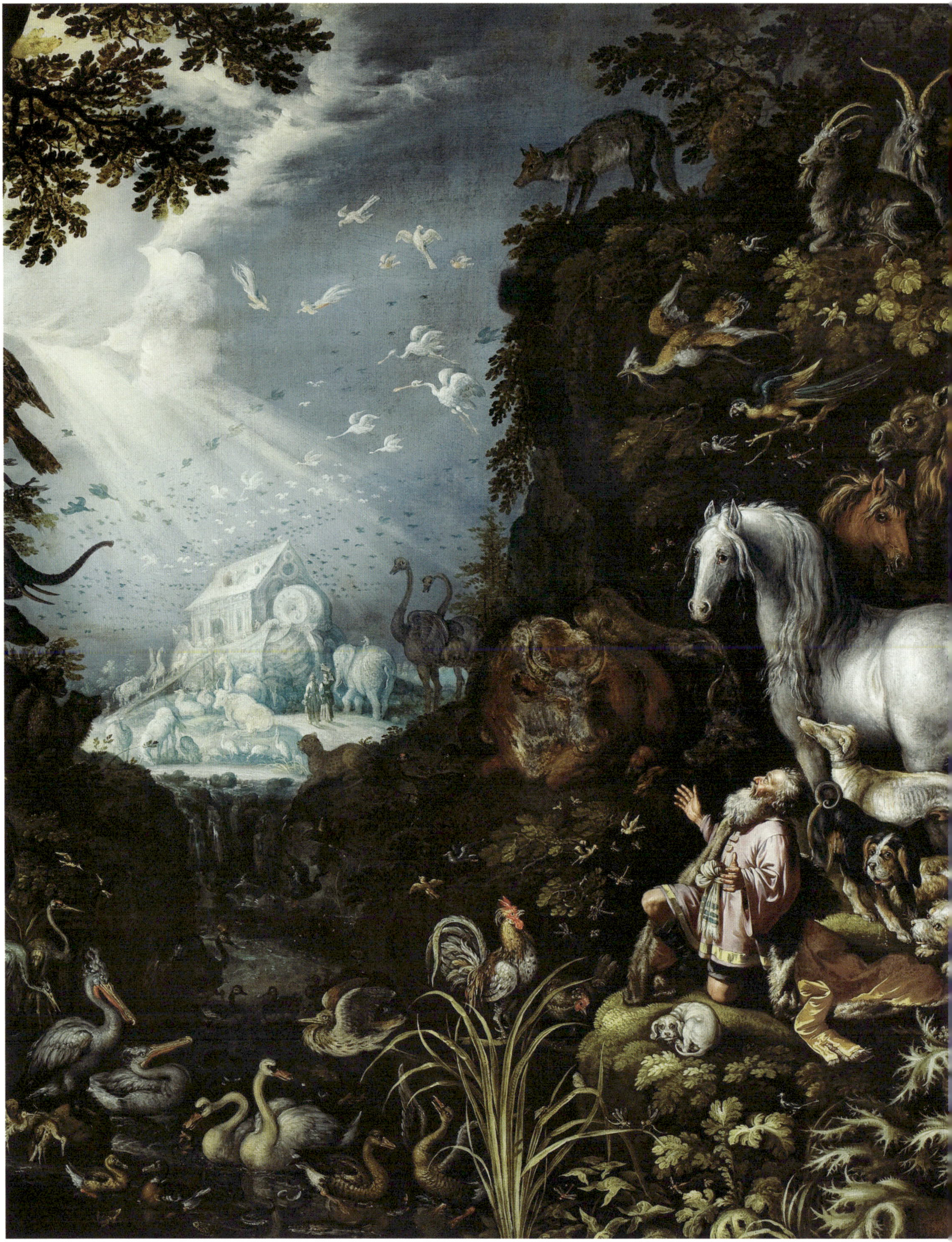

A visual treasury of Earth's living wonders

Stephen Ellcock

ARK

T&H

Foreword Robert Macfarlane
Introductory essay Sarah Perry

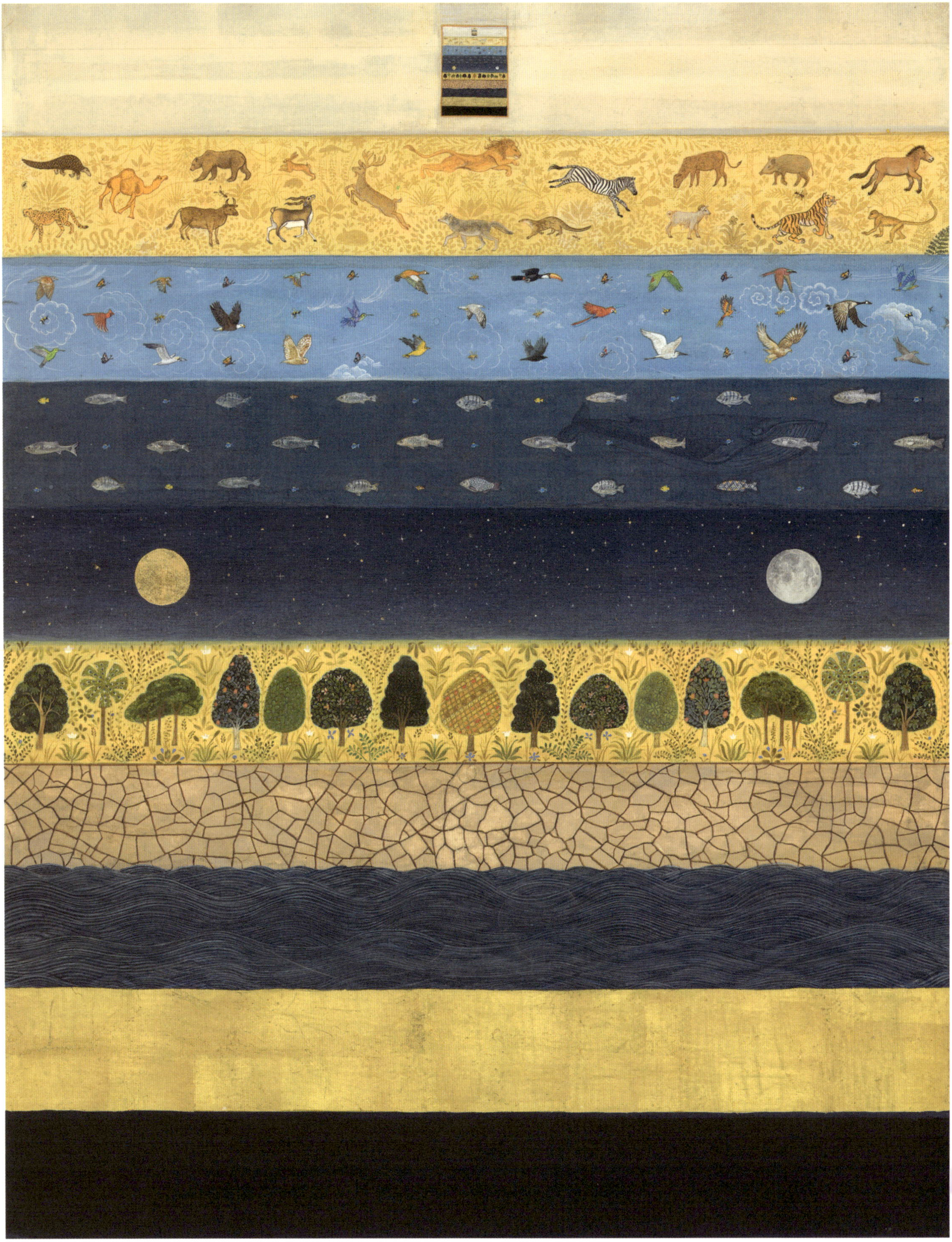

CONTENTS

FOREWORD

Robert Macfarlane

The English word 'ark' comes from the Latin *arca*, meaning a chest or box. An ark's first task is to contain. This function is what joins the two great arks of the Christian Bible: the Ark of the Covenant and Noah's Ark. The first holds the Ten Commandments and the second bears the blueprint for a repopulation of earth in the aftermath of apocalypse. It is an oddity of etymology that this strange, awkward little bark of a word should have come to mean a vessel of immense and vital capaciousness.

The book you hold in your hands is an ark. Like Noah's Ark – or Utnapishtim's before him in the *Epic of Gilgamesh*, which contains the earliest known written flood story – Stephen Ellcock has created a life-preserver, which seeks to gather and hold a representative fraction of what we inadequately call 'nature'. The wager of all such arks is that what they contain will allow the regeneration of what is to be lost. The hope of all such arks is that such loss never comes to pass. Noah and Utnapishtim were both warned by their god or gods that catastrophe was coming, and so they set to work with timber and pitch. Science is now our god, warning us in countless ways of the ecological calamity that is engulfing us – and so we set to work with paint, ink, photography and words to save life in its fantastical profusion, in its absurd and glorious diversity.

A mass extinction event occurs when existing species disappear far faster than new species are created; specifically, when more than 75% of the world's species are lost within a period of 2.8 million years or less. The earth is known to have gone through five historical mass extinction events, the most famous of which is the Permian-Triassic extinction event – also known as 'The Great Dying' – when around 95% of all species disappeared, probably due to the impact of a vast asteroid. The consensus of biologists is that we are currently living through the sixth mass extinction event, with the background rate of extinction now between 100 and 1,000 times higher than before humans existed.[1] We have now inaugurated our own 'Great Dying'. We are the asteroid.

'I don't know what "human nature" is,' wrote Ursula K. Le Guin (1929–2018) in her 1972 novella *The Word for World is Forest*, 'maybe leaving descriptions of what we wipe out is part of human nature…?'[2] The pages of *Ark* are testimony to that descriptive impulse. They demonstrate not only the staggering multiplicity of more-than-human life, but also the thrilling intricacies of humanity's own compulsive drive to represent that life. A second-order morphology is mesmerizingly in evidence here, as art itself mutates and shifts in its restless, futile, beautiful attempts to evoke the living world. Oak gall ink, gold leaf, silver gelatin plates, crushed cochineal, dyed threads, glass, clay, porcelain, carved wood: each new page showcases different modes and mediums of representation, from the iron oxide pigments used to paint bestiaries onto the limestone walls of Chauvet some 30,000 years ago, through to the latest microscopic imagery disclosing the wonder-worlds of fungi, algae, plankton and microbes. Cumulatively, a sense emerges of attention exercised as a kind of devotion, and of art as a form of worship directed towards what the great biologist Stephen Jay Gould (1941–2002) simply and memorably called 'wonderful life'.[3]

If there is a unifying motif to the diversity collected here, it is surely that of entanglement. Fungal mycelia stretch towards one another and fuse into ghostly labyrinths of hyphae. Tree roots tangle and weave. Moth-wings precisely mimic the species-specific patterns of tree bark: snow-white and coal-black wings for the moths who partner with silver birch; ash-grey and russet for those who live on oaks. In Kobayashi Kiyochika's (1847–1915) exquisite *c.* 1880 woodblock print, *Fireflies at Ochanomizu*, the lights of a pleasure boat floating on a night-time river are echoed by silver moon-glimmers upon the inky water, and by the star-chains of fireflies that pulse and flicker in the air around the boat. Here, luminosities which are variously human, lunar and creaturely in source echo one another.

→
Agusan del Sur is the largest inland wetland in the Philippines, and in 2024 fires raged through the area, caused by the highest-ever recorded temperatures in the region. The scene was captured by Gabriel Mejia, photographer and conservationist, in his image *Burning Lands, Talacogon, Agusan del Sur, Philippines*.

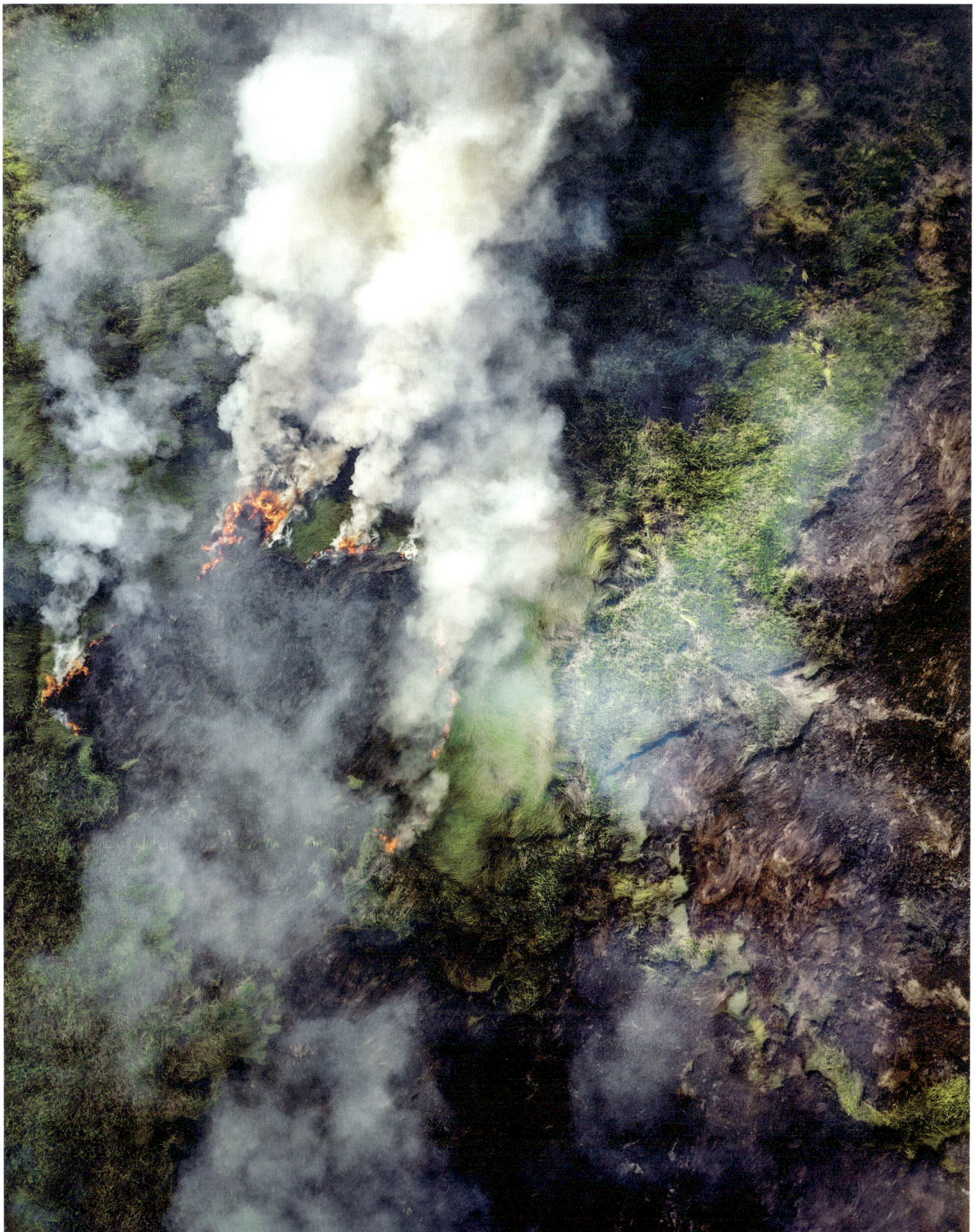

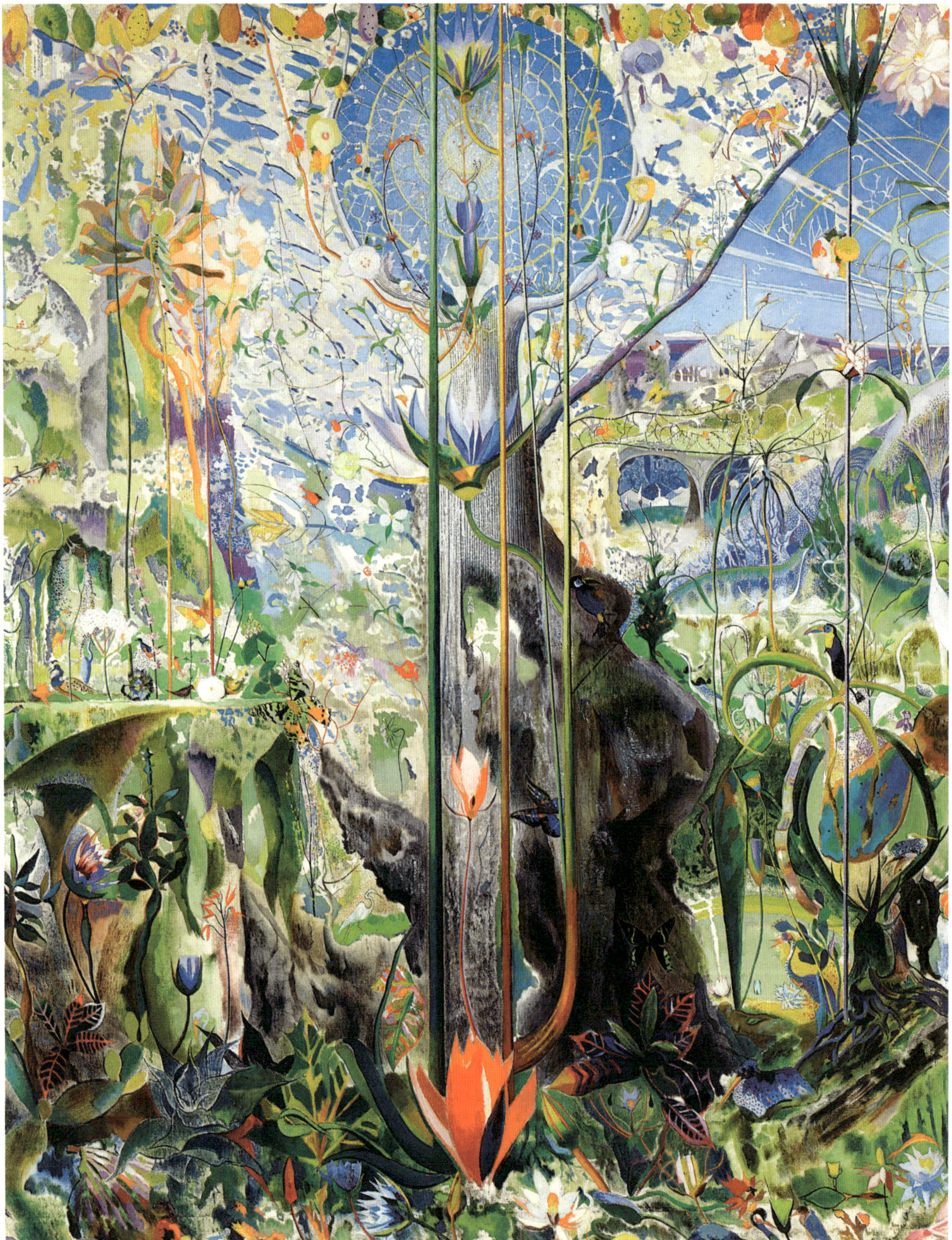

More broadly, although *Ark* is divided into seven chapters across five kingdoms, connections and rhymes proliferate between its parts with such intensity that they come to feel more like 'kindoms' than 'kingdoms'. In this respect, we might say that the ethos of the book's making is actively ecological – for the first rule of ecology is that 'everything is connected to everything else.' Or as the early environmentalist John Muir (1838–1914) put it in an 1869 journal entry: 'When we try to pick out anything by itself, we find that it is bound fast by a thousand invisible cords that cannot be broken to everything in the universe.'[4]

For more than 2,000 years, however, a very different model of life's organization has dominated the Western philosophical tradition – and, indeed, Western politics as it relates to the natural world. The 'great chain of being' or *scala naturae* ('ladder of nature') was set forth by Aristotle (384–322 BCE) in *Historia Animalium*,[5] his pioneering work of zoology, and enthusiastically installed within Christian doctrine from the 12th century onwards. According to the great chain, matter and life exist in a hierarchical structure, with God at its summit and then – in descending order of *anima* or 'life' – angels, humans, animals, plants and, at last, what Isaac Newton (1643–1727) later called 'inanimate brute matter'[6] (rocks, earth, water) at its base. The great chain was also what we would now refer to as a power structure, in which each level was shown to control the levels beneath it. As the King James translation of the book of Genesis put it, man shall 'have dominion over ... every living thing that moveth upon the earth.'[7] The concept of the great chain also influenced the basis of modern scientific taxonomy, Carl Linnaeus's (1707–1778) *Systema Naturae* (1735),[8] which organized the physical components of the world into three kingdoms – animal, vegetable and mineral – and initiated the Enlightenment project of life's description and systematization.

The models we use to imagine the world's organization matter deeply. We might even say that the difference between understanding ourselves to be entangled in an interdependent web of relations and believing ourselves to be standing aloof and supreme atop a pyramid of being, is the difference between survival and extinction. Life on our planet is a shimmering and finely woven fabric which has been millions of years in the making. As threads of this gossamer fray out, so others weave in to take their place. Existence is always *convivium* – a 'living-with'. We humans are the carriers and beneficiaries of this conviviality. Bacteria seethe on our skin and bloom in our guts. Our evolution has been shaped in part by our role as hosts to viruses, themselves ancient genetic elements that are capable of acquiring new genes from their hosts and even contributing to their hosts' own genomes by a process known as 'horizontal gene transfer', whereby viruses 'smuggle' genes from one organism to another.

Yet a powerful portion of humanity today lives in near-total denial of our interdependence, abides by a doctrine of human supremacism and proceeds as if we can live without limit on a finite planet. Never before in earth's history has one species been responsible for the deletion of so many other species. One by one, the planet's many other voices are falling silent; its innumerable more-than-human ways of being and sensing are blinking out. We are entering what the great biologist E. O. Wilson (1929–2021) chillingly called 'The Eremocene' – 'The Age of Loneliness'.[9]

So: let this glorious paper ark you are about to board stand as both a celebration and a warning. A warning of what is at risk if we do not change course, and if we cannot imminently find ways of halting the death-drive of terminal-stage capitalism. And a celebration of life in its teeming, tentacular, iridescent, intricate, inventive, glittering, spectacular marvellousness.

←
In this detail of Joseph Stella's painting, *Tree of My Life* (1919), a gnarled trunk is at the centre of a landscape teeming with plants and animals.

INTRODUCTION: UTOPIA OR OBLIVION?[1]

Stephen Ellcock

From their earliest days on earth, humans have often acted like the houseguests from hell – rude, messy, prone to violent outbursts – cursed with an alarming tendency to trash the fixtures and fittings, driven by an irrepressible urge to grab anything of the slightest conceivable value.

It took just ten generations from the creation of Adam and Eve for the God of the Old Testament to lose patience with humanity, the pinnacle of his creation. Adam and Eve were conceived in God's image and entrusted with the care of all that God had created, but from the moment they cast a covetous glance at the forbidden fruits of paradise they, along with their descendants, were doomed. Cast out of paradise, they and their offspring plunged headlong into a deadly spiral of sin and self-destruction. Faced with humanity's precipitous fall from grace, God resolved to teach his wayward children the ultimate lesson, sending the Flood to purge the planet. Only Noah – the one true righteous man – was spared, anointed as the last hope of creation and the guarantor of future life on earth.

The people of Noah's time succumbed to wickedness, corruption and violence; yet their transgressions pale in significance when compared to the evils, brutality and all-too-commonplace crimes against the planet of our own time. Now, scarcely ten generations on from the dawn of the modern industrial world, we stand on the brink of yet another cataclysm. However, unlike the Biblical Flood, the catastrophe approaching is entirely of our own making, the consequence of our own destructive impulses.

A ramshackle alliance of robber barons, ghouls and racketeers now hold the globe to ransom, forfeiting the future by hijacking and trashing the public realm, fracturing social bonds and poisoning the very earth herself. Pirates and profiteers, the shock troops of techno-feudalism, now run rampant, slashing and burning their way through our heritage, rights, the rule of law and the social contracts that once held us together. Compassion and respect are anathema to these brutes, who lay waste to everything that gives life value and meaning. Exploitation has replaced cultivation; mindless gratification and self-indulgence are considered virtues; while self-awareness and self-sacrifice are beneath contempt, the preserve of deadbeats and losers. Self-worth is measured in material wealth and the fruits of corruption. Unchecked greed, enforced by the extortionist practises of modern-day capitalism, has become the universal currency, accepted everywhere, no questions asked.

Today we could be living in respectful harmony with our neighbours and all other creatures on earth, but instead we bought the lie of endless progress. We could be the guardians of an unimaginably beautiful, endlessly fertile planet, a world of extravagant abundance whose natural resources, if carefully managed, could sustain and nurture every living thing for countless millennia to come. But the covenant with the natural world that is the birthright of humanity has been deliberately and violently shattered over the past few centuries, alienating us from our own planet – billions of orphans forcibly torn from the arms of Mother Earth.

As a result, we are facing multiple crises. These crises are real and are happening here and now.

Ark is my modest attempt to encapsulate the astonishing diversity and boundless variety of the living world, a catalogue of living wonders, from microbes to megafauna. I cannot hope to encompass the entirety and immensity of creation within the limits of a single volume. All I can offer is a small selection of the countless marvels I have discovered over many years spent as a hunter-gatherer of images, driven by a lifelong fascination with the natural world. Many of the species within these pages face grave challenges, while others may soon vanish forever, but all are part of the tree of life and all, without exception, should be valued, conserved and cherished.

This book is an affirmation rather than an elegy, a celebration of the richness of life that will inspire awe, astonishment and wonder together with a renewed appreciation and respect for the glories of the natural world and for every living thing with whom we share our fragile home. I hope it will encourage all of us to take up the role originally intended for humanity and become stewards of the earth and all its living creatures.

Throughout the book we have indicated the status of the species pictured according to the IUCN Red List, January 2026, from near threatened to critically endangered.

Critically endangered
Endangered
Vulnerable
Near threatened

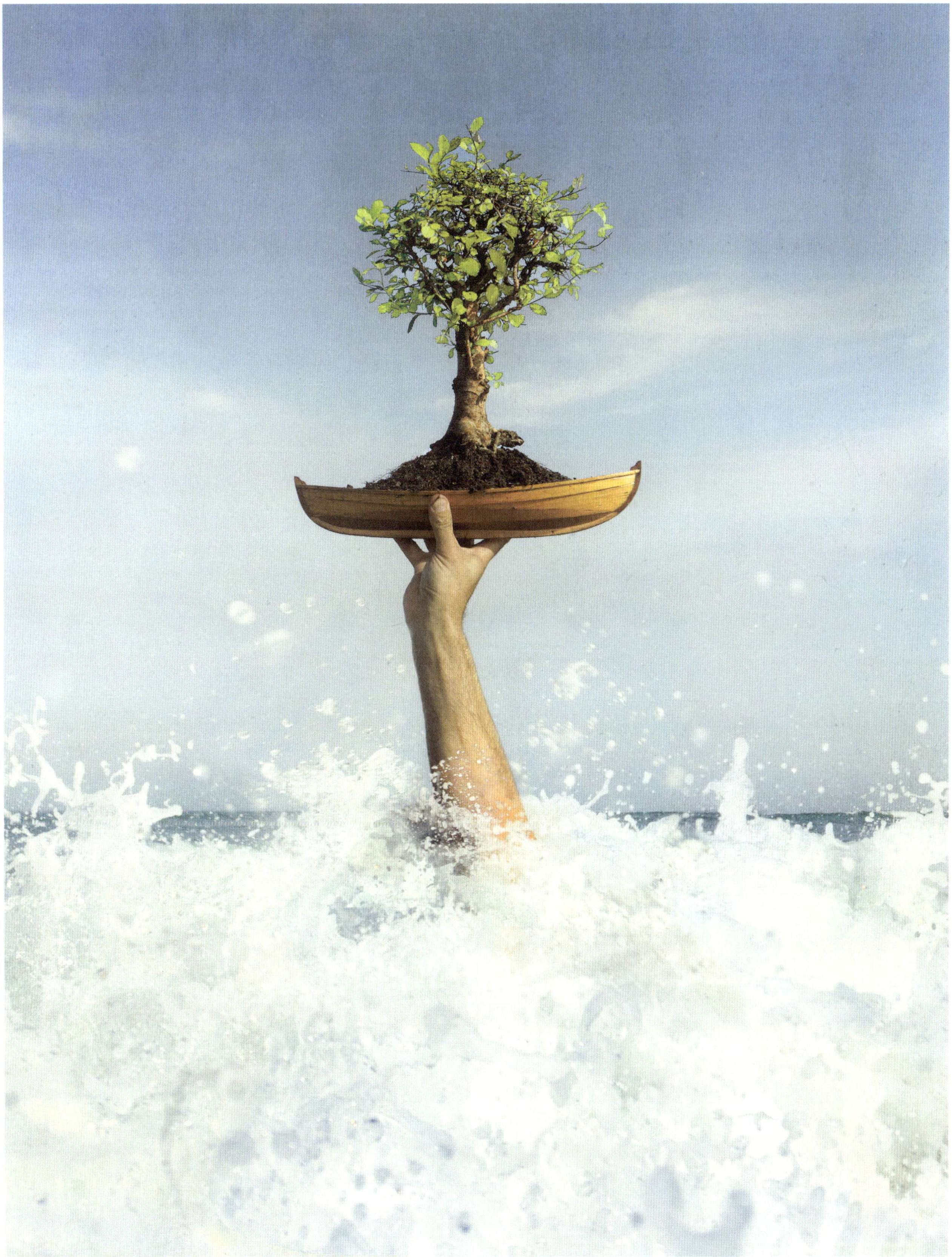

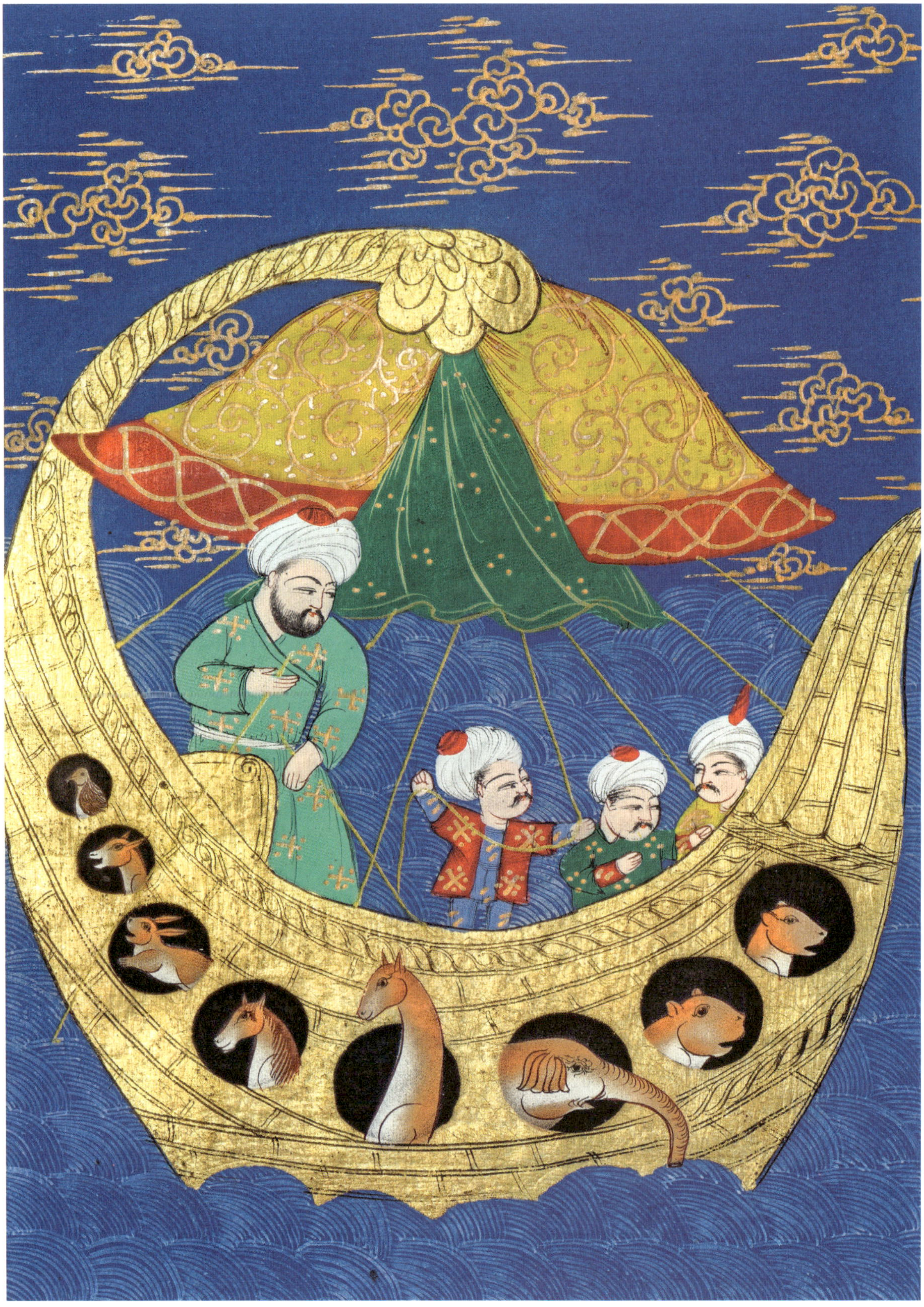

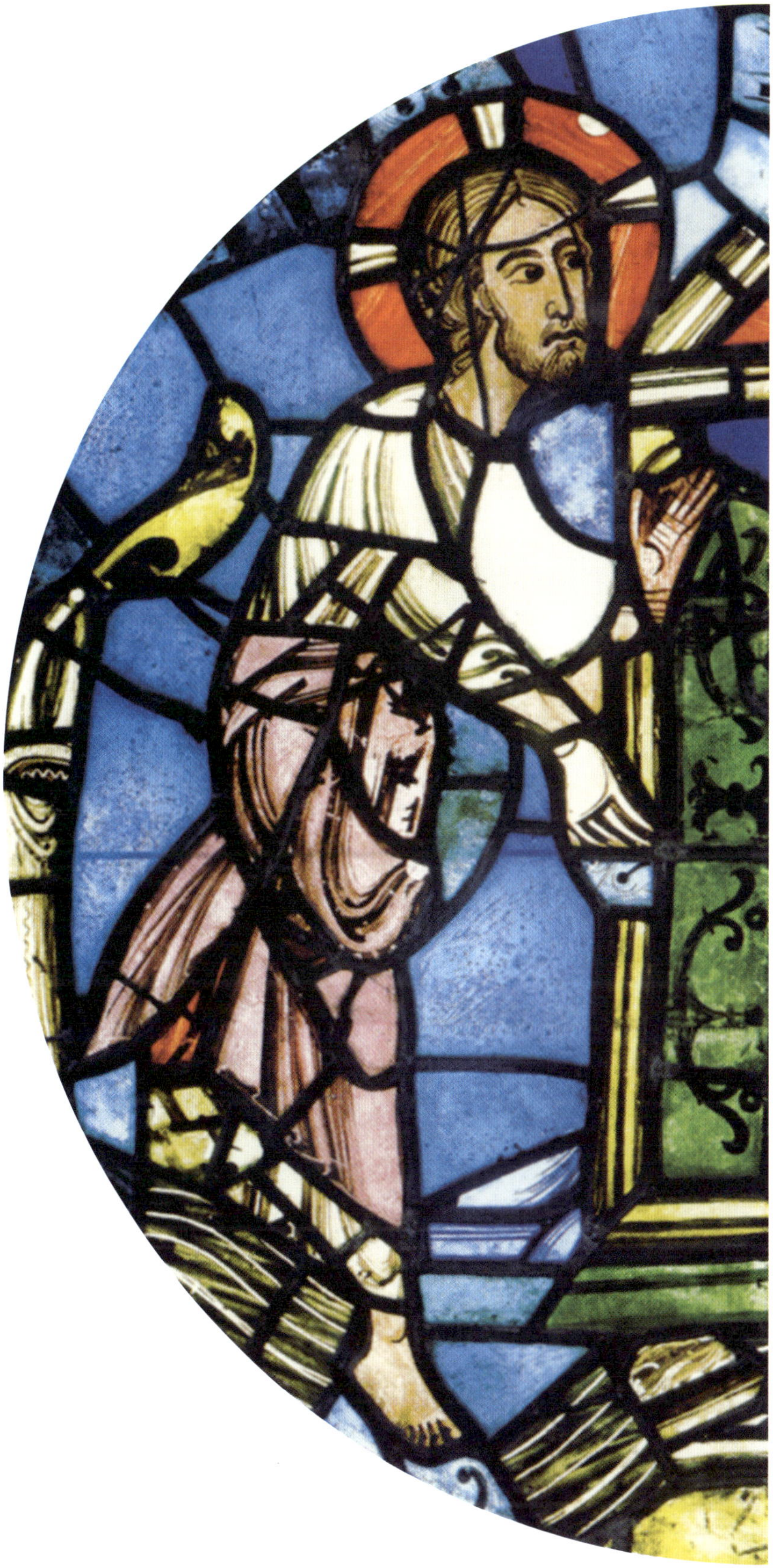

'And of every living thing of all flesh, two of every sort shalt thou bring into the ark, to keep them alive with thee; they shall be male and female. Of fowls after their kind, and of cattle after their kind, of every creeping thing of the earth after his kind, two of every sort shall come unto thee, to keep them alive.'

THE BIBLE, GENESIS 6:19–21

↲↲
In Kahn & Selesnick's 2016 to 2017 photograph, *The Augury of the Ark*, it is unclear whether the ark is being triumphantly saved from the flood or about to disappear under the waves.

↲
Richly coloured and decorated with gold leaf, this miniature of Noah's Ark, showing Noah with his three sons and a selection of animals, is from an Arabic manuscript from the Seljuk era (12th–14th century).

←
A French stained glass window of around 1190 illustrates the scene from the book of Genesis when God closes the door of Noah's Ark to protect its precious living cargo from the flood.

↳
This detail from the *Holkham Bible* (*c.* 1320–30) shows a raven – which was released by Noah, alongside a dove, to find out if land was near – feeding on bodies in the Flood waters.

ON THE ARK

Sarah Perry

Early one afternoon, on an iron-hard January day, I stood in an Elizabethan hall and looked out across the Norfolk lowlands. There was frost on the flats, and mist on the frost: the impression was of being on board a vessel drifting through winter. There was a table nearby, with the remains of Christmas on it – cards too lovely to be thrown on the fire, festive bottles opened and unfinished. One of those, barred and footed at the base, had a label that read ARARAT, with each letter A disproportionately sized to give the impression of mountain peaks. It was Armenian brandy, I was told, kept for years in casks of oak, its production and flavour more or less unchanged since it was first distilled. When I saw the label and the name, I thought at once of the story of Noah's Ark in the book of Genesis, which in its eighth chapter tells that 'the Ark rested in the seventh month, on the seventeenth day of the month, upon the mountains of Ararat'[1] – a dormant volcano in the far east of Turkey and on the border of Armenia. It seemed fitting to me that an Armenian brandy should take its name from the resting-place of the Ark, and in doing so call to mind an act of preservation against destruction, since so much of Armenian custom and culture was lost, in the early part of the last century, to genocide.

When I was a child, I took the legend of Noah's Ark to be a matter of fact. Once, while drying the dishes, I asked my mother what had become of it, unable to believe a boat so immense – and so real – could simply have disappeared. If this was only childish folly, it was not unshared. Over many centuries, faithful travellers have gone to Ararat seeking fragments of the gopher wood of which, the Bible says, the Ark was made (though 'gopher' itself is lost to us: the word is simply the transliterated syllables of unknown and untranslatable Hebrew script, and there is no such wood). The Talmud has it that the Assyrian King Sennacherib found a beam of the Ark, and worshipped it as an idol;[2] the 3rd-century Babylonian priest Berossus tells of visitors to fragments of the Ark who'd scrape off its protective pitch and wear the substance as a talisman against ill luck.[3] Amateur archaeologists have looked for the Ark, and so have oil engineers, and aviators, and Seventh-Day Adventist pastors; in 1883 a New Zealand newspaper reported that the Ark had been found,[4] and the story was disseminated across the world by publications that failed to notice the date, and that they'd been April-fooled.

When I asked my mother what had become of the Ark, I had no idea the Genesis Flood myth was only one of many – that in fact it's a tale told from Mesopotamia to Iran, from Ireland to Korea, handed from generation to generation over the course of many thousands of years. The first of these, preserved on fragments of Sumerian tablet, speaks of the god of an underworld sea instructing King Ziusudra to build himself a boat in which to ride out a seven-day storm that will destroy mankind. Meanwhile, in Hindu legend, Lord Vishnu in the guise of a fish warns a man that a flood catastrophe is coming, and so he must collect every kind of grain and preserve it on a boat. A Choctaw tale relates that an unnatural darkness fell on the earth, so that the people searched in despair for sight of the sun, until at last they rejoiced to see a shining in the north – but this was a rising tide of water that destroyed all those save a few who, suspecting the deluge, had built themselves a raft. Often the myths speak of a cleansing destruction, because of the wickedness of mankind, and a world made new – in the Lao flood myth, a crude and sinful people were punished by the gods for their faithlessness with a great flood, with a bare handful surviving in a gigantic and magical gourd.

How must we account for the universality of the flood myth and the idea of the ark, which seems as ancient and as common as the art of the story itself? It has been theorised, of course, that it speaks to historic fact: that a great flood, of natural origin or divine retribution, must once have covered the earth. Archaeological evidence has seemed sometimes to give credence to the notion, with distinct strata of sand and silt – interspersed between layers of evidence of human civilization – suggesting a widespread and catastrophic deluge. Contemporary understanding of geology dispels the idea, but science cannot shift the sense of wonder that the legend of the ark – whether a raft, or a gourd, or an immense vessel made watertight with pitch – has been told by so many household fires, and for so long.

Perhaps the myth persists because it speaks to what the ancient Sumerian held in common with the Choctaw storyteller, and the scribe of Genesis, and the Lao child

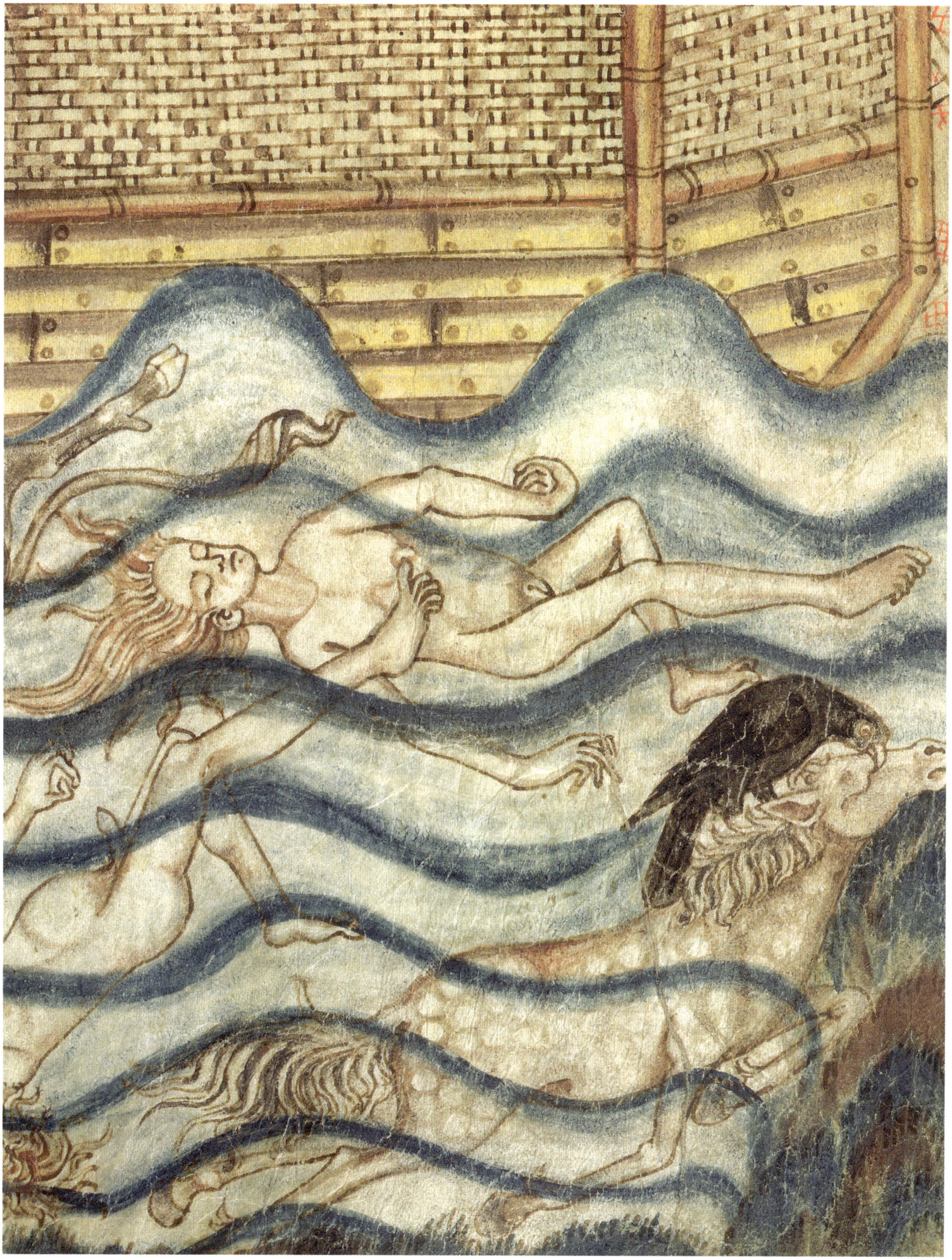

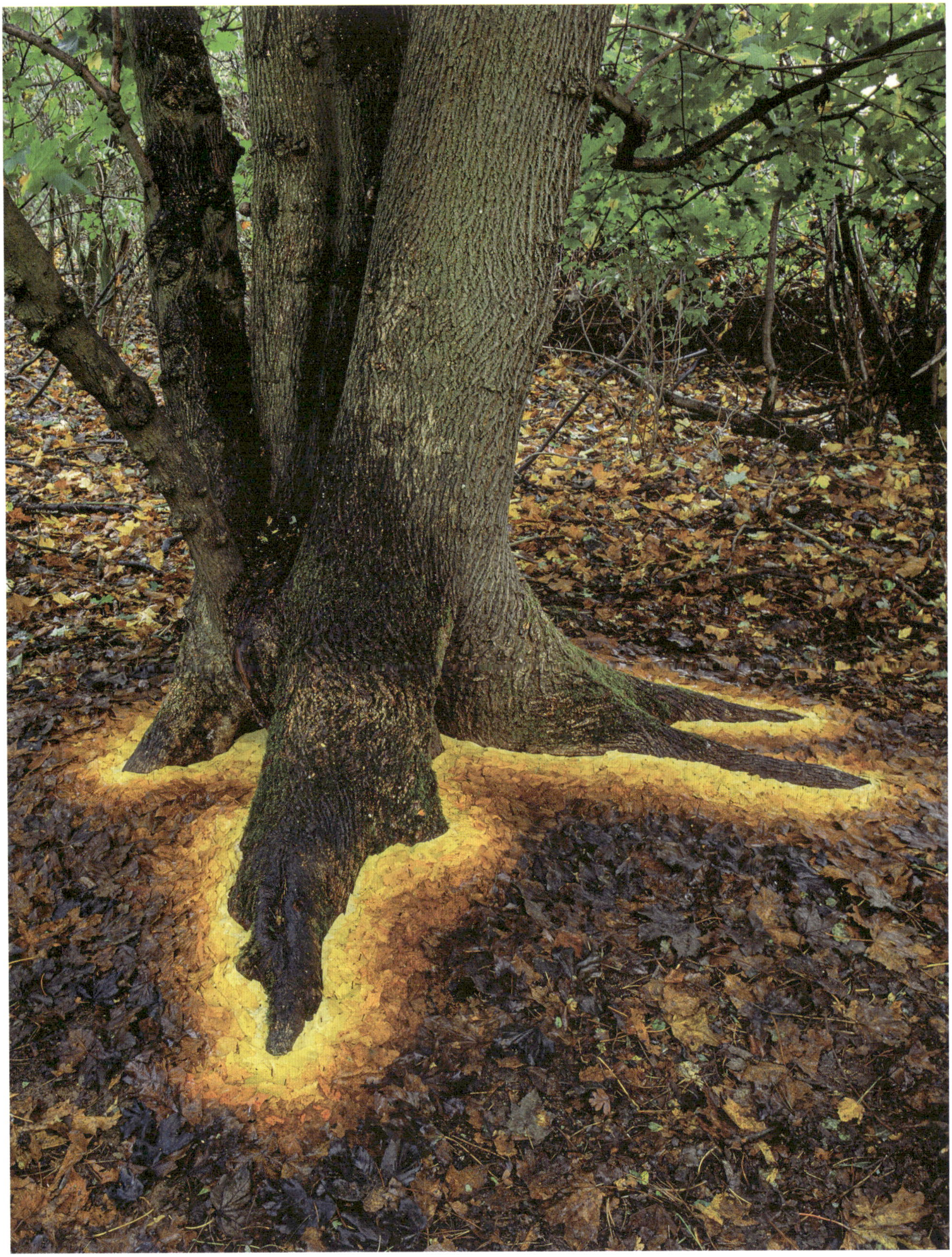

asking again and again to be told the tale of the flood and the magical gourd: the sense of the earth and its stores as an irreplaceable treasure threatened by human wickedness, and the hope of a means of preservation and escape. That is an essential and a unifying human instinct, which continues to find its expression in the written and the visual arts – the desire to create something that will withstand the wreckage of what we have done to ourselves, and to the world. The artist Joseph Beuys – once a member of the Hitler Youth, and disfigured by burns when his Stuka bomber crashed in 1944 – created a monumental work, which might be read as an act of cleansing as the tides of war receded, instituting in the German city of Kessel the planting of 7,000 oaks accompanied by basalt slabs: the city itself becoming an ark in which the natural world is preserved. Contemporary visual artists – seeing, like Noah, the coming destruction – turn to the matter of the earth, and in doing so display its terrible vulnerability, and insist on its preservation. In 2014, Olafur Eliasson's *Ice Watch* consisted of blocks of ice melting on stone pavements, to the indifference or horror of passers-by, the climate catastrophe in miniature; Andy Goldsworthy constructs intricate whorls and mandalas of fading leaves that could not outlast a passing shower. But the sense of the natural world in peril is not unique to this age of global heating and broken climate accords – Gerard Manley Hopkins, in his poem 'Inversnaid', wrote:

> What would the world be, once bereft
> Of wet and wildness? Let them be left,
> O let them be left, wildness and wet;
> Long live the weeds and the wilderness yet.[5]

This book represents that same response to the atomizing powers of industrial, military and political greed – it is an act both of defiance and of restitution, preserving what we hold in common, since what we most hold in common is both priceless and free: the natural world.

But if defiance, with its essential component of rage, is necessary, so is faith. Noah did not build his Ark because he despaired, but because he hoped. Despair is a diminishing and a stultifying force, and one we cannot afford – faith meanwhile will build an ark despite the appearance of madness and folly, and believes it possible a dove can outlast a deluge. I've sometimes wondered if, against the thunder of anguish at climate breakdown, and at the loss of species and habitat, the still, small voice of hope is lost. It can be heard, if it is sought out – there was a time, for example, when I could not recall having ever seen a red kite: now they're so numerous I barely trouble to look up when they pass over with their forked tails outspread. That is a still, small voice, local and particular, but all the same – it speaks.

There is another Biblical ark – small, local, particular, and shaped by hope. In the book of Exodus, the Egyptian pharaoh, fearful of the growing population of enslaved Hebrew people, gave the order that every male Hebrew child should be slaughtered at birth. A young mother named Jochebed concealed her infant son until, fearing he'd be discovered, she wove a basket of the papyrus reeds that grew on the banks of the Nile, made it watertight with pitch, and set the child sailing in his little ark downriver while his sister watched. It is deeply touching – and it ought to be instructive – to picture the woman's hands silently weaving, and the tidal waters of the Nile bearing the child away: human endeavour and the natural world's provision, uniting against wickedness. It may have seemed to Jochebed a futile gesture, and a frail hope – a handwoven basket set against the vengeance of a pharaoh – but she lived to see her child become the prophet Moses, who led the Hebrew people to a promised land.

I no longer have the child's faith that Noah's Ark is still concealed on Ararat; Jochebed and Moses now seem the stuff of fairytales to me. But I take from the Ark what I have found in this book – the belief that we hold in common a native instinct to preserve what is wonderful and irreplaceable and good, setting our hope against apparently insurmountable forces of destruction. This is Tennyson in *In Memoriam A. H. H.*, fending off his despair:

> Behold, we know not anything;
> I can but trust that good shall fall
> At last – far off – at last, to all,
> And every winter change to spring.[6]

←
Andy Goldsworthy, *Sycamore leaves edging the roots of a sycamore tree*, Berrydown Foundation, Hampshire, 1 November 2013.

↳
Heritage (2013) is a reflection of artist Cai Guo-Qiang's idea of Queensland, Australia, as a 'last paradise', where a harmonious society has been – for the moment – protected and preserved.

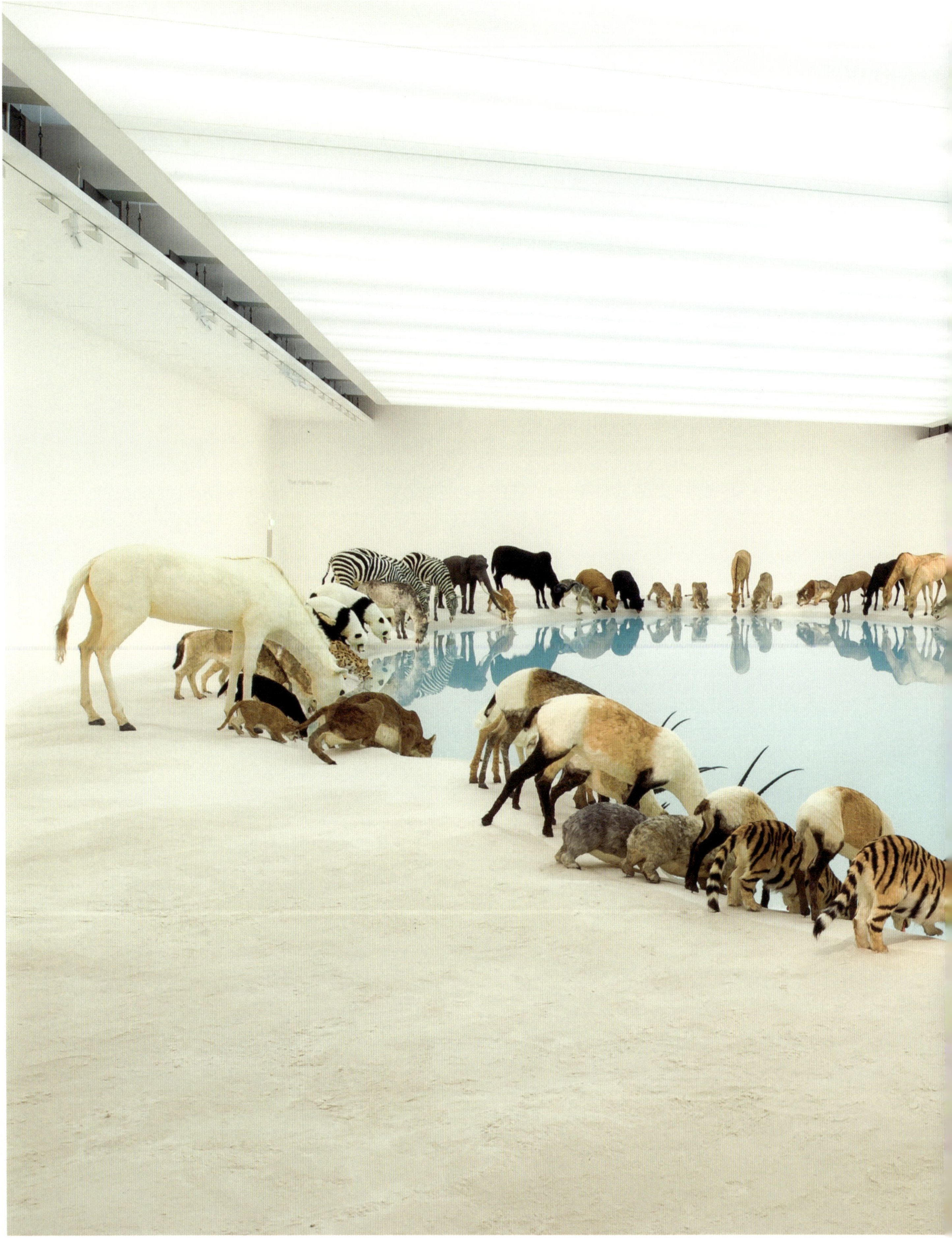

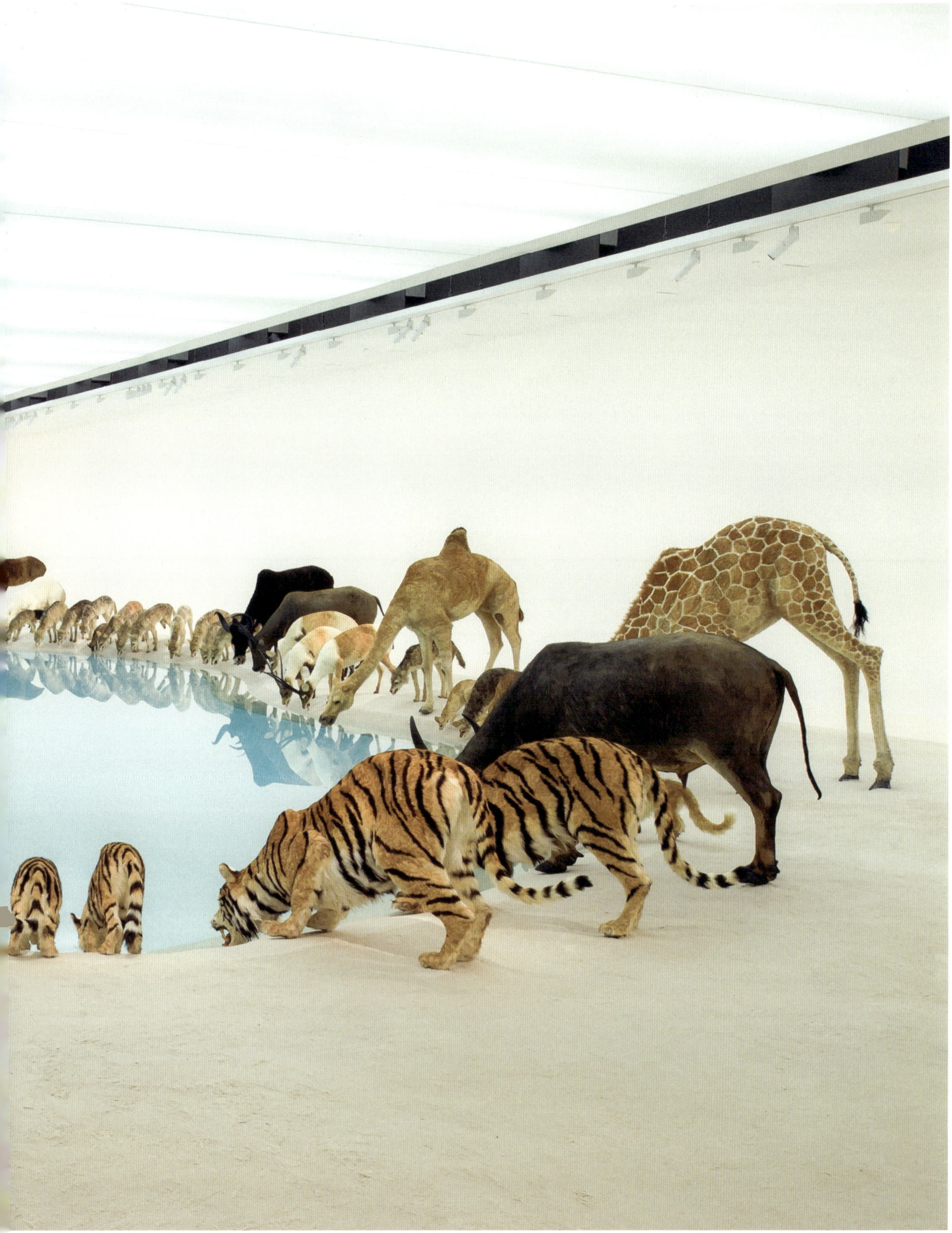

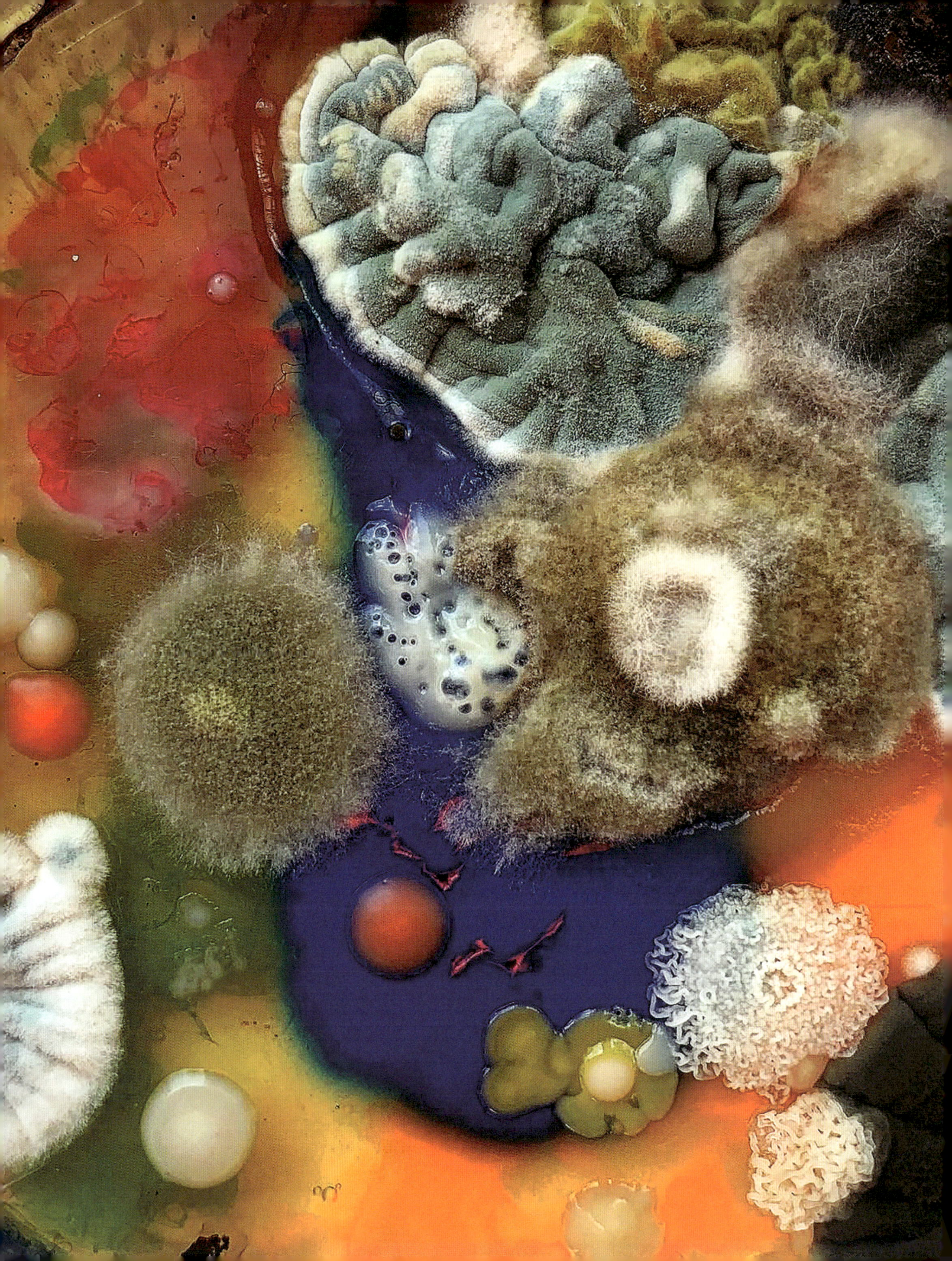

(I) MICRO-ORGANISMS & FUNGI

DINOFLAGELLATE	RADIOLARIA
FORAMINIFERA	DIATOM
CHROOCOCCUS	NOSTOC LINCKIA
CHROOCOCCUS TURGIDUS	PHYTOPLANKTON
SLIME MOULD	TRUE ICELAND LICHEN
LATTICE STINKHORN	SCARLET ELFCUP
CHANTERELLE	BITTER BOLETE
BUTTER-FOOT BOLETE	RUDDY BOLETE
PEPPERY BOLETE	WINTER RUSSULA
DESTRUCTIVE PHOLIOTA	FLY AGARIC
WOOD DECAY FUNGUS	WOOD EAR
TIGER SAWGILL	WITCHES' BUTTER

MICROORGANISMS & FUNGI play vital roles in nutrient cycling, plant growth and the production of antibiotics. Microorganisms may be single-celled or exist in a colony of cells and are often invisible to the naked eye. They include bacteria, radiolarians and many species of algae. Fungi include slime moulds and mushrooms, and range in size from the microscopic to multicellular large mushrooms. They have a unique cell structure, are able to digest organic matter and can form symbiotic relationships with plants.

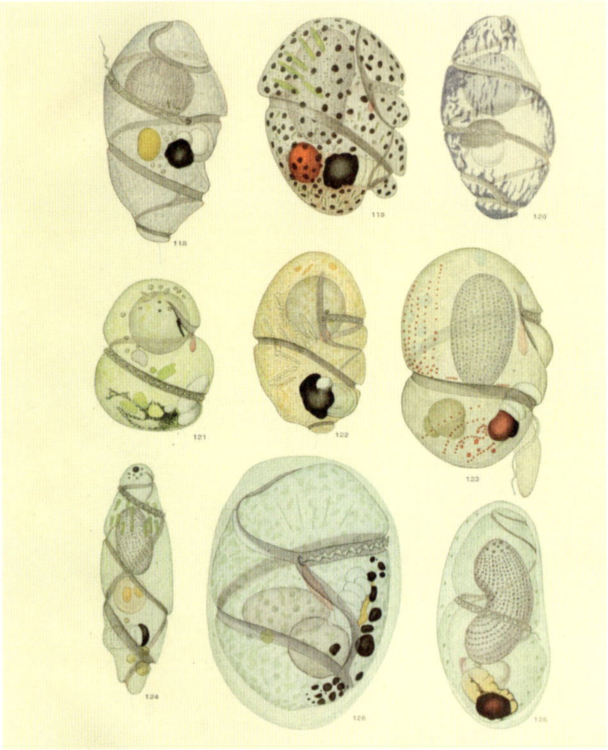

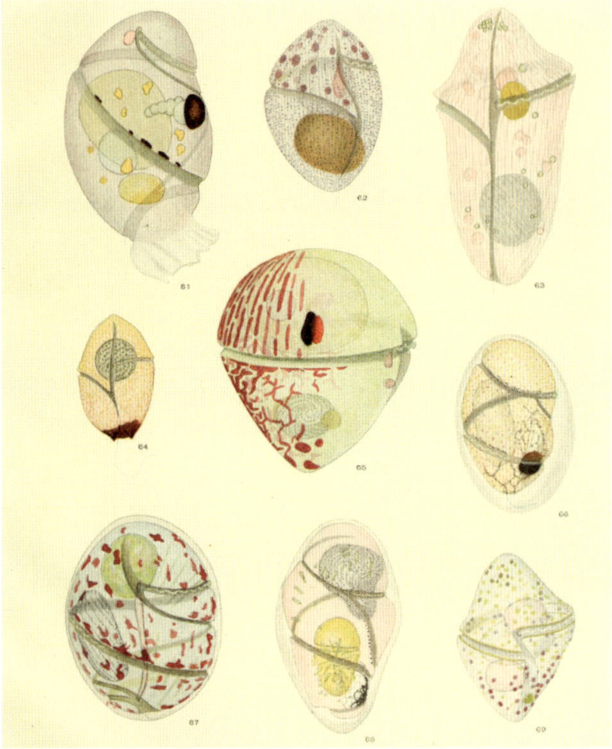

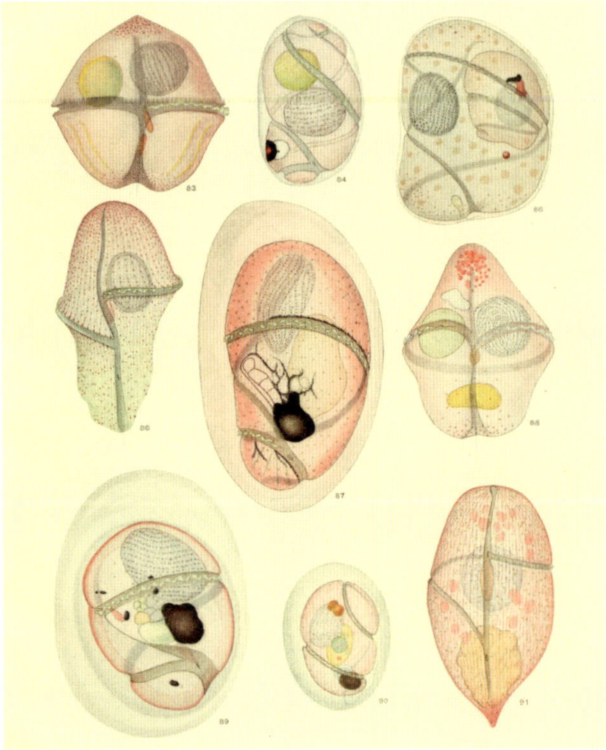

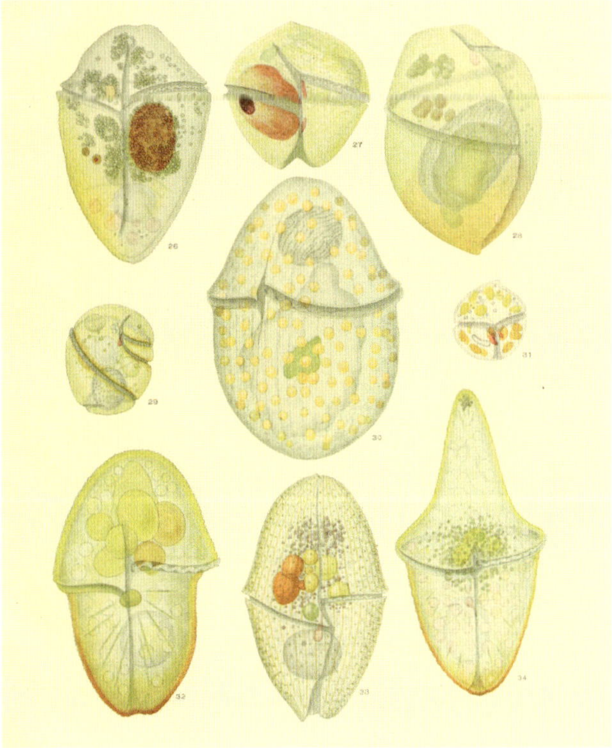

↑
Under certain conditions dinoflagellates, such as the ones shown in these 1921 illustrations, can rapidly reproduce to form toxin-producing algal blooms.

→
Radiolaria, pictured here in a micrograph by Christian and Noé Sardet, are types of zooplankton that feed on even smaller species of plankton.

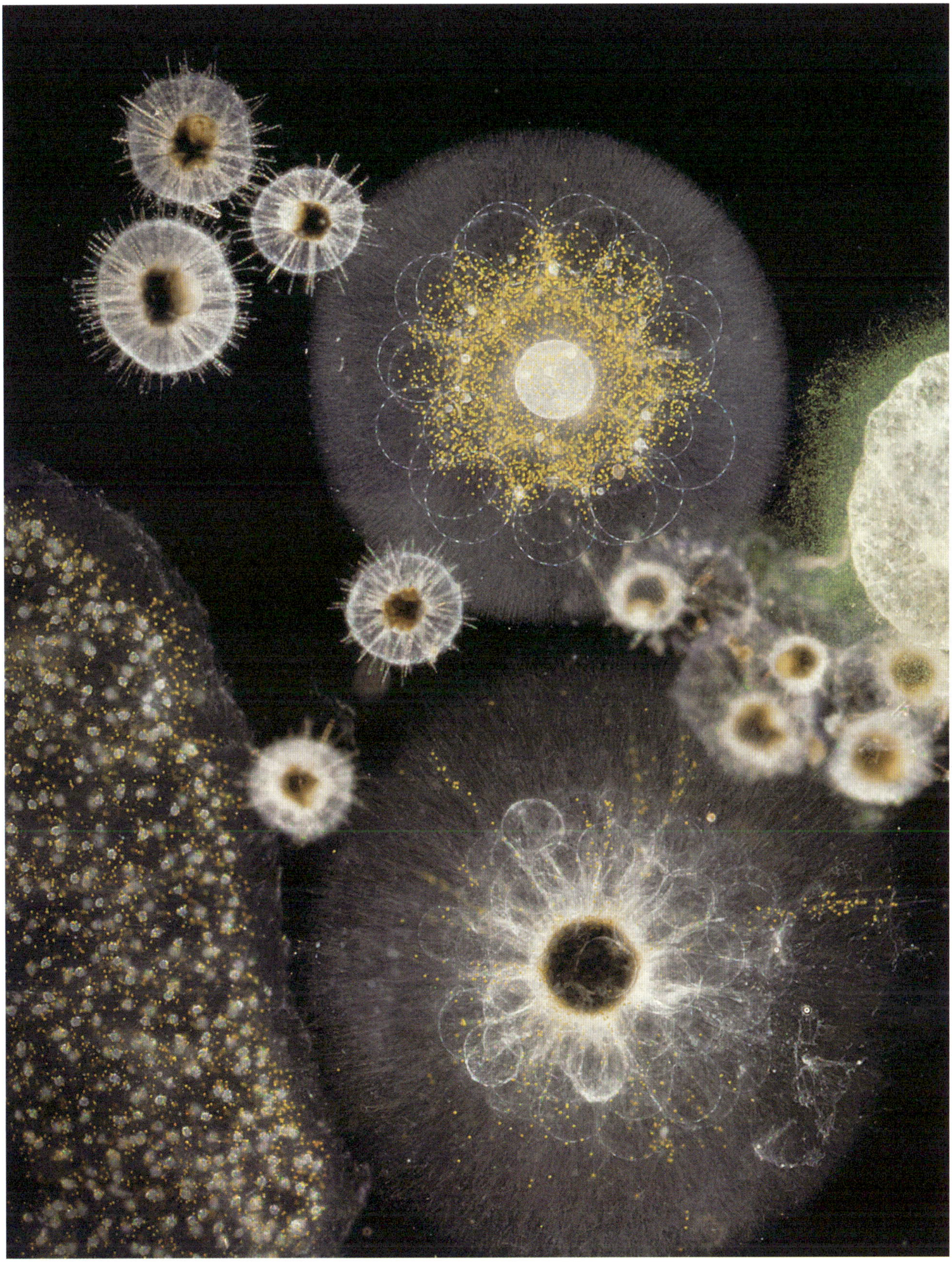

'Diatoms can manipulate silica in ways that nanotechnologists can only dream about. If we understood how they can design and build their patterned frustule as part of their biology, perhaps this could be adapted by humans.'

DAN ROKHSAR, QUOTED IN 'TINY DIATOM HAS GLOBAL ENVIRONMENTAL IMPACT', *UW NEWS*, 2004

(1)

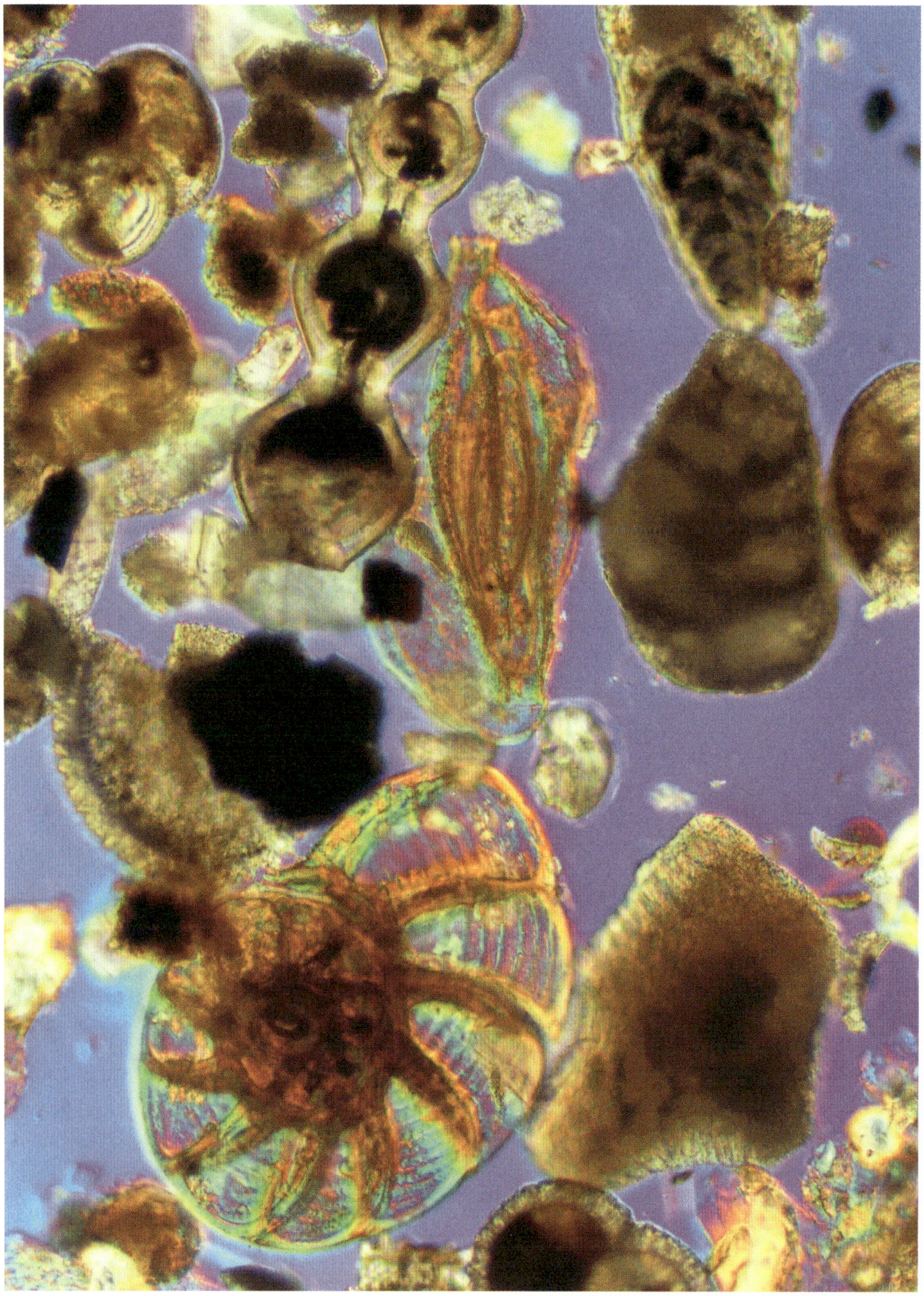

←
Foraminifera are single-celled microorganisms that have been around for more than 500 million years. Each individual foraminifera builds astonishingly complex, intricate shells around themselves that are called tests. These examples, photographed by Josef Reischig, demonstrate some common test shapes: a series of interconnected chambers, a planispiral shell (like the shell of a snail) and a shell with two rows of interconnected chambers. By studying the chemistry of fossilized tests, scientists can gain insights into long-term patterns in earth's climate history.

→
Diatoms are a group of single-celled, photosynthetic algae that contains many species – some scientists estimate that there could be two million species. Through photosynthesis, diatoms produce 20–50% of the air we breathe. Known as a frustule, the diatom's cell wall is unique because it is made of glass-like silica and can be geometrically patterned, often resembling snowflakes, mandalas or stained glass windows. These examples are photographed by microscopists Frank Fox (top right) and Jan Hinsch (right).

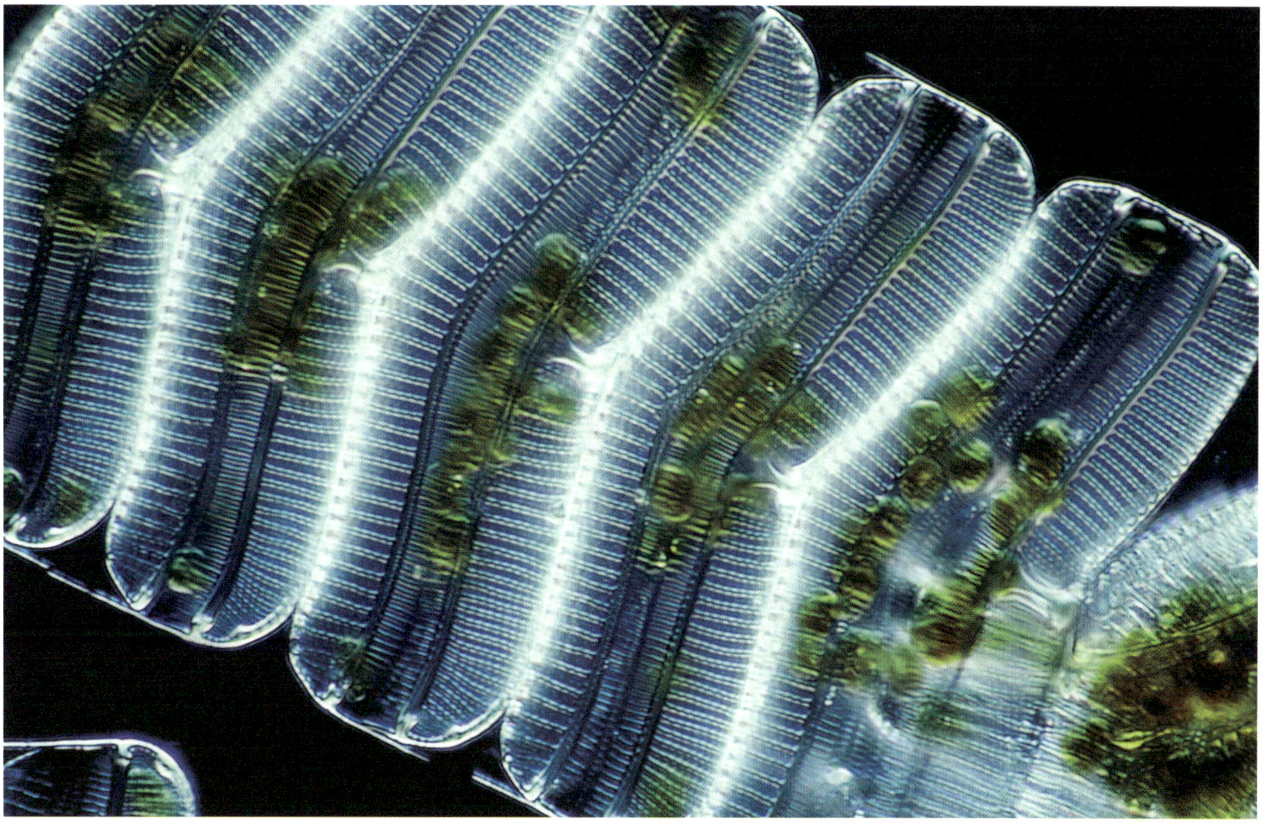

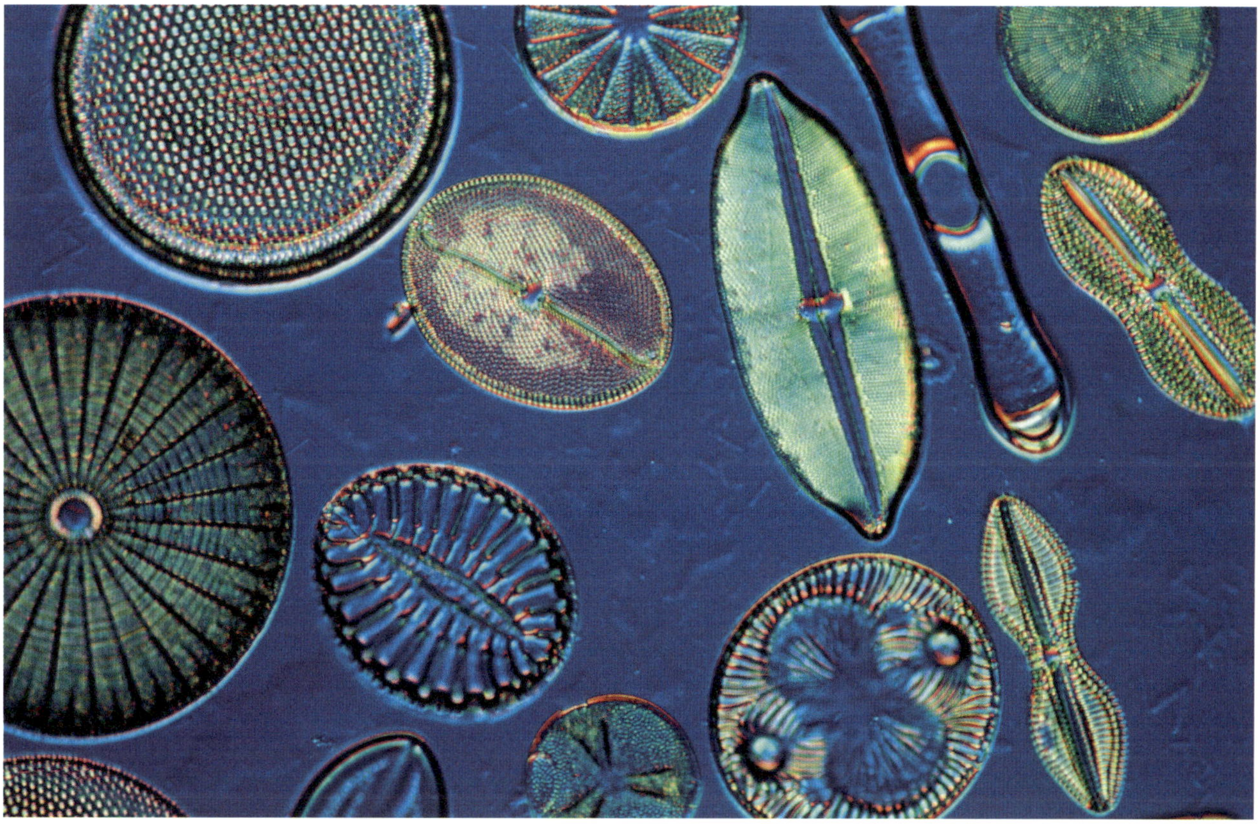

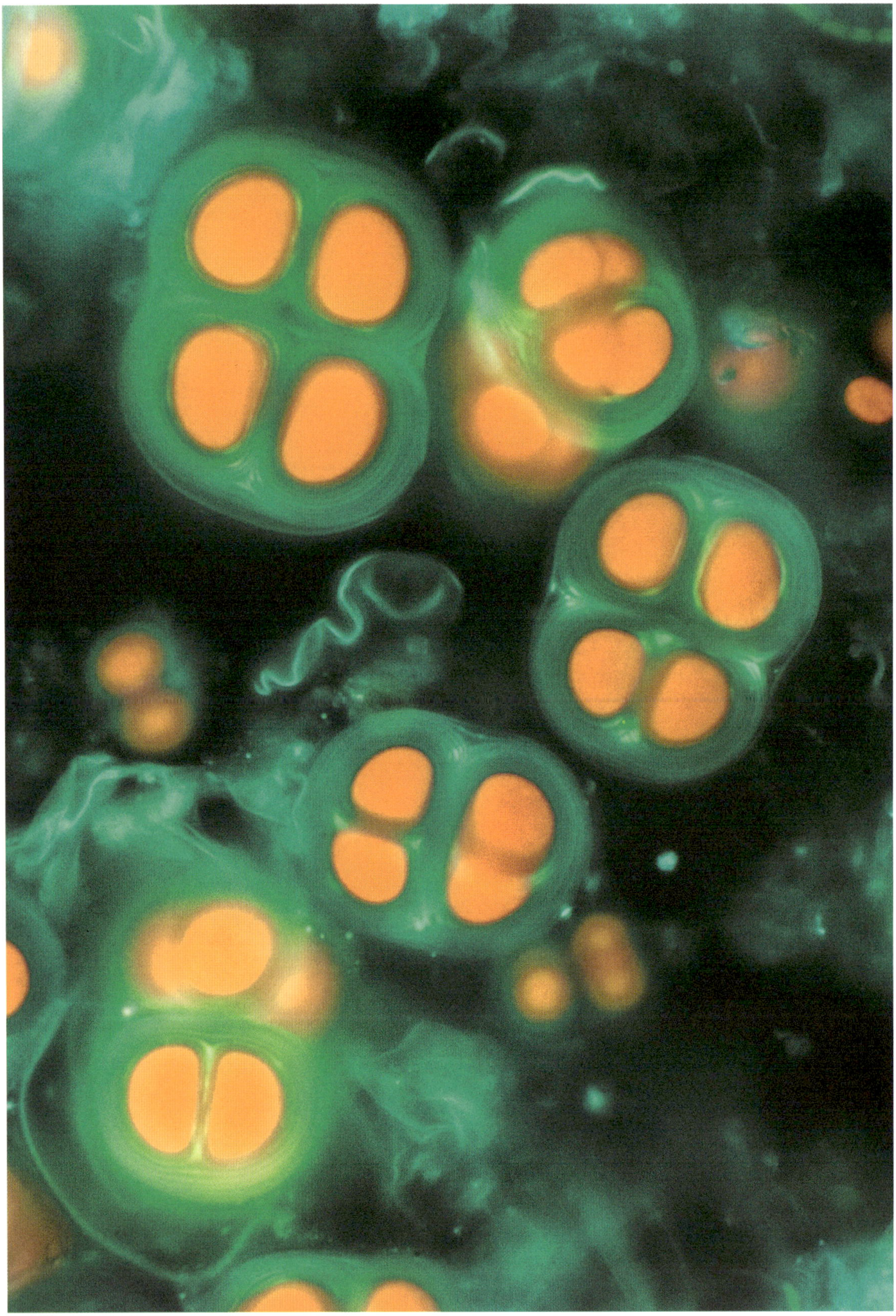

(1)

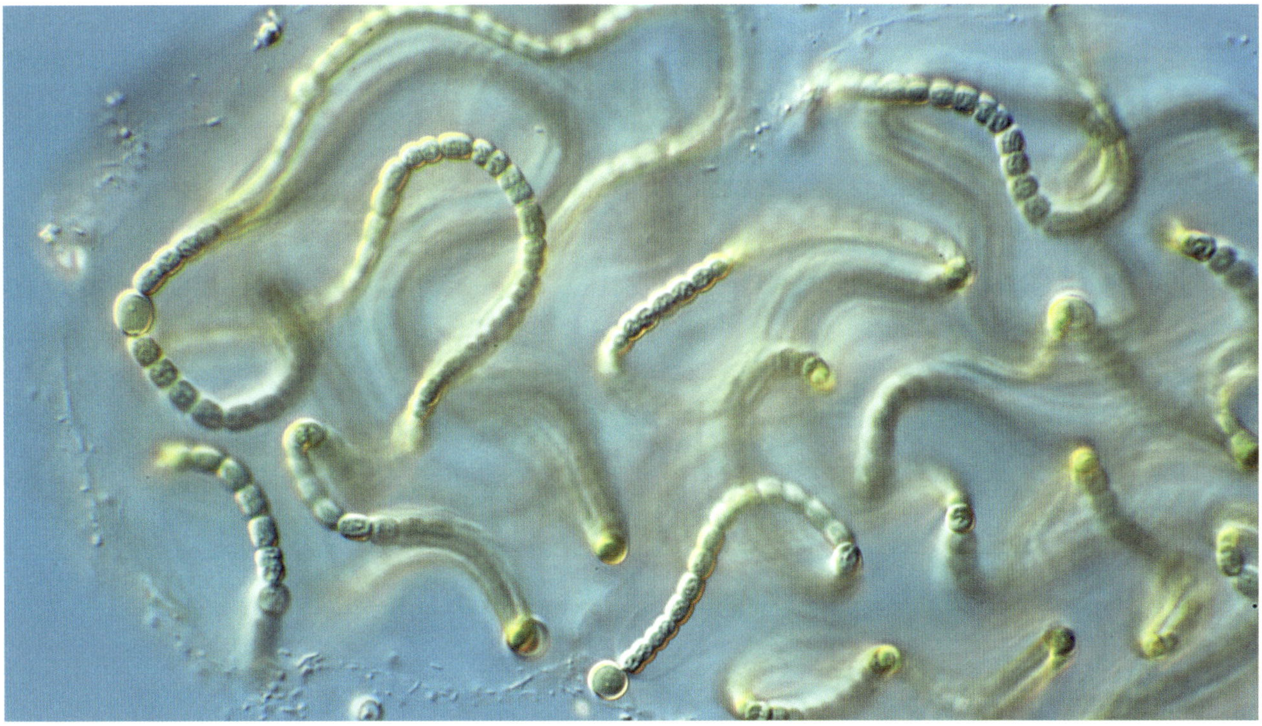

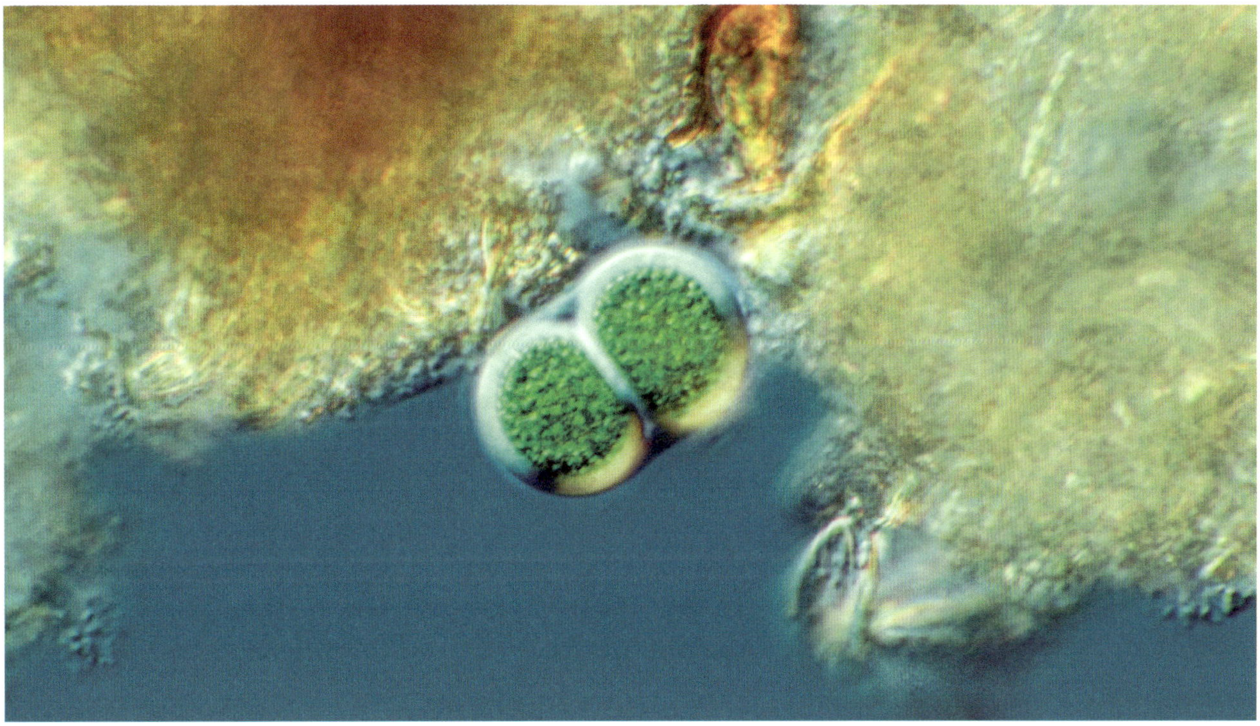

←↑
Chroococcus (left and above) are a genus of cyanobacteria, or blue-green algae. These single-celled aquatic organisms often join together in colonies or filaments made up of two, four, eight or sometimes more cells. Cyanobacteria, including the *Chroococcus* genus and *Nostoc linckia* species (top), use photosynthesis to create energy, absorbing carbon and producing oxygen in the process.

(1)

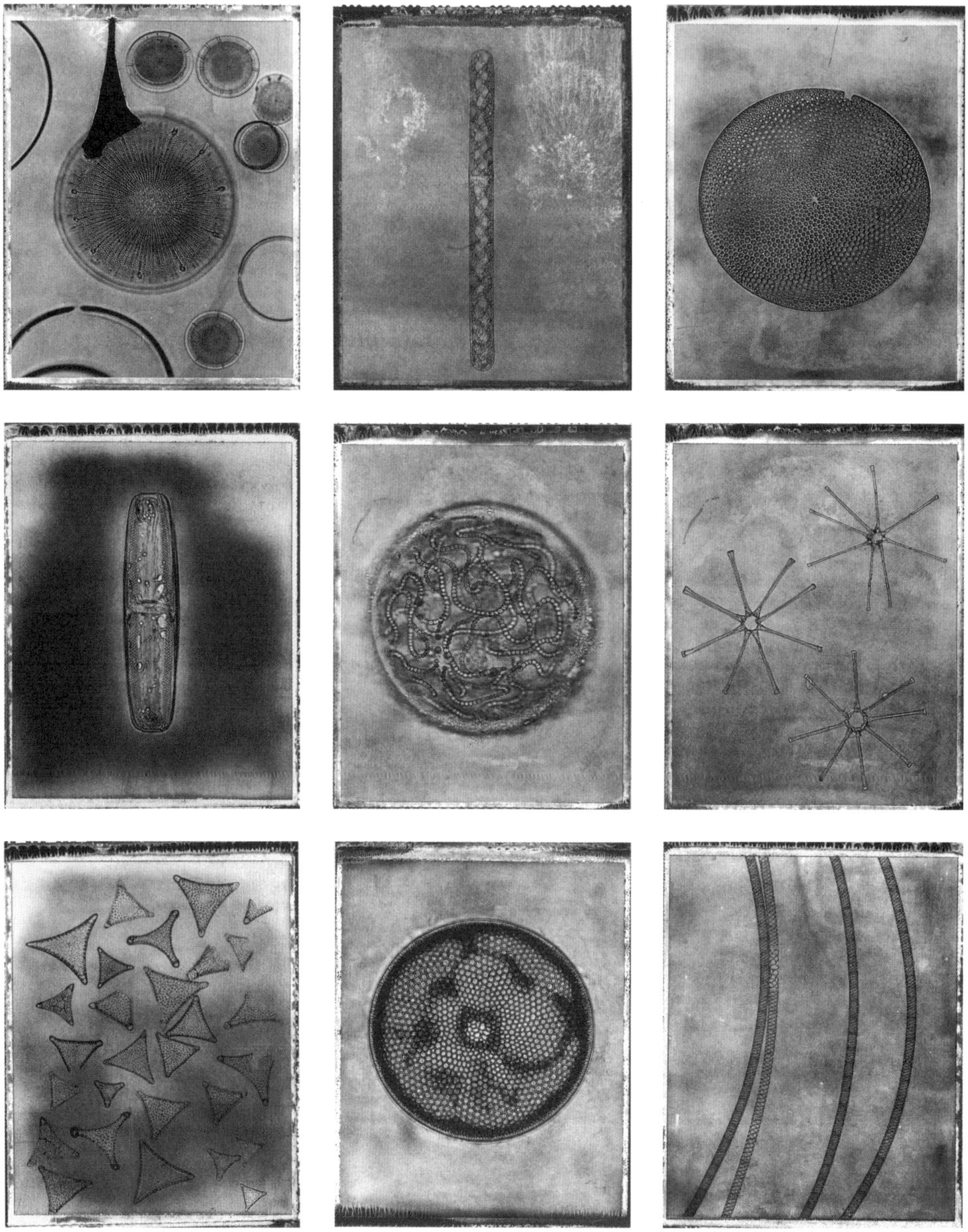

'About two-thirds of the oxygen in our atmosphere is produced in the surface waters of the sea by phytoplankton ... which initiate the entire food web of the ocean.'

JACQUES-YVES COUSTEAU, 'OCEAN POLICY AND REASONABLE UTOPIAS', 1981

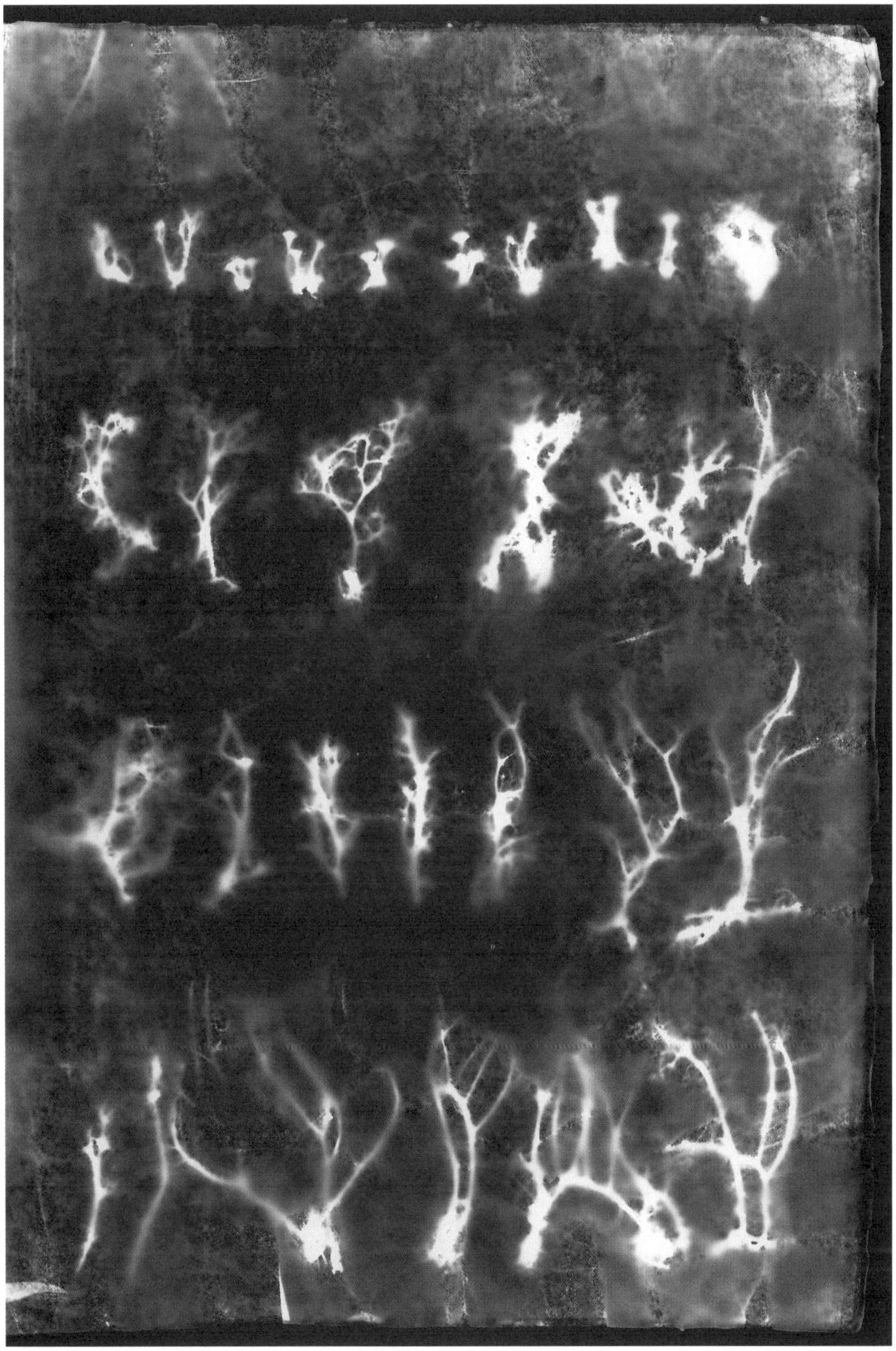

↖
Nine photographs of phytoplankton from Jan C. Schlegel's *Life on Earth* series (2024) reveal the forms of these microscopic aquatic organisms. Phytoplankton contain chlorophyll to capture sunlight for photosynthesis. In the process, they absorb carbon dioxide and generate more than half of the world's oxygen.

↑
Anne Arden McDonald made *Lichen Alphabet* (2014) by 'taking an antique ledger book, full of transparent pages, and glueing a few of the pages together with botanical objects sandwiched inside of them,' then 'printing through them to make a photograph.' Lichens are sensitive to air quality, and their presence or absence can act as a guide to the health of an environment.

↓
Katie Scott's artwork shows stages of development for various types of slime mould. The oscillation of slime moulds in nature is the basis of a population-based optimization algorithm, known as SMA (slime mould algorithm), that has been applied in diverse fields including astrophysics, where it was used to aid the visualization of the cosmic web.

→
Eliot Porter's *Lichen Behind Skogar, Iceland, July 6, 1972* illustrates the true Iceland lichen or Iceland moss (*Cetraria islandica*). The orange lichen could be *Xanthoria parietina*. Lichen isn't a single organism, but a symbiotic relationship that can be formed between a fungus and algae or cyanobacteria.

(1)

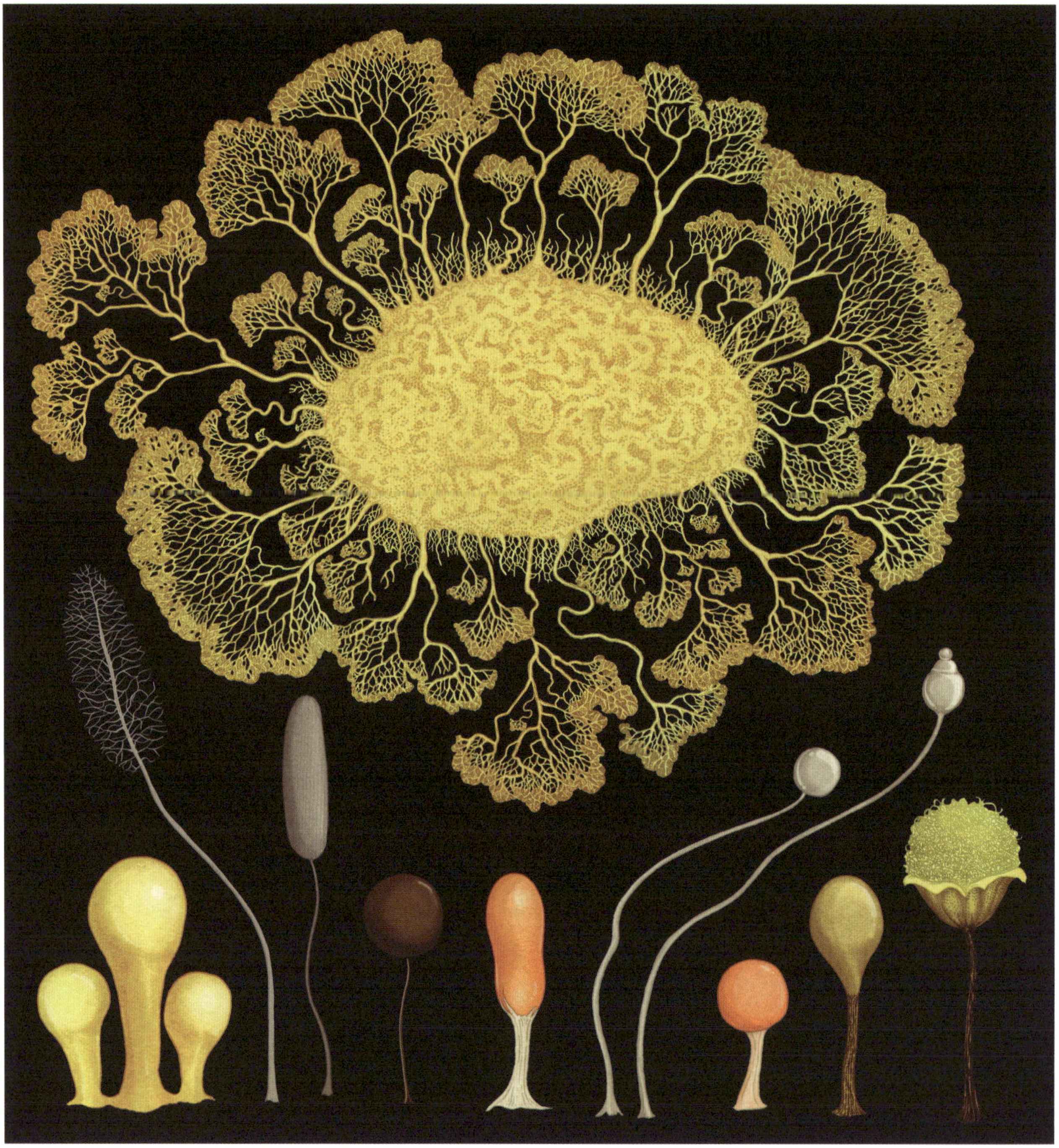

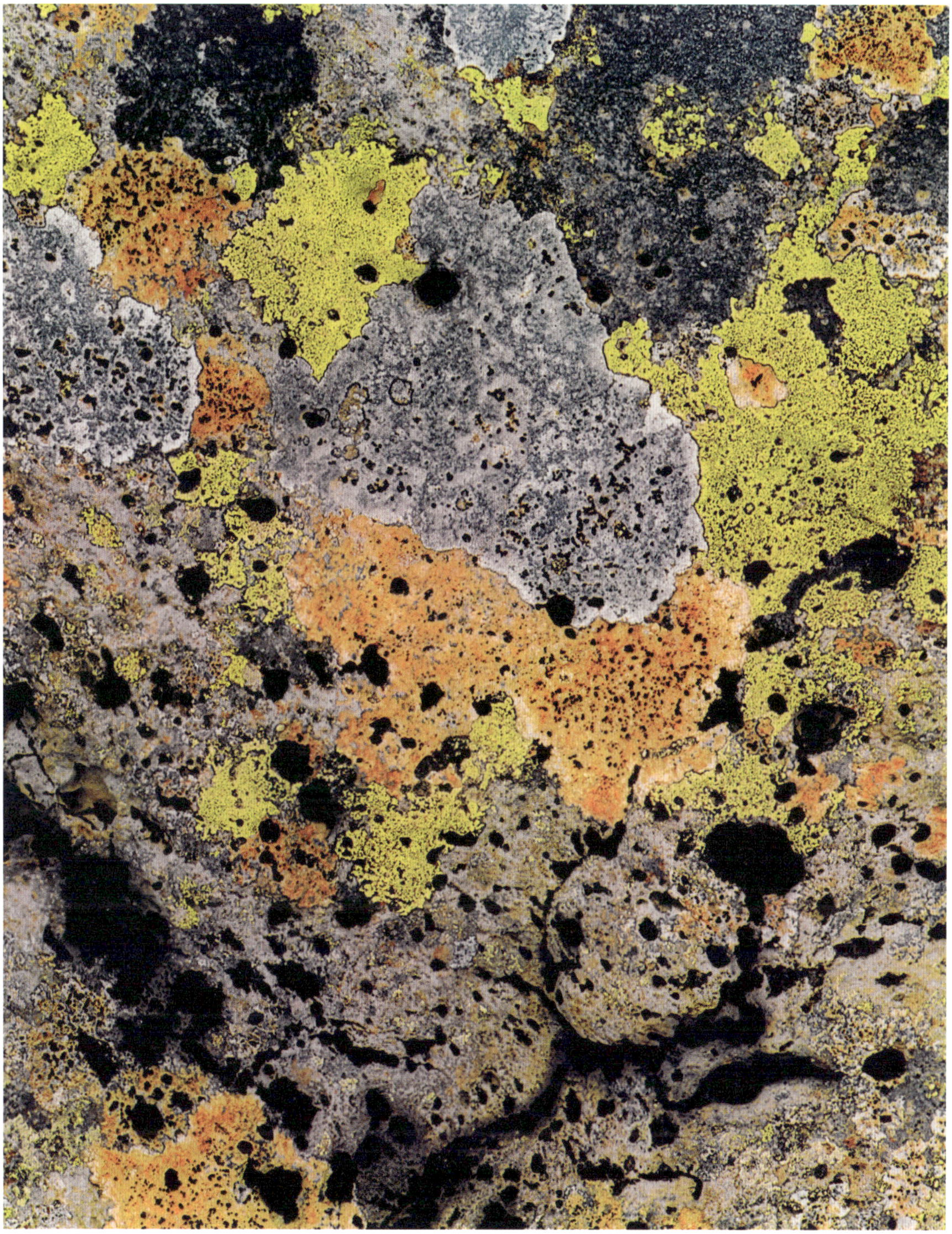

'One could speculate that lichens would be among the last inhabitants to succumb on a dying earth.'

STEVEN L. STEPHENSON, *THE KINGDOM FUNGI: THE BIOLOGY OF MUSHROOMS, MOLDS AND LICHENS*, 2010

↓
Theo Carreras's *The Lattice Fungus and Orange Elf-Cup* (1912) portrays two woodland fungi. The lattice stinkhorn (*Clathrus ruber*) smells like rotten meat. It breaks down mulch and dead wood, enriching the soil. In the foreground is the scarlet elfcup (*Sarcoscypha austriaca*), a species that feeds on decaying wood and leaf litter.

↘
Viennese designer Koloman Moser's *Schwämme* ('Mushrooms') textile design, dated from 1899, shows chanterelle (*Cantherellus*) mushrooms. There is a red species, known as *Cantherellus cinnabarinus*, but the colour of Moser's pattern is more likely to have been chosen for its aesthetic value rather than its accuracy.

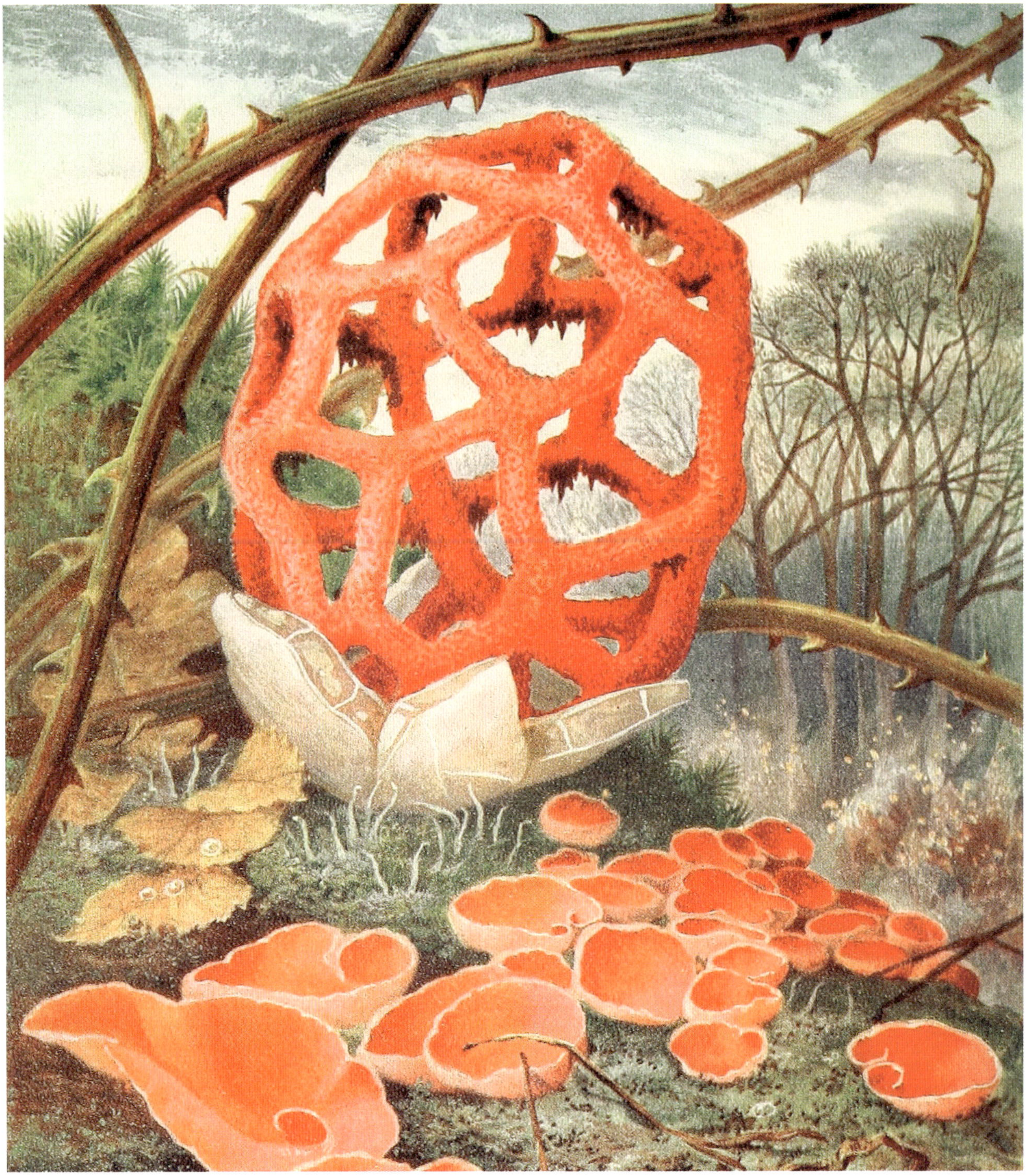

'When we try to pick out anything by itself, we find it hitched to everything else in the universe.'

JOHN MUIR, *MY FIRST SUMMER IN THE SIERRA*, 1911

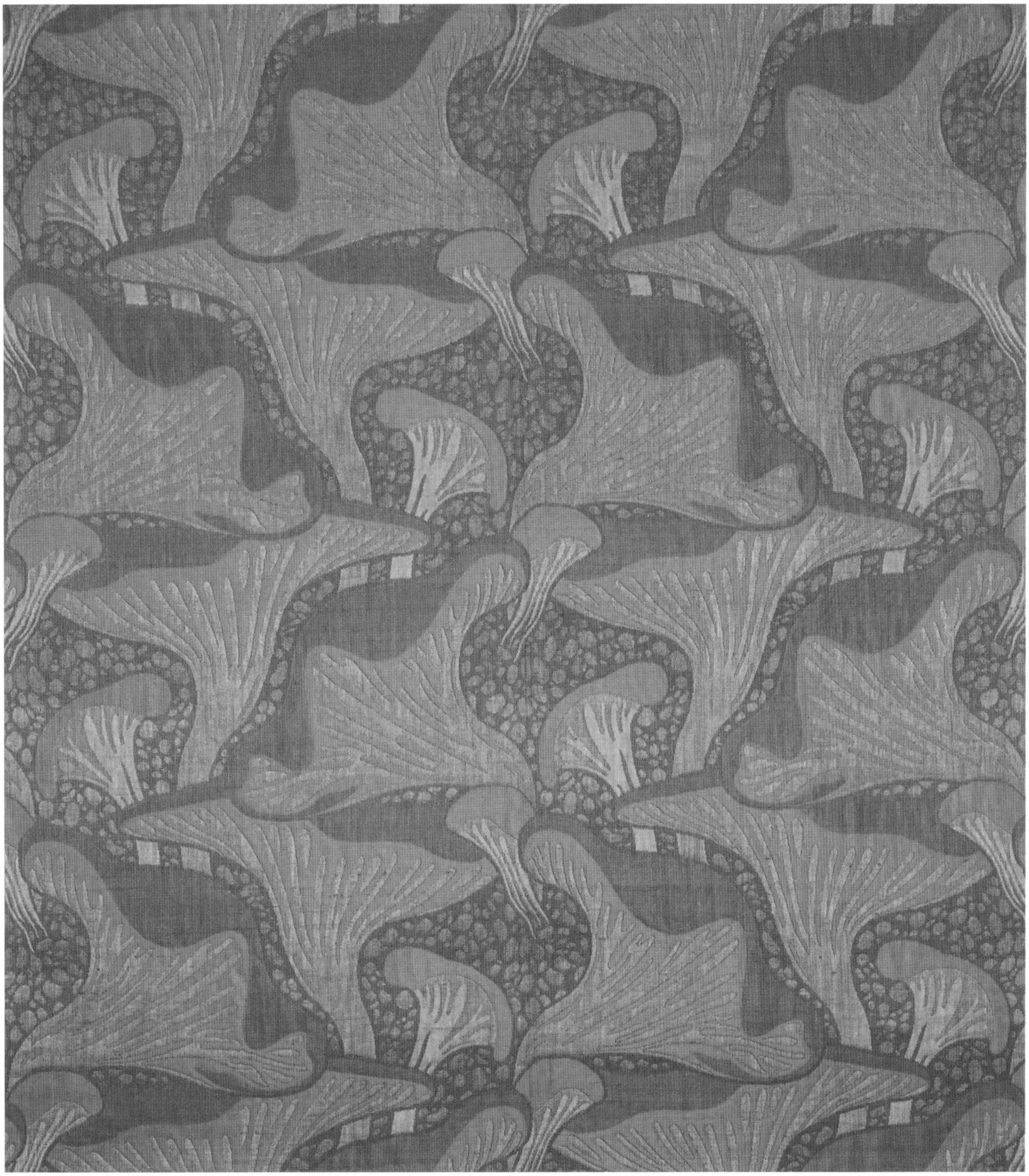

'The simplest and most lumpish fungus has a peculiar interest to us – compared with a mere mass of earth – because it is so obviously organic – and related to ourselves – however mute.'

HENRY DAVID THOREAU, JOURNAL ENTRY, 10 OCTOBER 1858

(1)

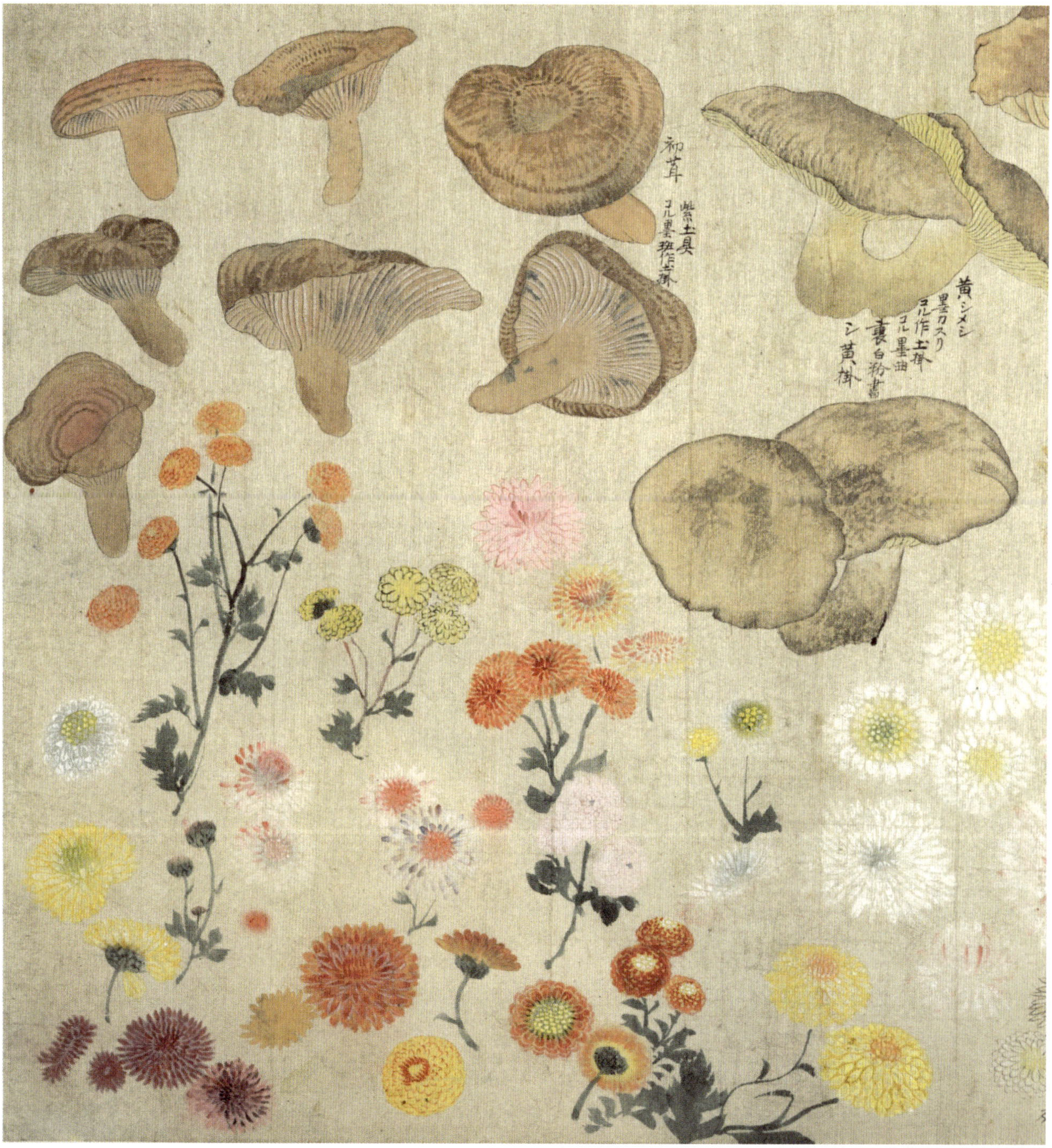

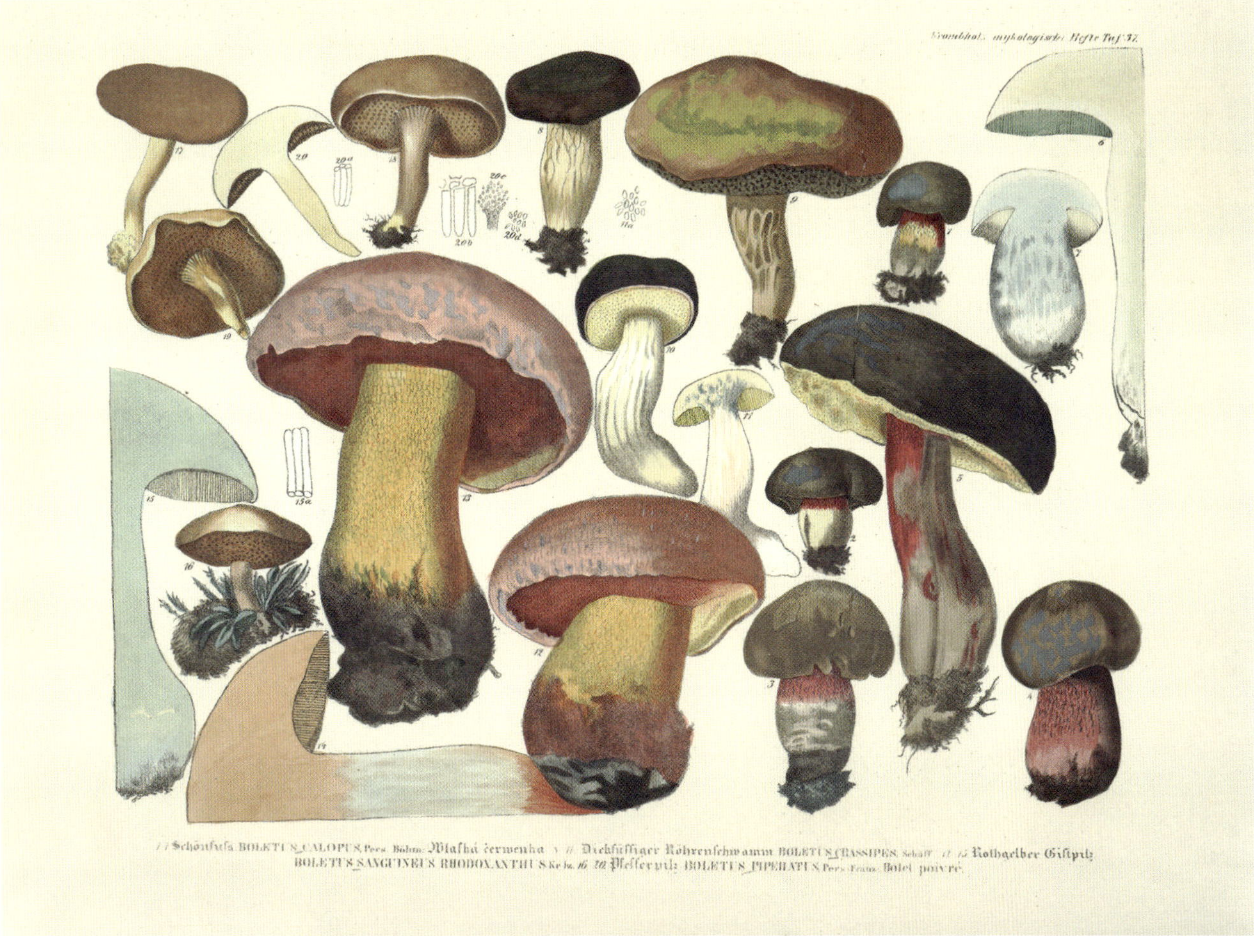

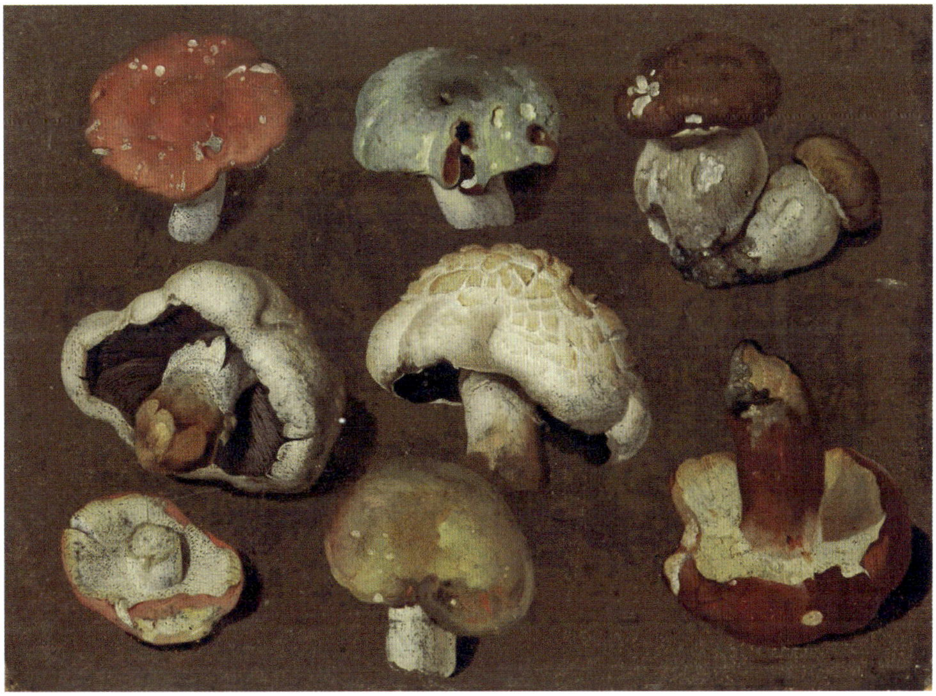

←
Chanterelle (*Cantherellus*) mushrooms, shown in this image from Japanese artist Maruyama Ōkyo's sketchbook (*c.* 1770), can't be farmed as they rely on their mycorrhizal relationships to the roots of specific species of trees in order to survive.

↑
⊖ This 19th-century study is by J. V. Krombholz, who sought to identify and illustrate edible, harmful and cautiously named 'suspicious' species of mushrooms. The mushrooms grouped in the bottom left of the image are identified as what is commonly known ruddy bolete.

→
This oil painting (*c.* 1736) is just one example of many studies of mushrooms produced by Philippe Ferdinand de Hamilton.

(1)

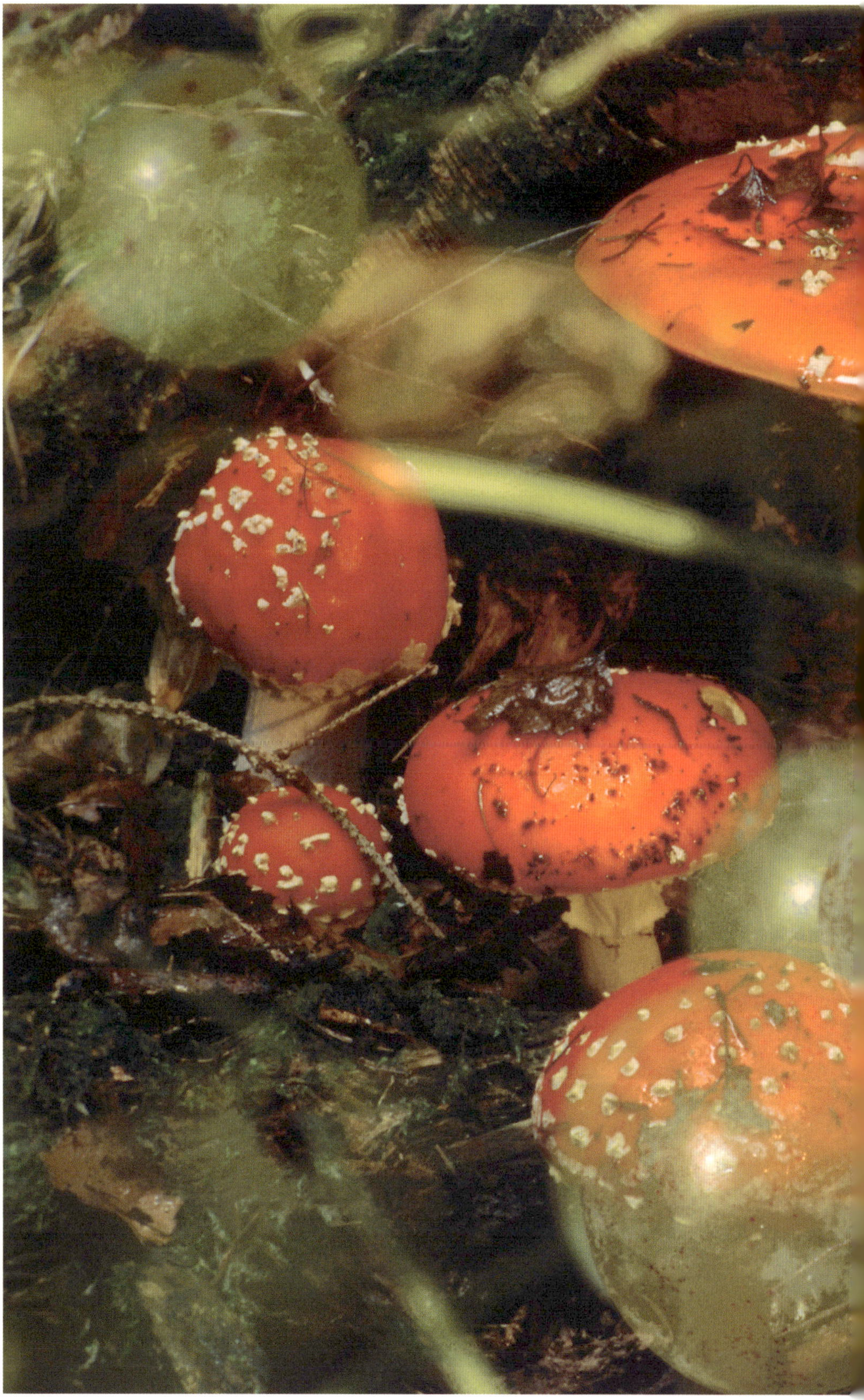

→
This image from Peter Fischli and David Weiss's 1997 to 1998 series of flower portraits, *Untitled (Flowers)*, shows the fly agaric (*Amanita muscaria*), a highly toxic and hallucinogenic fungus. In the past it was used to poison flies.

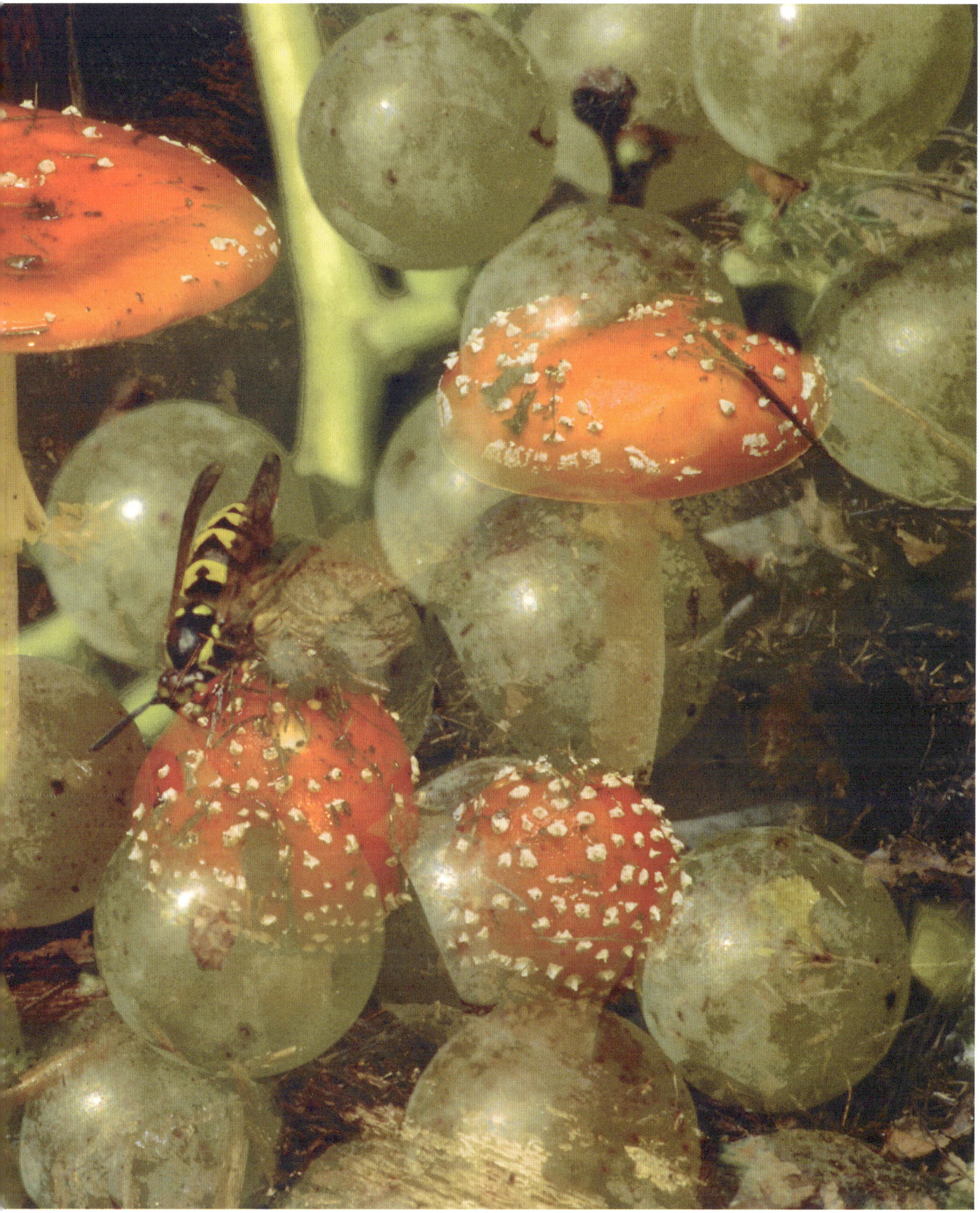

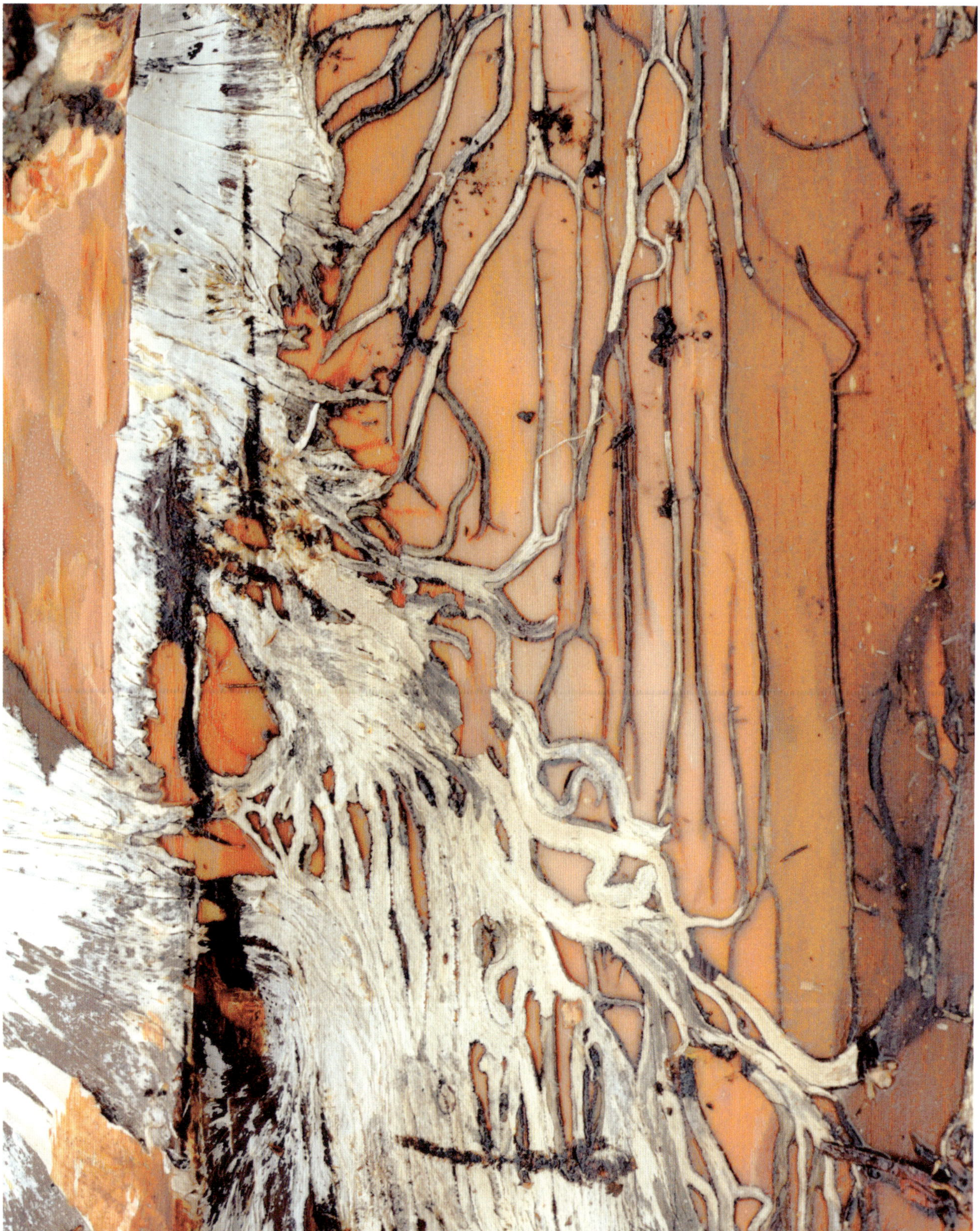

(1)

'The ever-present spores of some of the fungi have germinated ... and have sent out their delicate mycelia over the wood, the threads penetrating any season cracks or fissures, thence piercing and growing in the wood-cells; the manner of growth is not only interesting but wonderful, and almost leads one to think, for an instant, that the fungi, like animals, have instincts to protect them in their development.'

P. H. DUDLEY, 'WOODS AND THEIR DESTRUCTIVE FUNGI II', *POPULAR SCIENCE MONTHLY*, SEPTEMBER 1886

←
The mycelium of a wood-decay fungus resembles a root-like network of white filaments on the underside of bark. It can cause wood in living tress to rot in a short space of time.

↓
Sixteen species are included in this *c.* 1827 lithograph by A. Cornillon, including, centre left, the lattice stinkhorn (*Clathrus ruber*); bottom left, the tiger sawgill (*Lentinus tigrinus*); and bottom right, the jelly-like witches' butter (*Tremella mesenterica*).

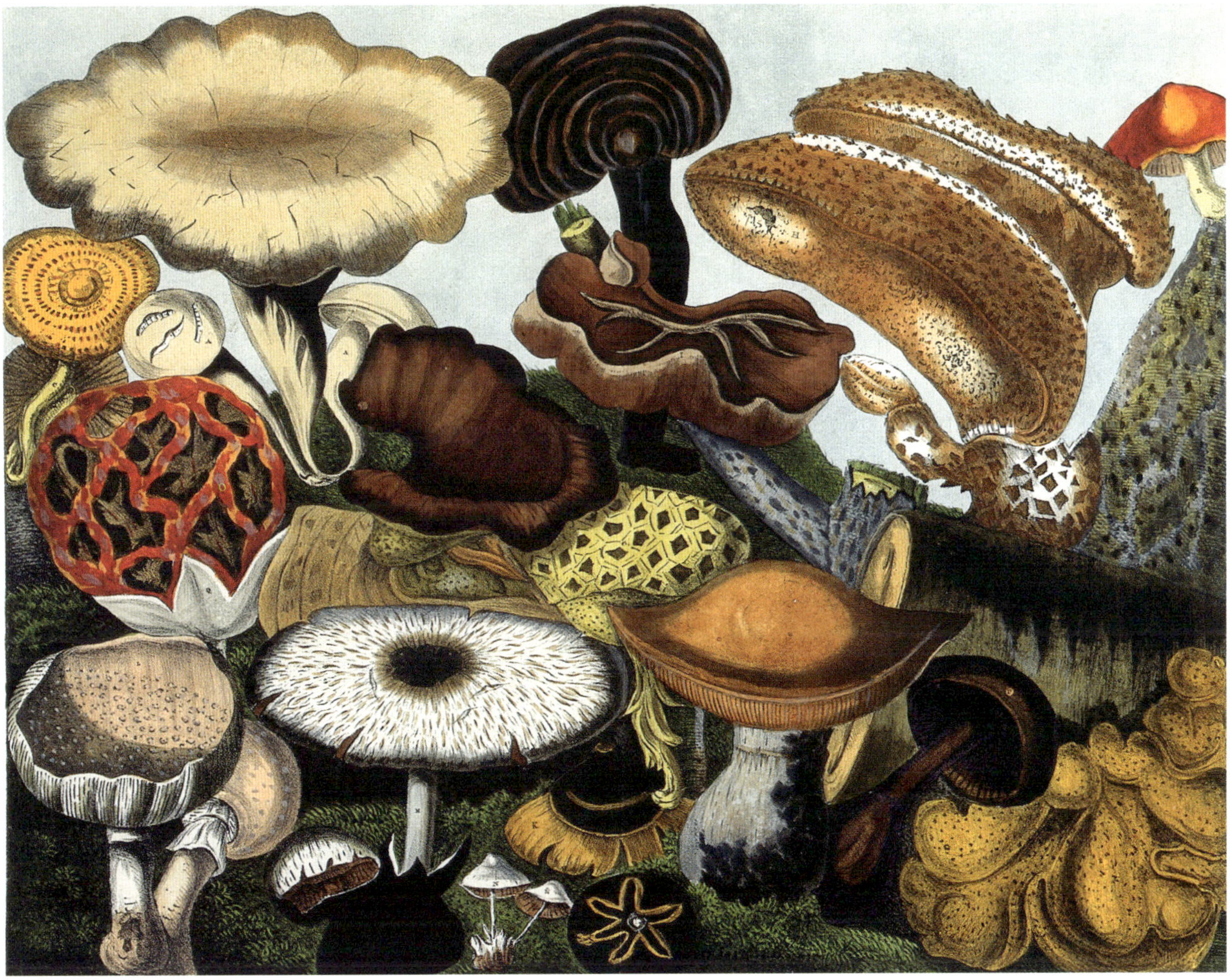

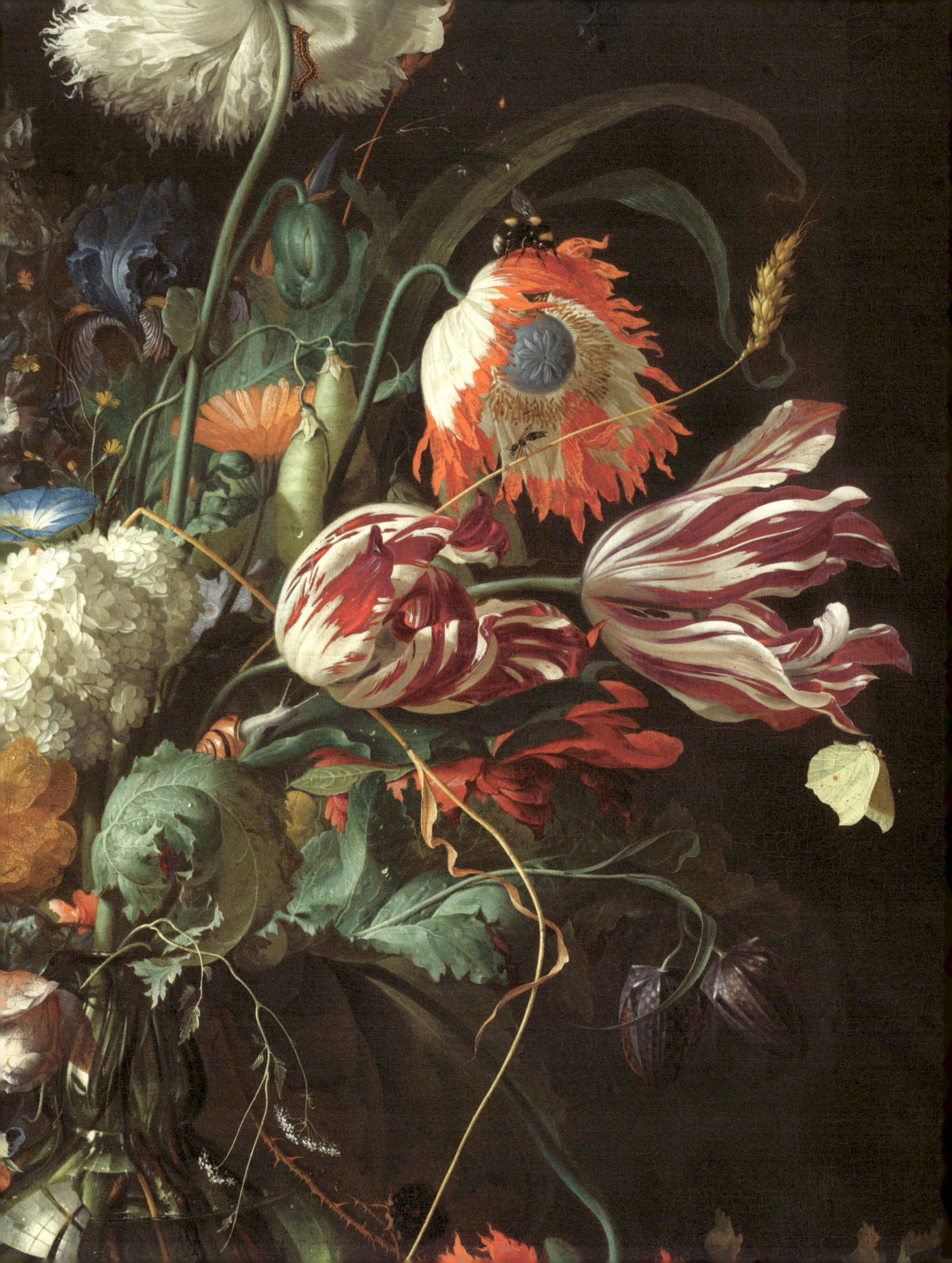

(II) PLANTS

MORNING GLORY	TULIP
HYDRANGEA	SNAKE'S HEAD FRITILLARY
LEGUME	SQUASH
SWEET POTATO	BUTTERNUT WALNUT
GRASS	VOODOO LILY
BLACK SALSIFY	MADONNA LILY
SUNFLOWER	WATER LILY
DANDELION	PEACH
BELLEVUE PLUM	BROWN BEURREE PEAR
GOLDEN BEURREE PEAR	COLMAR PEAR
PHOENIX APPLE	NORROWAY'S BEAUTY APPLE
ROMANO MELON	POTATO
SYCAMORE	BLACKBERRY
KUMQUAT	ASIAN CULTIVATED RICE
BUTTERBUR	GRAPE
LOTUS	PURPLE LAVER
WINGED KELP	FAN WEED
AMARYLLIS	ORCHID
JAPANESE SACRED LILY	NIGHT-BLOOMING CEREUS
ORIENTAL POPPY	BREADSEED POPPY
CABBAGE ROSE	PEONY
GIANT SEQUOIA	BEAR'S BREECHES
SMALL SCABIOUS	AMERICAN MAIDENHAIR FERN
NASTURTIUM	POMEGRANATE
RED CABBAGE	BUTTERCUP
QUEEN OF THE NIGHT CACTUS	THISTLE
CABBAGE	CHRYSANTHEMUM
ENGLISH OAK	CABBAGE PALM
GREATER PLANTAIN	GRANDIDIER'S BAOBAB
MANGROVE	CROCUS

PLANTS are mostly photoautotrophs, meaning they produce their own food through photosynthesis, which uses the pigment chlorophyll, combined with sunlight, minerals, water and carbon dioxide. Their defining features are that they have rigid, cellulose cell walls, are essentially capable of unlimited growth, do not have a nervous system and are more or less stationary. They come in many forms, including trees, flowers, mosses, grasses and ferns, and can be found in nearly every environment on the planet.

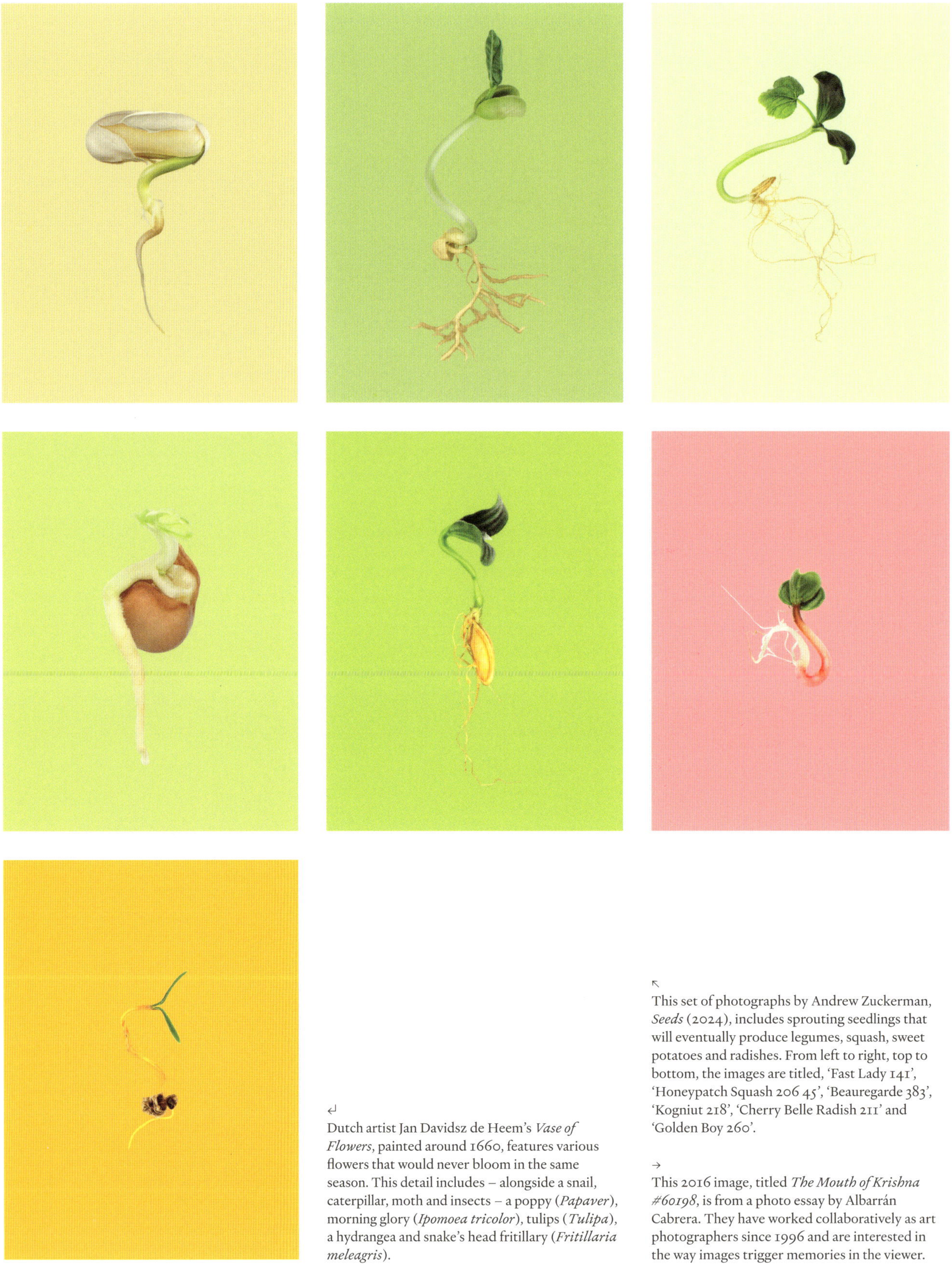

↩
Dutch artist Jan Davidsz de Heem's *Vase of Flowers*, painted around 1660, features various flowers that would never bloom in the same season. This detail includes – alongside a snail, caterpillar, moth and insects – a poppy (*Papaver*), morning glory (*Ipomoea tricolor*), tulips (*Tulipa*), a hydrangea and snake's head fritillary (*Fritillaria meleagris*).

↖
This set of photographs by Andrew Zuckerman, *Seeds* (2024), includes sprouting seedlings that will eventually produce legumes, squash, sweet potatoes and radishes. From left to right, top to bottom, the images are titled, 'Fast Lady 141', 'Honeypatch Squash 206 45', 'Beauregarde 383', 'Kogniut 218', 'Cherry Belle Radish 211' and 'Golden Boy 260'.

→
This 2016 image, titled *The Mouth of Krishna #60198*, is from a photo essay by Albarrán Cabrera. They have worked collaboratively as art photographers since 1996 and are interested in the way images trigger memories in the viewer.

'Aye, flowers! The very name of flowers,
That bloom in wood and glen,
Brings Spring to me in Winter's hours'

JOHN CLARE, 'THE TELL-TALE FLOWERS', 1837–64

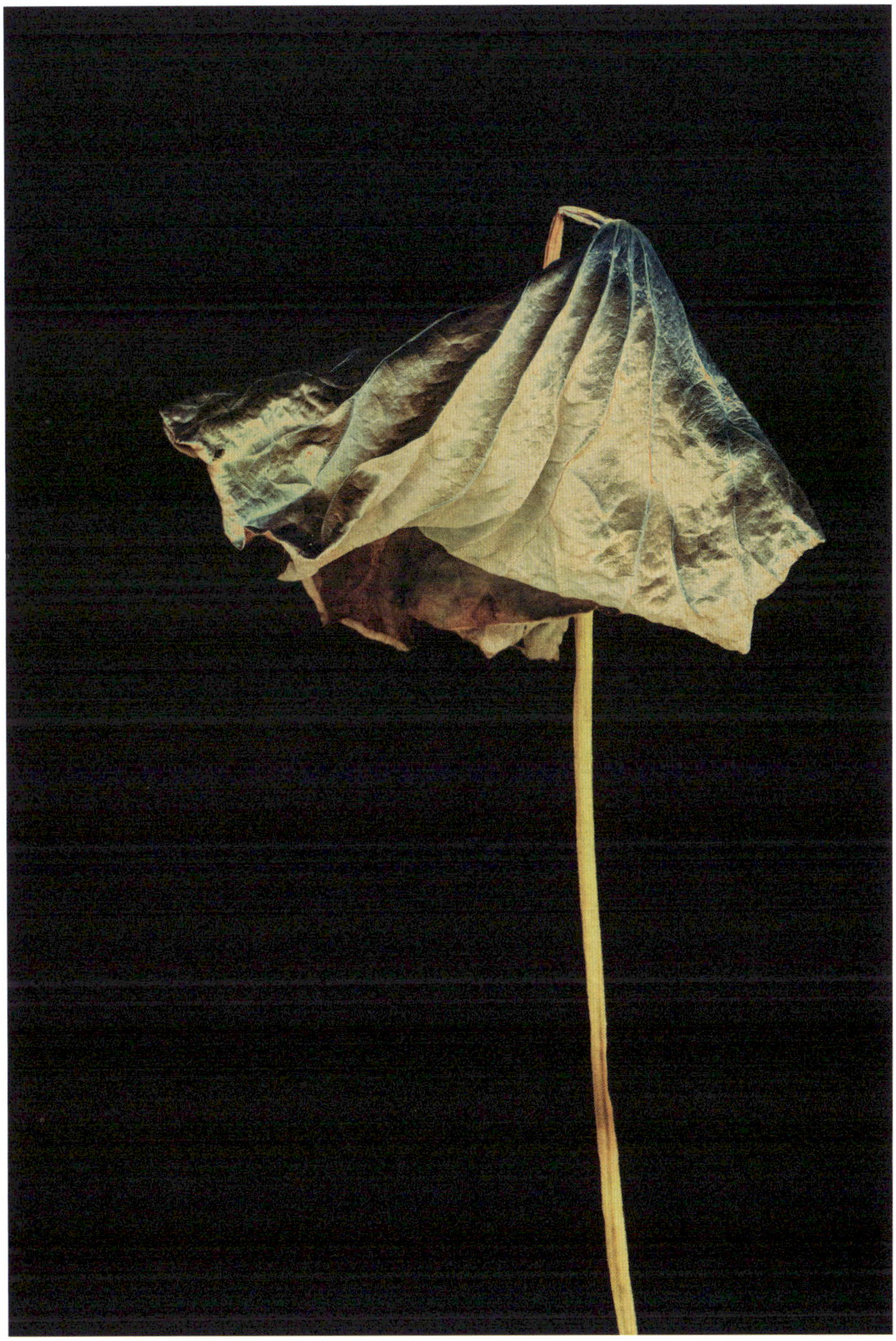

'The ear of wheat returns a hundredfold the grain from which it grew. The surface of the earth offers to us far more than we can consume.'

RICHARD JEFFERIES, *THE LIFE OF THE FIELDS*, 1884

(II)

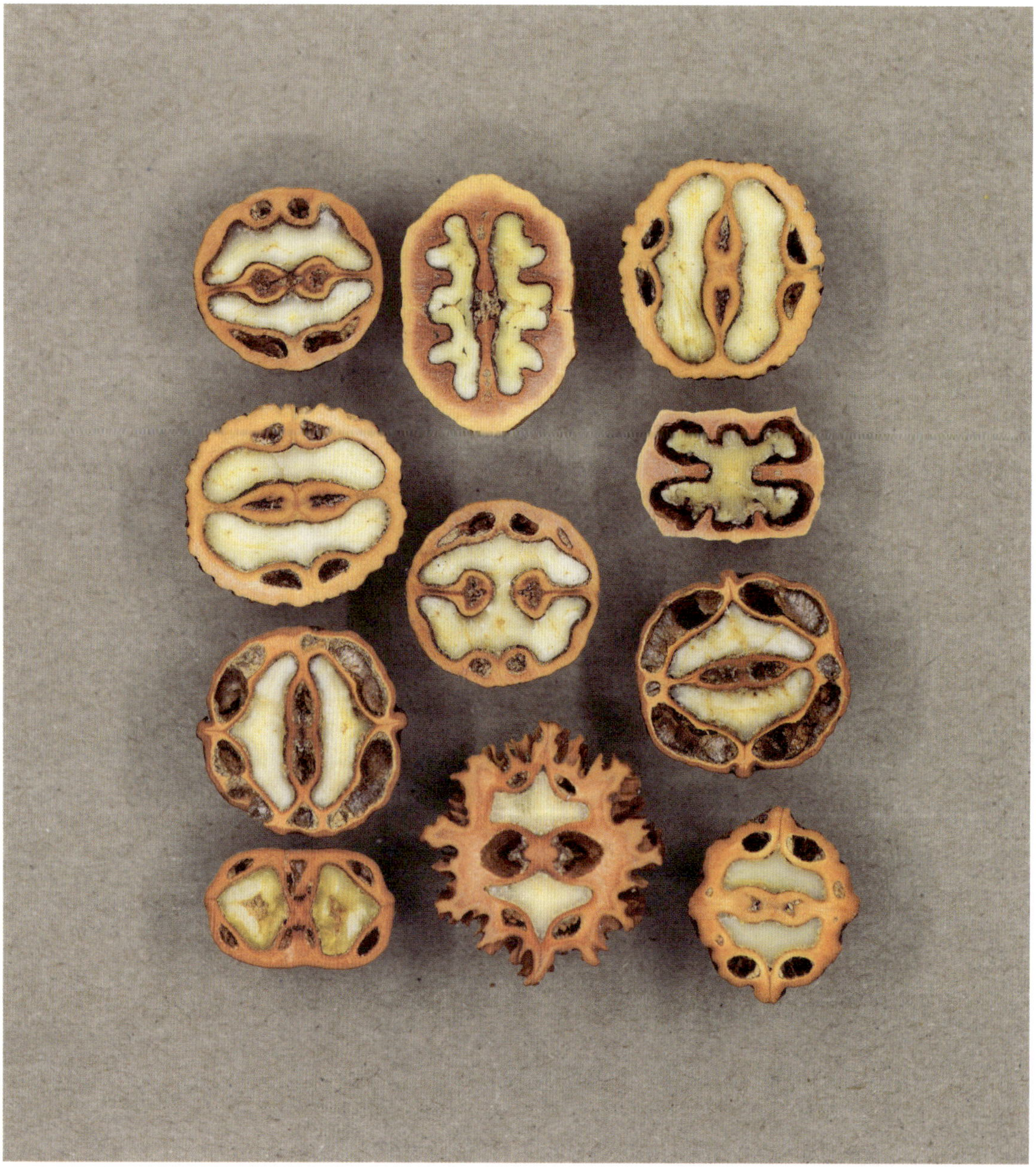

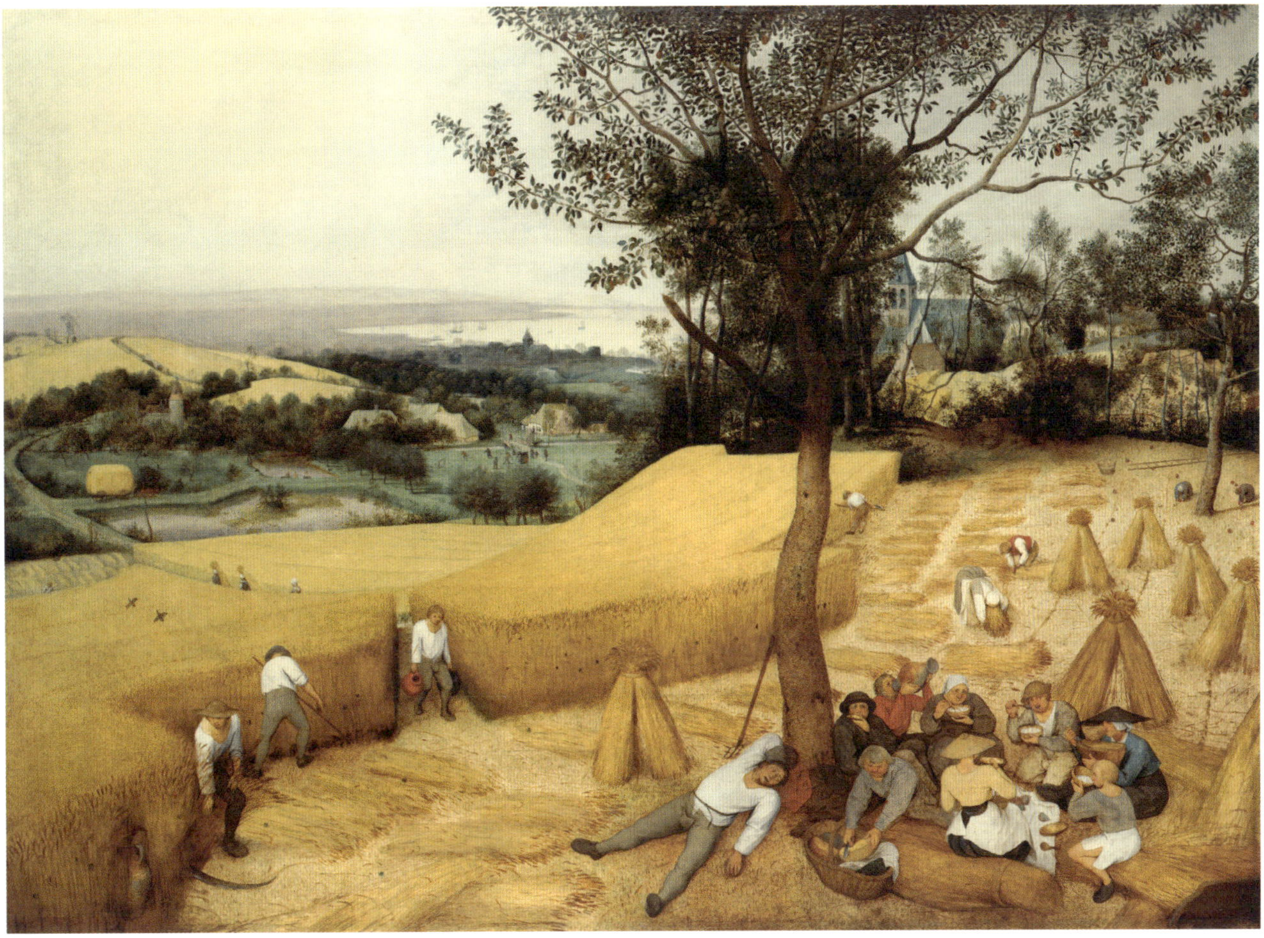

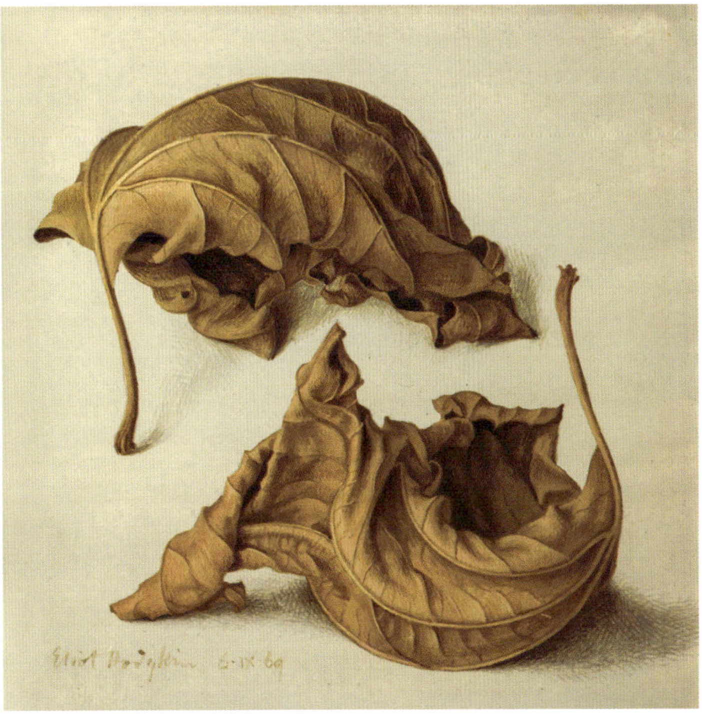

←
◎ One of the species of walnut selected by Jonas Frei for this 2024 artwork is endangered: the butternut walnut (*Juglans cinerea*). The bark and nut rinds from this tree can be used to dye cloth in shades ranging from light yellow to dark brown, a technique that was used in parts of North America in the 19th century. During the American Civil War, some of the Confederate troops from remote areas of the South used this dye for their homemade uniforms, leading to their nickname: butternuts.

↑
Pieter Bruegel the Elder's 1565 painting *The Harvesters* depicts late-summer country life: peasants harvesting wheat, and eating bread, cheese and pears. It is one of a series of six paintings that were commissioned by an Antwerp merchant for his country home.

→
This tempera painting, *Two Dead Leaves No. 2* (1969), exemplifies Eliot Hodgkin's artistic goal, as he described it to *The Studio* magazine in 1957, to 'show the beauty of natural objects which are normally thought uninteresting or even unattractive.... People sometimes tell me that they had never really "seen" something before I painted it, and I should like to believe this.'

'When the pollen is of full size ... out the pollen flies. The growth of the plant begins here, with the pollen.'

JULIA MCNAIR WRIGHT, *NATURE READERS: SEA AND THE WAY-SIDE*, NO. 3, 1888

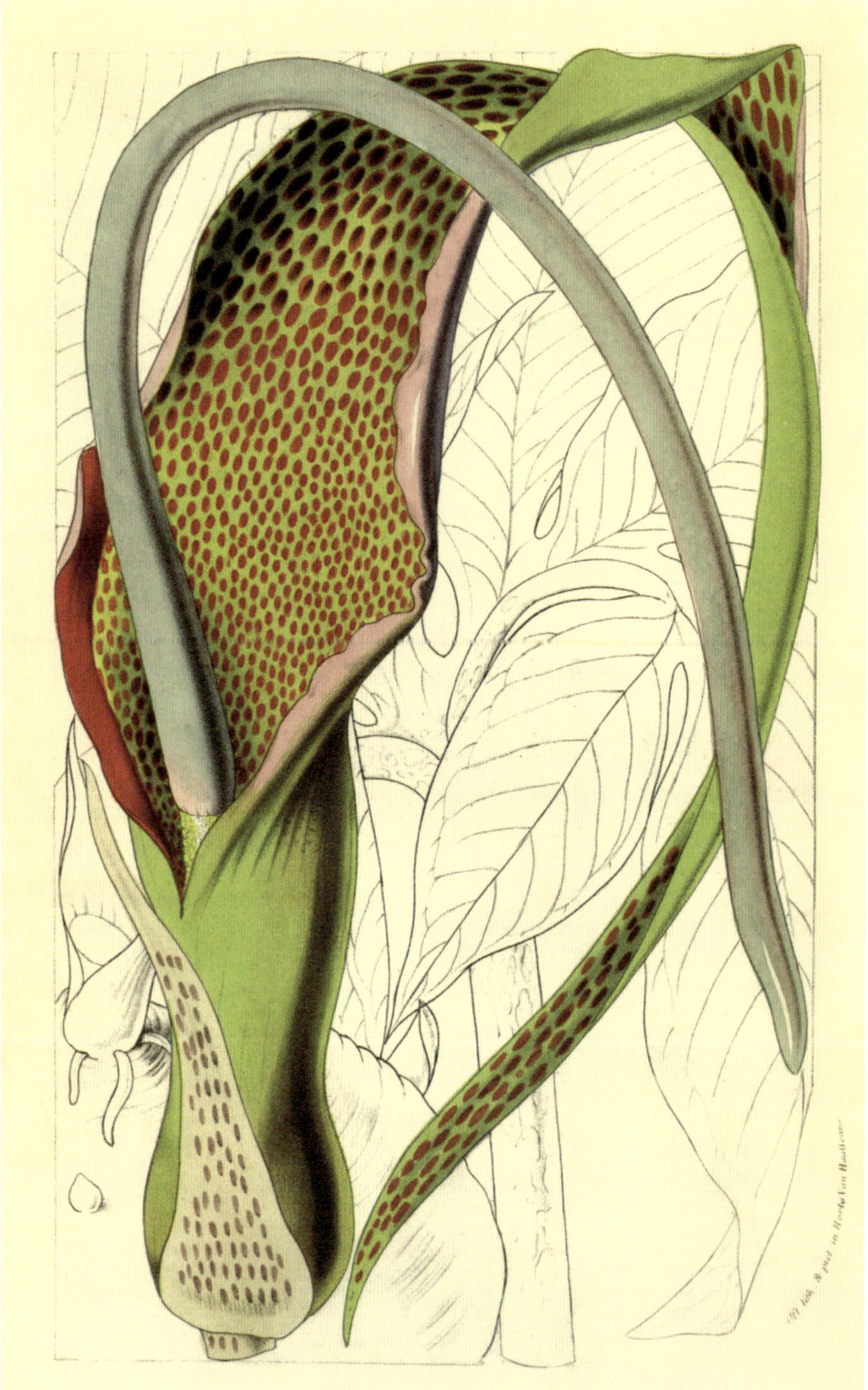

←
This illustration of a voodoo lily (*Sauromatum venosum*, formerly *S. guttatum*), a native to Asia and Africa, is from *Flore des serves et des jardins de l'Europe* ('Flora of the Meadows and Gardens of Europe') by the 19th-century Belgian horticulturist, Louis van Houtte.

→
⊖ Carl Julius Fritzsche used a microscope with a magnification of 500x to make these studies of the pollen grains of flowering plants, in order to understand their structure. The illustrations, published in *Uber den Pollen* ('About Pollen') in 1837, include, top left, black salsify (*Scorzonera*); and, top right, the Madonna lily (*Lilium candidum*), which is near-threatened on the IUCN Red List.

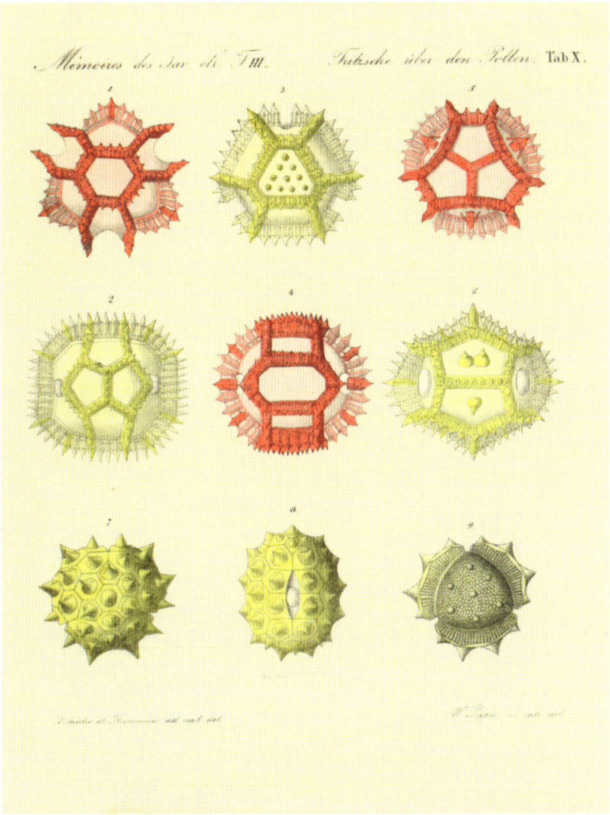
Tab X.

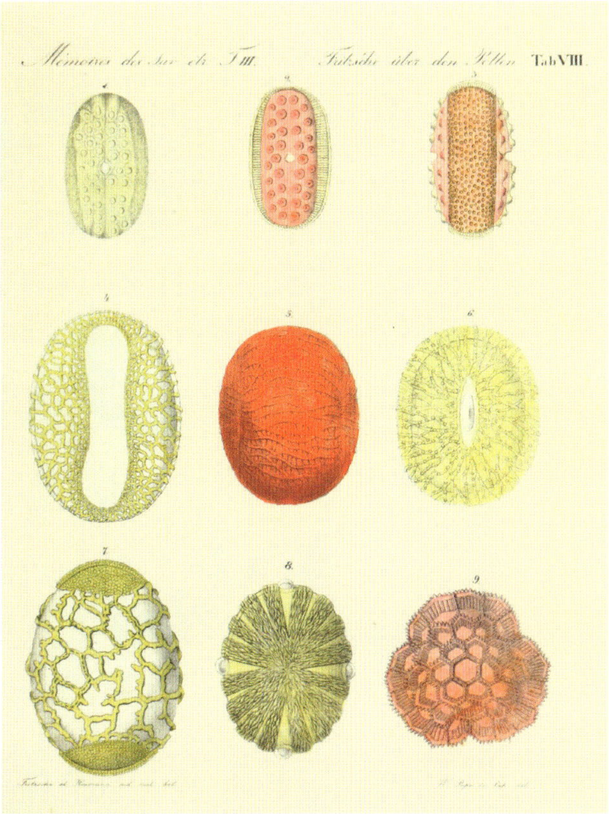
Tab VIII

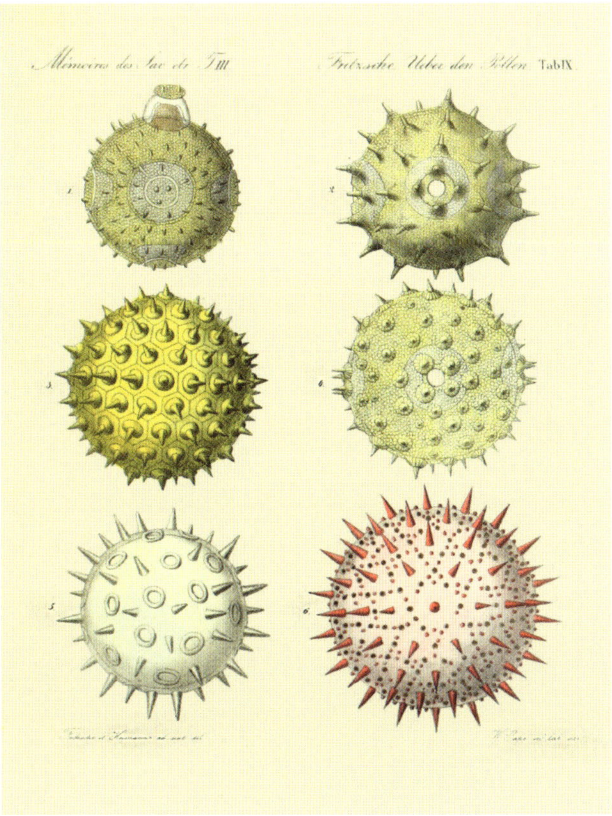
Tab IX.

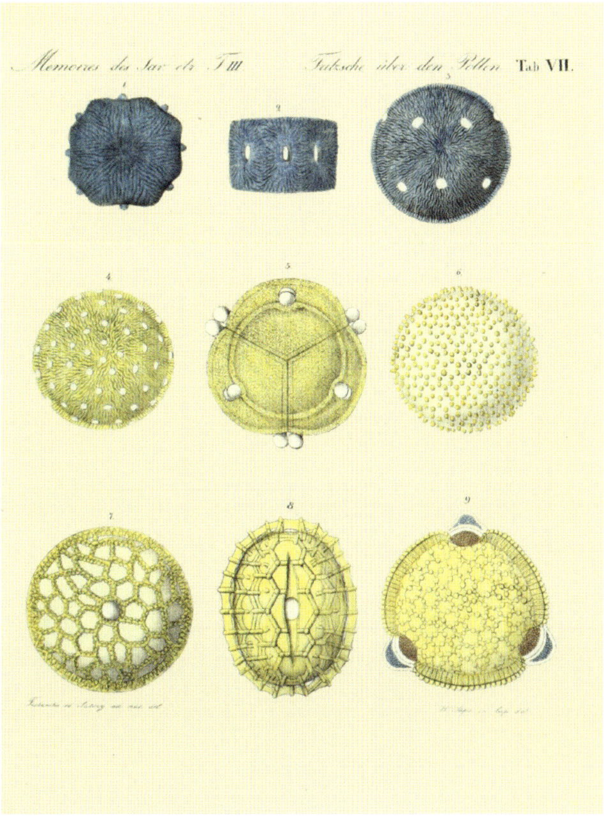
Tab VII.

'I am not a lover of lawns. ... Rather would I see daisies in their thousands, ground ivy, hawkweed, and even the hated plantain with tall stems, and dandelions with splendid flowers and fairy down, than the too-well-tended lawn.'

WILLIAM HENRY HUDSON, *THE BOOK OF A NATURALIST*, 1919

(11)

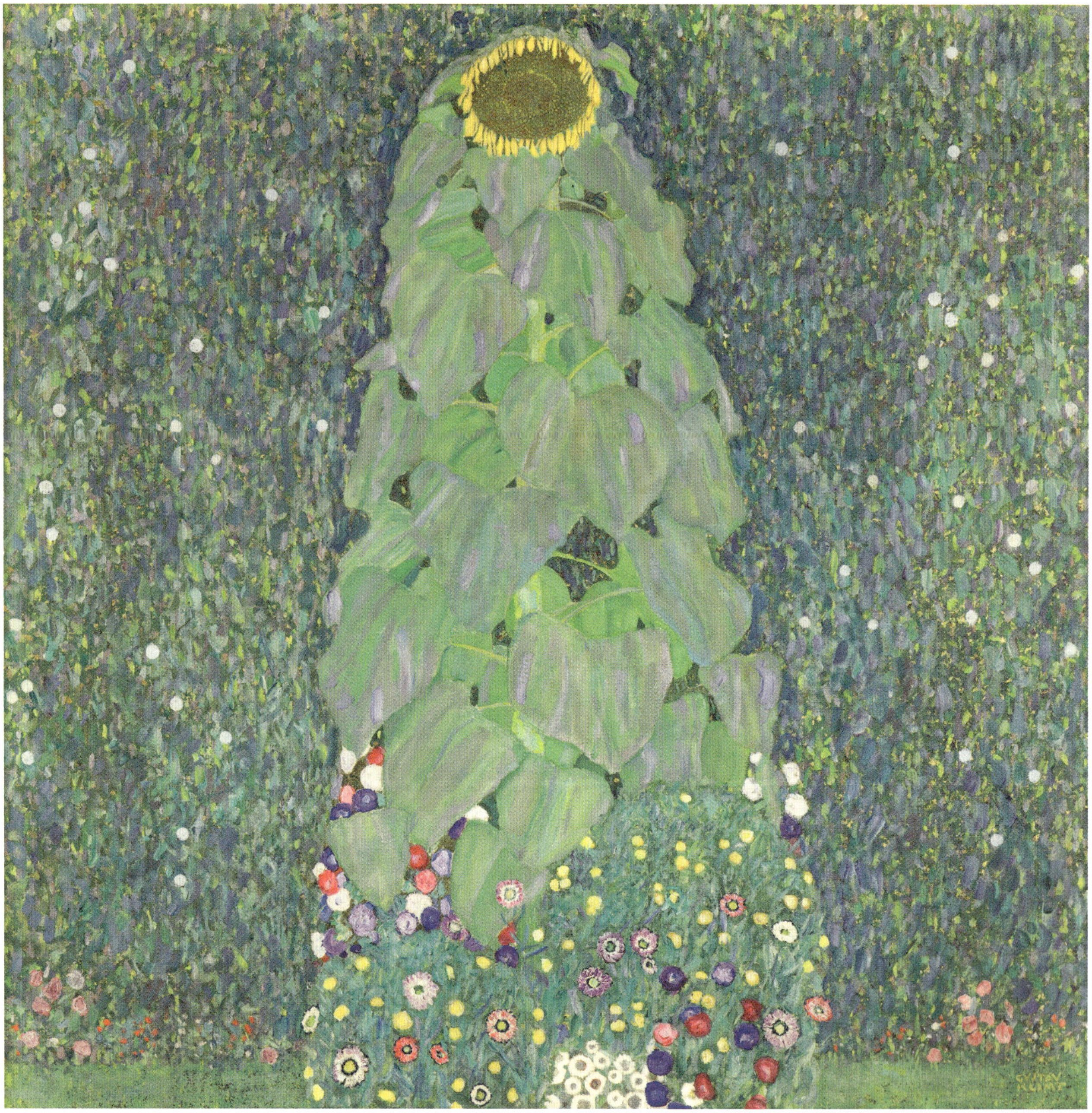

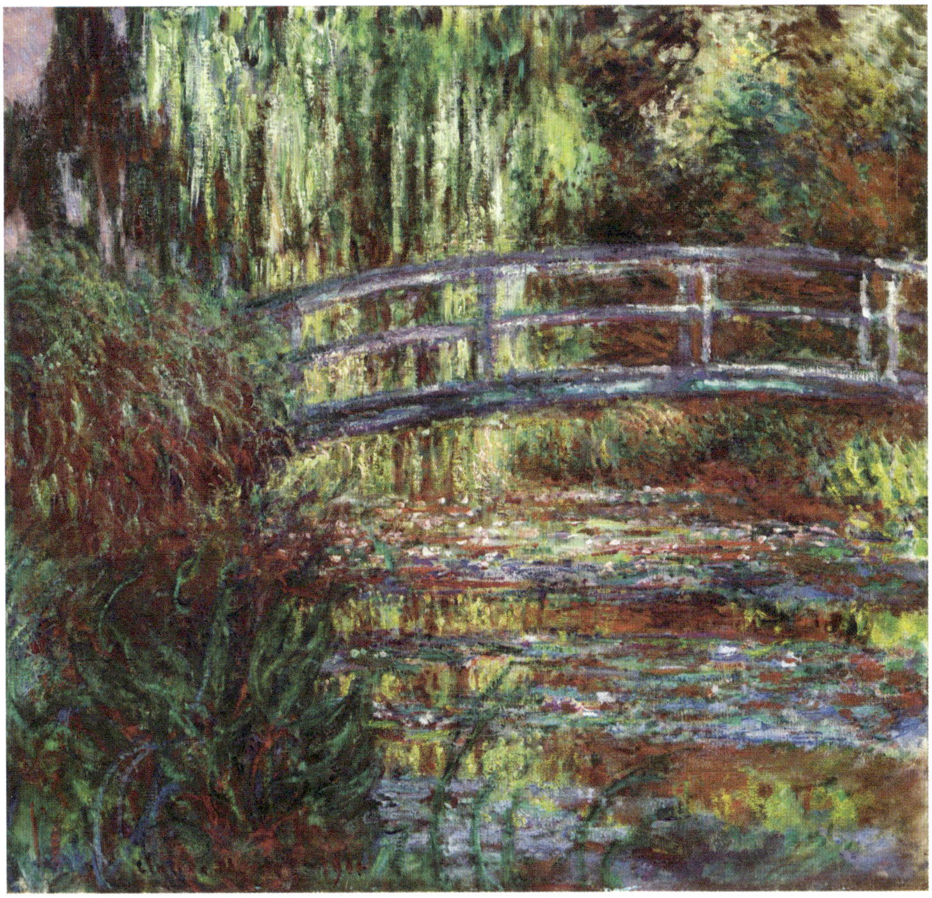

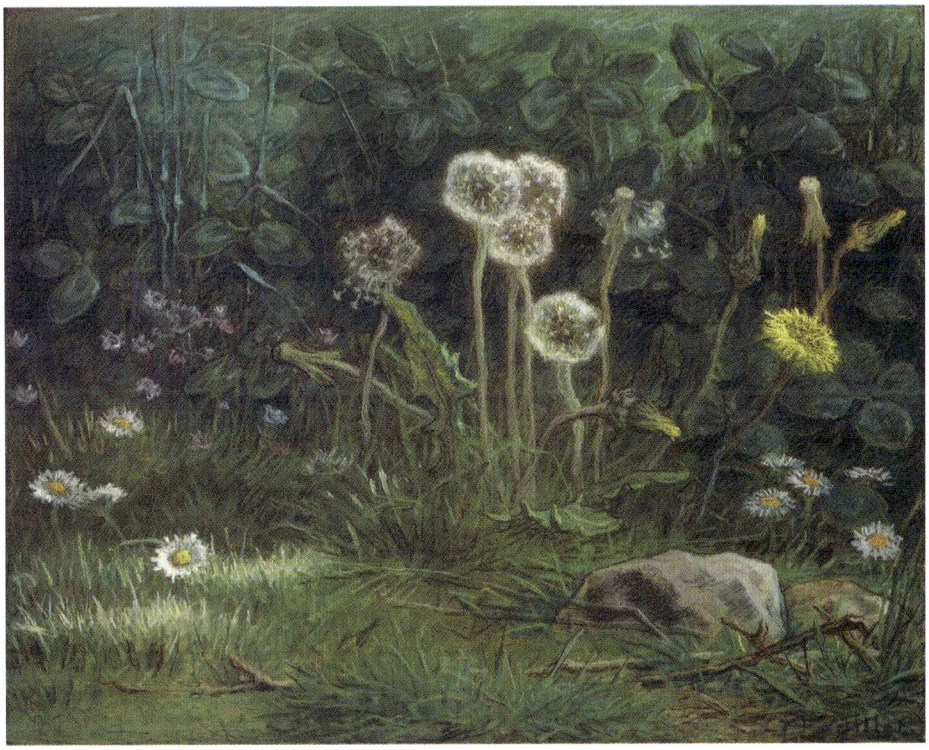

←
Typically, sunflower (*Helianthus*) heads follow the trajectory of the sun when young, before maturing to face east. Unusually, the head of Gustav Klimt's *Sunflower* (1907–8) droops to face the earth.

↗
The Water Lily Pond (1900) is one of many paintings by Claude Monet depicting his water garden at Giverny. Monet often depicted the new water lily (*Nymphaea*) varieties – hybrids of European lilies and tropical species from America – that were introduced to Paris in the late 19th century.

→
Jean François Millet's *Dandelions* (1867–68), which shows the different stages of growth of the common dandelion (*Taraxacum*) flower, celebrates the beauty of humble plants that might be considered weeds.

'Buy me no meat, but mellow
Apricots, melons yellow,
Cream, and strawberries.
These have the sweetest savor'

PIERRE DE RONSARD, 'SUMMER'S REVEL', 1524–85,
TRANSLATED BY CURTIS HIDDEN PAGE

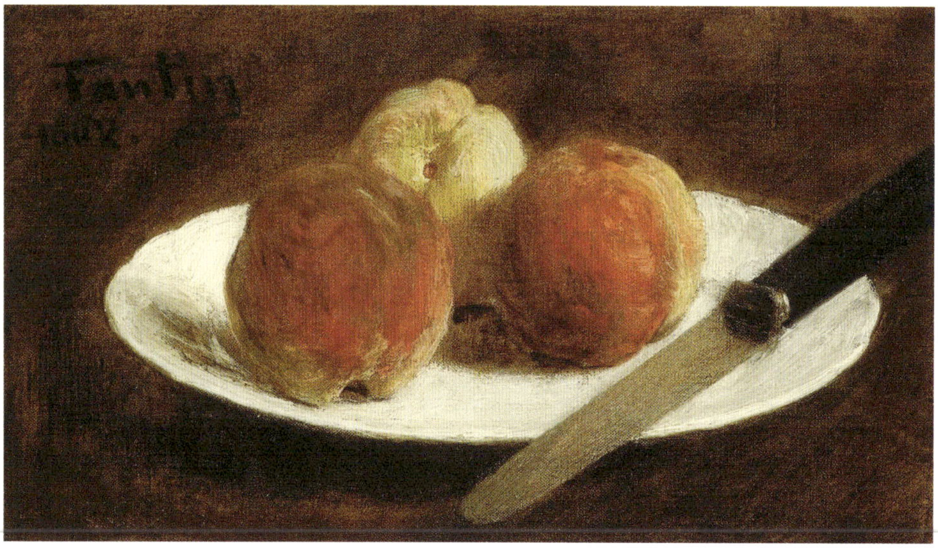

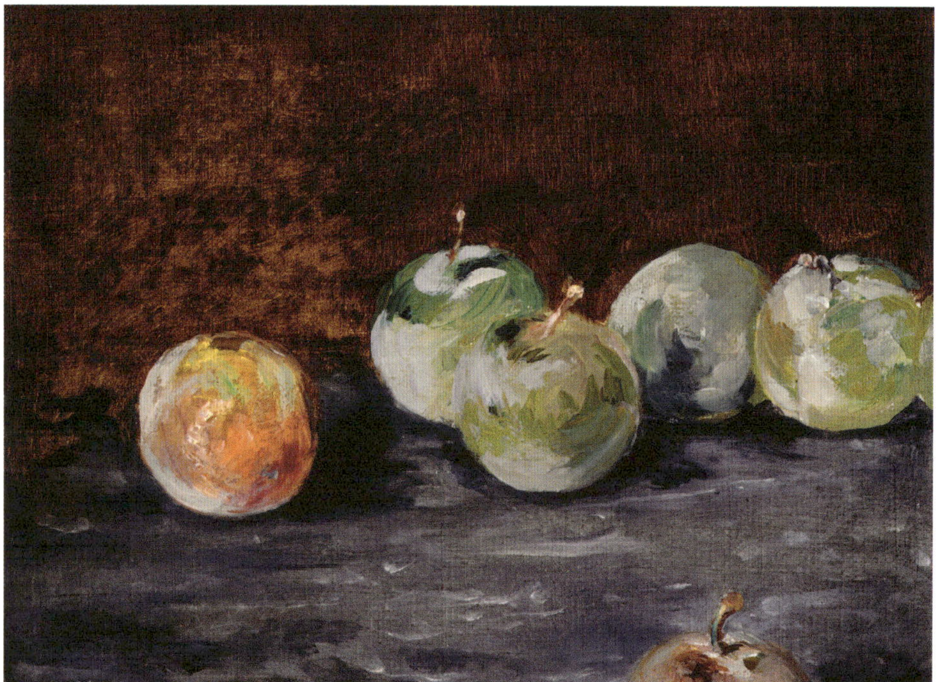

↖
One of several still lifes by Henri Fantin-Latour that focus on peaches (*Prunus persica*), this 1862 example could show the Téton de Vénus cultivar, which was grown near to Paris around the time of the artist.

←
In 1880 Édouard Manet spent five months in Bellevue, to the west of Paris, where he underwent a course of hydrotherapy. Prevented from working outside by bad weather, Manet turned his attention to still life and often illustrated his letters to friends with fruits, including these 'Belle de Louvain', or 'Bellevue', plums (*Prunus domestica*). He likely created this oil painting during, or upon his return from, this stay.

→
The US Department of Agriculture holds a remarkable collection of thousands of pomological watercolours painted between 1888 and the 1930s, some of which are included here. The works are by a variety of artists who were employed by the USDA to document the species grown in the US, including Ellen Isham Schutt, Elsie E. Lower, Deborah Griscom and Amanda Newton.

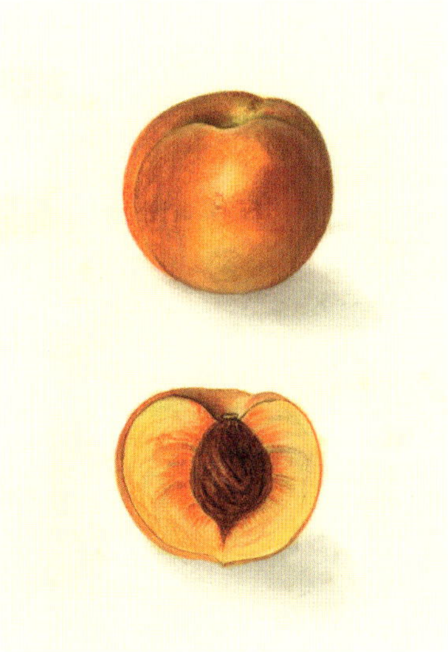
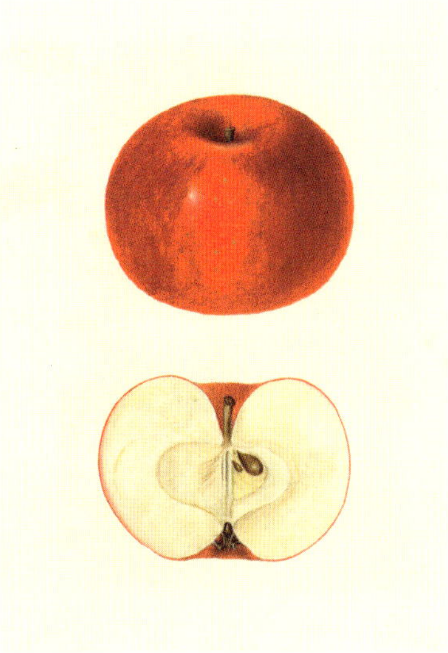
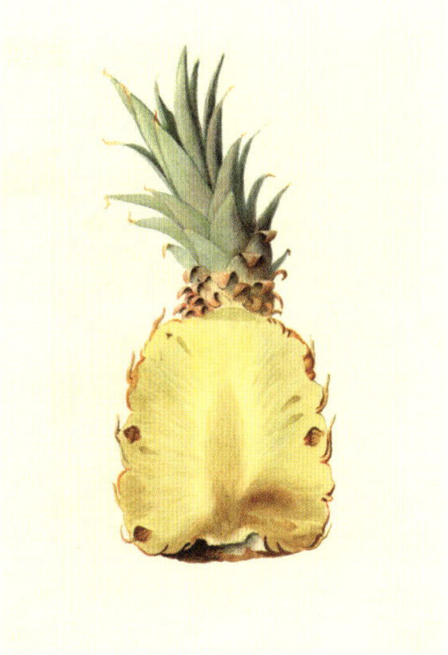
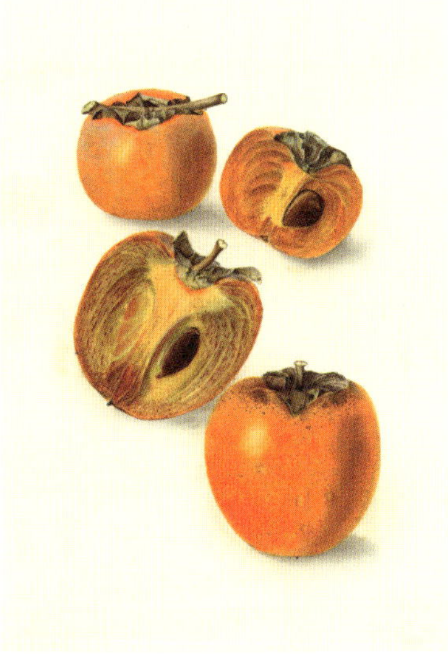
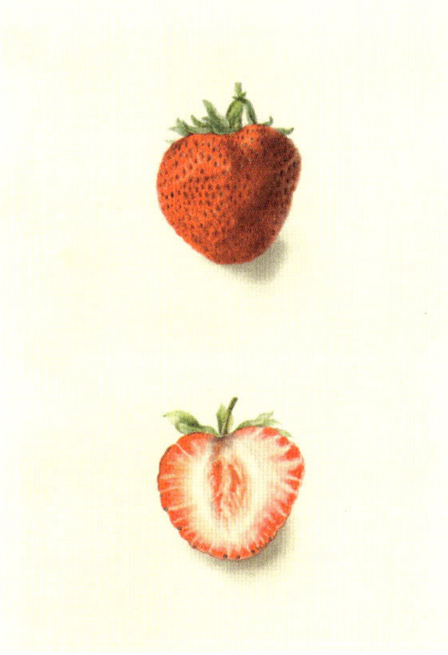
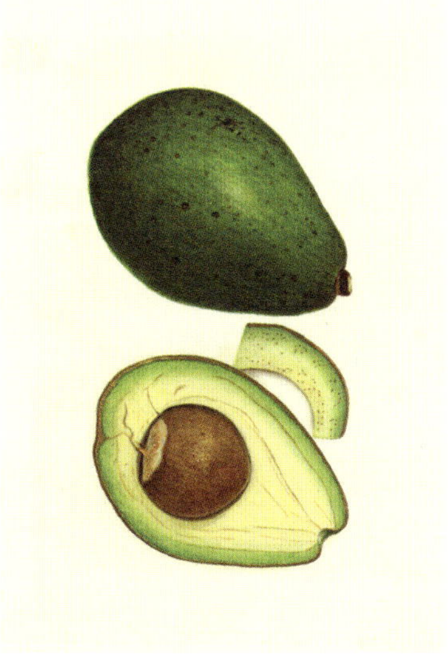
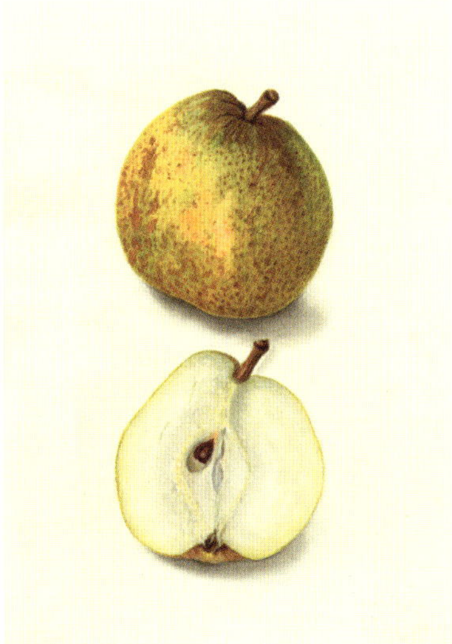
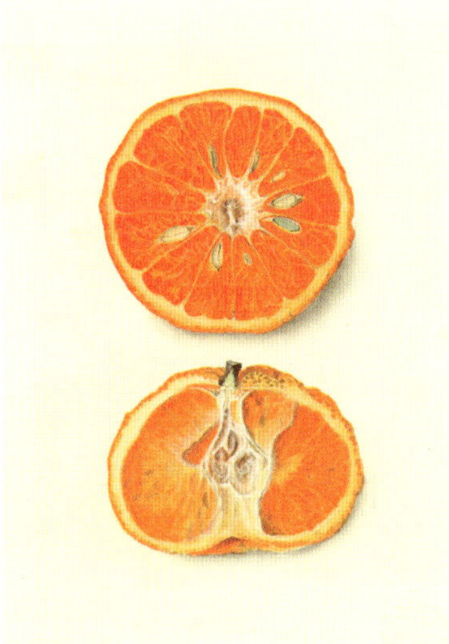
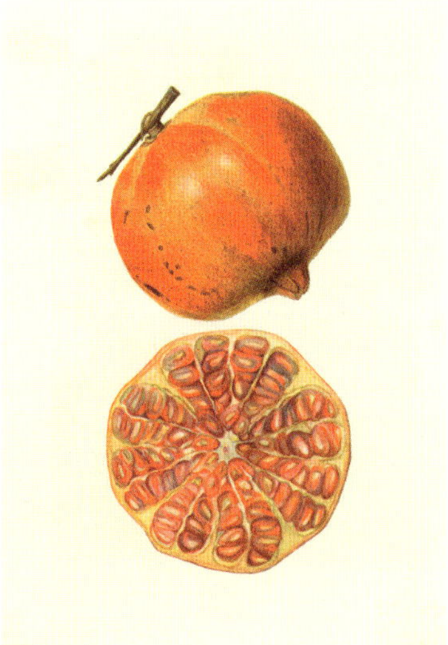

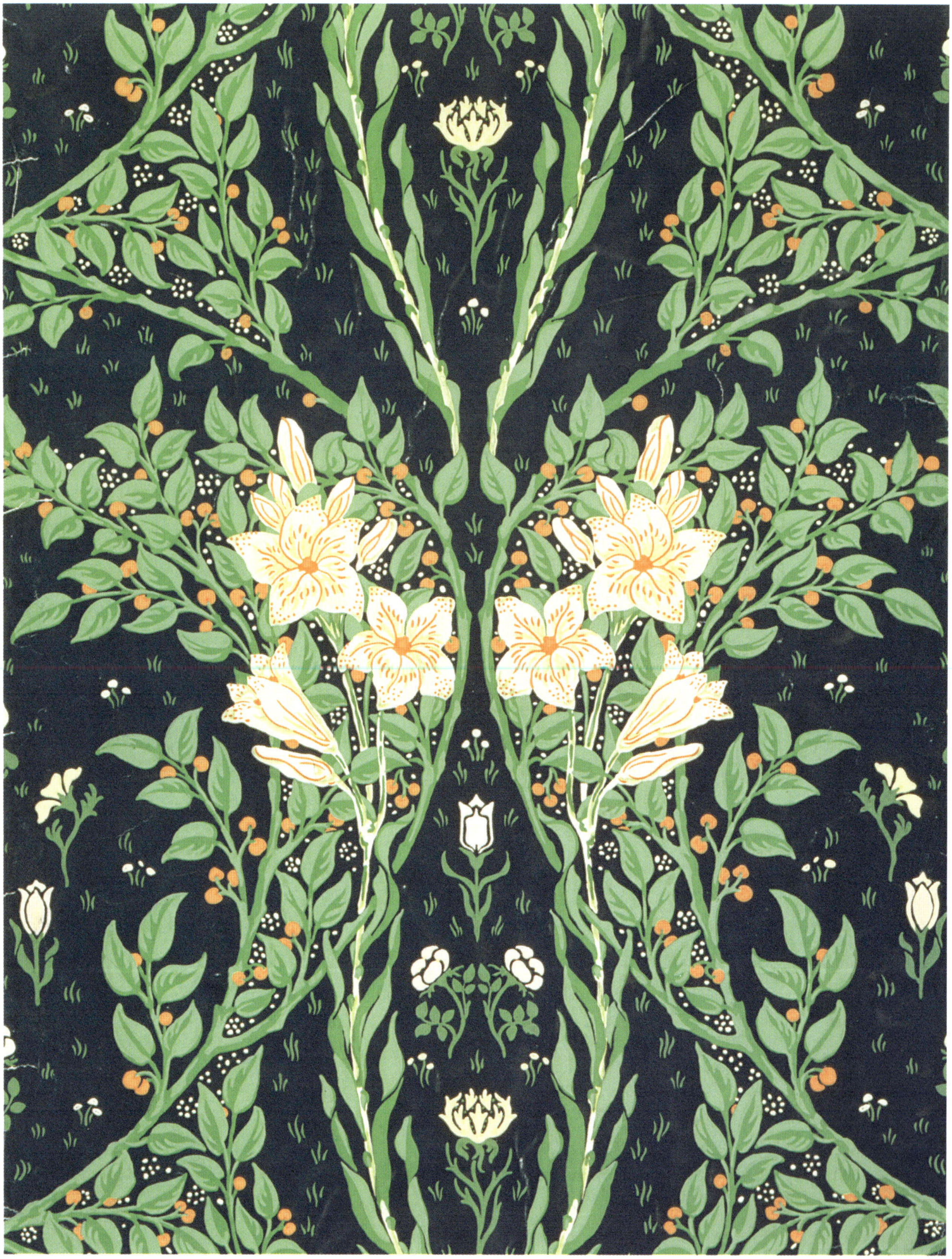

(11)

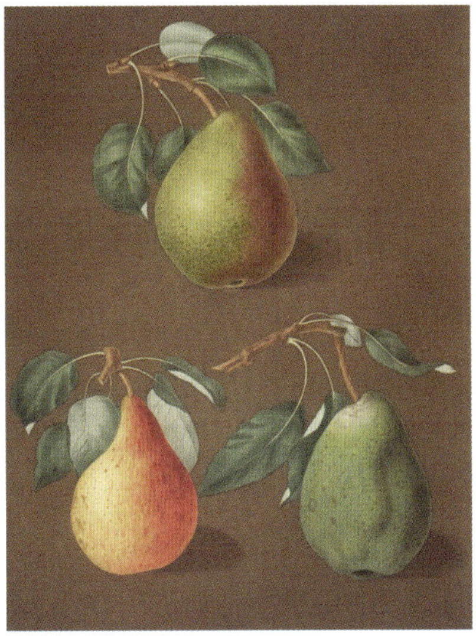
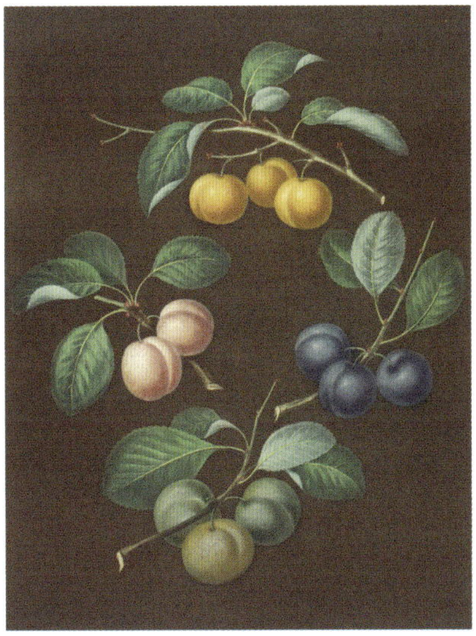
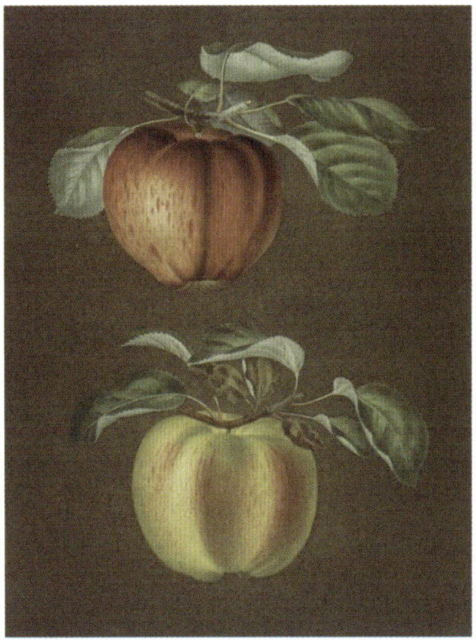

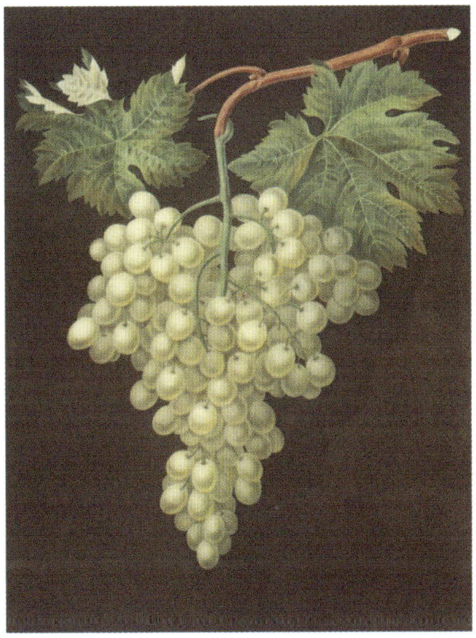
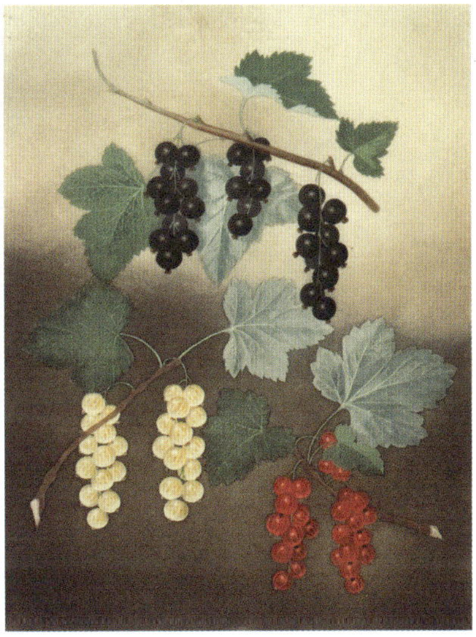
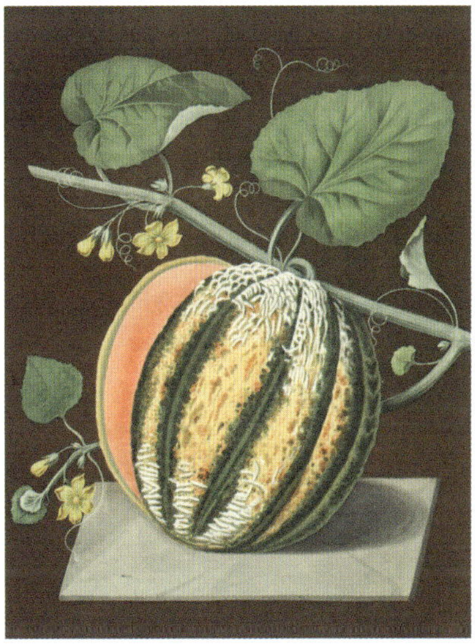

←
⊖ Painter, illustrator and designer, Walter Crane's 'Francesca' wallpaper design (1902) is based on lilies, possibly the Madonna lily (*Lilium candidum*).

↗
Many of the illustrations in George Brookshaw's *Pomona Britannica* (1812), a collection of ninety plates illustrating British fruit, were based on specimens in the gardens of Hampton Court Palace. Several varieties here, such as the 'Brown Beurree', 'Golden Beurree' and 'Colmar' varieties of pear (top left), the 'Phoenix' and 'Norroway's Beauty' apples (top right), and the scarlet flesh 'Romano' melon (centre right), are no longer cultivated.

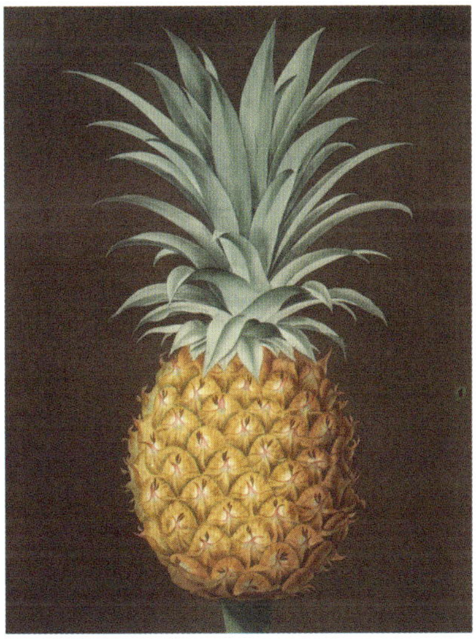
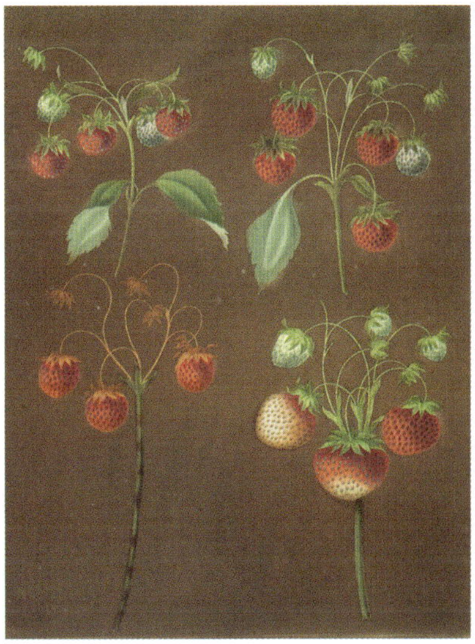

(11)

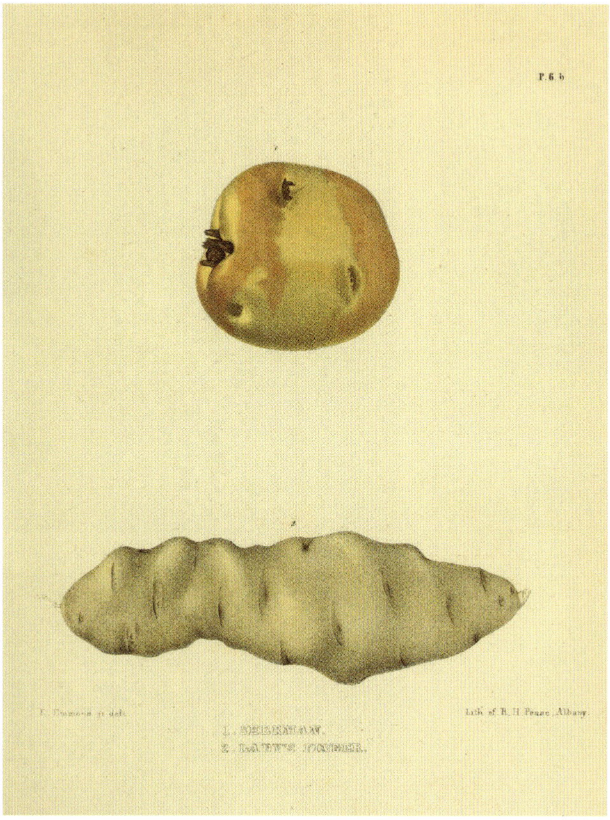
P. 6. b
Lith. of R. H. Pease, Albany.
2. LADY'S FINGER.

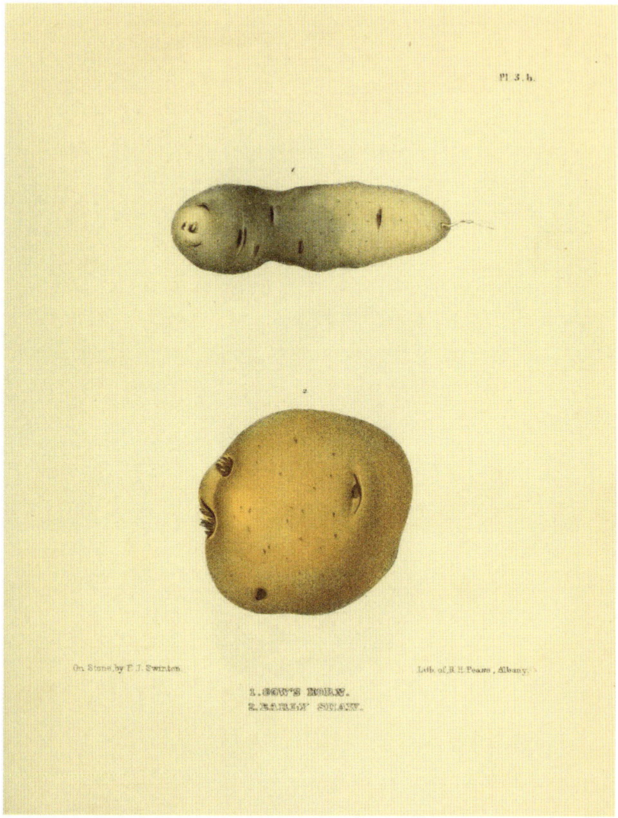
Pl. 3. b.
On Stone, by F. J. Swinton.
Lith. of R. H. Pease, Albany.
1. COW'S HORN.
2. EARLY SHAW.

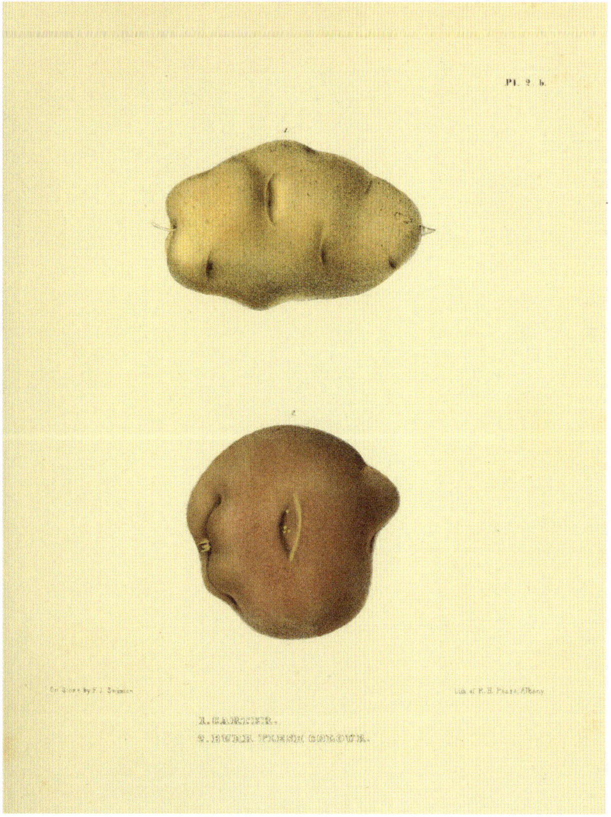
Pl. 2. b.
1. CARTER.

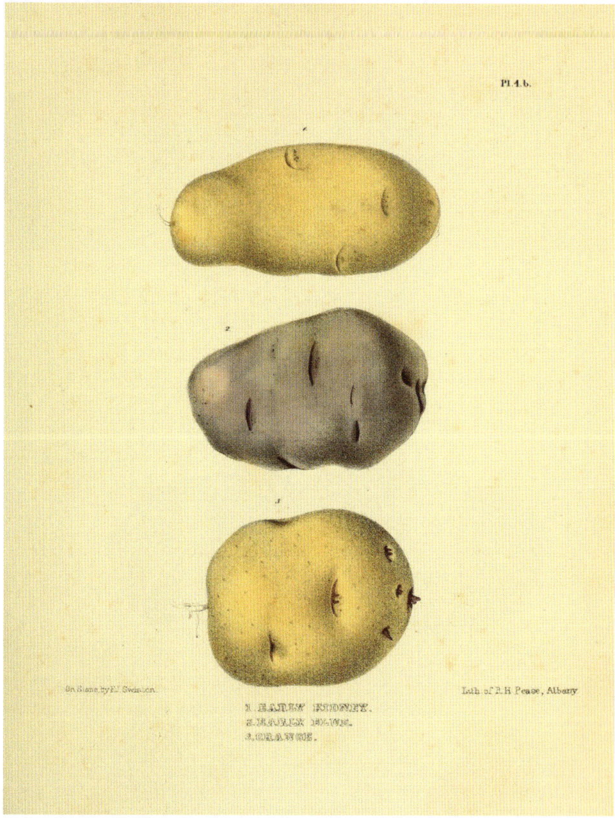
Pl. 4. b.
On Stone by F. J. Swinton.
Lith. of R. H. Pease, Albany
1. EARLY KIDNEY.
3. ORANGE.

'It has been demonstrated that an acre of potatoes will feed double the number of people that can be fed from an acre of wheat.'

MRS BEETON, 'THE POTATO AS AN ARTICLE OF HUMAN FOOD', *THE BOOK OF HOUSEHOLD MANAGEMENT*, 1861

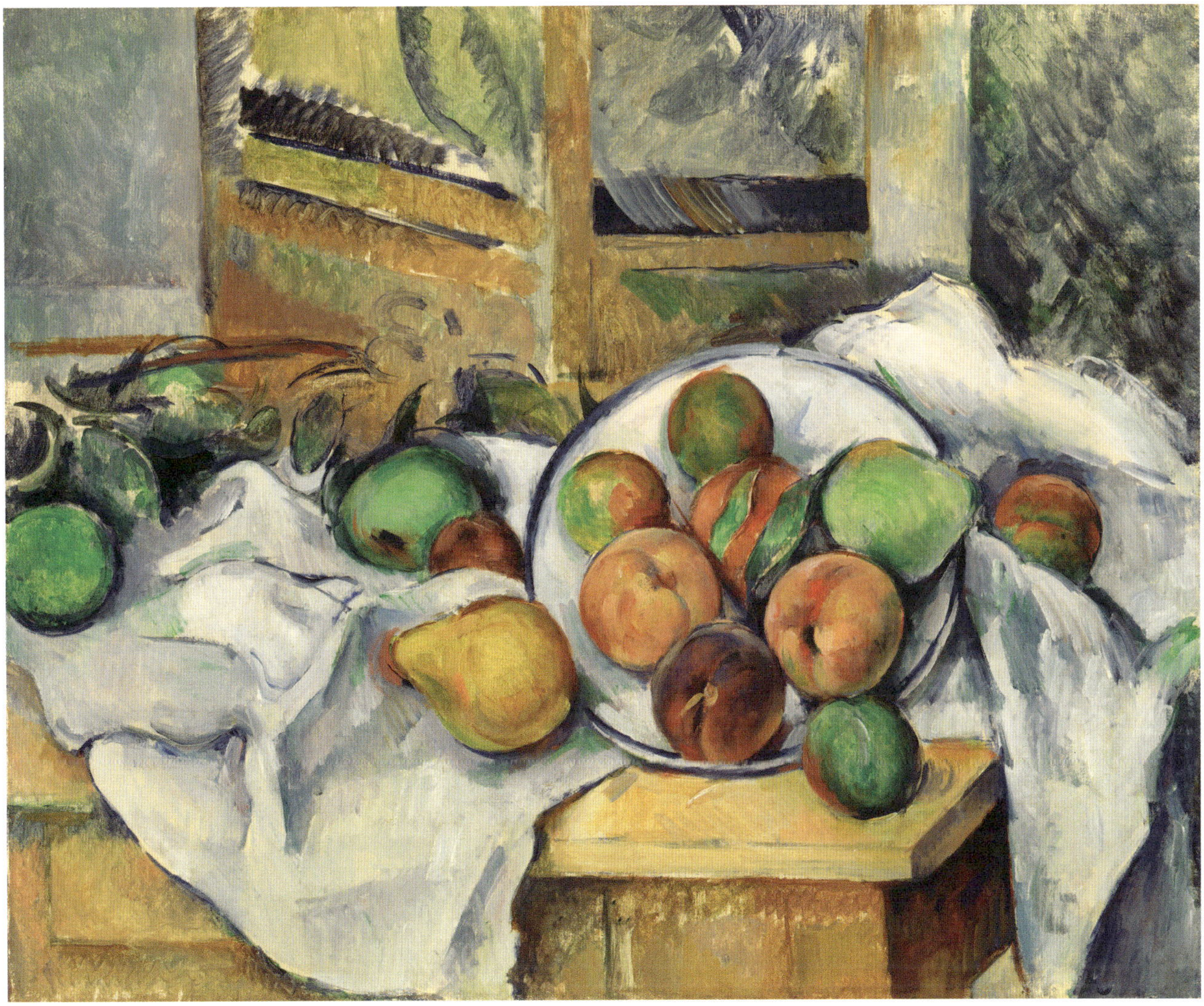

←
These scientifically accurate plates depicting heirloom and extinct potato (*Solanum tuberosum*) varieties by Ebenezer Emmons (1799–1863) were published in a survey of produce grown in New York State. Clockwise from top left, they show 'Sherman' and 'Lady's Finger'; 'Cow's Horn' and 'Early Shaw'; 'Early Kidney', 'Early Blue' and 'Orange'; and 'Carter' and 'Burr's Flesh Colour'.

↑
Paul Cézanne's *Un coin de table* (*A Table Corner*, *c.* 1895) features peaches (*Prunus persica*), pears (*Pyrus*) and apples (*Malus*), one of his favourite subjects to paint.

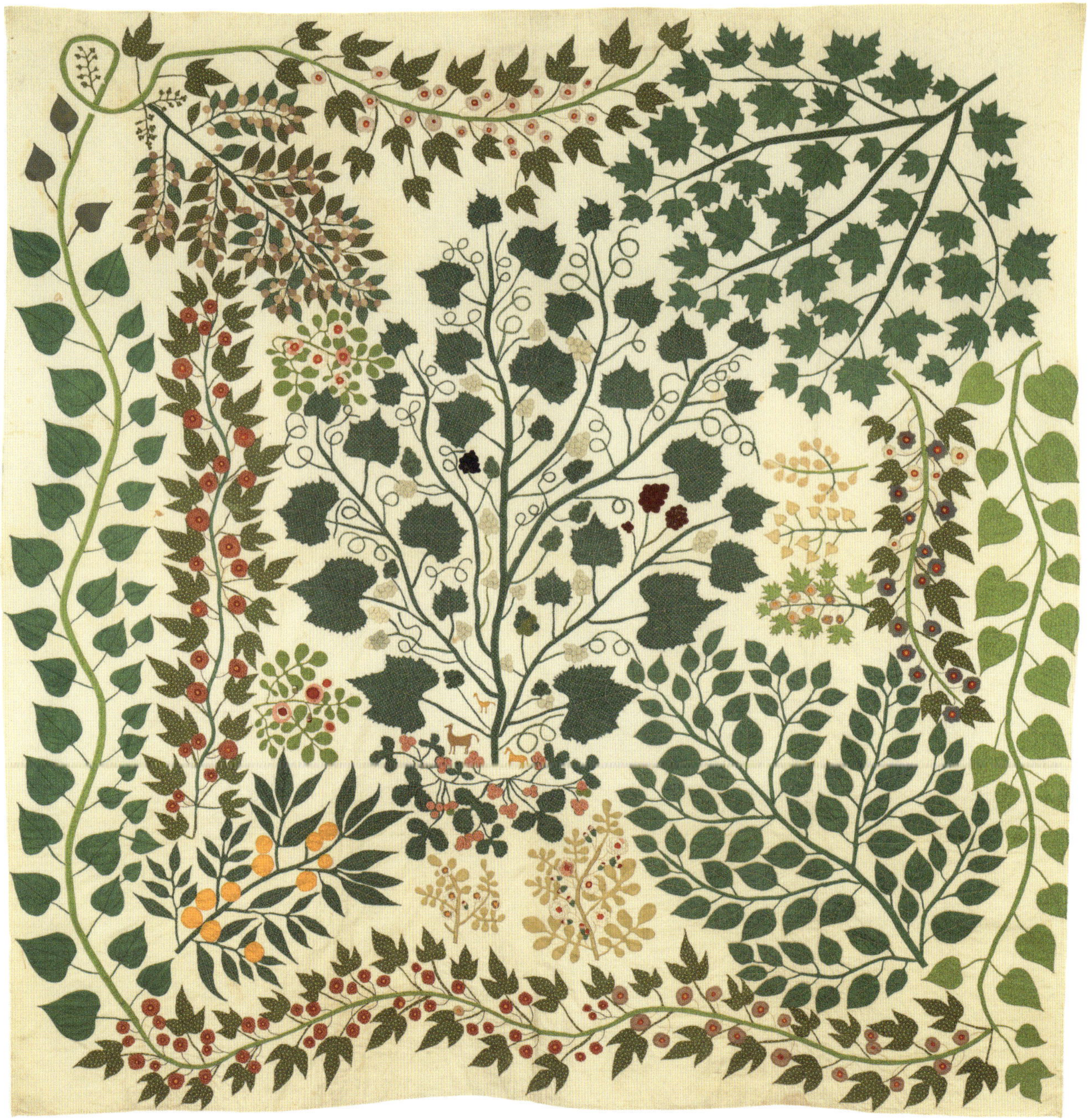

(11)

↑
Ernestine Eberhardt Zaumseil's *Branches and Vines Quilt* (*c*. 1875) is decorated with a Tree of Life pattern incorporating, top right, sycamore leaves (probably *Platanus occidentalis*), blackberry (*Rubus*) leaves in the centre, and, in the bottom left corner, the kumquat (*Citrus japonica*).

↗
Rice Plant and Butterburs is by the 19th-century artist Kubo Shunman. Asian cultivated rice (*Oryza sativa*) is one of only two cultivated types of this staple food. In the UK, the butterbur (*Petasites hybridus*) provides an early source of food for bees.

→
Winemaking scenes appear in ancient Egyptian tomb paintings from the Old Kingdom onwards. This facsimile detail of *Nebamun Supervising Estate Activities* (dated 1928) from the Tomb of Nebamun (1928 CE, based on the *c*. 1400–1352 BCE original) shows workers gathering the harvest, pressing the grapes (*Vitis vinifera*) and pouring the wine into large pottery jars while Nebamun looks on.

'The vineyard flourishes in the district of the Noble One. ... with their load, they bear fruit, with grapes more than sand.'

WINE OFFERING LITURGY FROM PTOLEMY X (*C.* 140–*C.* 88 BCE) TO THE GODDESS HATHOR, TEMPLE OF EDFU, EGYPT, TRANSLATED BY MU-CHOU POO

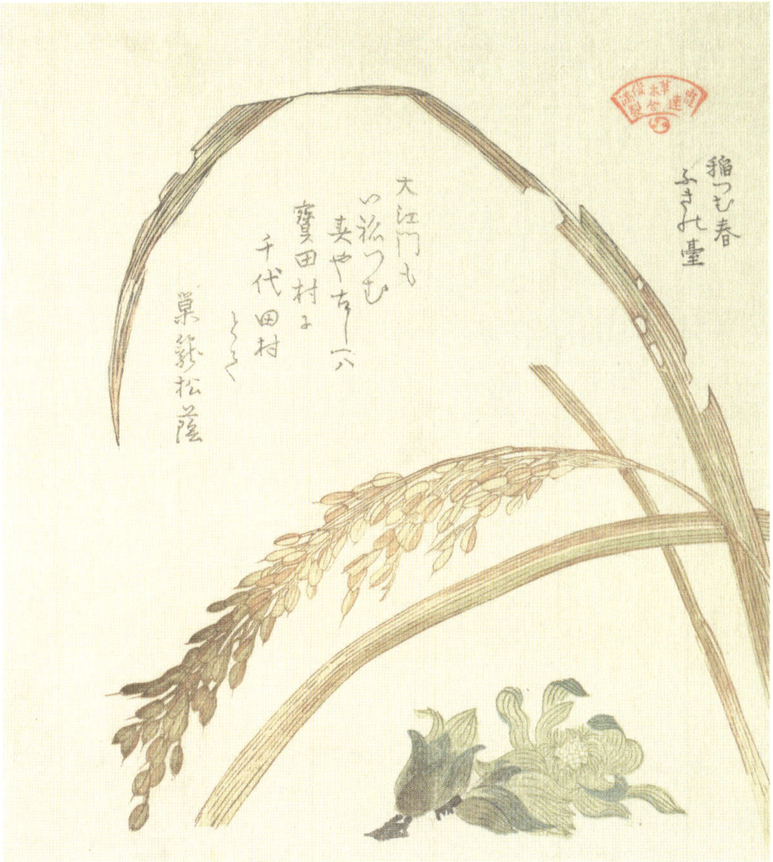

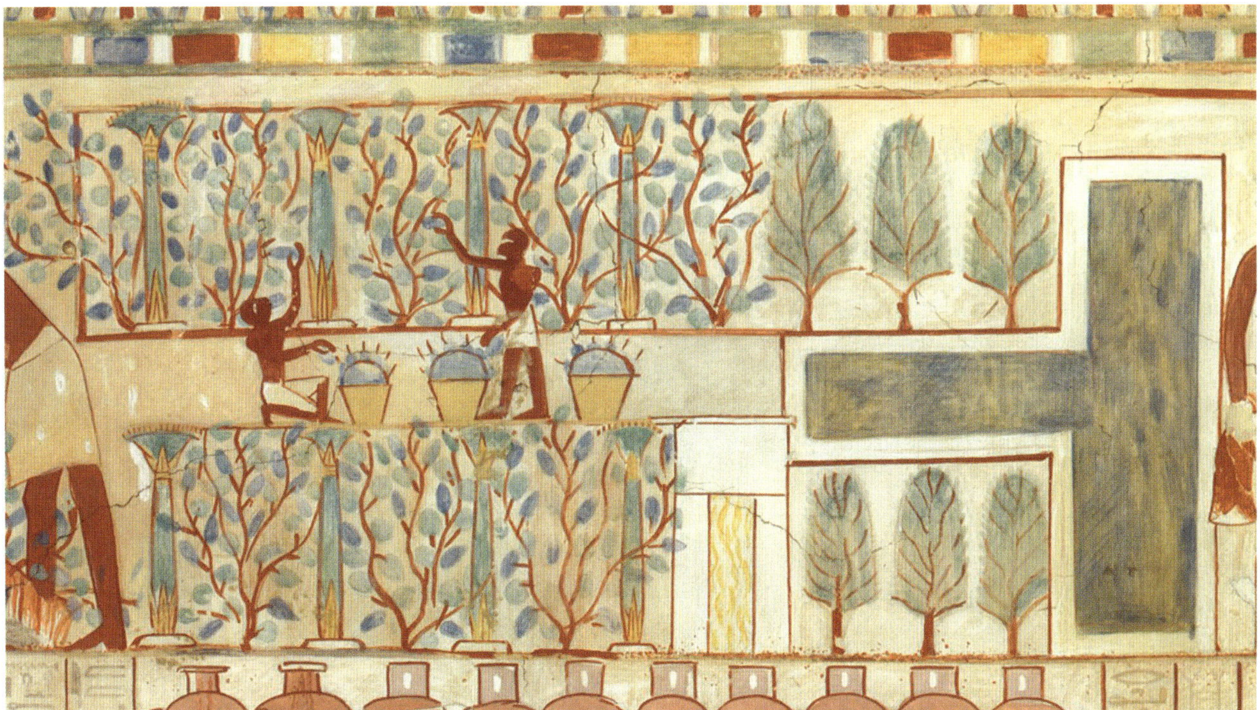

(11)

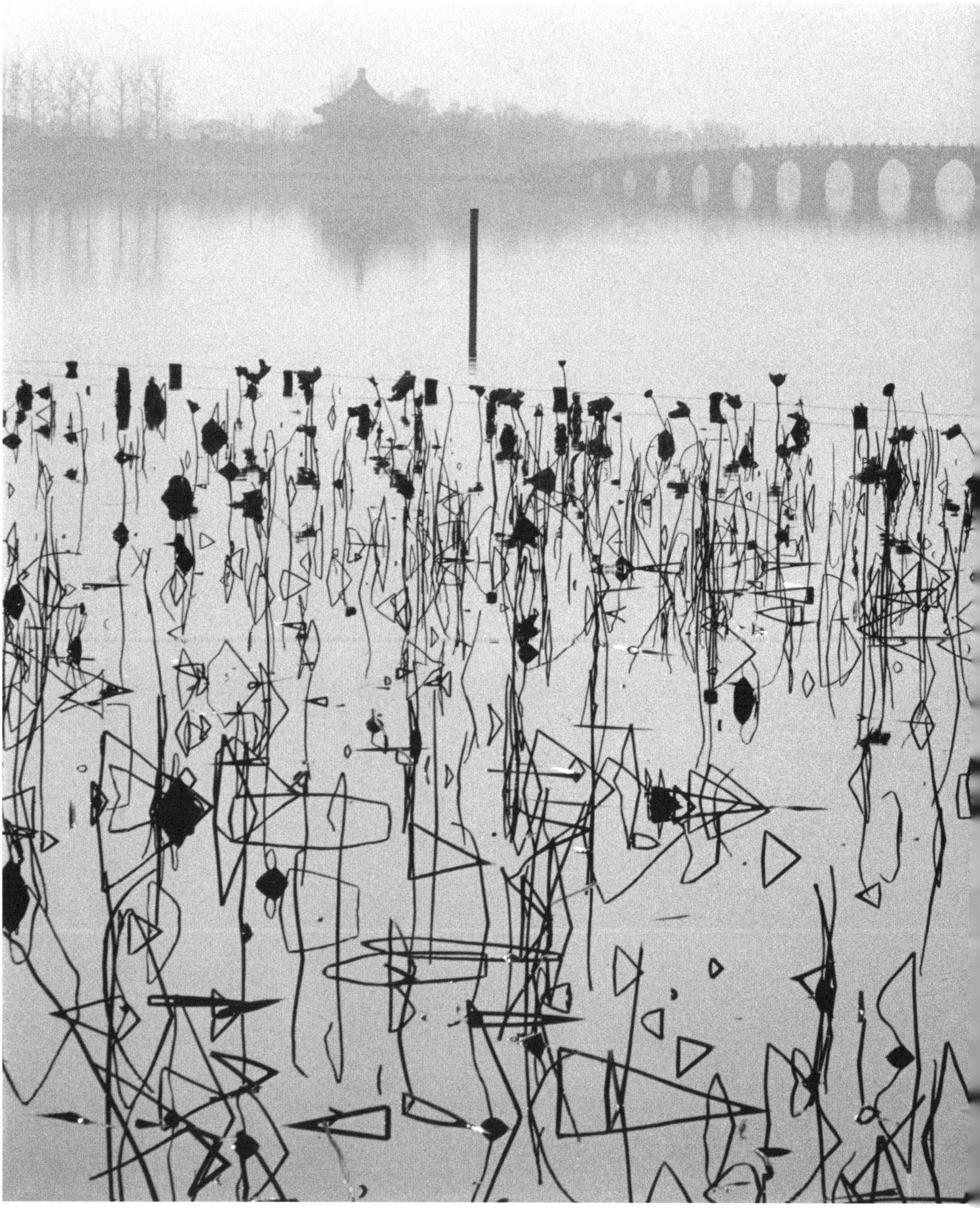

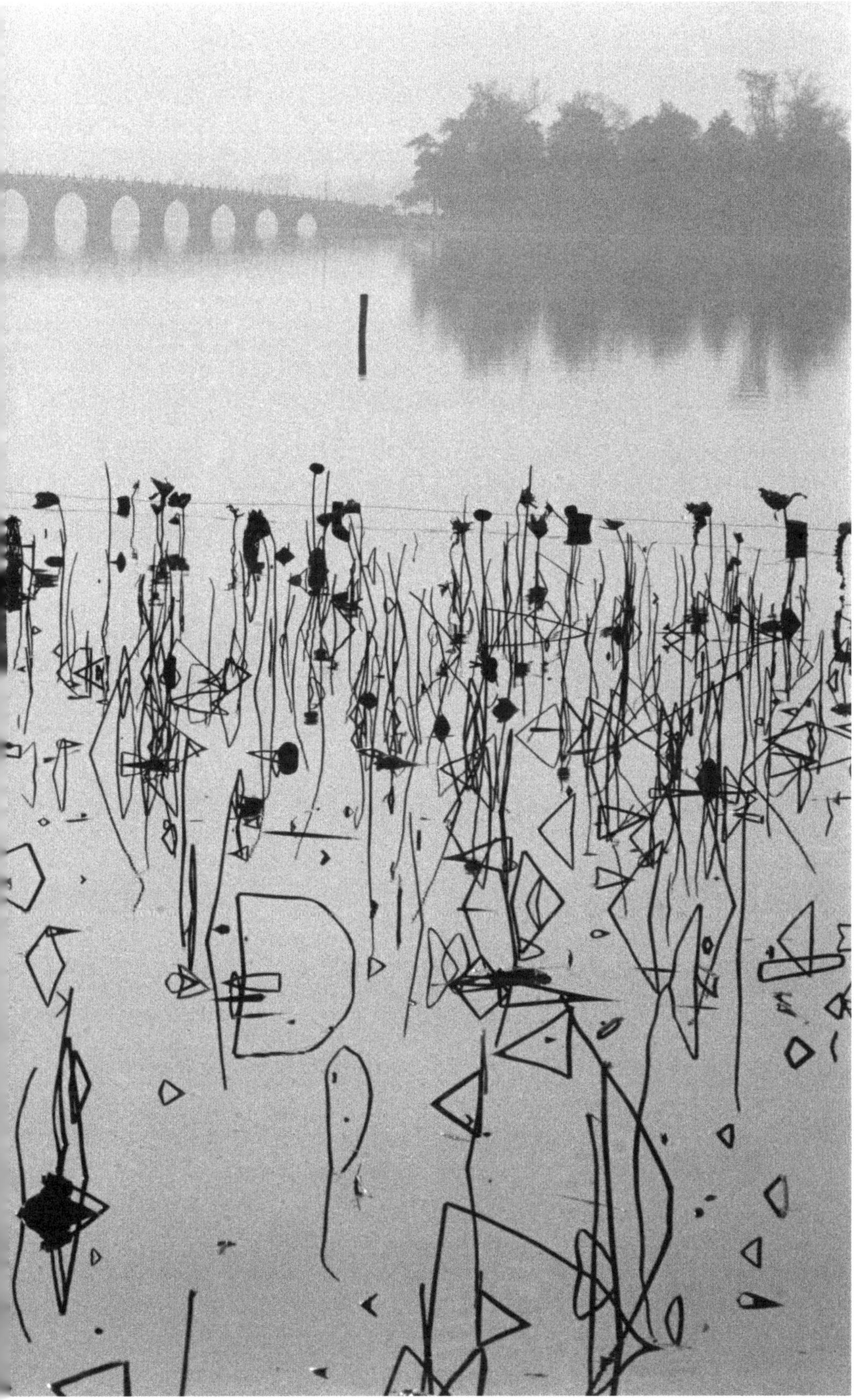

←
The lotus flower (*Nelumbo nucifera*) opens to the sun each morning and disappears underwater at night, re-appearing again next morning. In Hinduism and Buddhism, it is a symbol of purity and rebirth. In René Burri's photograph, *Former Summer Palace. Dead Lotus Flowers on Kunming Lake* (Beijing, China, 1964), the dead flowers and their reflections combine to form angular, abstract shapes.

(11)

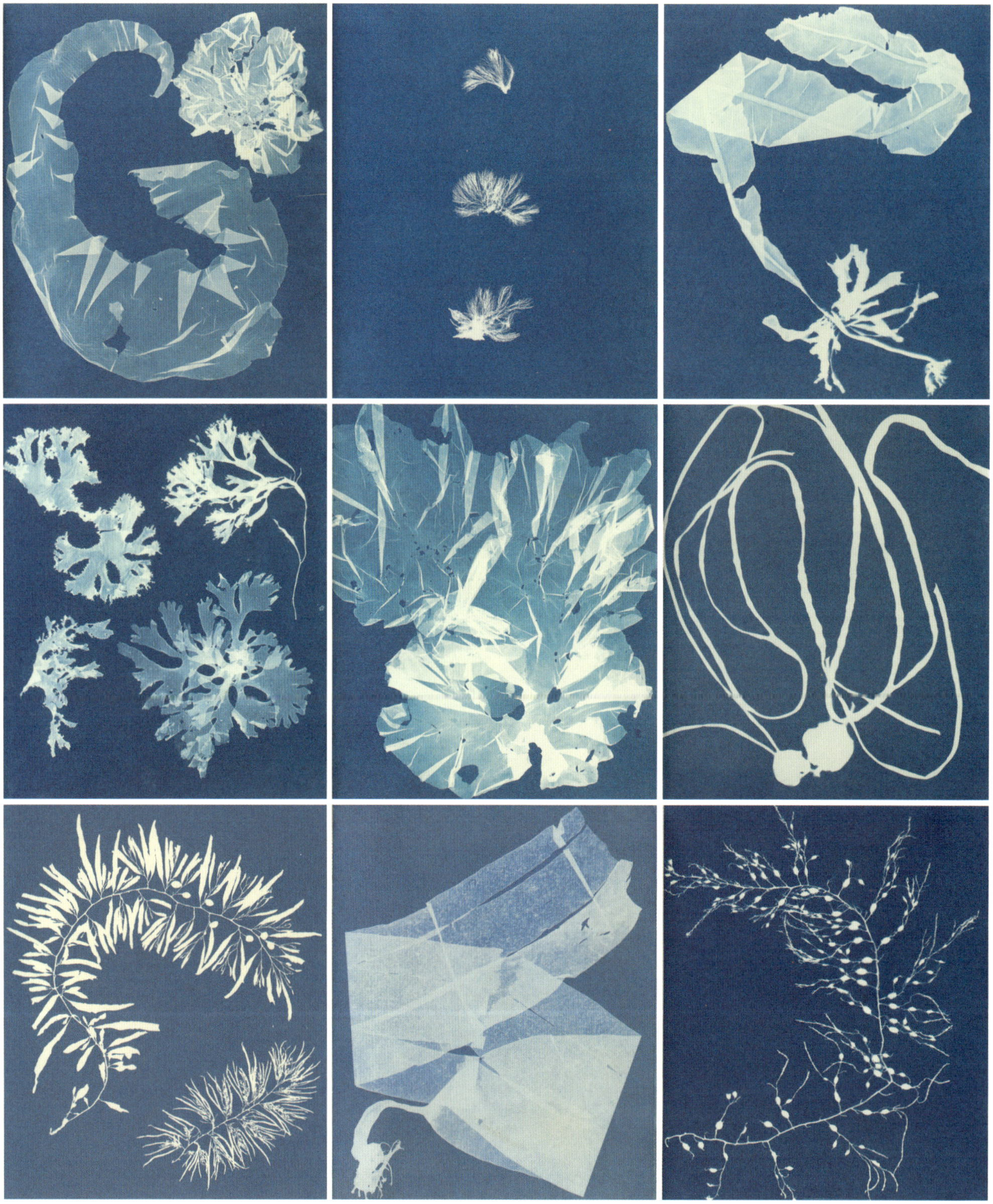

↑
Anna Atkins was the first person to use photographic images to illustrate a book. *Photographs of British Algae* (1843–53) used cyanotypes to display species including, top left, purple laver (*Porphyra laciniata*); top right, winged kelp (*Alaria esculenta*); and, centre left, fan weed (*Rhodomenia laciniata*).

→
Better known for his abstract paintings, Dutch artist Piet Mondrian also painted flowers, such as *Red Amaryllis with Blue Background* (*c.* 1907). The cultivar could be 'Red Lion' or 'Red Pearl', which are available in the Netherlands.

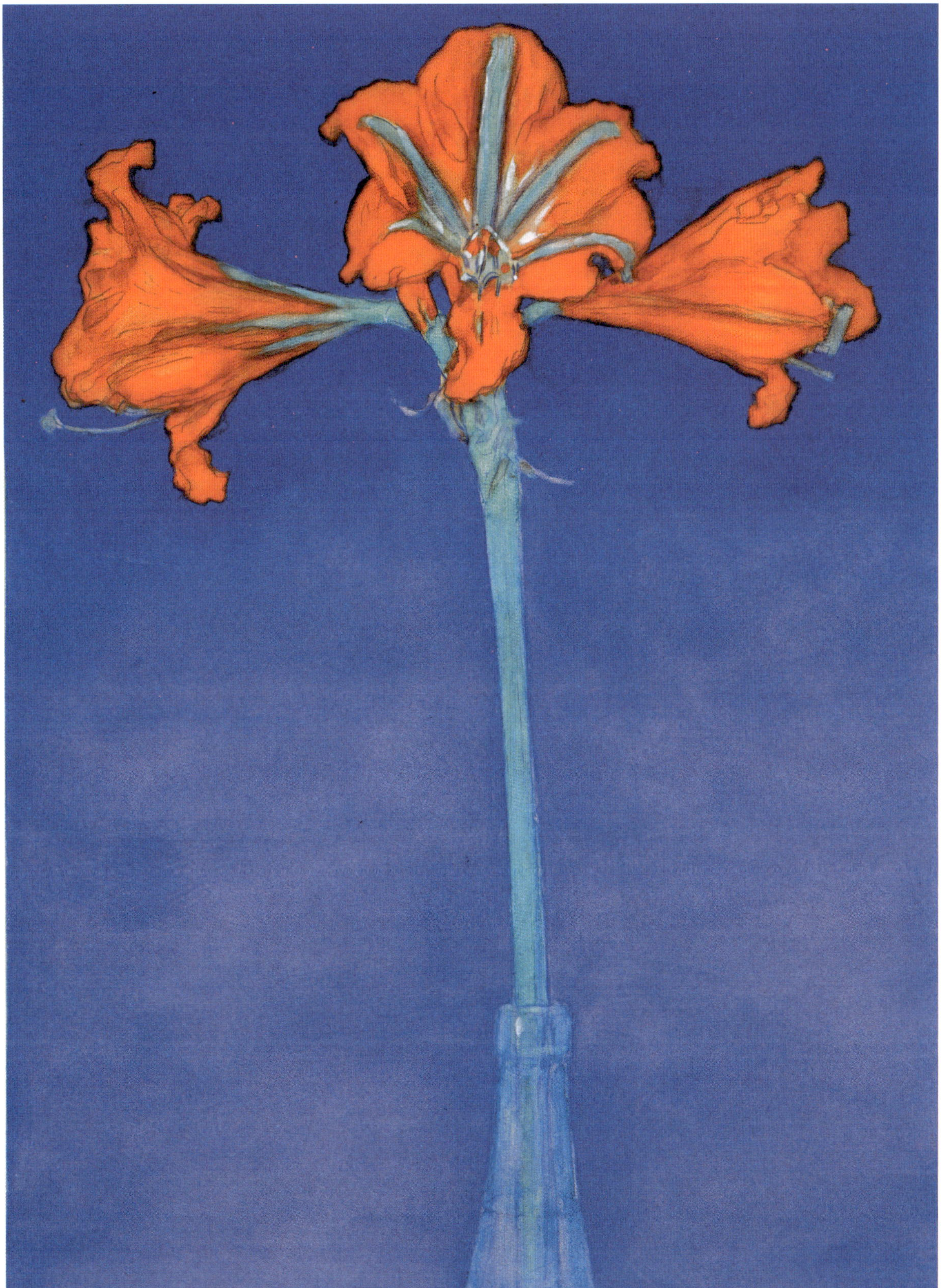
PIET MONDRIAAN

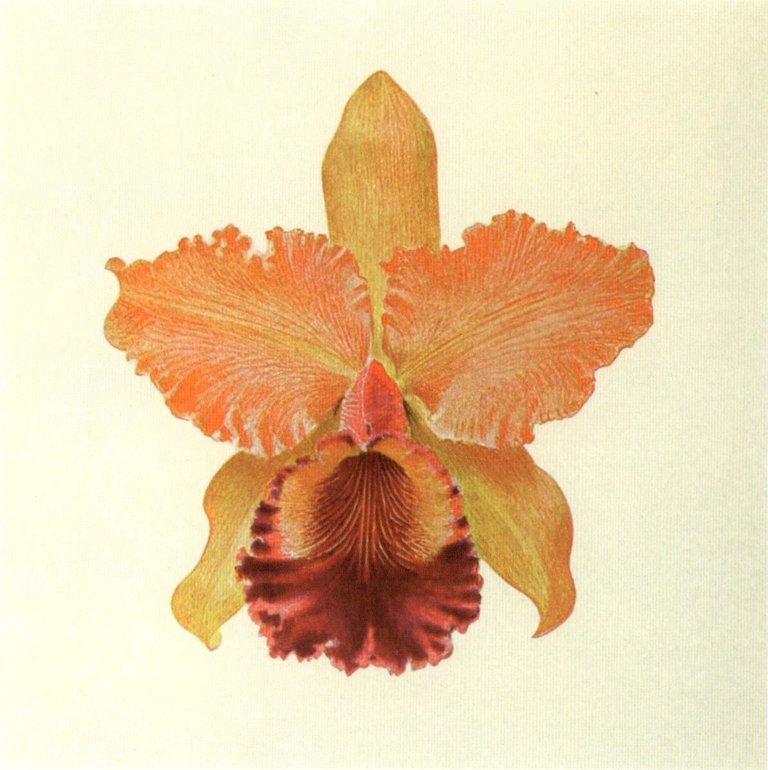

←
The orchid *Laeliocattleya firminii* is a *Cattleya* hybrid, illustrated here in *The Orchid World* magazine in 1910.

↓
This detail of Sekine Untei's 1832 illustration shows several species of the Japanese sacred lily (*Rohdea japonica*). This small, cultivated, ornamental lily is used in traditional Chinese medicine.

→
A *Nowruz* (New Year) display of exotic fruit is the subject of this 19th-century painting, signifying the beginning of spring. Still-life paintings were a relatively new genre in Iran at the time and were hung in reception rooms and garden pavilions.

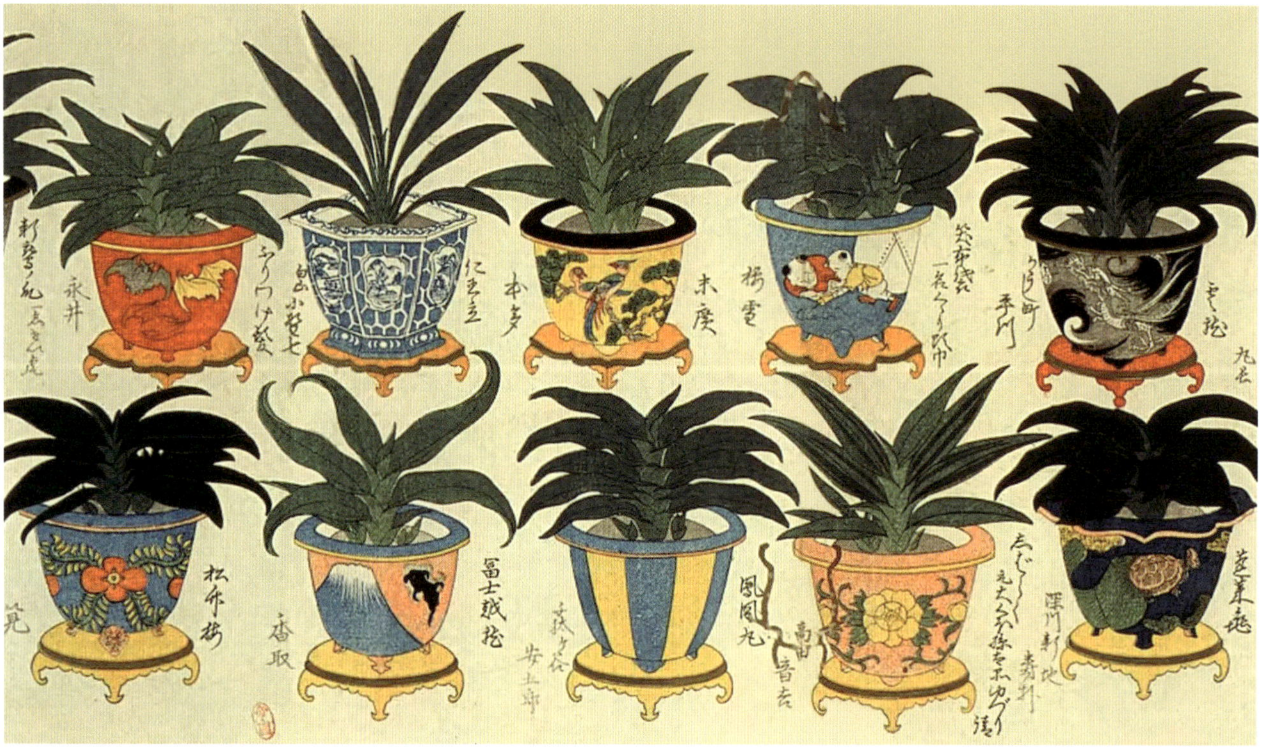

'Earth laughs in flowers, to see her boastful boys
Earth-proud, proud of the earth which is not theirs'

RALPH WALDO EMERSON, 'HAMATREYA', 1846

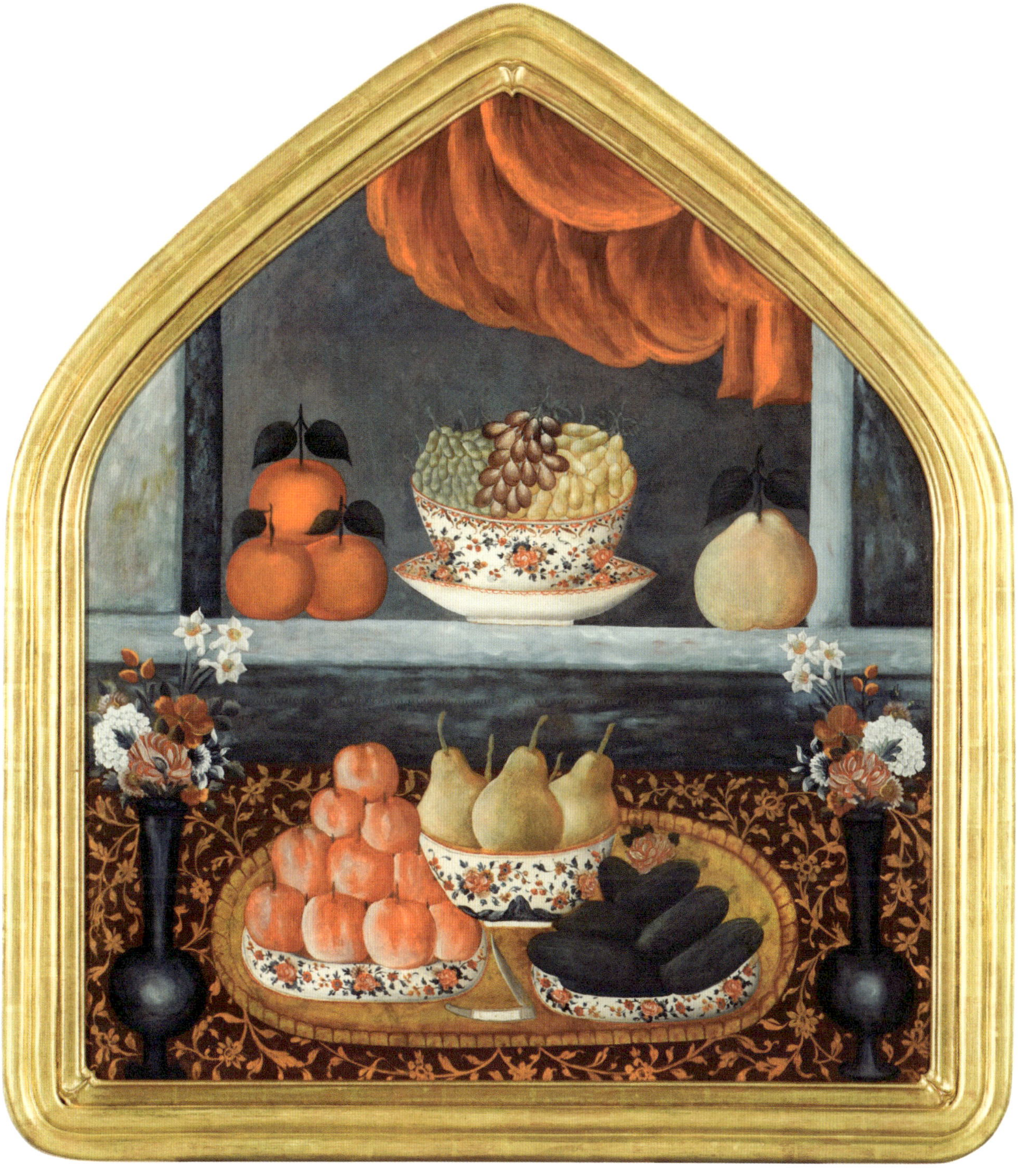

(11)

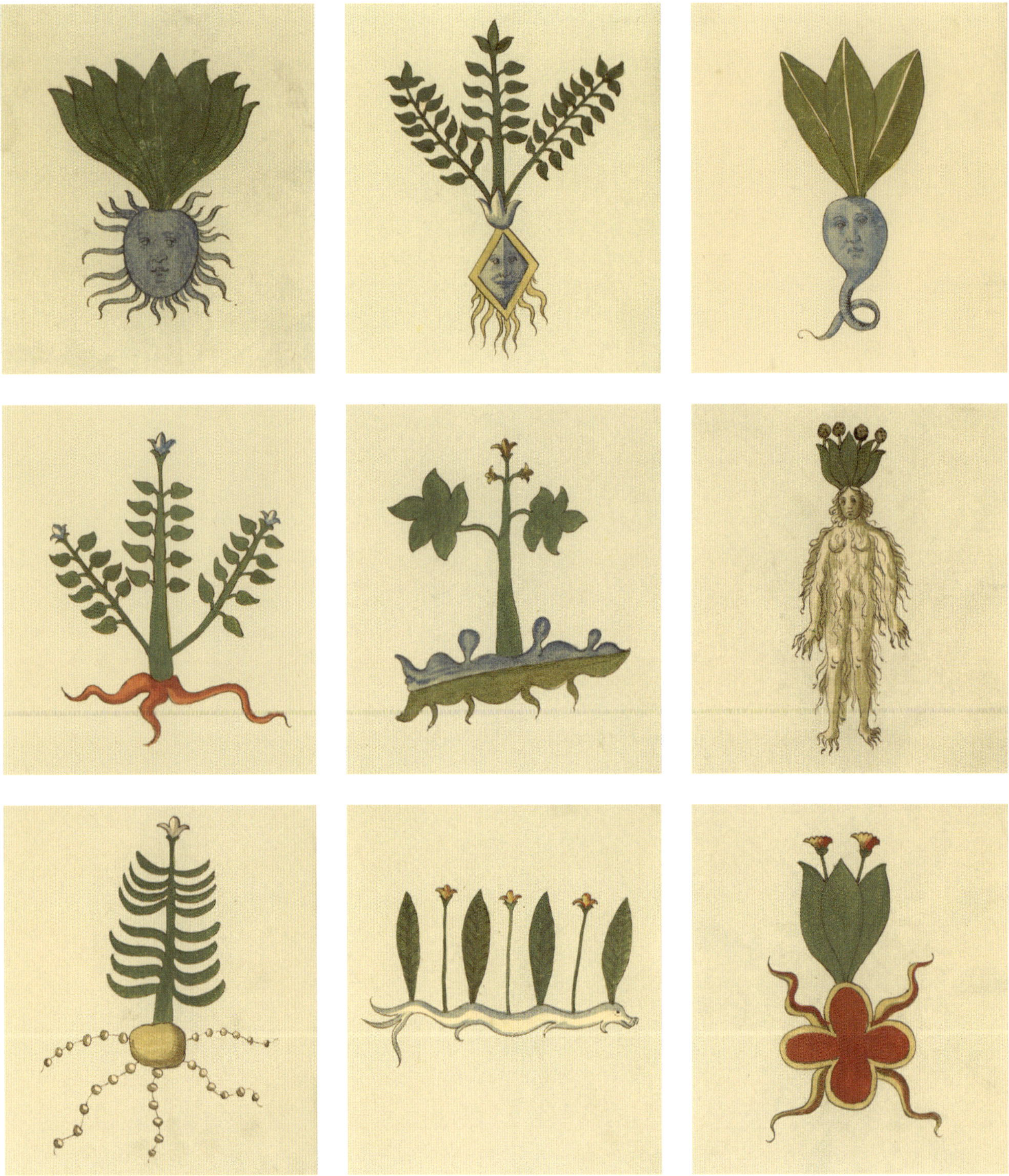

↑
The *Erbario* was a herbal created in Italy in around 1400. Some illustrations from it follow medieval conventions, incorporating fantasy elements, such as human faces. Others are more naturalistic representations of plants, showing roots, leaves and flowers.

→
'The Night-Blowing Cereus', from Robert Thornton's *The Temple of Flora* (1807), is a hand-coloured mezzotint after a painting by Philip Reinagle. This cactus, known as the night-blooming cereus (*Selenicereus grandiflorus*), blooms for just one night and its flowers wilt before dawn.

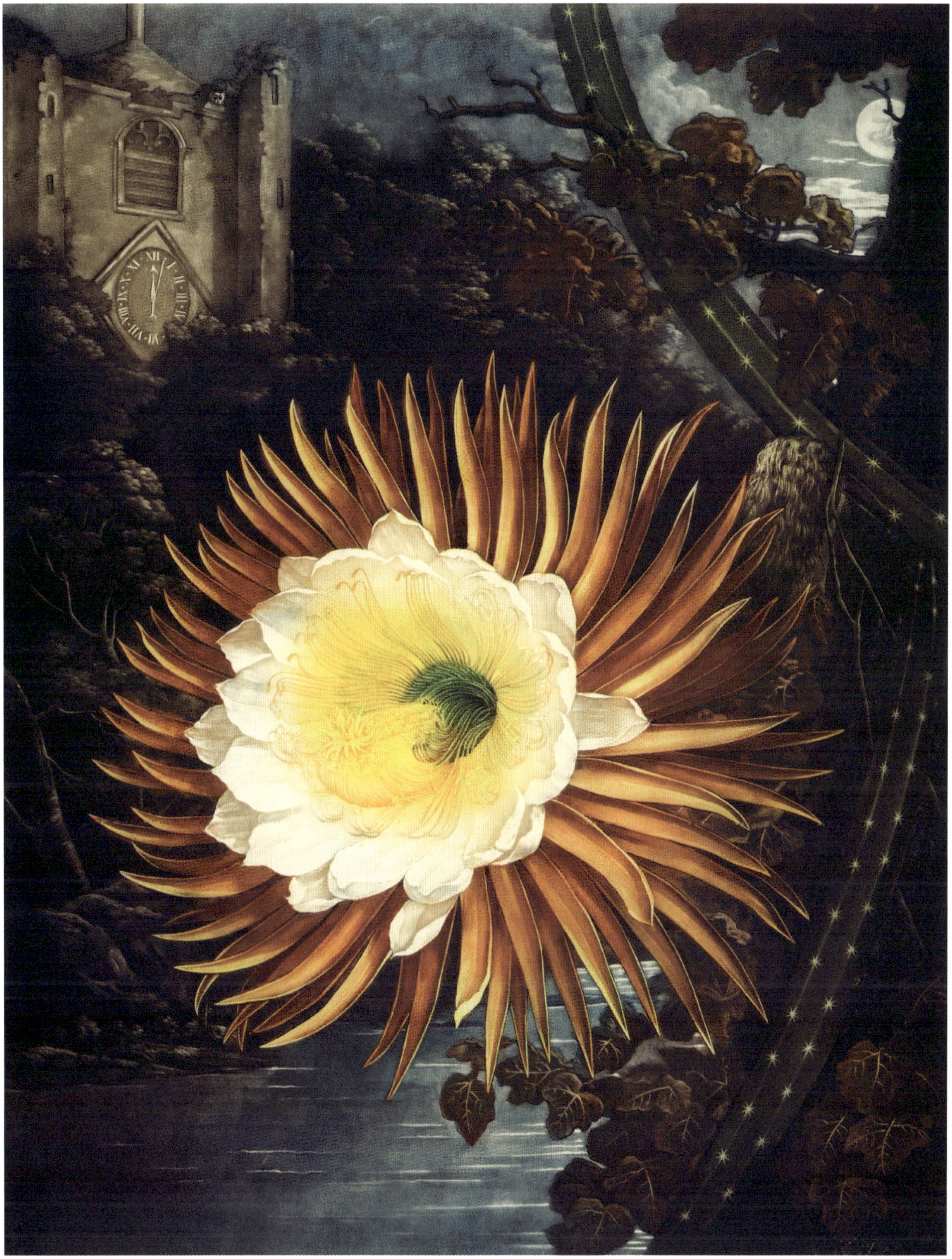

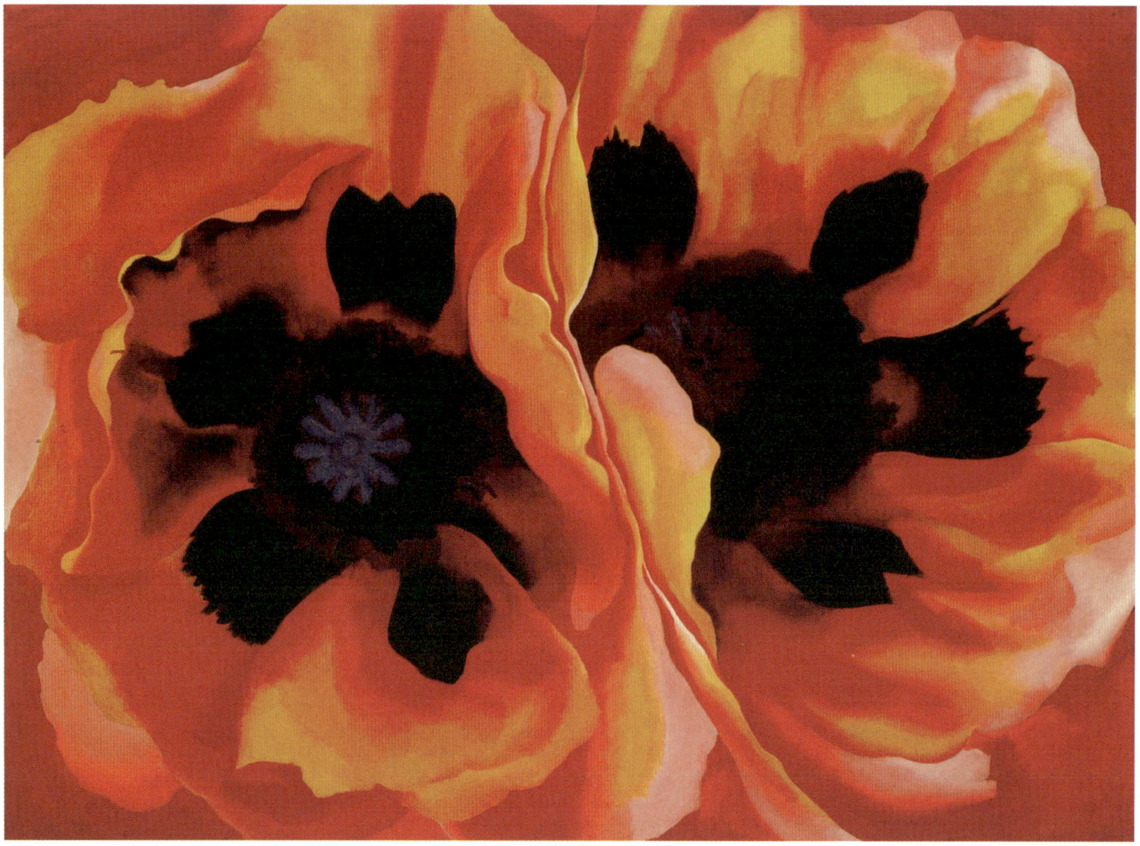

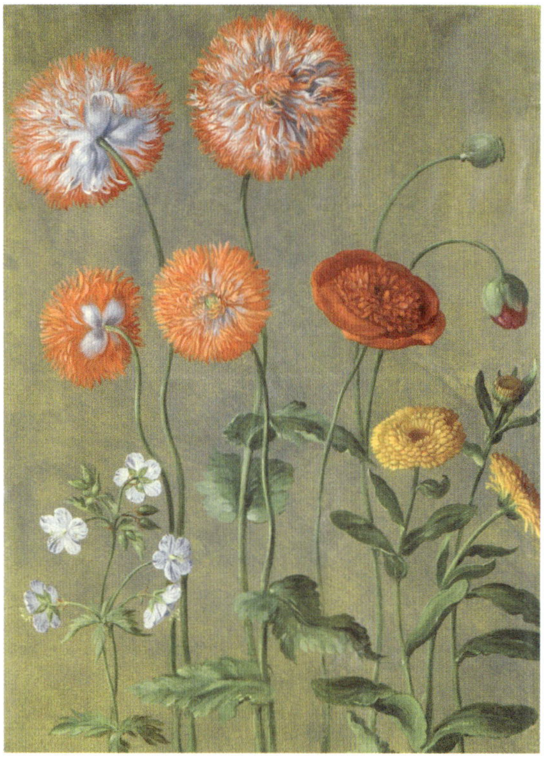

↑
In *Oriental Poppies* (1927), American artist Georgia O'Keeffe has cropped in close on the subject. Poppies are a big draw for pollinators such as bees and butterflies.

←
The flowers in Johann Jakob Walther's watercolour *Poppies* (17th century) are similar to the opium species (*Papaver somniferum*), also known as breadseed poppies. Walther produced a collection of flower illustrations, reflecting the fashion for collecting and cultivating exotic plants among the wealthy at the time.

→
Rachel Ruysch's *Flowers in a Glass Vase* (1704) is composed of a number of species, including tulips (*Tulipa*), a cabbage rose (*Rosa × centifolia*), lilies (*Lilium*) and peonies (*Paeonia*), along with a variety of insects that have been attracted by the scent of the flowers.

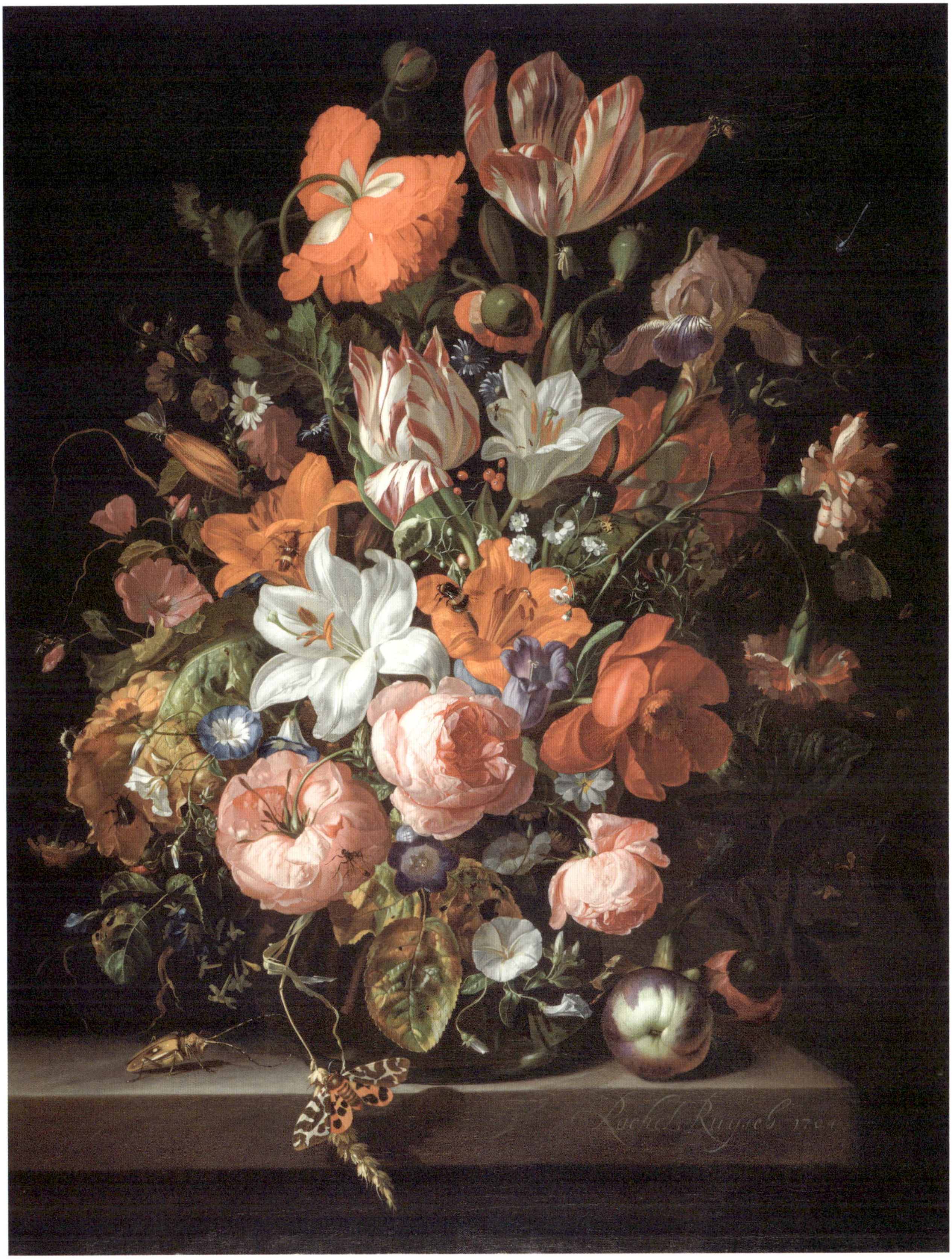
Rachel Ruysch 1704

(11)

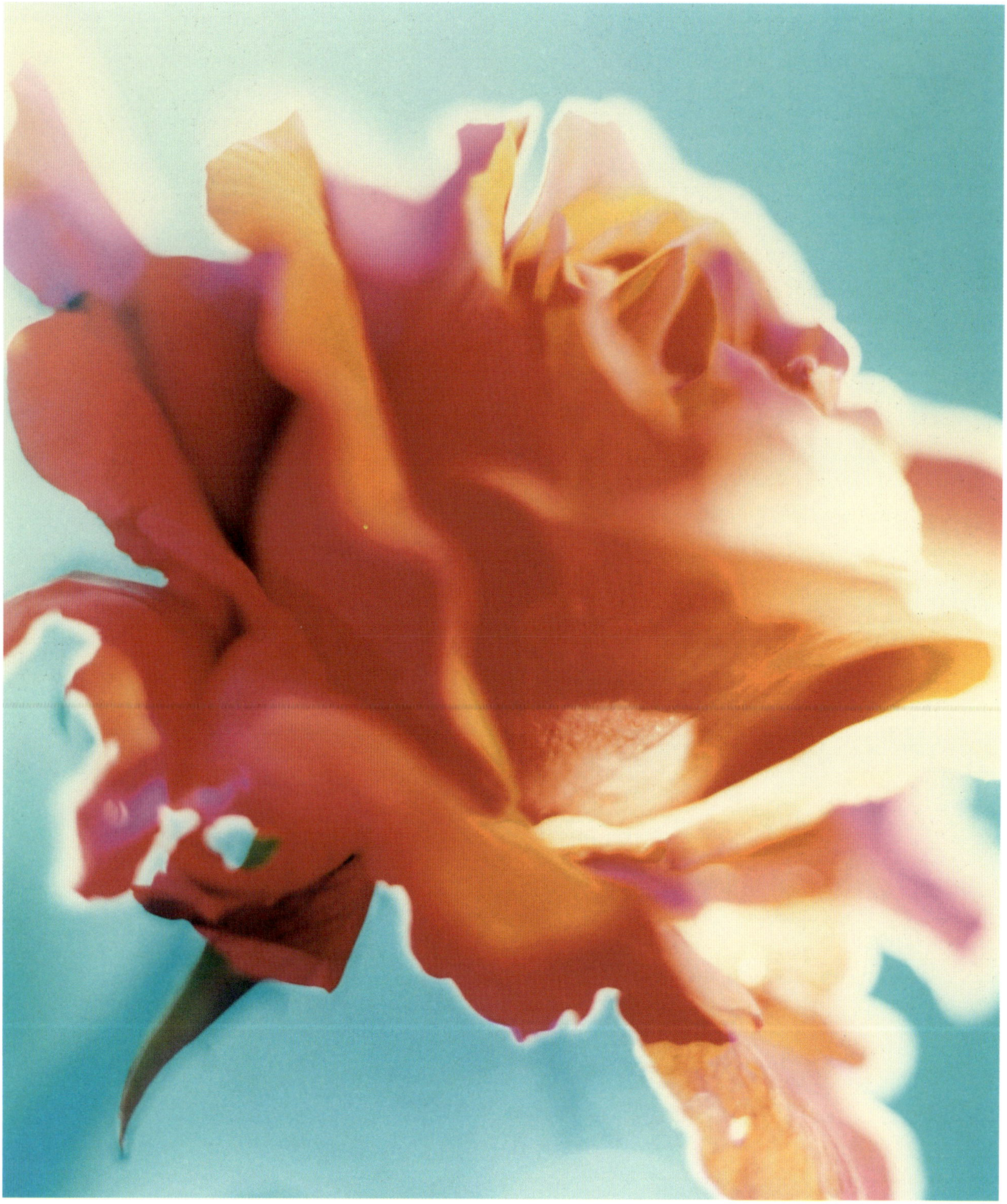

↑
For his series that includes *Dream Meadow LI* (above, 2024) and *Dream Meadow XXIV* (opposite, 2021), Gareth McConnell used long exposures to create psychedelic-inspired imagery that is reminiscent of the double-exposed works in Peter Fischli and David Weiss's *Flowers & Mushrooms* (1997–98).

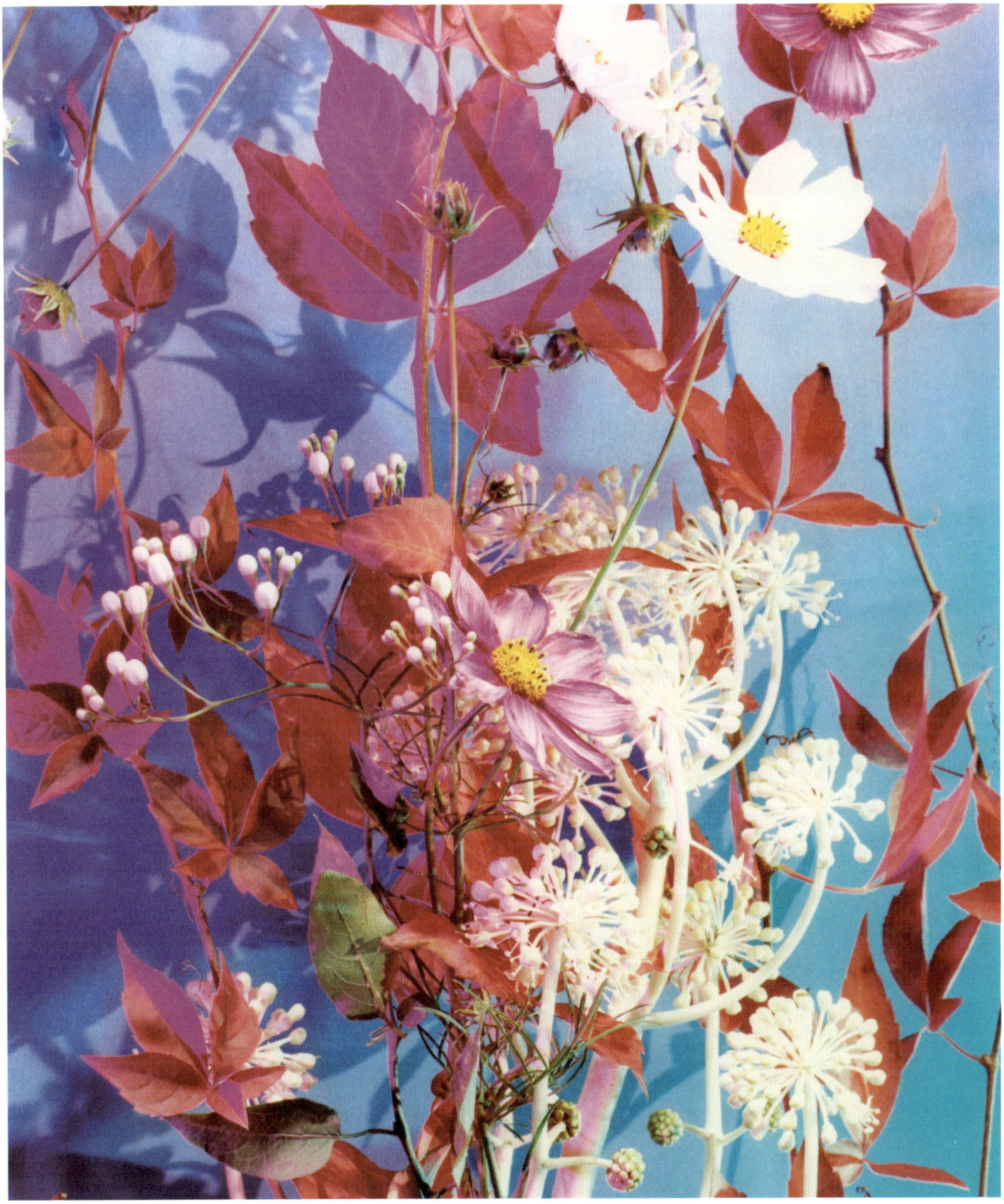

'During a man's life only saplings can be grown, in the place of the old trees – tens of centuries old – that have been destroyed.'

JOHN MUIR, 'THE AMERICAN FORESTS', *THE ATLANTIC*, AUGUST 1897

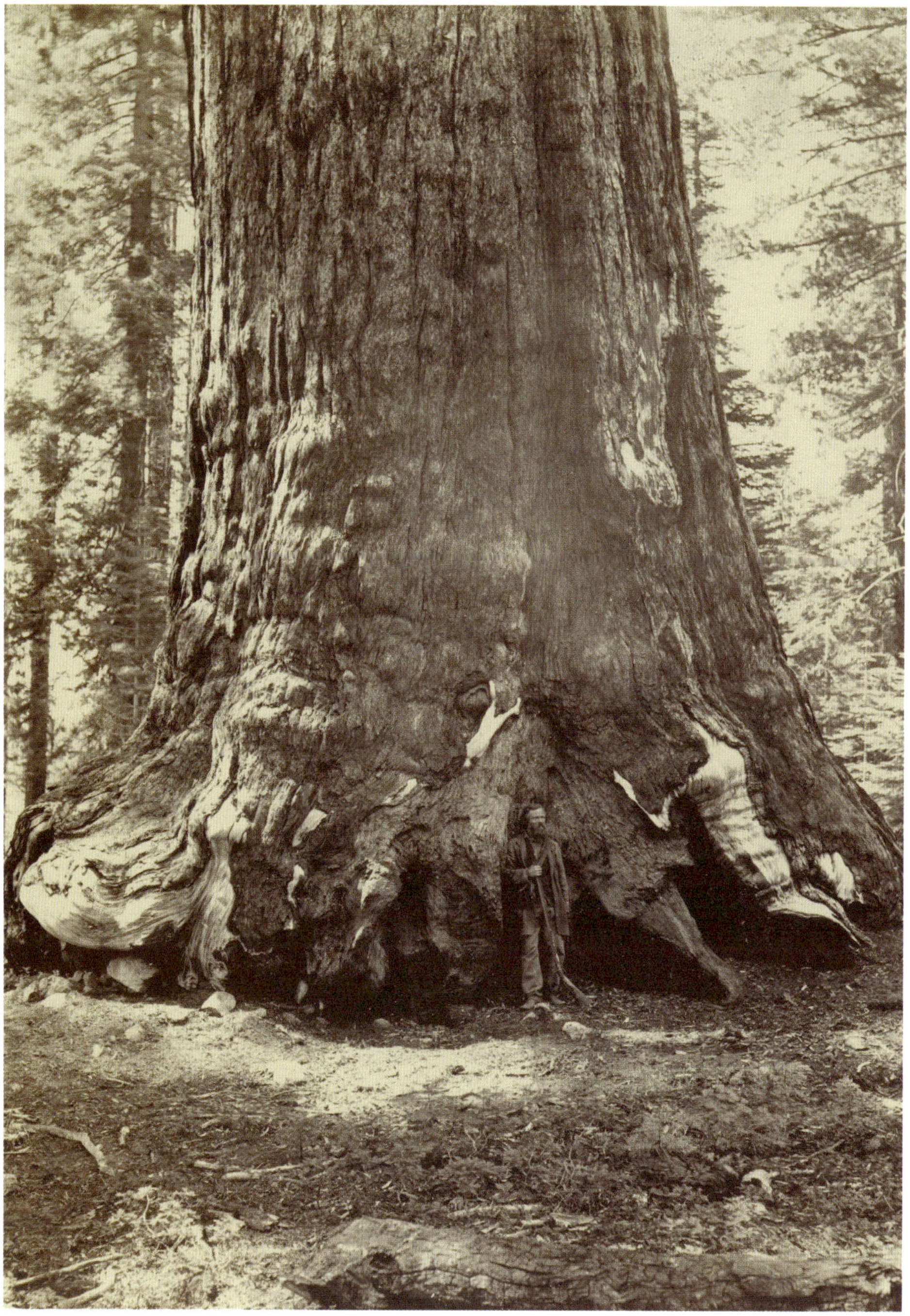

←
◎ Photographer Carleton E. Watkins spent twenty years photographing Yosemite Valley. For this image of a 3,000-year-old giant sequoia (*Sequoiadendron giganteum*) nicknamed the Grizzly Giant, taken in 1865–66, Watkins posed conservationist Galen Clark against the tree to demonstrate its size.

→
These images of bear's breeches (*Acanthus mollis*), small scabious (*Scabiosa columbaria*), oriental poppy (*Papaver orientalis*) and American maidenhair fern (*Adiantum pedatum*) are from a series of close-up studies of plants, flowers and seed heads by German sculptor and photographer Karl Blossfeldt, published in his *Urformen der Kunst* (*Art Forms in Nature*) in 1928.

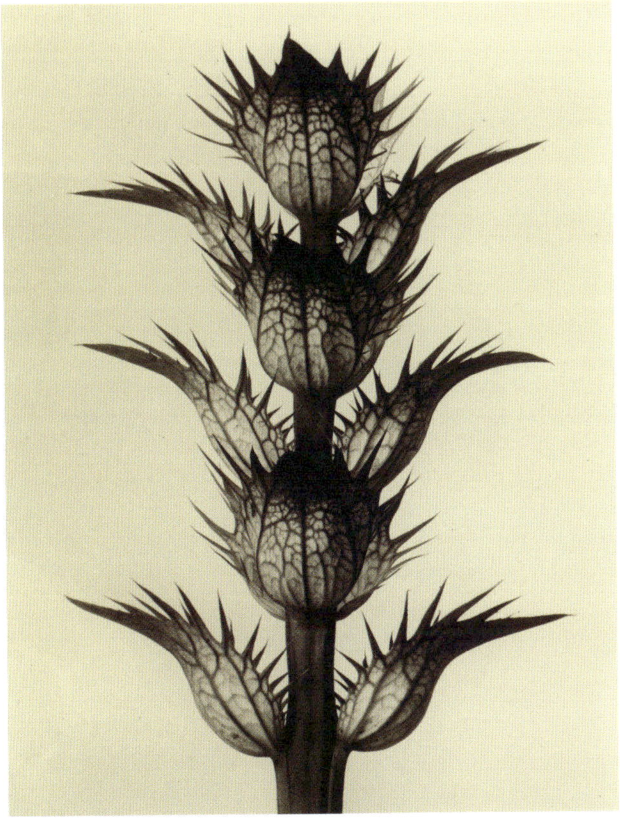

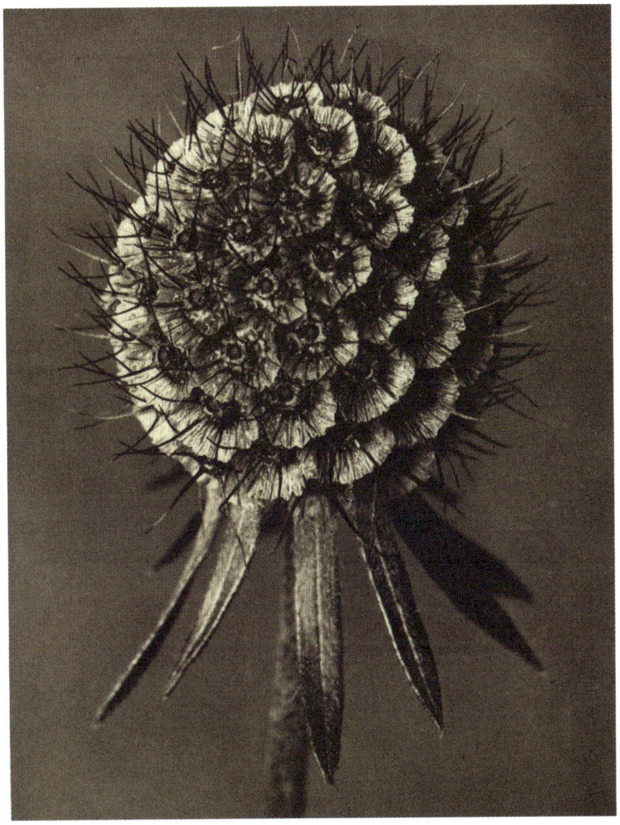

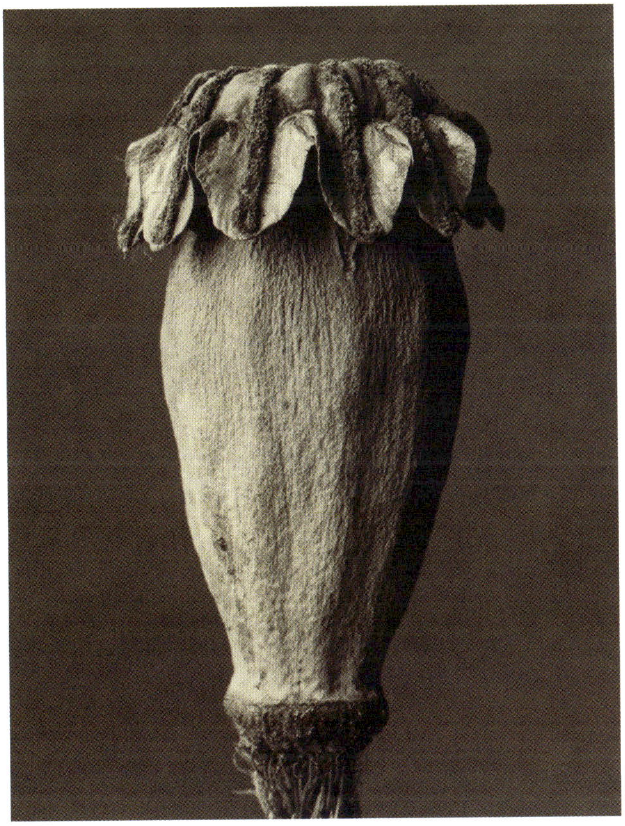

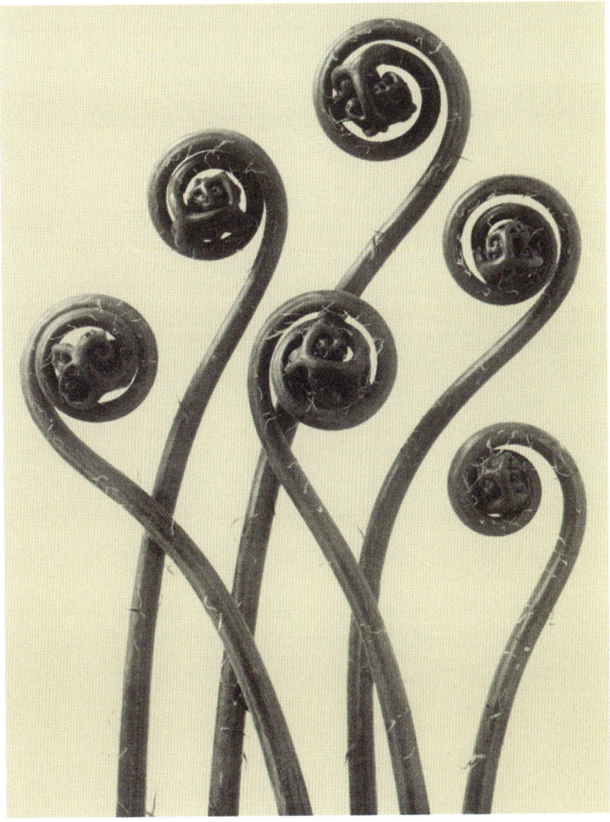

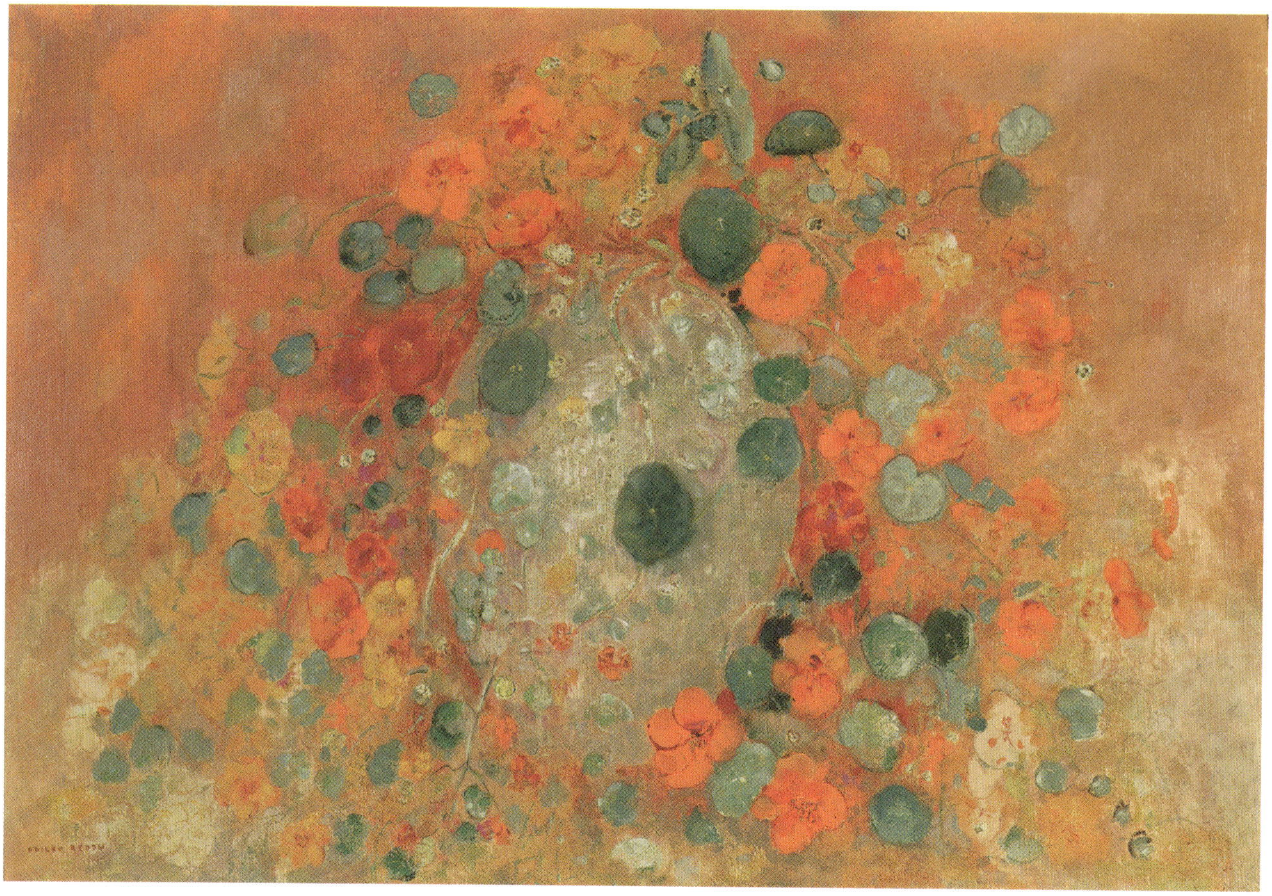

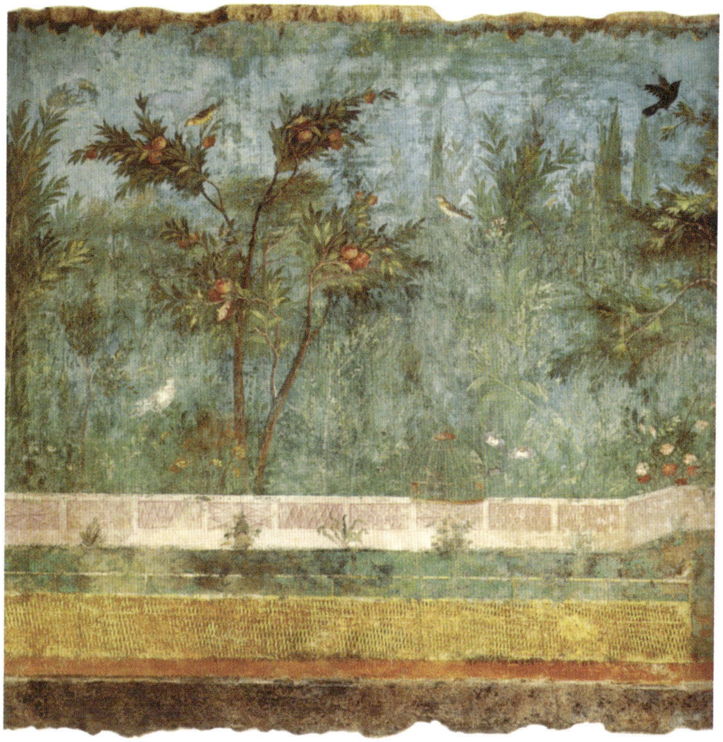

↑
Though French Symbolist painter Odilon Redon's early works were almost entirely monochromatic, his later works, including *Nasturtiums* (1905), utilize vibrant colours.

←
This detail from a fresco in the Villa of Livia, near Rome, dates from between 38 and 20 BCE. The fresco covers all four walls of the villa's dining room and is filled with an array of accurately portrayed trees, plants and birds, including the pomegranate tree (*Punica protopunica*).

→
In Swiss artist Paul Klee's 1917 work, titled *Sky Blossoms above the Yellow House (The Chosen House)*, natural forms can be seen to link sky and earth.

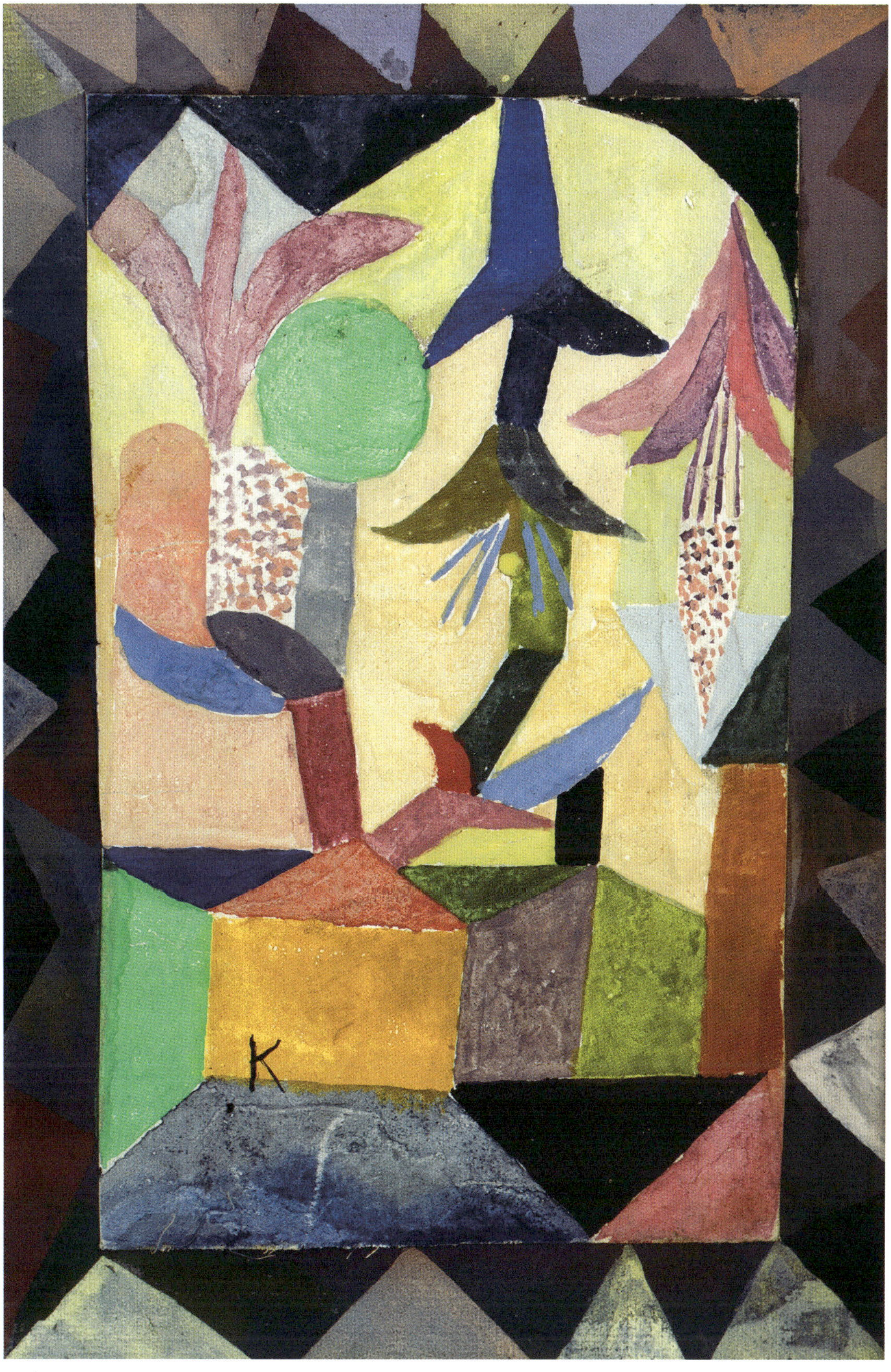
K

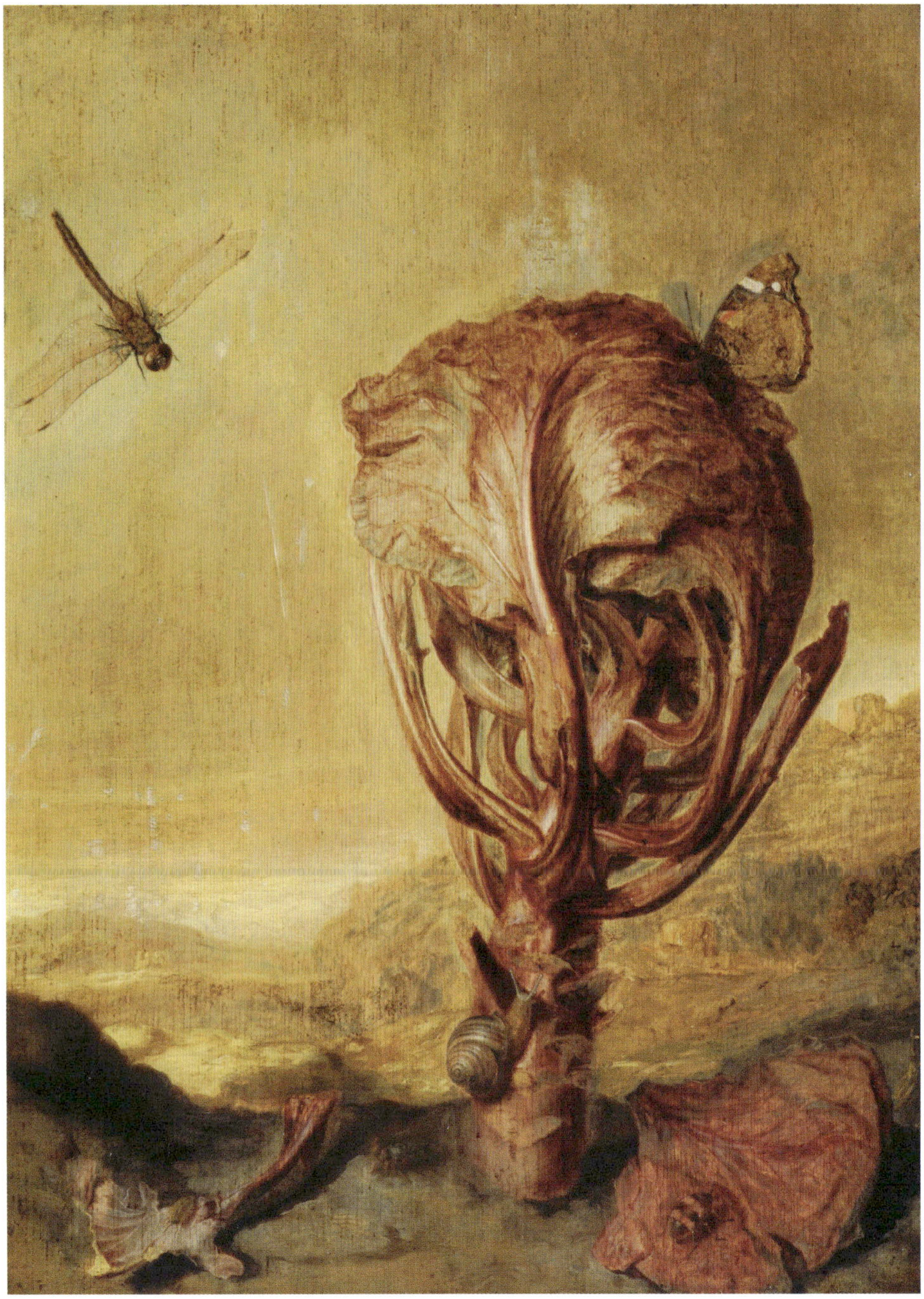

←

17th-century Dutch artist Margaretha de Heer specialized in paintings of plants and insects, such as *A Red Cabbage, a Snail, a Butterfly, a Dragonfly, a Bee and a Wood Louse, in a Landscape*.

→

Bavarian botanical artist Barbara Regina Dietzsch was one of a family of artists, several of whom specialized in natural history drawing. Her illustrations combine plants and insects, putting each plant at the centre of an ecosystem that revolves around it. Shown here are (clockwise from top left): buttercup (*Ranunculus*) with a butterfly and an insect; Queen of the Night cactus (*Selenicereus grandiflorus*) (attribution uncertain); thistle (*Cirsium*) with butterfly, dragonfly, ladybug and spider; and cabbage head (*Brassica oleracea*) with pearl-bordered fritillary butterfly (*Boloria euphrosyne*) and caterpillar.

'Some see nature all ridicule and deformity ... and some scarce see Nature at all. But to the eyes of the man of imagination, Nature is Imagination itself.'

WILLIAM BLAKE, LETTER TO REV. DR TRUSLER, 23 AUGUST 1799

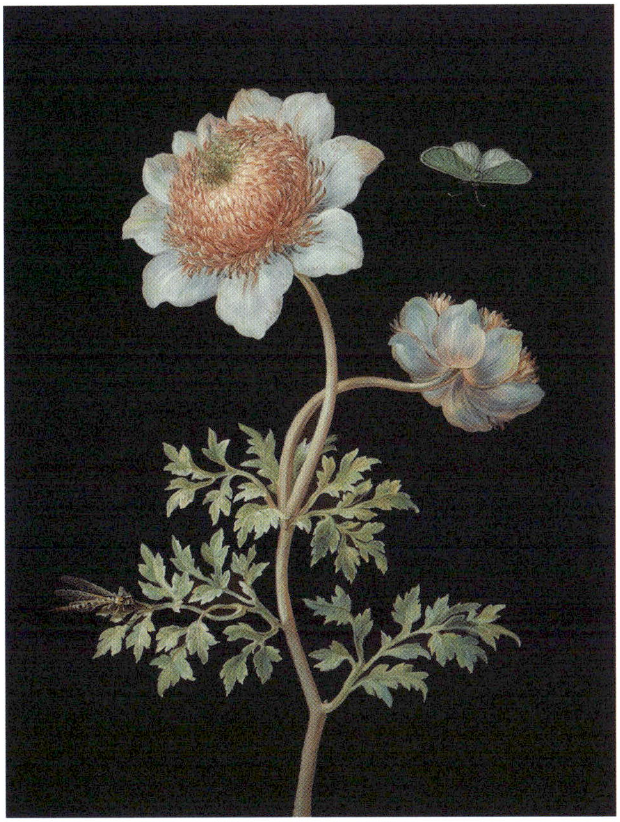
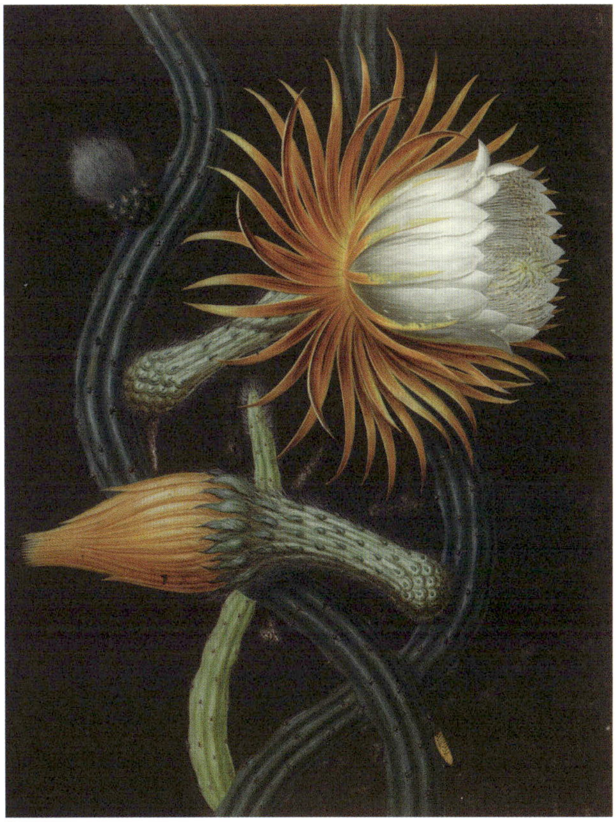

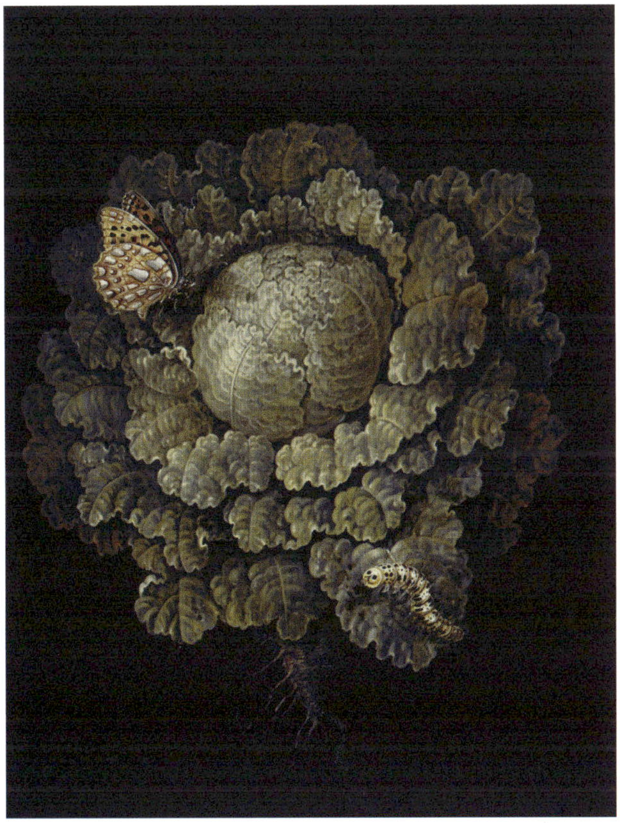
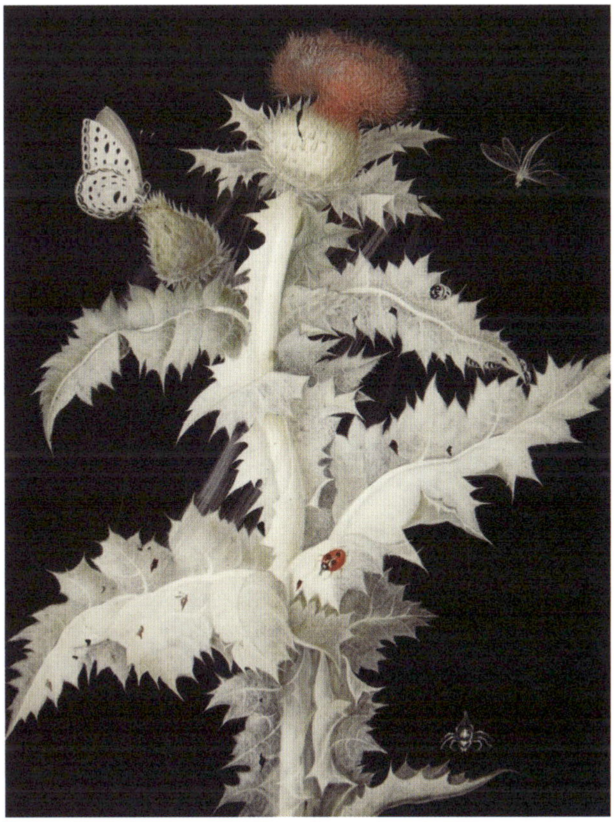

'The tree is beautiful; but that's not everything; it has its own right to live, it is as necessary as water, the sun, or the stars. Life on earth is inconceivable without trees.'

ANTON CHEKHOV, LETTER TO A. S. SUVORIN, 18 OCTOBER 1888, TRANSLATED BY S. S. KOTELIANSKY AND PHILIP TOMLINSON

(11)

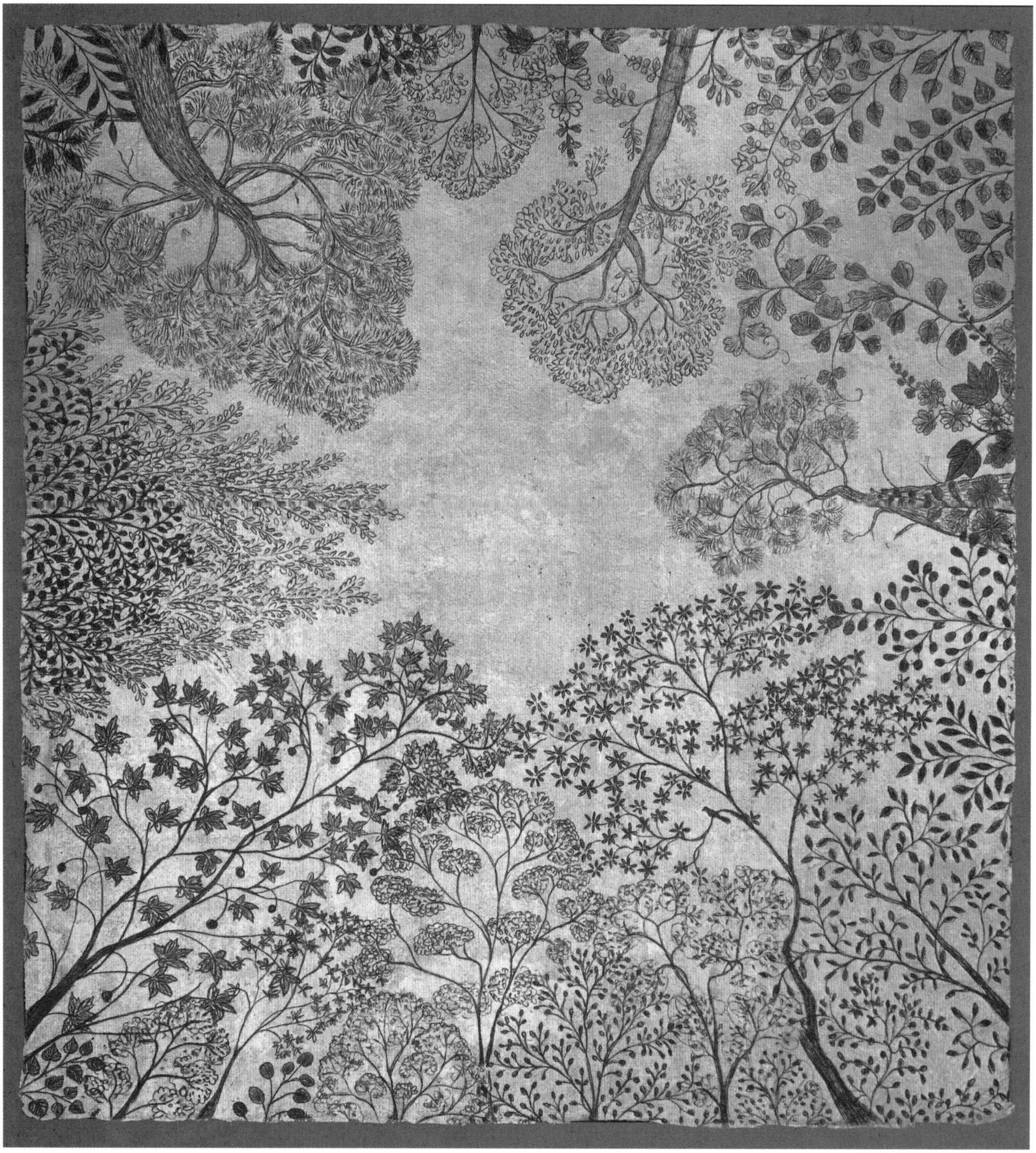

↙
Jethro Buck's *These Great Trees are Prayers* (2020) uses – like all his work – traditional techniques inspired by the skill and craftsmanship of Indian miniature paintings to explore the richness of the natural world.

→
Chrysanthemums arrived in Japan from China in around the 8th century, and became the symbol of the Japanese Imperial Family. Published in the 17th century, *The Picture Book of Chrysanthemums* was the first book on this flower to appear in Japan. It covers one hundred varieties and contains hand-painted illustrations.

↓
Samuel Palmer's *Oak Trees, Lullingstone Park* (1828) celebrates this familiar and symbolic species, which can live 500 years or more. The English oak (*Quercus robur*) supports more life forms than any other tree species in the UK. Insects, birds, bats, mammals and fungi are among the creatures that live in and feed off the tree.

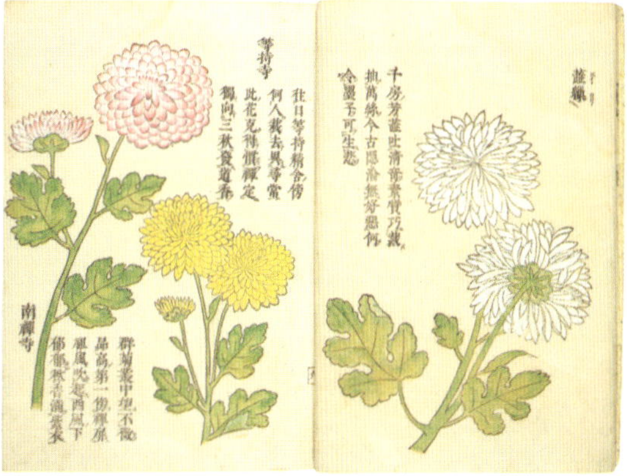

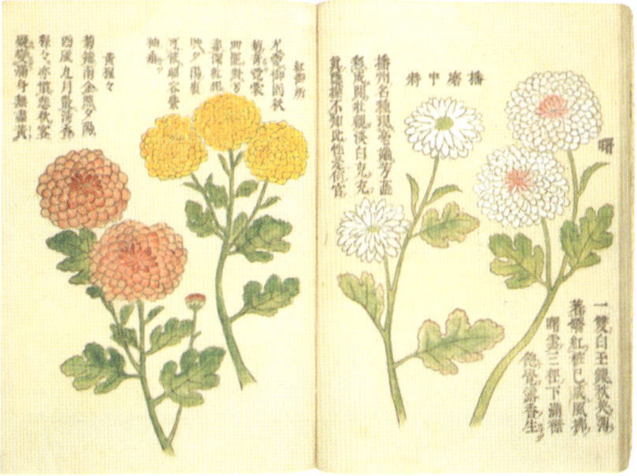

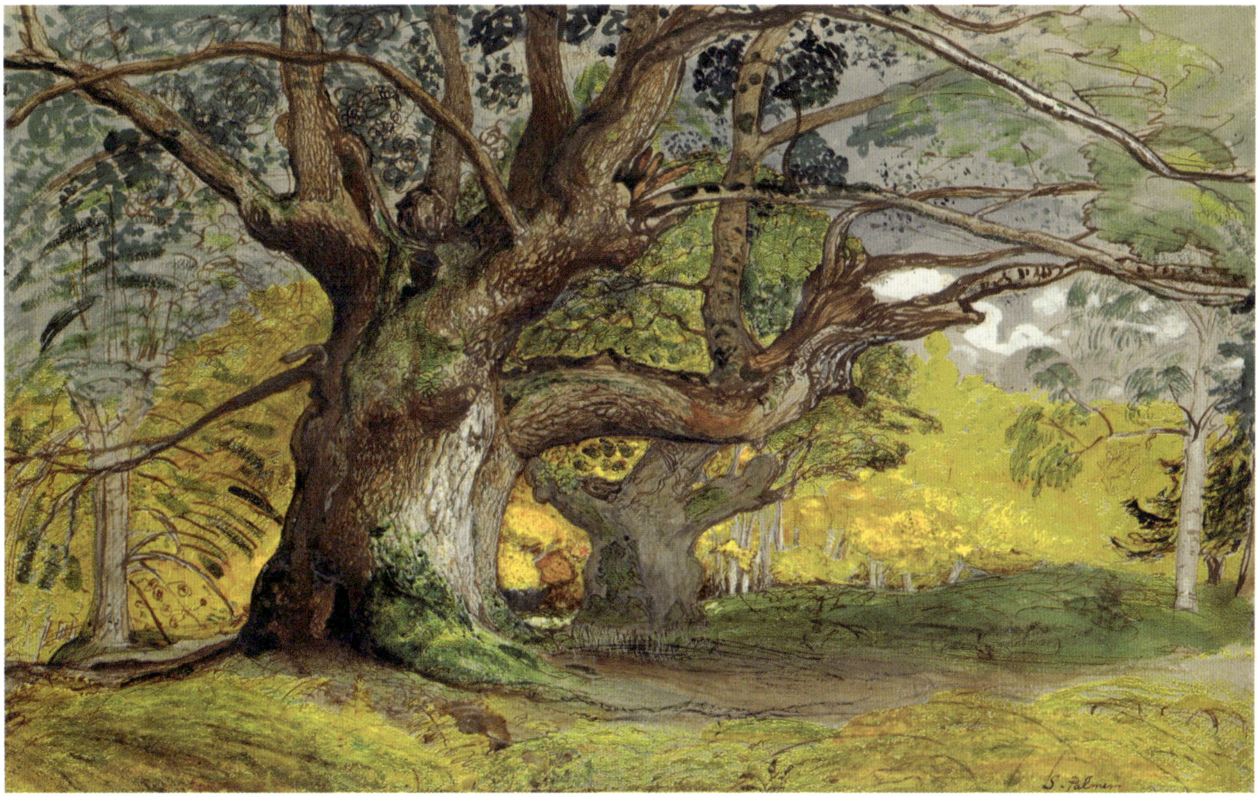

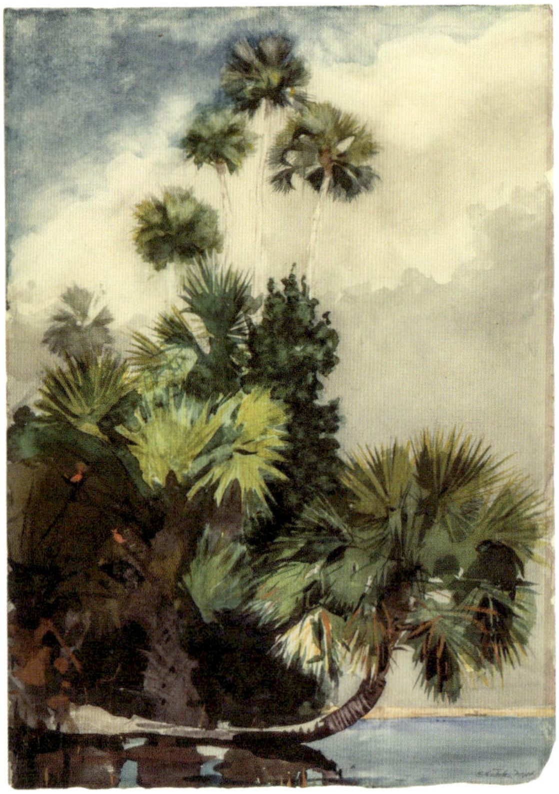

←

Winslow Homer's 1904 watercolour, *Palm Trees, Florida*, includes a variety of palm trees. Palms that are native to Florida include the cabbage palm (*Sabal palmetto*) and the Paurotis palm (*Acoelorrhaphe wrightii*). The state's palm trees are threatened by a bacterial infection known as lethal bronzing disease, which is spread by insects that feed on tree sap.

↓

Sous-bois ('Undergrowth', *c.* 1870–90), a watercolour by the French painter Pierre Puvis de Chavannes, features plants that thrive when they are left to grow wild in undisturbed conditions.

→

The subject of Albrecht Dürer's *Great Piece of Turf*, painted in 1503, was a slice of ground dug from a meadow near his home. It shows numerous species of wild flowers and grasses – such as dandelion (*Taraxacum*) and greater plantain (*Plantago major*), and possibly meadow grass (*Poa pratensis*), germander speedwell (*Veronica chamaedrys*) and hound's tongue (*Cynoglossum officinale*) – growing together in their natural environment.

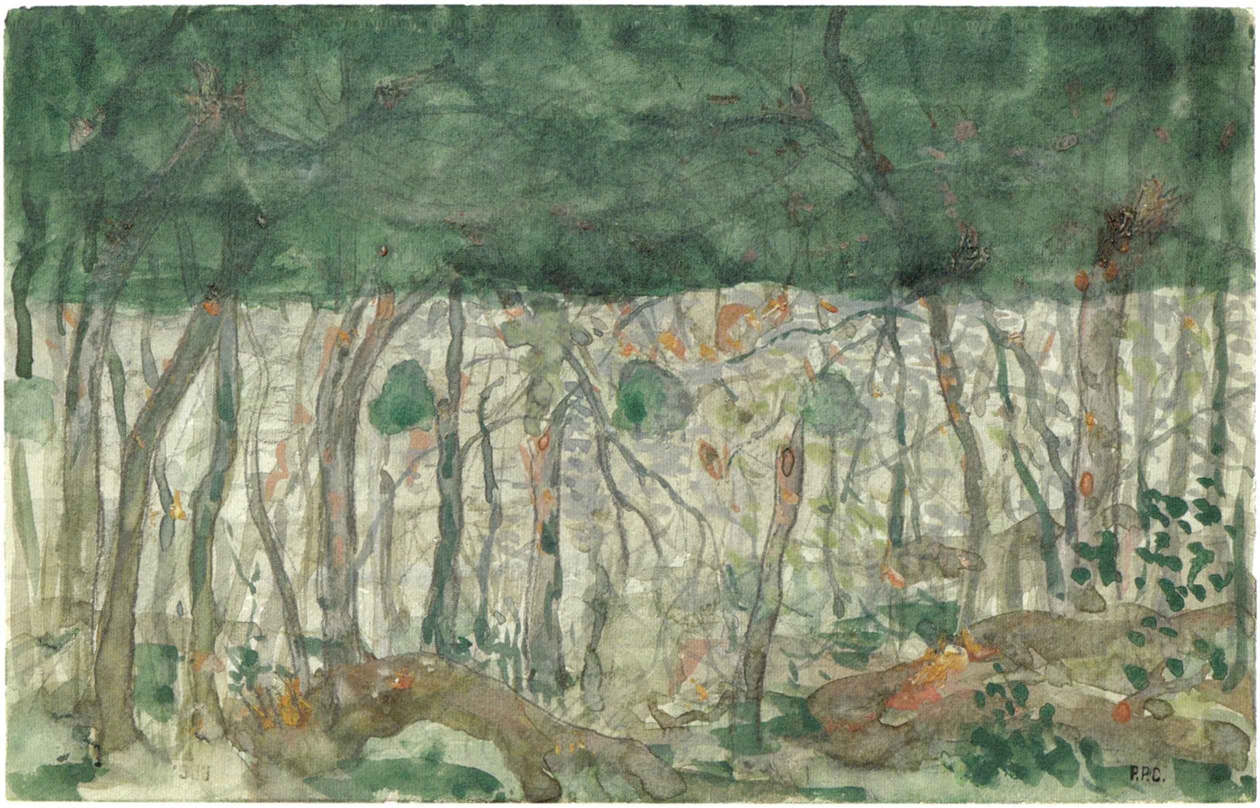

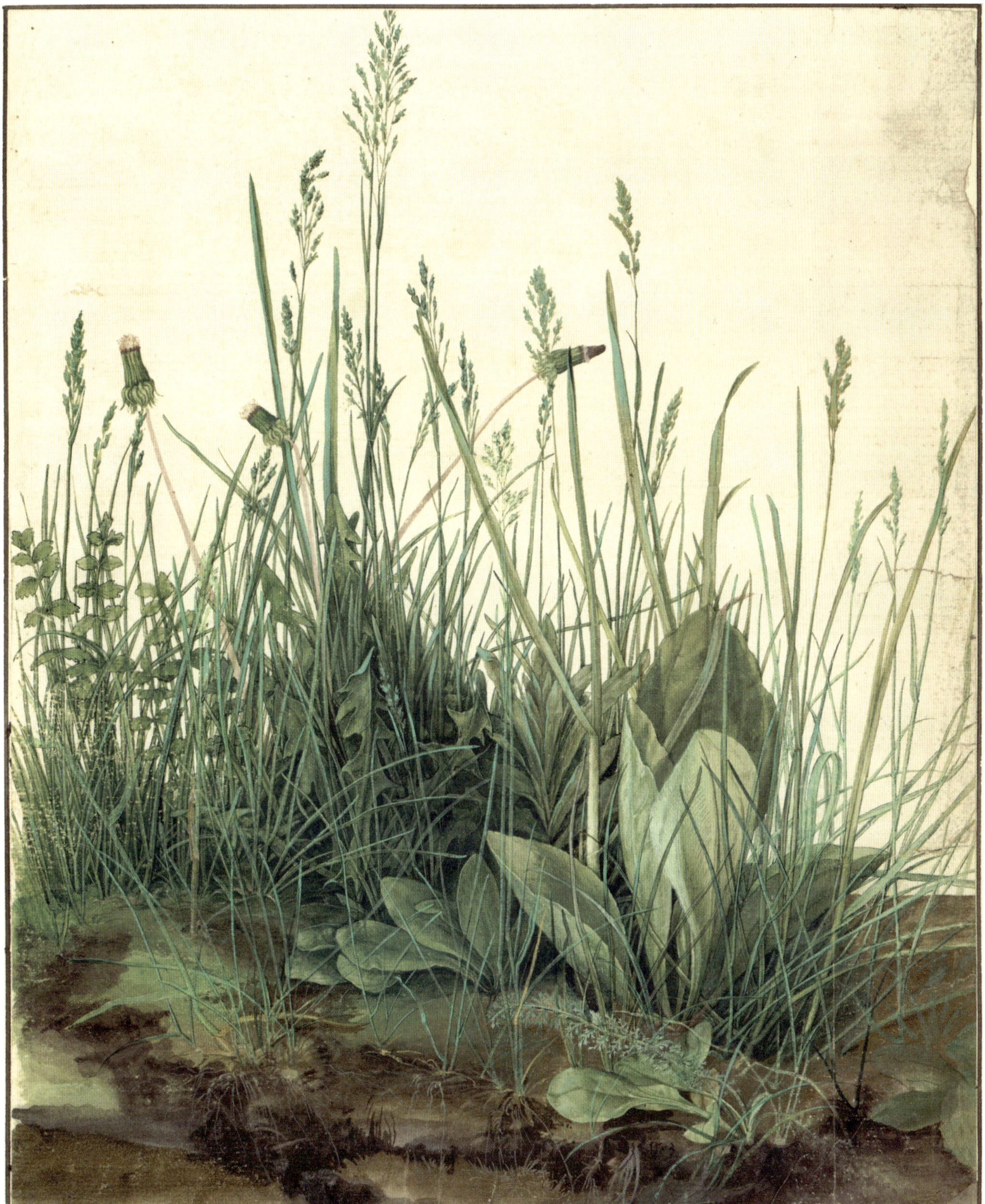

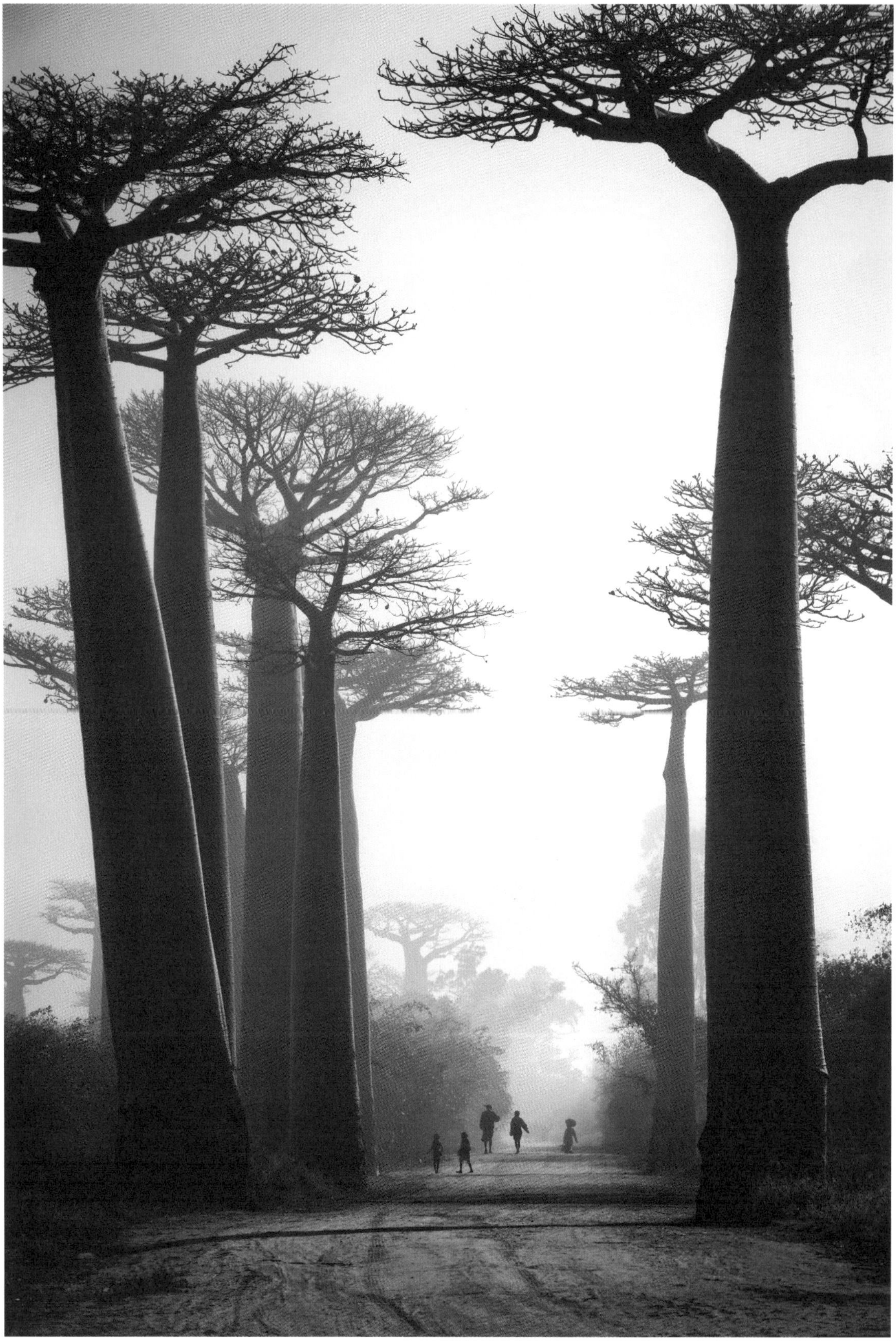

(11)

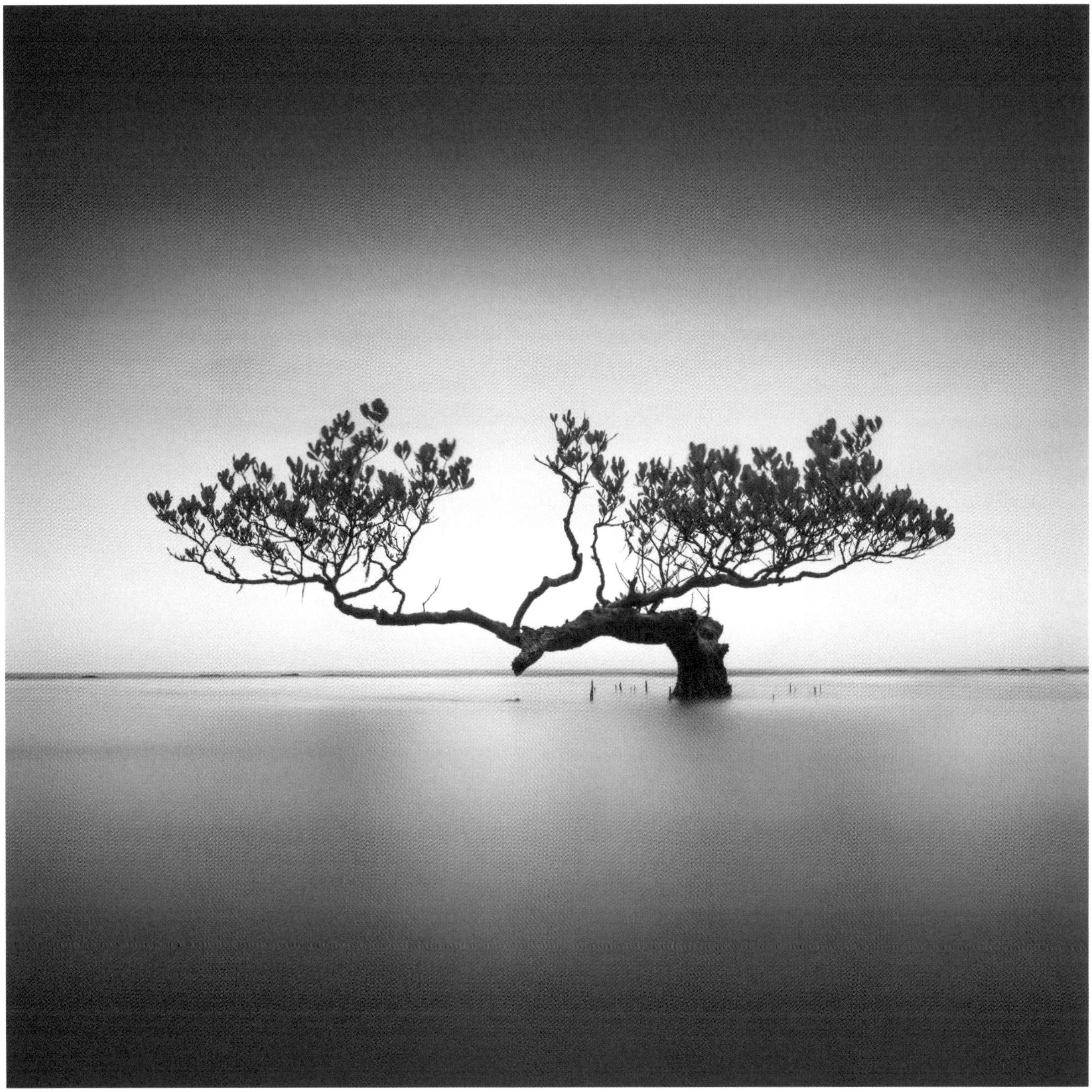

←
◎ Madagascan photographer Pierrot Men's *Allée des Baobabs, Morondava, Madagascar*, 2012, features Grandidier's baobabs (*Adansonia grandidieri*). This avenue of giant trees used to be surrounded by dense forest, which has been cleared for farming.

↑
Indonesian photographer Hengki Koentjoro's *Mangrove* (2017) draws attention to the importance of mangrove ecosystems, one of the most threatened tropical marine ecosystems on the planet. Mangove forests are ten times more efficient at storing carbon than traditional forests, they provide buffer zones between land and sea, protecting coastlines, and they shelter an array of rare and protected species.

(11)

TREES: ARCHITECTS OF THE NATURAL WORLD

Harriet Rix

Trees built the ark, but were they on it? We know that they must have been, because without trees there is no way life could survive on earth; it's more absurd to think of trees being absent from an oasis of life than it is to imagine animals could survive cataclysms by floating around in a wooden boat. It took millions of years for trees to weave stability across dry land and enable animals to live there. At the very beginning of their development, trees were frequently wiped out by small tidal waves or floods roaring down mountains. Eventually, their roots, branches and trunks gradually knotted together rivers and stabilized mountains, and in their slipstream came detritovores, herbivores and freshwater fish. Now their presence provides the safe niches that we rely on, and all the strange and bizarre glories of the world, from luminous fireflies to three-toed sloths, exist under their shadows.

Of course, not all animals live in forests. Emperor penguins of the Antarctic will never see a tree, and it's likely that neither the Saharan silver ant nor the large-eared pika of the Himalayas – a charming species of rabbit that lives on high mountain peaks – will encounter one. However, these are the exceptions. Even in deserts or grasslands or tundra, the overwhelming majority of life on earth relies on trees because trees have dominated terrestrial ecosystems for 370 million years.[1]

Energy from the sun is the basis of all human and animal life, and nothing on earth can convert sunlight into food quite like trees. They are the bottom cog of the wheel, grinding away whenever there is sunlight to join together 16 billion tonnes of the carbon dioxide in the air into solid carbon chains every year.[2] These chains become the building blocks of the elephants and giraffes, butterflies and fungi that consume them, or are broken down to release the energy they store. Looking at the size and mass of a tree gives us a direct indication of how much carbon dioxide and energy has gone into the making of it.

Trees are architects of the visible world, whether animal, vegetable or mineral, and implicit in tree architecture is the construction of habitats for other life forms. Trees are never uniform: branches will all be of a different thickness, trunks may be hollow or smooth, leaves can grow very differently even on the same tree. This means that there is an almost unlimited number of different worlds, shapes and environments that can be created by trees. In tropical environments a large tree can look like a waterfall of green, a cascade of multilayered branches that can produce micro-climates like the rooms in a house – shady and warm – and support orchids and monstera, ferns and prayer plants. The oaks of the UK support 716 species of lichen and 1,178 different species of invertebrate, and those that have been pollarded support different beetle species from those that have not.[3] Even in arctic climates trees can be smothered with mosses, lichens and ferns.

Trees are also architects of the invisible world. When we think of a tree with a trunk and branches and leaves and roots, we are seeing only the skeleton of the tree. Clothing trees in the air and under the ground is a world of unseen chemicals, and these chemicals act as a proxy through which trees can react with the rest of the world. With enough water, air and sunlight, trees have huge amounts of energy to create with, and they use this to twirl carbon into a plethora of shapes and sizes. The smallest molecules they create, one or two carbons long, often make up the tree scents that we know – the smell of orange blossom and *Elaeagnus*, cinnamon and myrrh. Larger rings give us the sugars in maple syrup and mangos. The largest matted chains and rings of carbon are the resins for cello bows and the wood pulp that this page is made from.

The visionary architect Sou Fujimoto (b. 1971) writes that 'the essence of architecture is to enclose a particular territory while opening it up to the world at the same time.'[4] Trees do exactly this in myriad ways. Plants have become massive and woody at least seven times in evolutionary history, each time lignifying their cells to make a stiff palisade of wood, shooting up huge canopies to project leaves into the domain of air, and drilling roots deep down into the domain of water. In growing tall, trees solve a problem; they need both air and water, so they use their trunks as a pipeline that can mix these two notoriously irreconcilable elements. As trees build these tall, wide structures, they have created an enclosed territory, but air and water are sucked in as a result. In a similar way, tree bark and wood provides a physical barrier around the territory of the tree, hard for a beetle to chew into, but trees' tolerance of mutualism acts as a magnet to lichens, plants, birds and insects. The enclosed territory of the tree becomes a magnet for the world.

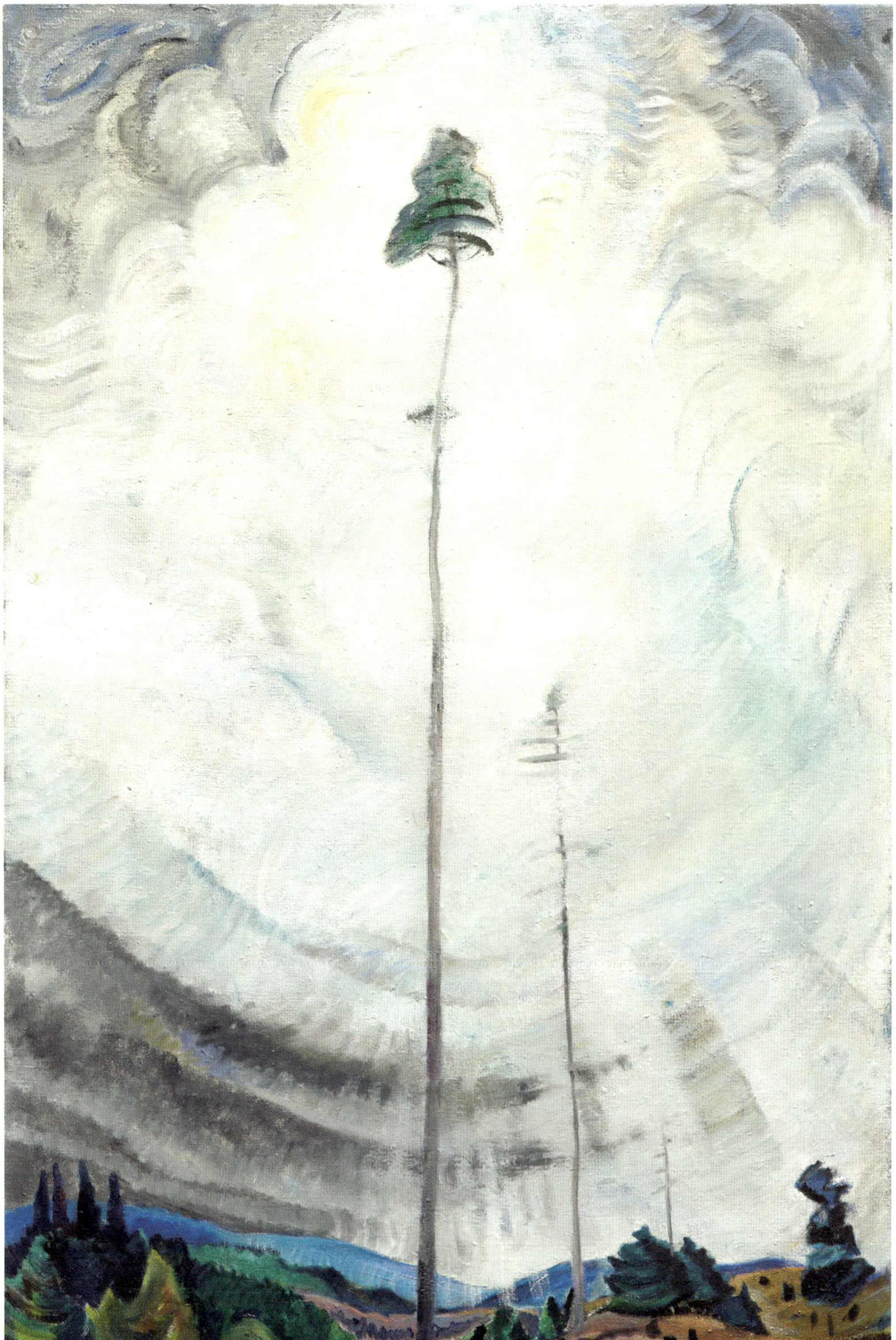
M.E.CARR

(11)

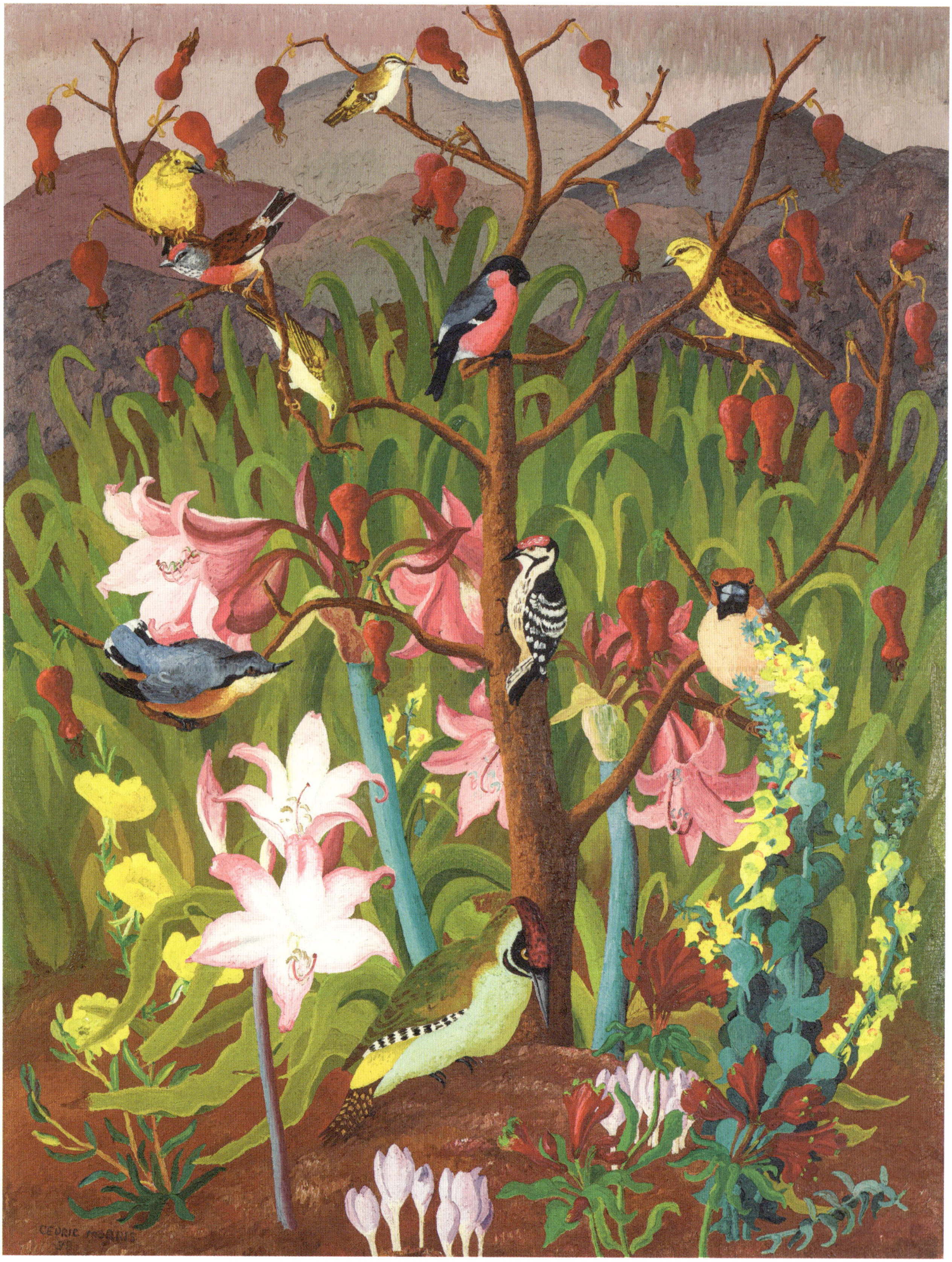

Nowhere is this more evident than in the relationships between trees and fungi, an entangled, complex, give-and-take of sugar and some of the rarest metals in the soil. When trees put out their very early roots, fungi will often try to enter the root and form a link with it. In some cases, the tree will resist by thickening the cell wall or poisoning the fungus, in others it will let the fungus in, and in doing so merge its root system with the mycelium network of the fungus, which can then stretch out ten times further than the roots.

Trees can survive without fungi, but with fungi they become so much more. Trees provide the sugar, fungi provide access to places that tree roots cannot reach and immensely efficient digestion of dead matter, contributing nitrogen and rare metal ions as a result. It's an ambivalent relationship, of course; fungi are the only life forms that can digest the toughest part of a tree, the lignin, which makes up a major part of the tree, and as such they are a threat. However, as trees and fungi have evolved together the balance has been maintained well enough that some fungi will co-exist only with specific trees. The delicious saffron milk cap, for example, will only grow with pines and no other tree. One fungus may link several pines together with its mycorrhizal network, and they may exchange sugars, as well as nitrogen and other nutrients, and be more productive as a result.

It's not just fungi that connect trees to the wider world. In the air, we can smell the scent of the pines as we walk into a forest, feel the change of air as we walk into a beech wood in spring, smell the sweetness of a fig in an autumn garden; a shifting cloud of scents and chemicals attracts us even through the barriers we have set up between us and the natural world. Trees have created an actively open boundary that is centripetal to plants and insects and humans. Life is attracted into the orbit of the tree.

Trees move huge distances while remaining firmly rooted, by the simple expedient of using animals (or sometimes air) to do their moving for them. The Brazil nut tree, for example, can grow to 60 metres (197 feet) tall in the Amazon if it finds the right niche, and after the rainy season its canopy above the rainforest will be covered with creamy white flowers which only flower for a day. Only one glossy, black bee species is strong enough to get into the flower, and this bee can fly for kilometres over the canopy to find the right tree and the scented orchids that grow on it. After the nuts develop, it is only a small mammal, the agouti, that can get through the extraordinarily hard shell to eat the nuts or bury them for later consumption, spreading the Brazil nut tree species as a result, sometimes by kilometres.

There are myriad examples of co-evolutions like this, the shaping of animal and plant together like a hand to a glove. Counter-intuitively, these moments of intense change and adaptation are often powered by extinctions. Plants like the Brazil nut tree and the magnolia, the oak and the honeysuckle – in fact all the flowering plants we see on earth today – only evolved after the apocalyptic impact of the Chicxulub asteroid sixty-six million years ago. The asteroid released the energy of ten billion atomic bombs. In the two years of darkness that followed, everything changed; the great open conifer forests waned and shrunk and were eventually replaced by the hugely productive broadleaved forests such as the Amazon. Fragments from the time before the asteroid are still found around the Pacific in the giant sequoias, the kauri of New Zealand and the *Cryptomeria* of Japan, but they are a still shrinking minority fitting into a new world. Even before that, cataclysms ended the great and strange trees that made coal, and the earliest trees – with their huge root systems and beautiful, feathery, ginkgo-like leaves – were probably victims of their own success, possibly producing so much oxygen that they burnt to extinction.

In each cataclysm thousands of tree species have become extinct, but in each case the response of those remaining has been to diversify and, in turn, diversify the animals they live with. This diversity – there are about 80,000 taxa – together with the adaptations of species to all the niches of the earth, means that trees did not really need the ark. Trees do so much to find a good niche and stay there, and they also create niches – both for themselves as individuals and collectively. Perhaps the myth of the ark was created by someone who was oblivious to the symbolism of wood supporting all of the rest of life, but I don't think so. I think that those great ribs and planks were symbolic of a cradle – a tree-cradle supporting then, as now, the whole of life on earth as we know it.

↲
Canadian artist Emily Carr's *Scorned as Timber, Beloved of the Sky* (1935), features a pine tree that has been rejected by loggers because of its spindliness. The painting reflects her concerns over the impact of forestry and mining on coastal landscapes.

←
All the flowers in Cedric Morris's *Crisis* (1938) – including lilies, crocuses, a Dalmatian toadflax (*Linaria dalmatica*) and a shrub that may be a common alpine rose (*Rosa pendulina*) – were painted from nature.

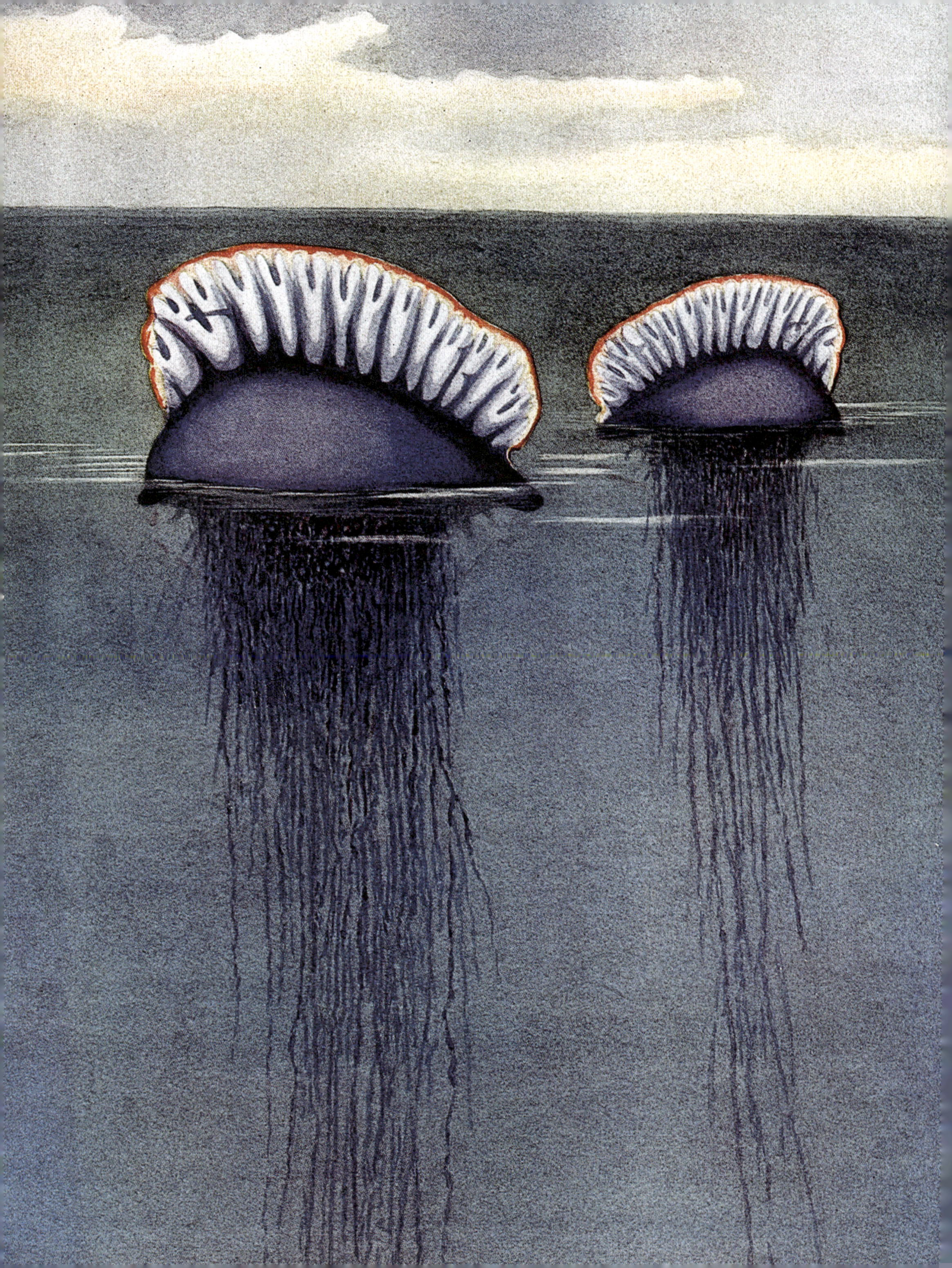

(III)

INVERTEBRATES

PORTUGUESE MAN O' WAR	BEE
PRAYING MANTIS	FLY
SMALL WHITE BUTTERFLY	PAINTED LADY BUTTERFLY
SMALL HEATH BUTTERFLY	MEADOW BROWN BUTTERFLY
COMMON BLUE BUTTERFLY	RED ADMIRAL BUTTERFLY
LARGE YELLOW UNDERWING MOTH	DRAGONFLY
NARROW-BORDERED FIVE-SPOT BURNET	BUFF-TIP MOTH
PEPPERED MOTH	ANGLE SHADES MOTH
LOBSTER MOTH	STICK INSECT
SCARAB BEETLE	BLUE MORPHO BUTTERFLY
GREATER STAG BEETLE	CICADA
GRASSHOPPER	FIREFLY
SPIDER	GIANT PEACOCK MOTH
ANT	SNAIL
SILKWORM	EARTHWORM
JAPANESE HORSESHOE CRAB	PYURA MICROCOSMUS SEA SQUIRT
QUEEN CONCH	CYANEA ANNASKALA JELLYFISH
KNOBBLY ANEMONE	ELEPHANT EAR SPONGE
BRITTLE STAR	GIANT BASKET STAR
CAKE URCHIN	CALCAREOUS SPONGE
COMB JELLY	STARFISH
MADREPORA PULCHRA STAGHORN CORAL	SEA BAMBOO
ATLANTIC WHITE-SPOTTED OCTOPUS	FLYING SQUID

INVERTEBRATES are animals that do not have a backbone or internal bony skeleton and include insects, worms, sponges and crustaceans. They are found on land and in water, and make up more than 90% of existing animal species. Without a rigid skeleton, some invertebrates, such as earthworms and octopuses, are soft-bodied. Many other terrestrial and aquatic invertebrates, though – including all insects, crustaceans, arachnids and shelled molluscs – have developed hard exoskeletons for the protection of internal organs and structural support of the body.

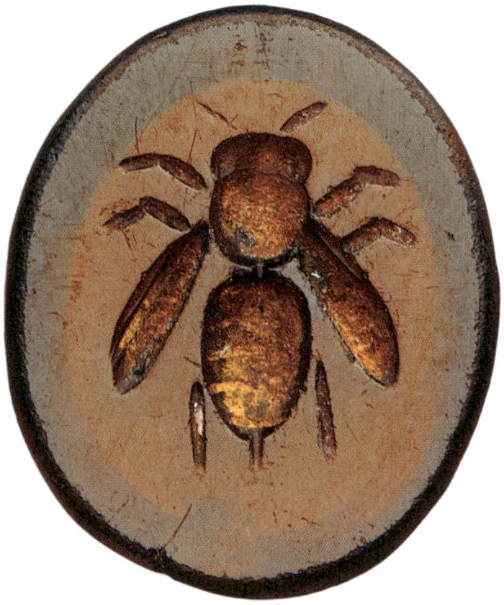

'The pedigree of honey
Does not concern the bee;
A clover, any time, to him
Is aristocracy.'

EMILY DICKINSON, *POEMS*, BOOK III: NATURE, POEM V, PUBLISHED 1890

(III)

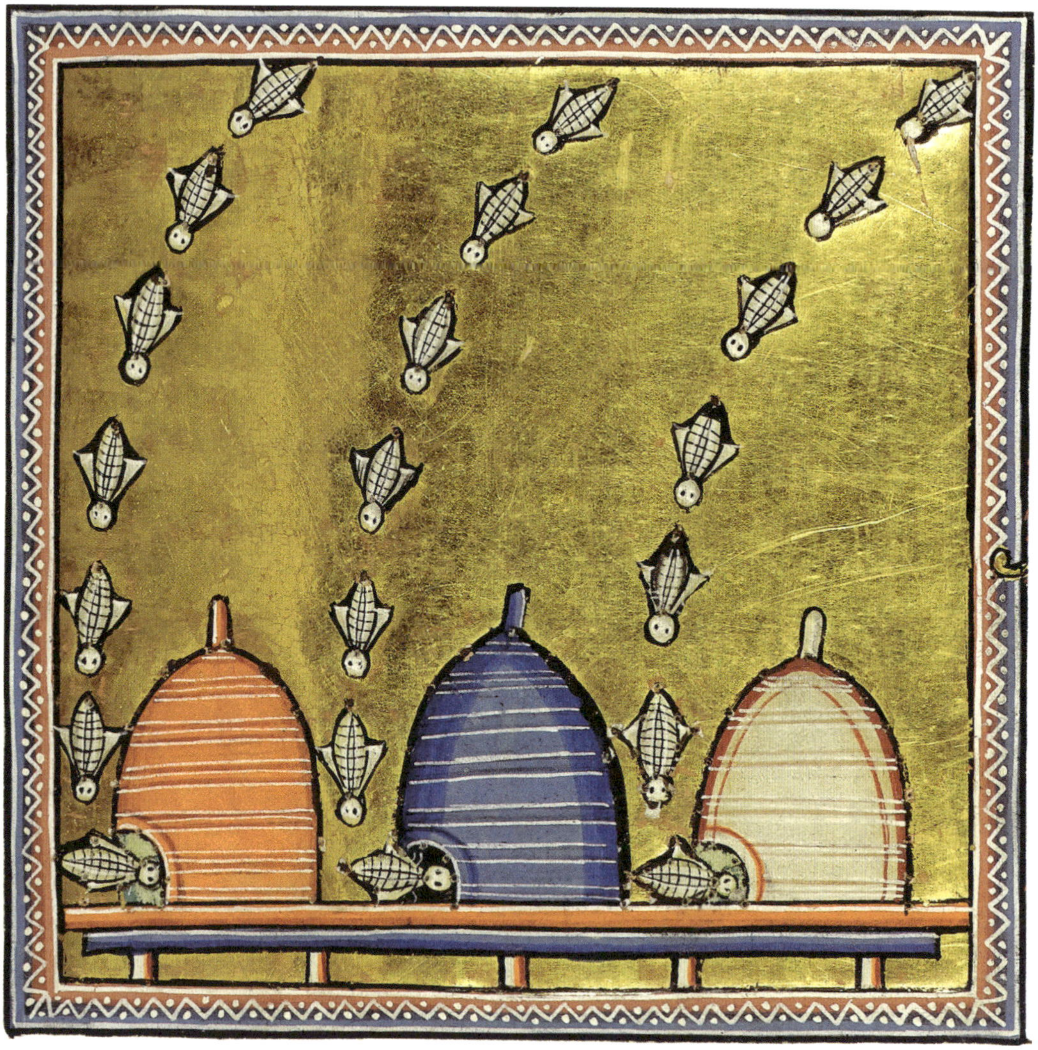

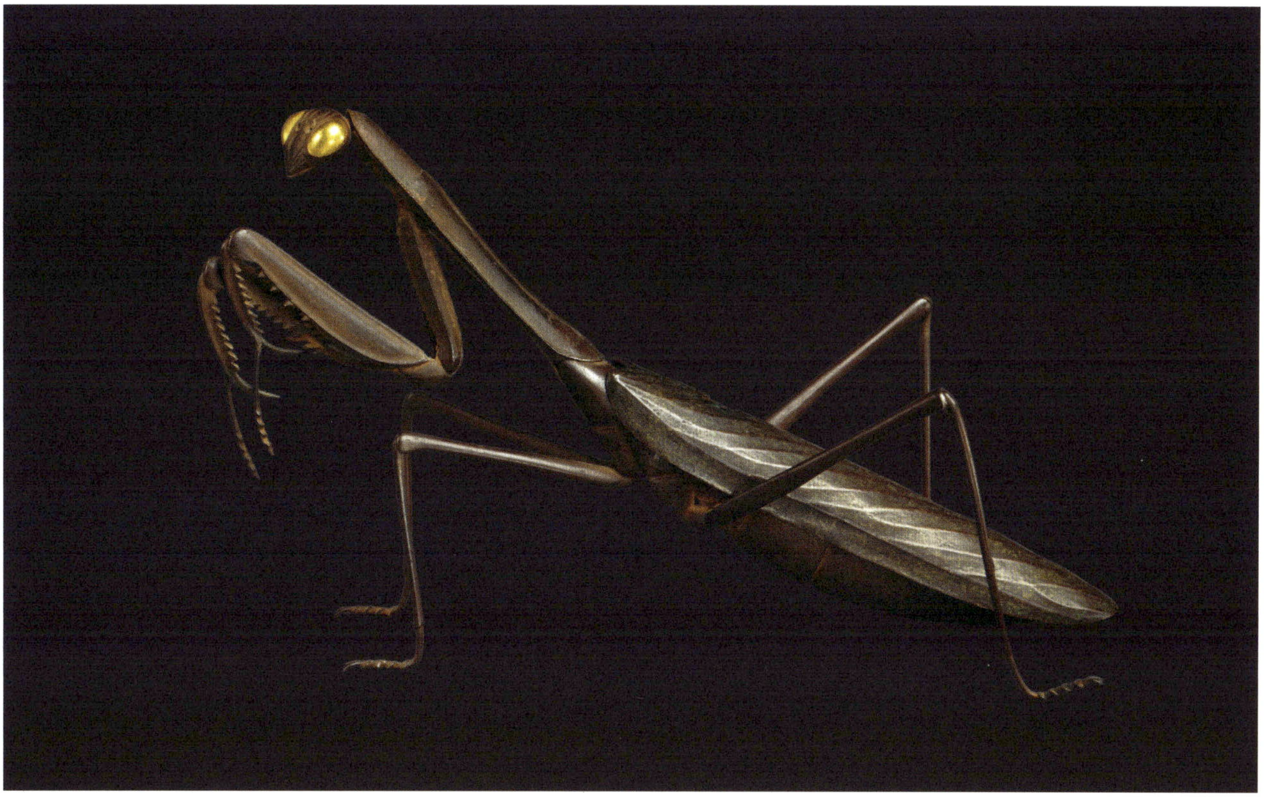

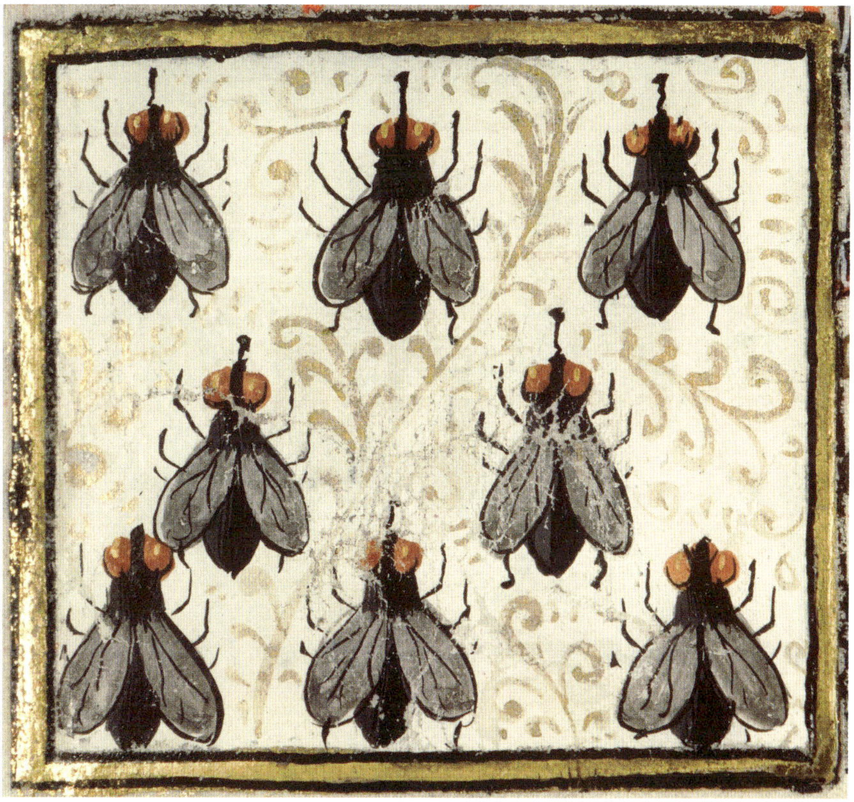

←↖
The ancient Romans were highly knowledgeable beekeepers. This Roman carnelian engraved with the image of a bee (opposite, top), dating from between the 1st and 3rd centuries CE, may have been used in a signet ring. The industriousness of bees is celebrated in an illustration from the *c.* 1200 *Aberdeen Bestiary* (opposite), in which an orderly stream of bees enters each hive, or skep.

↑
The praying mantis fulfils a vital role in the food chain as it feeds on insects such as aphids, cockroaches and mosquitoes while acting as a food source for creatures such as birds, frogs and spiders. This 19th-century metal incense burner in the form of a praying mantis is from Japan.

→
The humble fly is celebrated in a beautifully decorated plate from a 1447 French manuscript of *On the Properties of Things*, a popular medieval encyclopaedia written by Bartholomaeus Anglicus, or Bartholomew the Englishman, between 1220 and 1240.

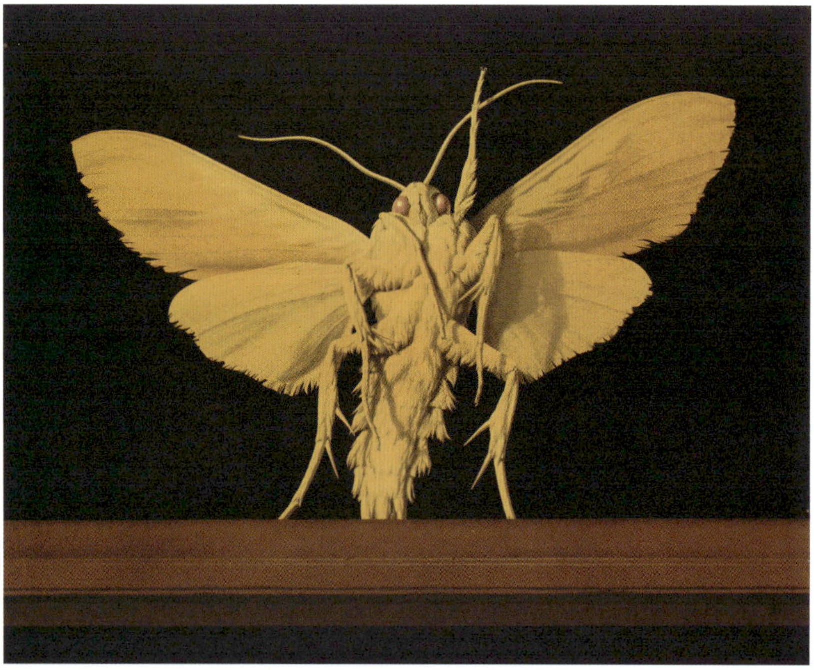

'Butterflies and birds
restlessly they rise up
a cloud of flowers'

MATSUO BASHŌ (1644–1694), TRANSLATED BY JANE REICHHOLD

(III)

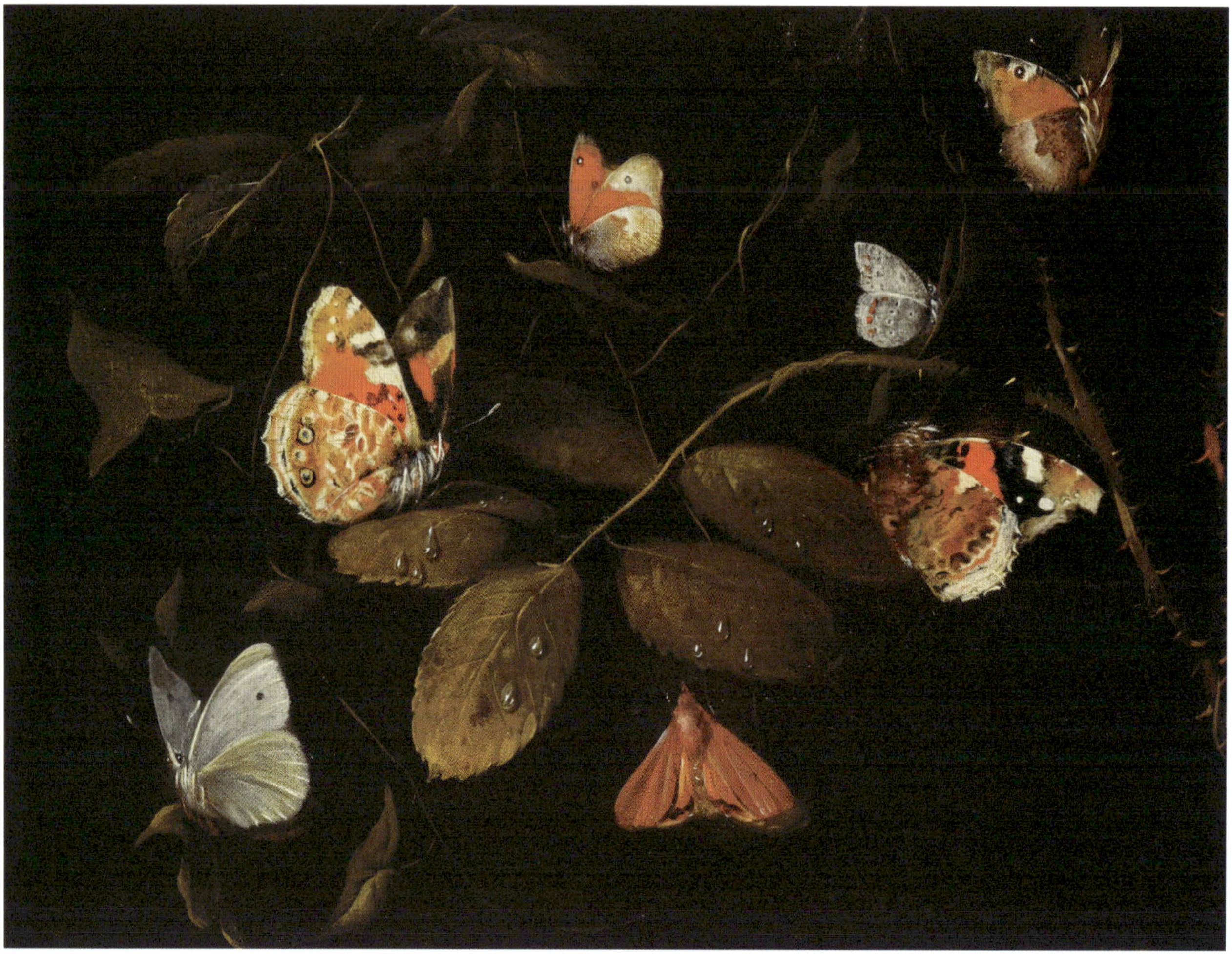

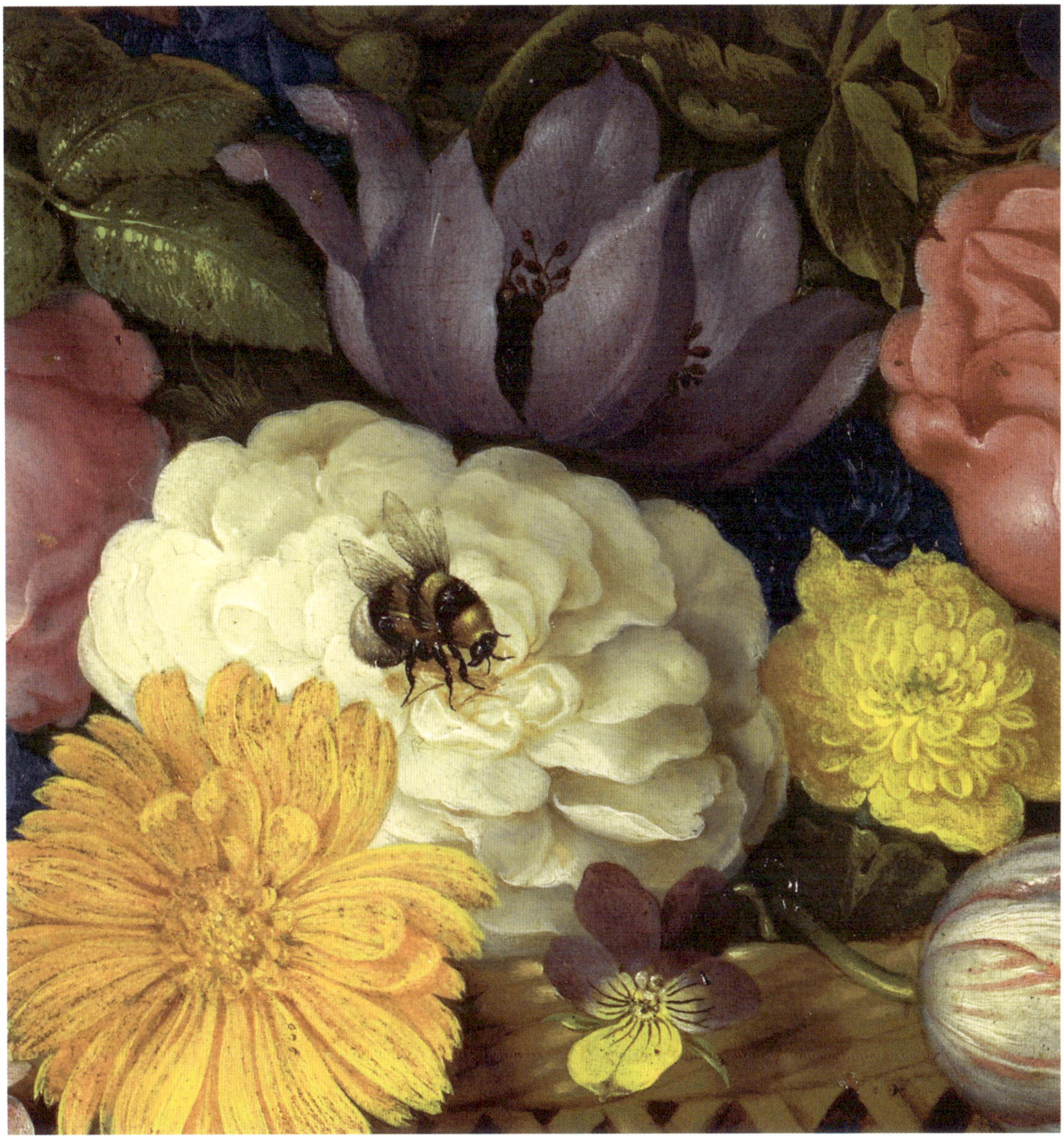

←
Six Butterflies and a Moth on a Rose Branch (*c.* 1690) by the Scottish still-life painter William Gouw Ferguson illustrates several well-known British butterfly species: a small white (*Pieris rapae*), painted lady (*Vanessa cardui*), small heath (*Coenonympha pamphilus*), meadow brown (*Maniola jurtina*), common blue (*Polyommatus icarus*) and red admiral (*Vanessa atalanta*). The moth (bottom, centre) is a large yellow underwing (*Noctua pronuba*).

↖
Peter Birkhäuser's watercolour and crayon drawing, *Moth* (1944–45), was inspired by his seeing a moth fluttering against the window after dark. He later interpreted this as symbolic of his state of mind at the time: the moth, representing his soul, struggles towards the light, symbolizing his consciousness.

↑
Flower Still Life (1614) by Ambrosius Bosschaert the Elder is filled with lifelike renditions of flowers and insects, as can be seen in this detail. In Dutch art at the time, these subjects were a reminder that life is short and beauty transient.

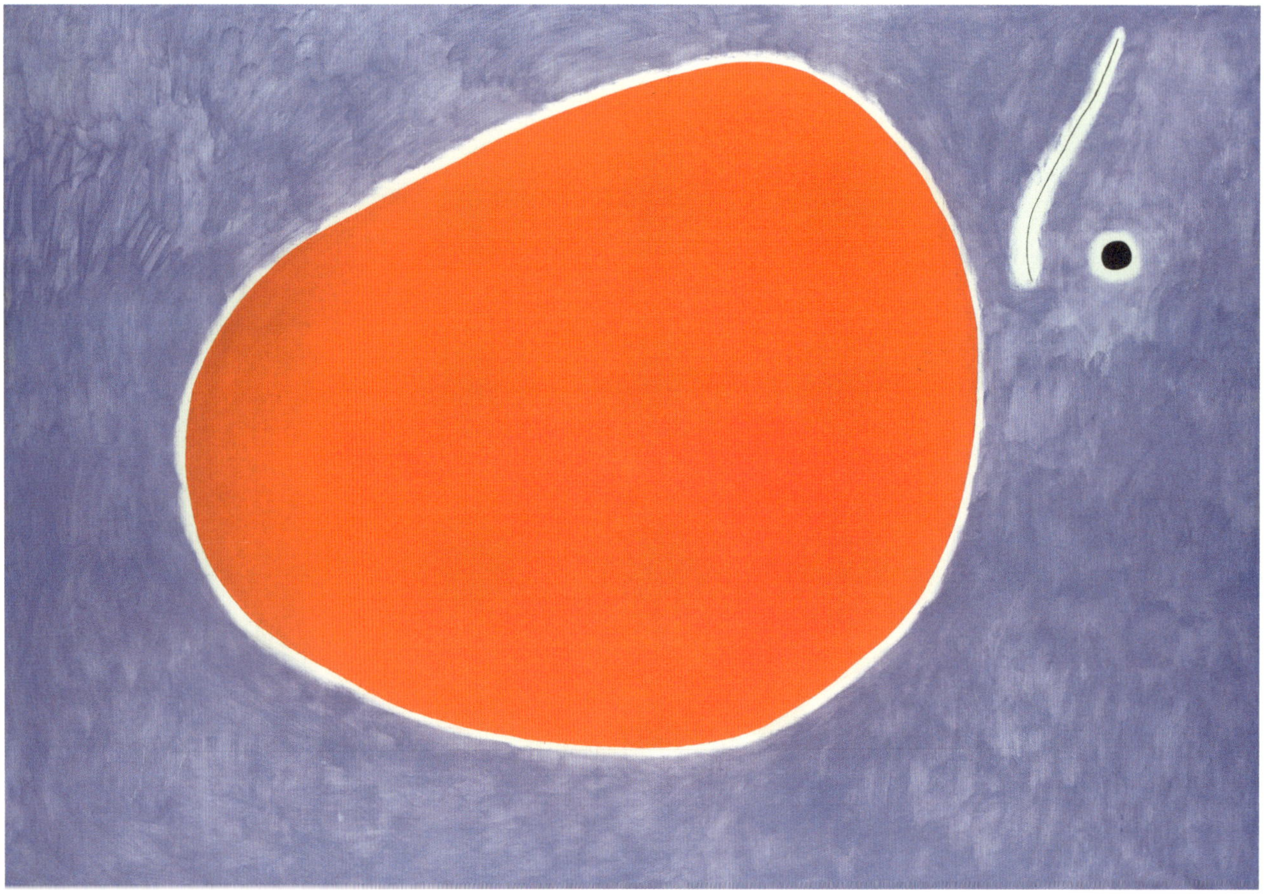

'Well, out comes the dragonfly, all whole, and he sits in the sun – sort of coming alive, you know; and he gets strong, and then he spreads his pretty wings, and flies up in the air, and never is a grub anymore.'

LOUISA MAY ALCOTT, *LITTLE MEN*, 1871

↑
In Joan Miró's *Flight of the Dragonfly in Front of the Sun* (1968), one small but eye-catching line represents the flight of the dragonfly set against the huge form of the sun. Miró has used minimal means to express a balance in nature between large and small.

→
This watercolour illustration of the narrow-bordered five-spot burnet (*Zygaena lonicerae*) shows the brightly coloured moth foraging in the grass. It appears in *Les Papillons dans la Nature* ('Butterflies in Nature', 1934) by the Swiss entomologist and illustrator Paul-André Robert.

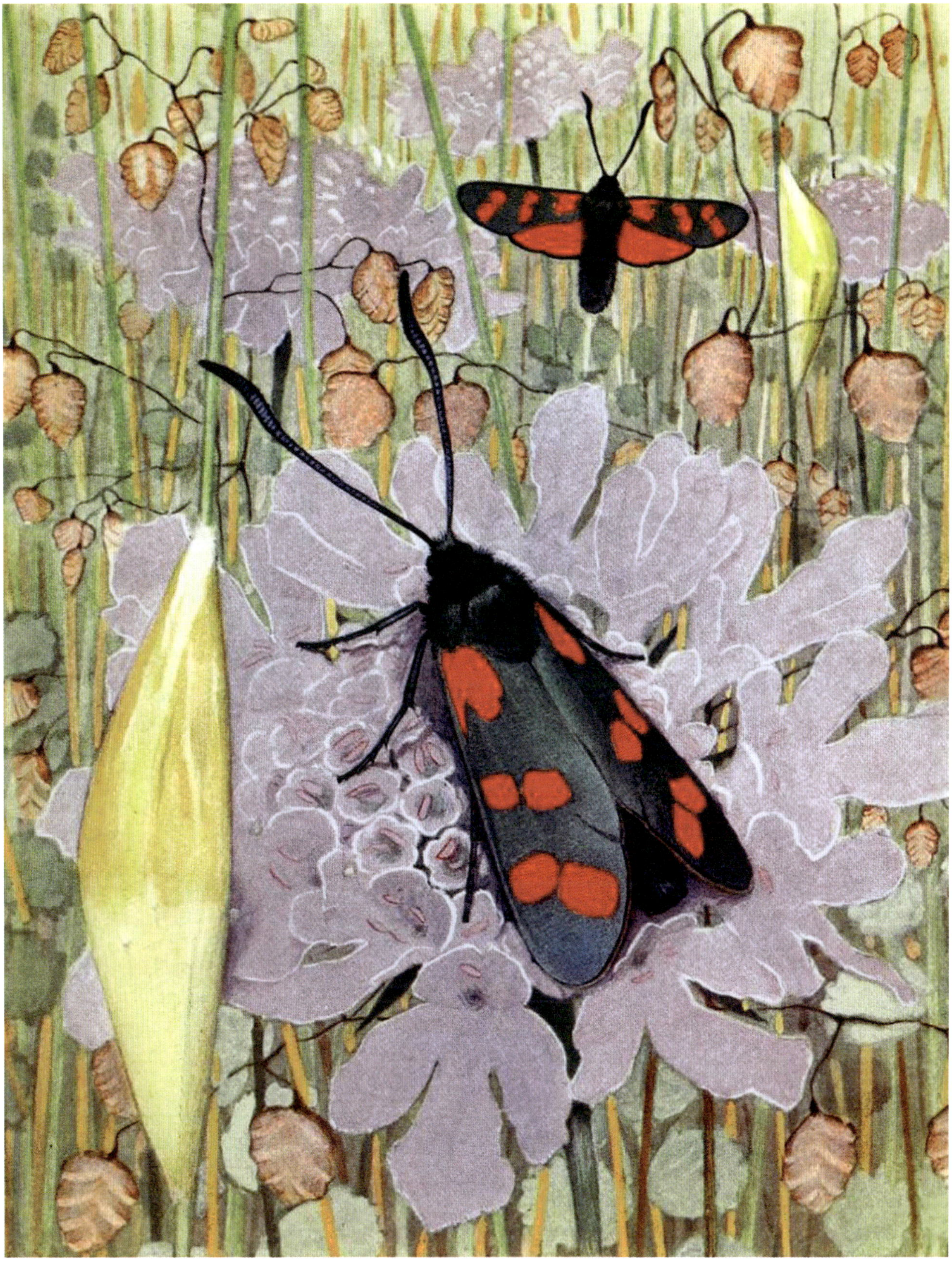

(III)

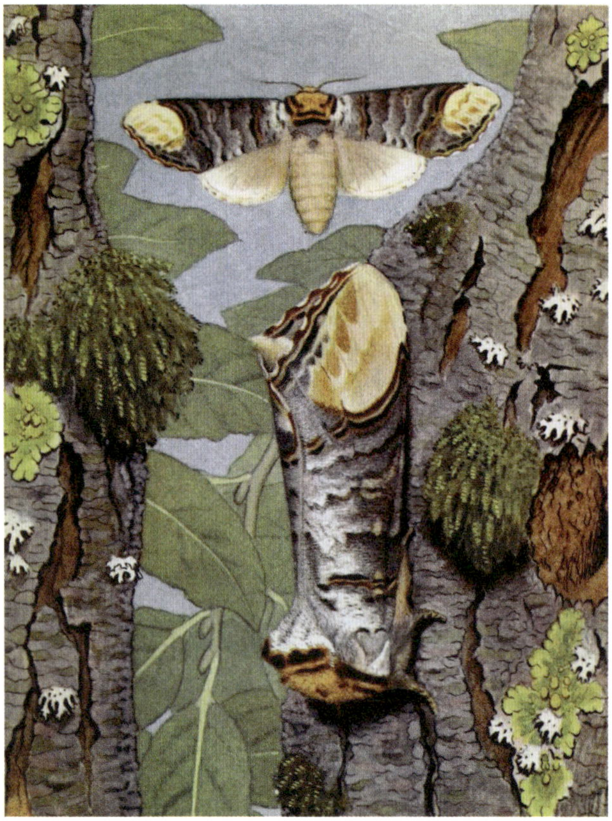

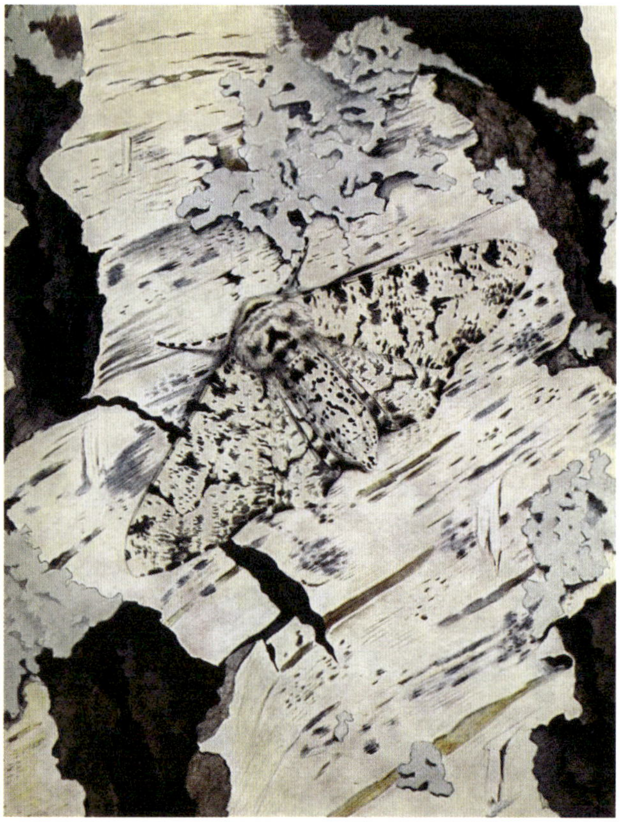

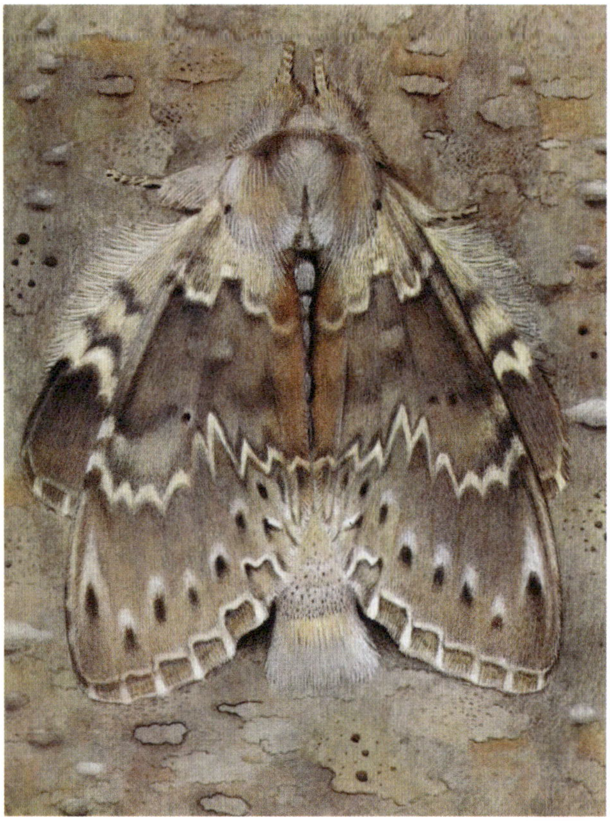

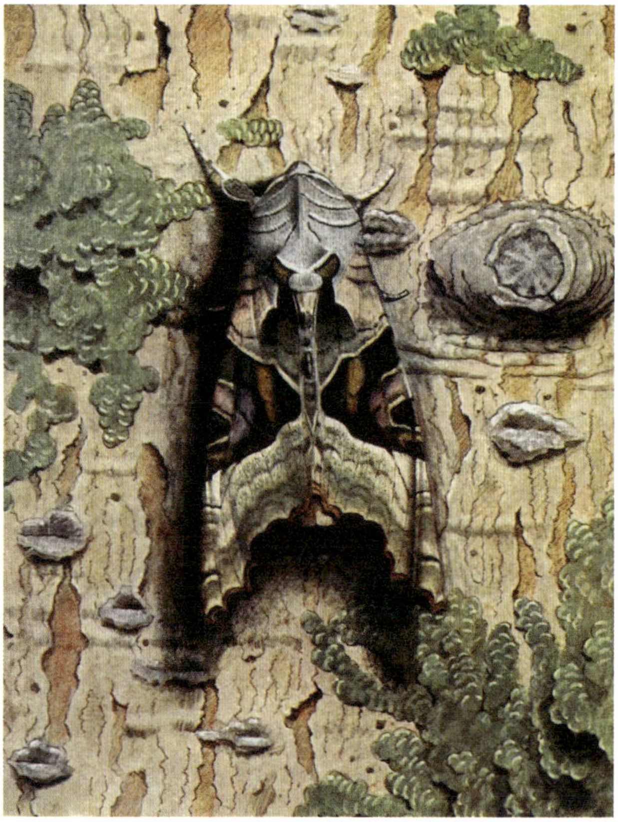

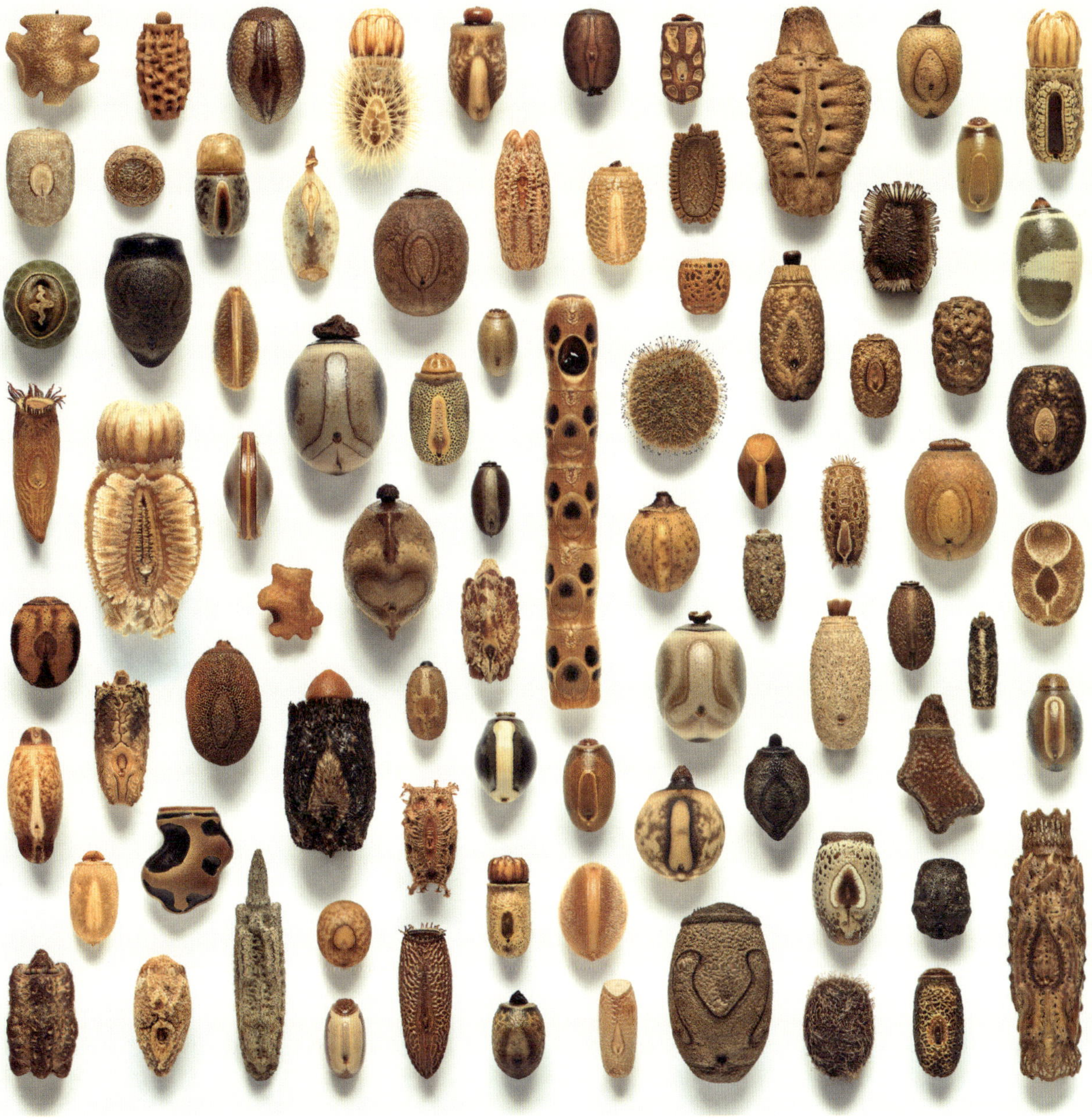

←
Paul-André Robert 1934 book, *Les Papillons dans la Nature* ('Butterflies in Nature'), includes beautiful illustrations of both butterflies and moths, but the four species illustrated here are all moths. Clockwise from top left: buff-tip moth (*Phalera bucephala*), peppered moth (*Biston betularia*), angle shades moth (*Phlogophora meticulosa*) and lobster moth (*Stauropus fagi*).

↑
For *Phasmid (Stick Insect) Eggs* (2022), Levon Biss meticulously cleaned dozens of eggs under a microscope with a single-haired paintbrush. Each egg pictured is less than 3 millimetres (⅛ inch) long. The shapes and designs of stick insect eggs are completely unique to each species; many imitate the appearance of plant seeds in an effort to disguise themselves from predators.

↓
In Jean-Henri Fabre's *Book of Insects* (1921), illustrated by Edward Julius Detmold, two scarabs roll a ball of dung across the ground. In Fabre's words, the beetle is shown working 'in partnership with a friend'. Scarabs lay their eggs in the dung balls, and the ancient Egyptians believed this represented the cycle of life, death and rebirth.

→
American artist Martin Johnson Heade painted *Blue Morpho Butterfly* (*c.* 1863–64) following a trip to Brazil. The blue morpho (*Morpho peleides*) is one of the largest butterflies in the world, with a wingspan of 12 to 20 centimetres (5 to 8 inches). It is threatened by the deforestation of its rainforest home and habitat fragmentation.

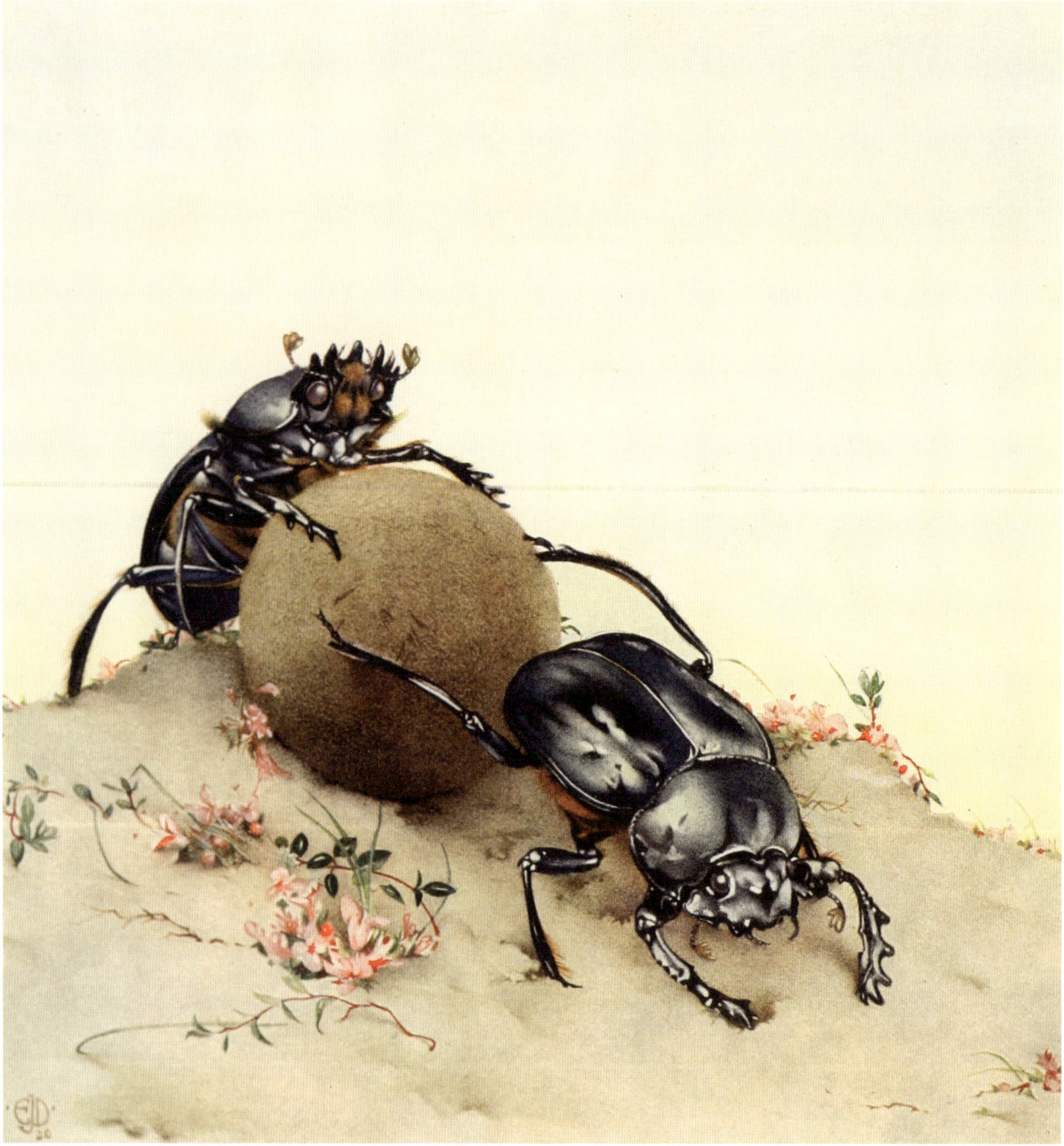

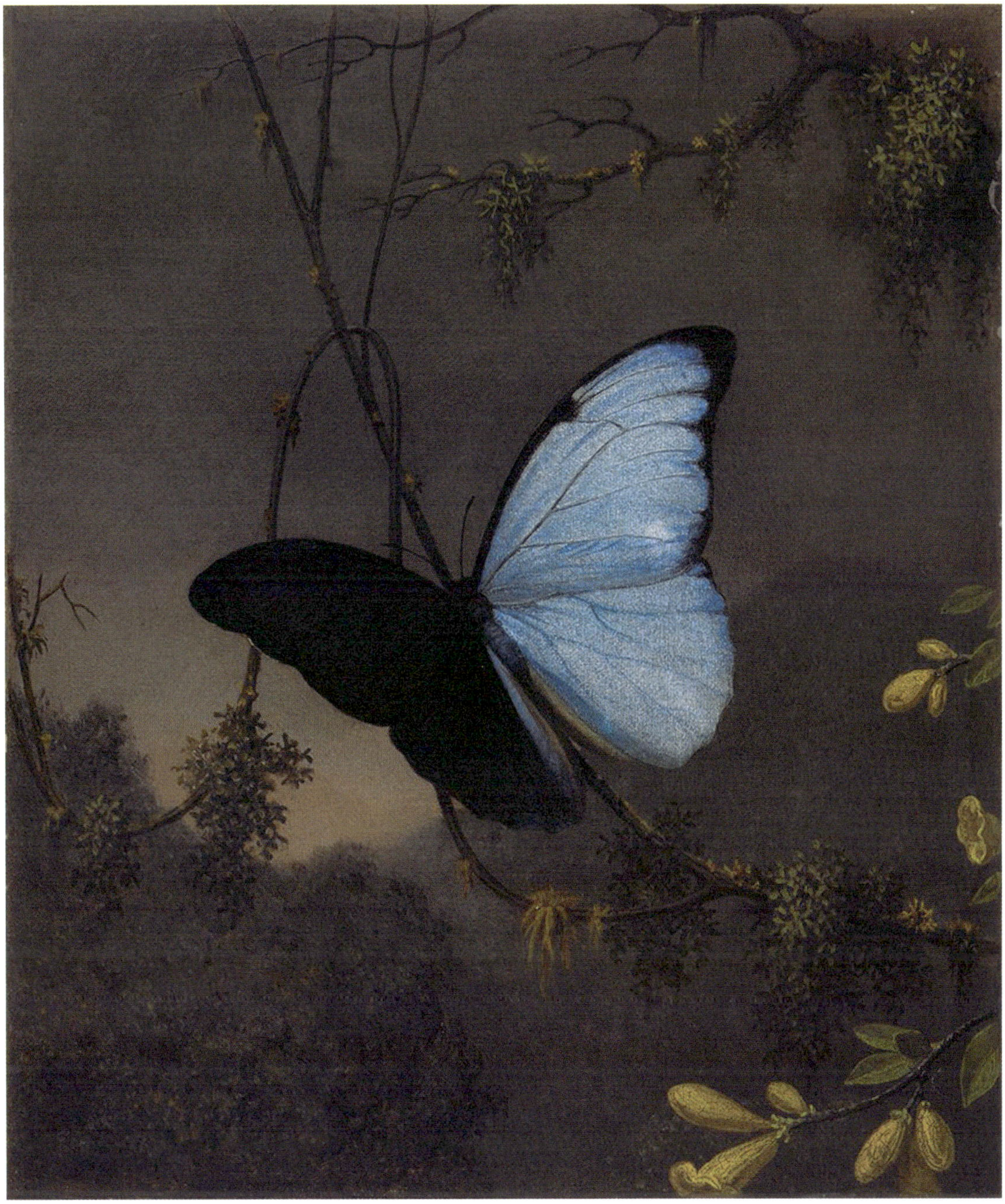

'If [invertebrates] were to disappear ... many of the plants would no longer be pollinated. Lots of animals, amphibians, reptiles, birds, mammals would have nothing to eat. And our fields and pastures would be covered with dung and carrion.'

SIR DAVID ATTENBOROUGH, IN THE BBC TV SERIES *LIFE IN THE UNDERGROWTH*, 2005

'The leaf that taps, the pod that breaks,
The breeze on the grass-tops bending by,
The shrill quick sound that the insect makes.'

KATHERINE MANSFIELD, 'VOICES OF THE AIR', 1916

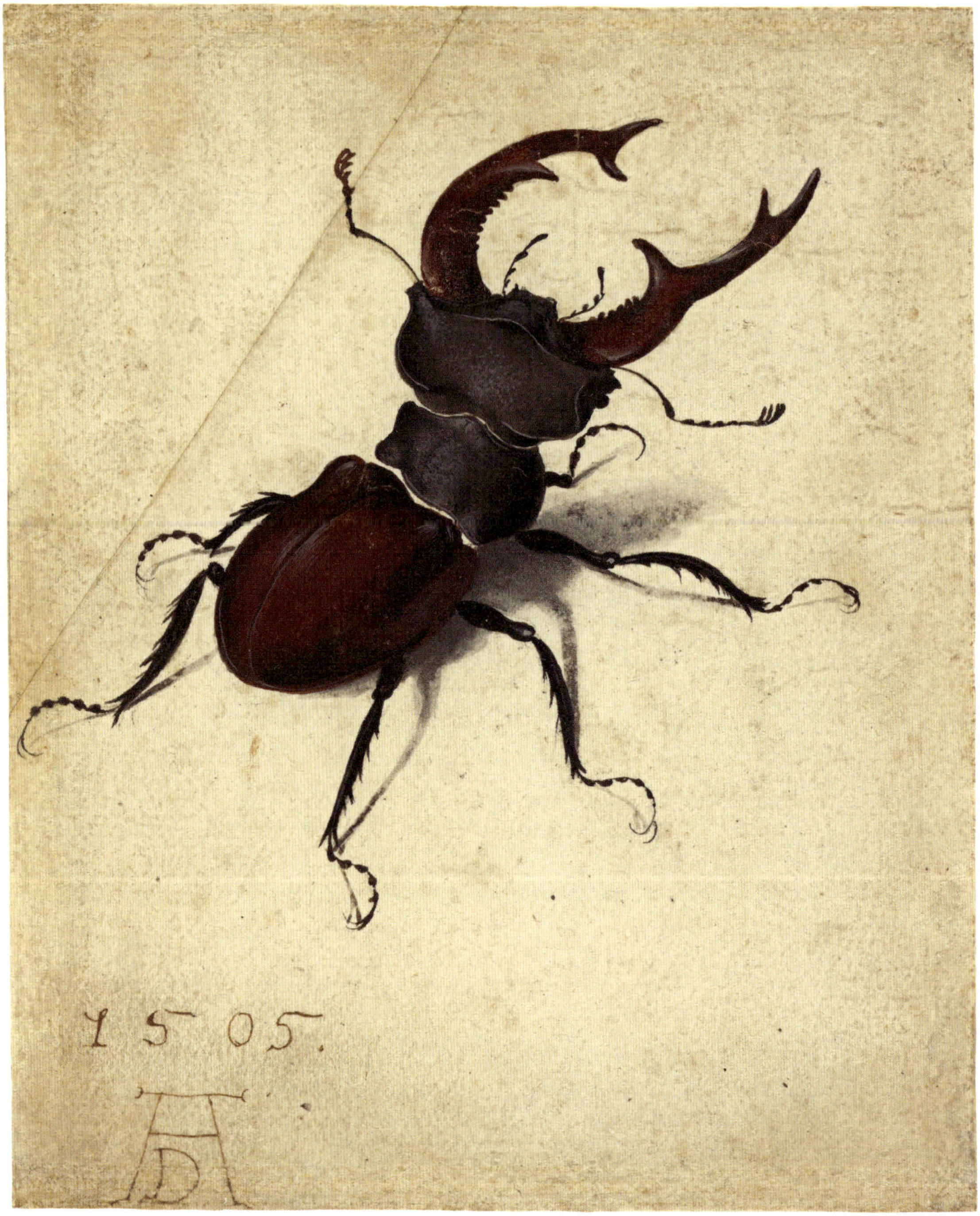

↙
⊖ The greater stag beetle (*Lucanus cervus*) is given star treatment in this 1505 watercolour by Albrecht Dürer. The adult beetle feeds on tree sap and, particularly in Europe, its woodland habitat is under threat in many areas.

→
Vincent van Gogh's *Sketch of Three Cicadas* (1889), from a letter to his brother Theo, captures one of the world's loudest insects. The males sing to attract females by contracting and relaxing ribbed membranes in their stomachs.

↓
Chinese artist Qi Baishi, in whose album of plants and insects (1924) this illustration of a grasshopper appeared, showed insects in their habitats, rendered in a spontaneous but accurate style and bright colour.

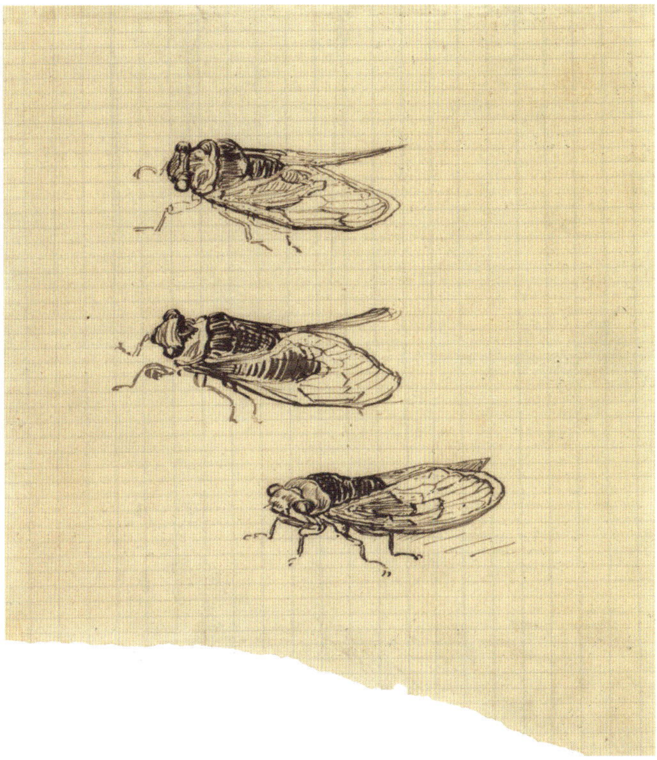

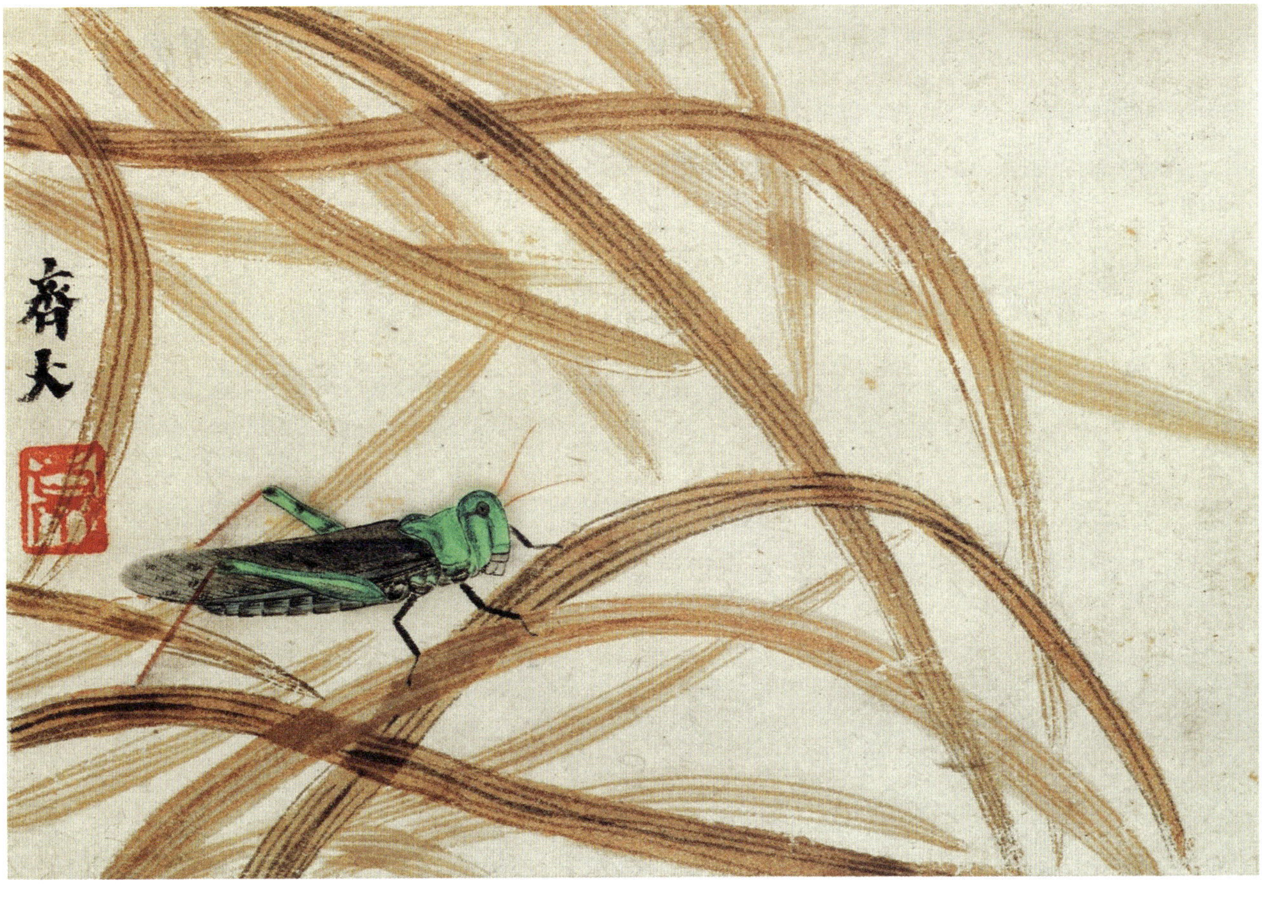

(111)

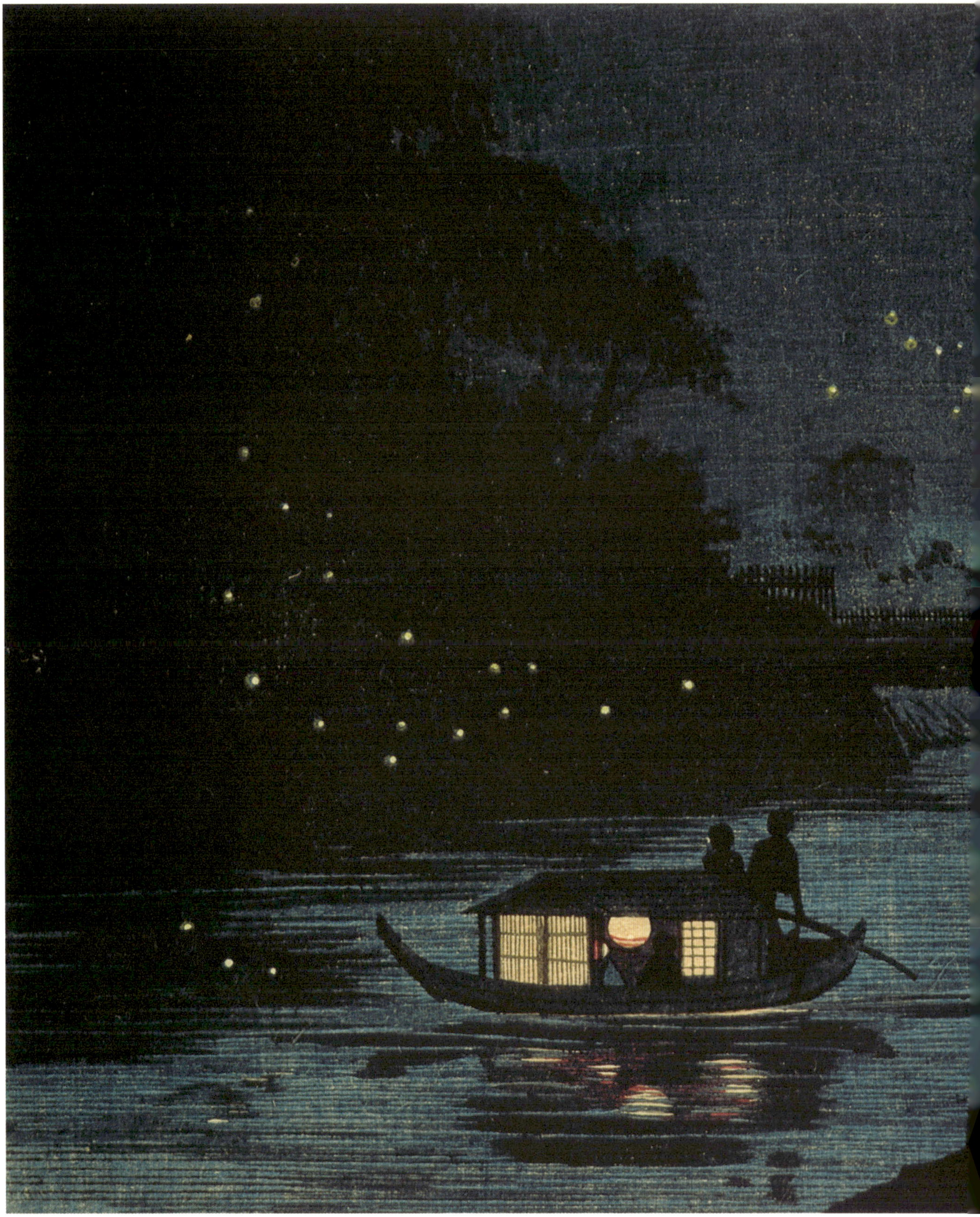

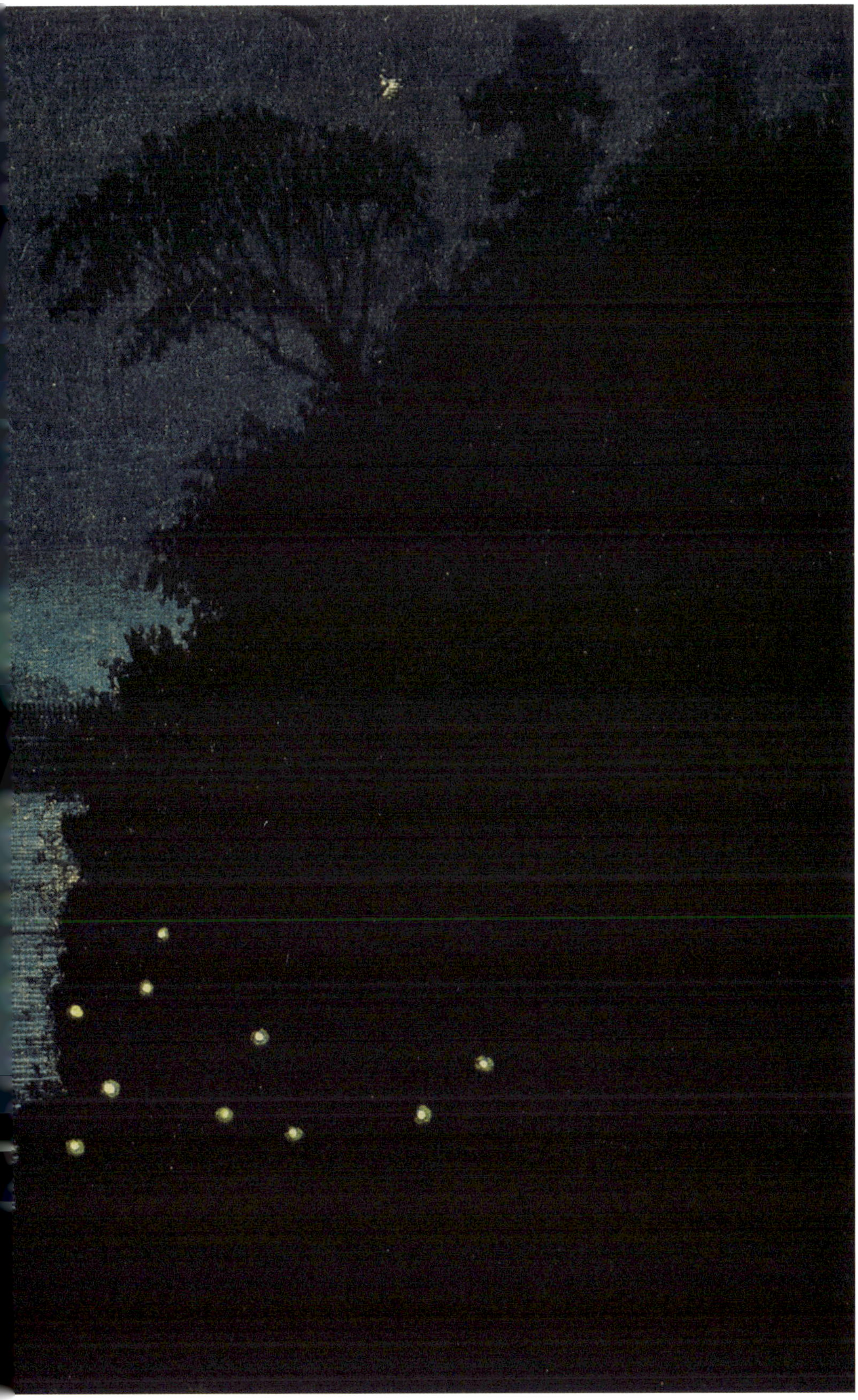

←
As they hatch, fireflies produce light to find mates, and each species has its own system of signals. Kobayashi Kiyochika's woodblock print *Fireflies at Ochanomizu* (*c.* 1880) conveys the beauty of this natural display. In Japan large crowds still turn out to watch this event each spring.

'We can allow satellites, planets, suns, universe, nay whole systems of universe to be governed by laws, but the smallest insect, we wish to be created at once by special act.'

CHARLES DARWIN, ENTRY IN NOTEBOOK N, NOVEMBER 1838

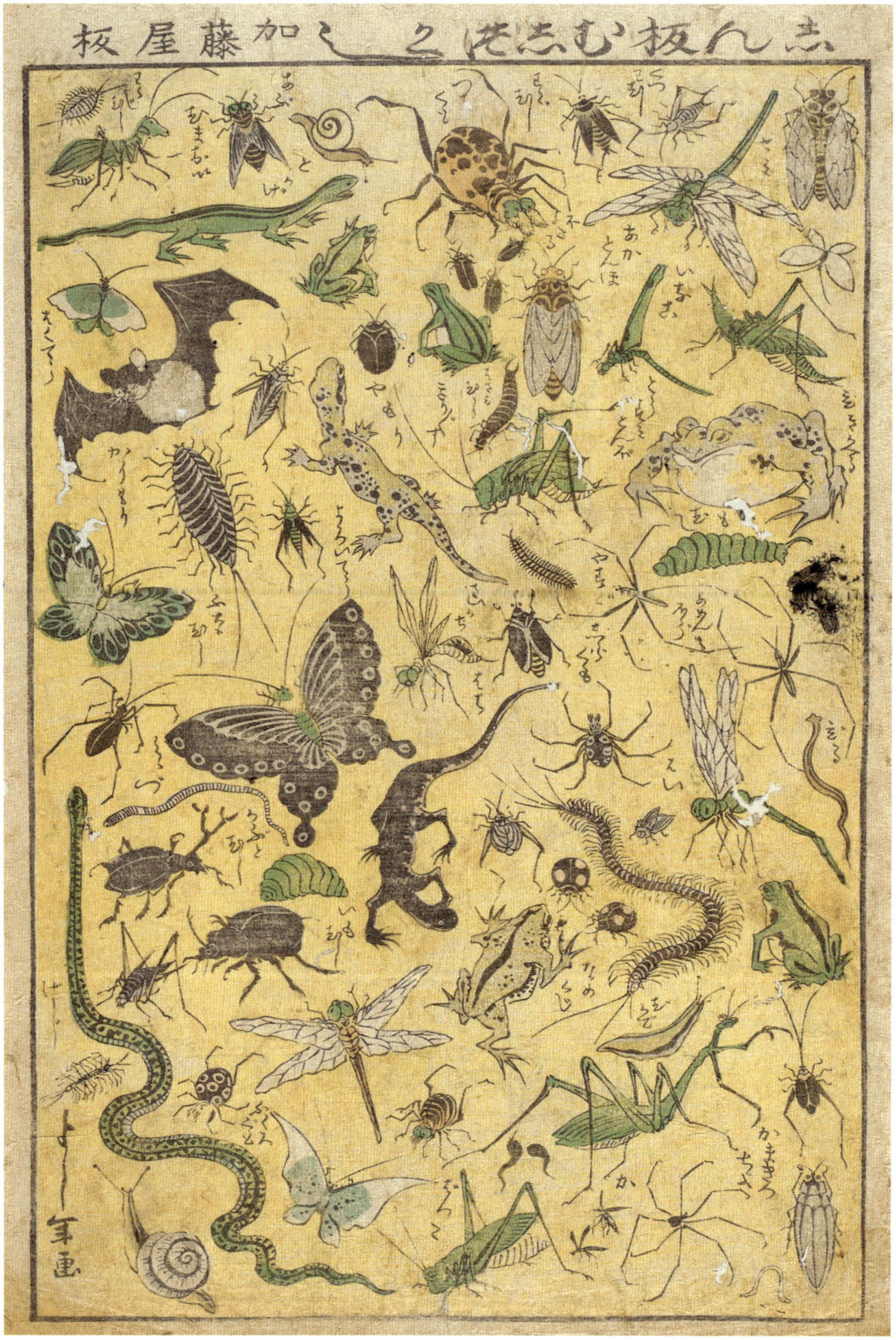

←
An array of insects in different stages of their life cycles is illustrated in the *New Publication of Selected Insects* (19th century) by Tsukioka Yoshitoshi, Japan's last great master of the *ukiyo-e* style of woodblock printing.

→
The French designer Émile-Allain Séguy was fascinated by insects, which he described as 'mechanic wonders'. In 1925, he published *Insectes*, a collection of twenty plates of accurate, decorative illustrations produced using the pochoir, or stencilling, technique.

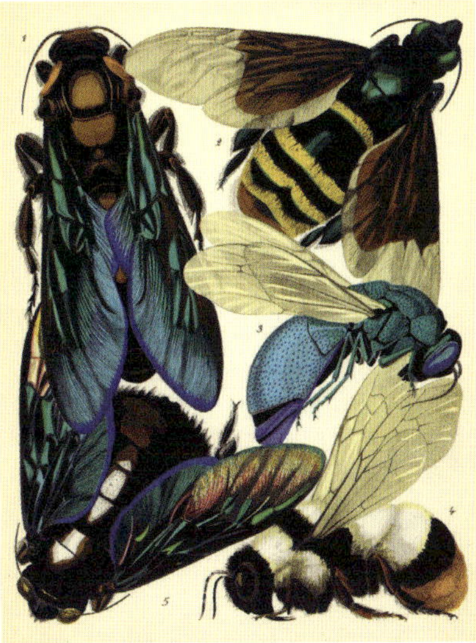
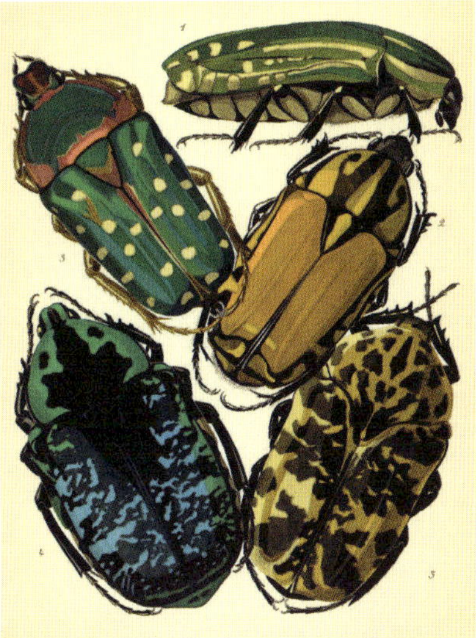
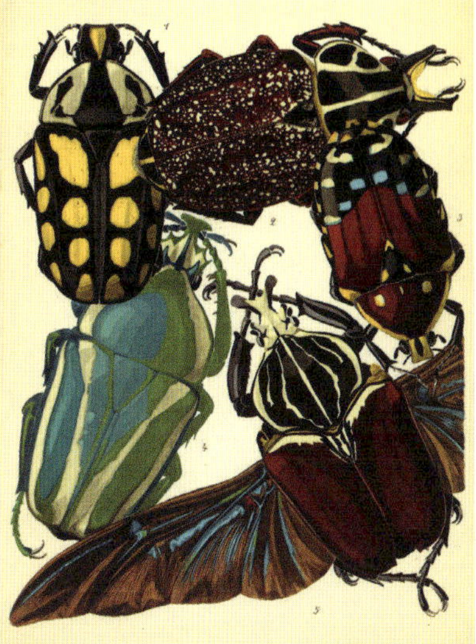
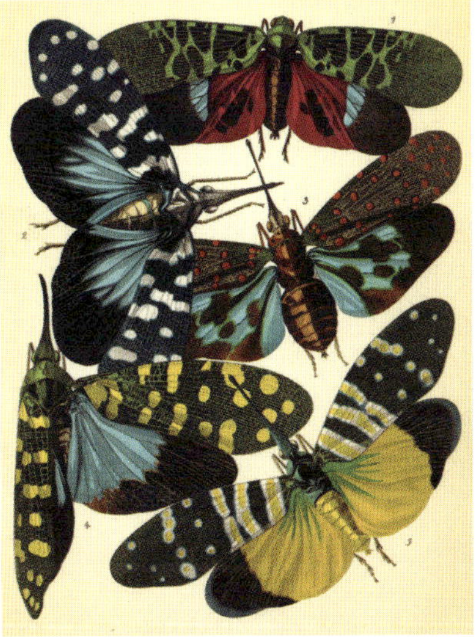
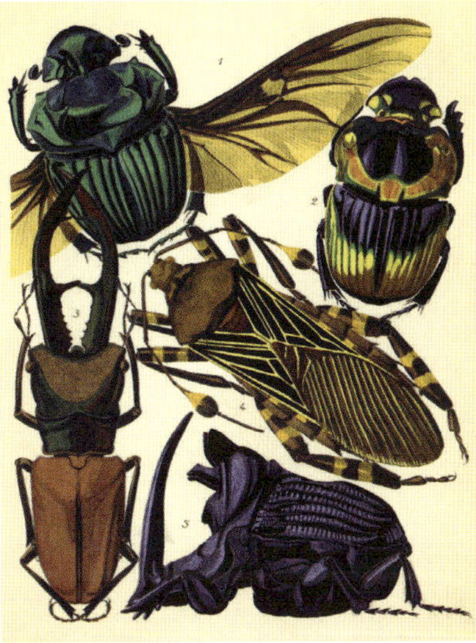
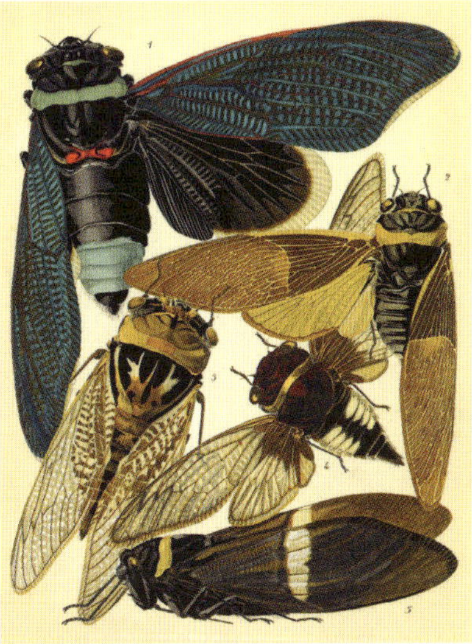
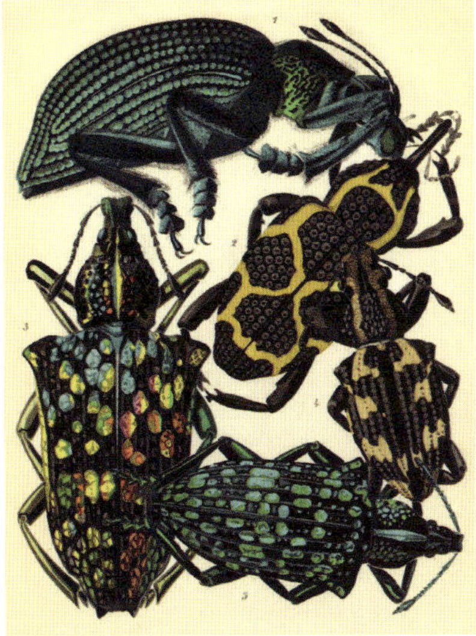
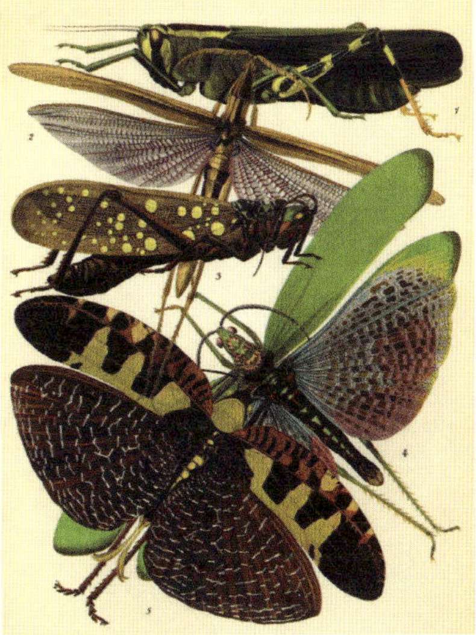
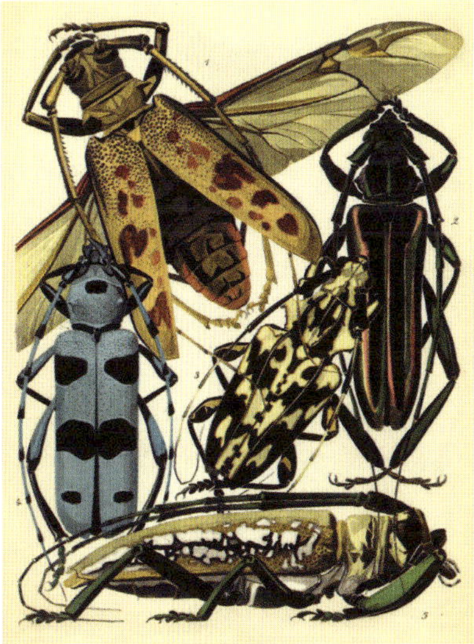

(111)

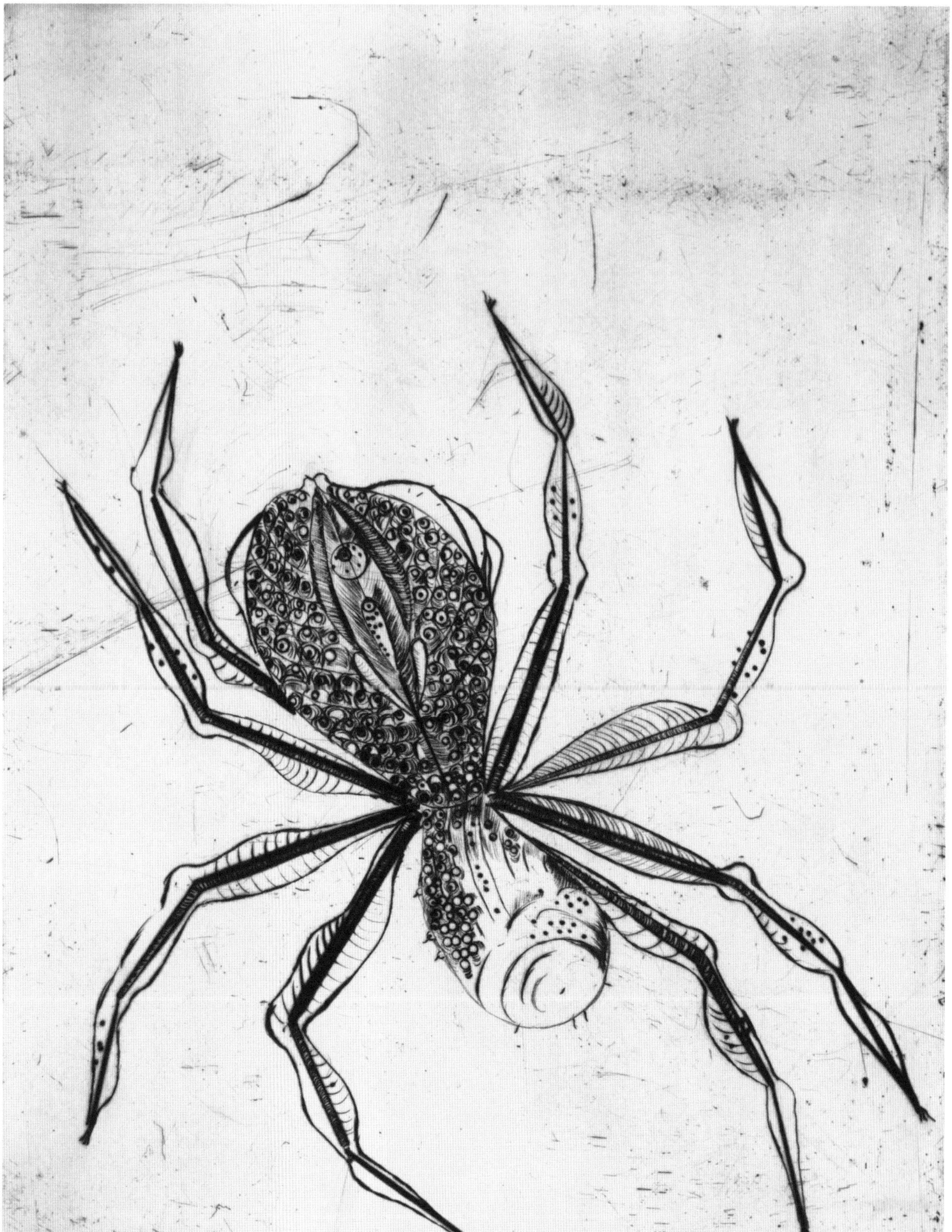

'Spiders were around long before humans, and it is likely they will be around in some number far into the future, even if humans are not. When most land and sea animals died out during the Permian mass extinction, spiders survived.'

LESLIE BRUNETTA AND CATHERINE L. CRAIG, *SPIDER SILK*, 2010

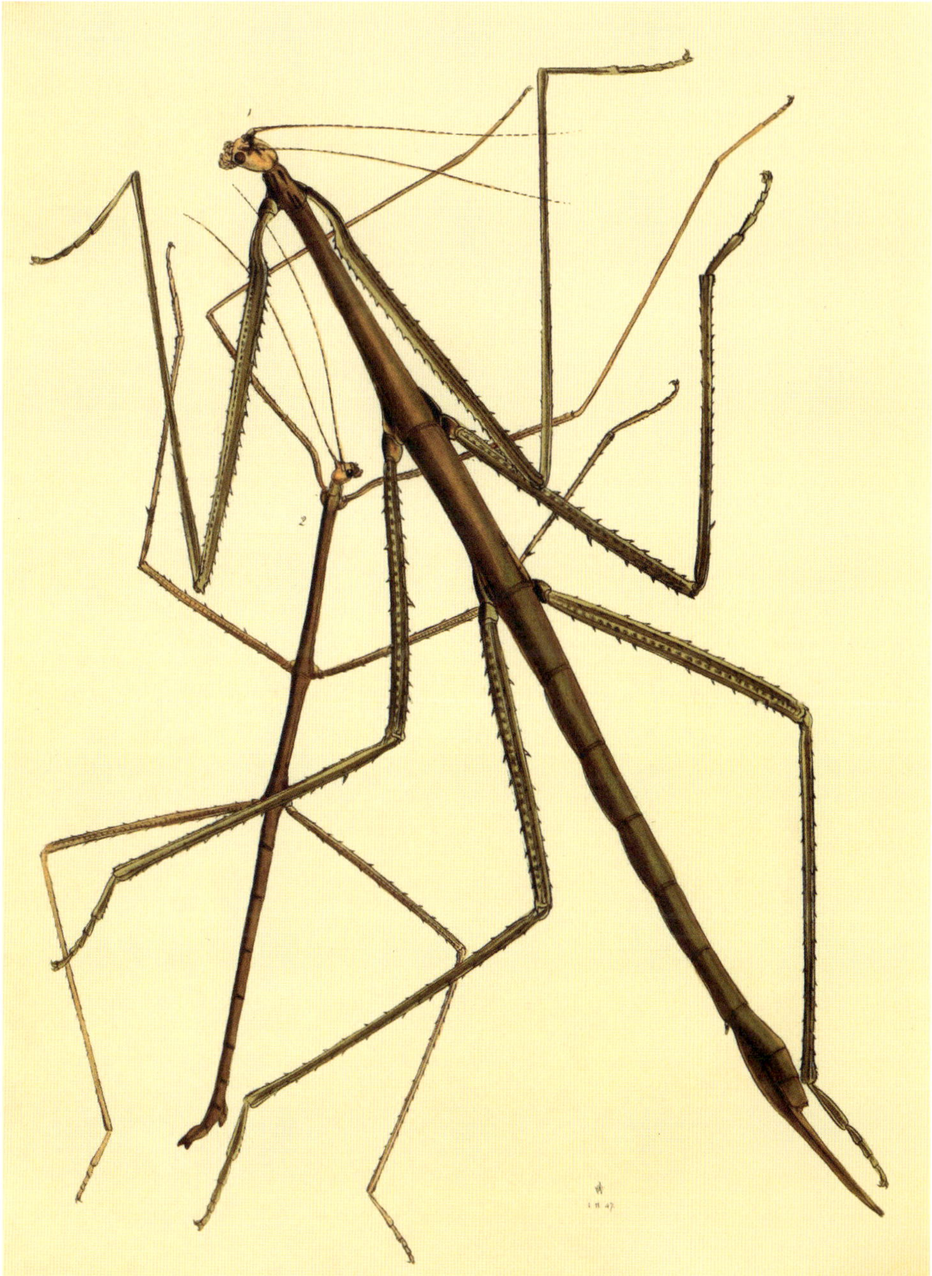

←
The spider was a recurring motif in the work of the French-American artist Louise Bourgeois, appearing in every form from drawings and prints, such as this drypoint print from 1995, to large sculptures. The spider is a wily predator that preys on other insects, but Bourgeois also associated it with the maternal qualities of protecting and nurturing its young.

→
Masters of camouflage and with a voracious appetite, stick insects use their powerful mandibles to feed on fresh leaves. This image is from J. O. Westwood's *The Cabinet of Oriental Entomology* (1848), a collection of illustrations of insects from India and nearby islands, many of which had not been illustrated previously.

↓
The giant peacock moth (*Saturnia pyri*) is the largest European moth with a wingspan of 15 to 20 centimetres (6 to 8 inches). Vincent van Gogh referred to the insect as a 'death's-head moth' and depicted it with a skull-like pattern on its back in his painting *Giant Peacock Moth* (1889). The skull and wing pattern are both inaccurate.

→
Illustrator Jan Augustin van der Goes's *Spider* (1690–1700) captures the jewel-like creature. The artist has used gouache to render the arachnid in minute detail.

↘
In an illustration from the *Northumberland Bestiary*, an English illuminated manuscript from *c.* 1250–60, ants are shown marching in formation between their nest and a wheatfield, harvesting the grain to store for winter.

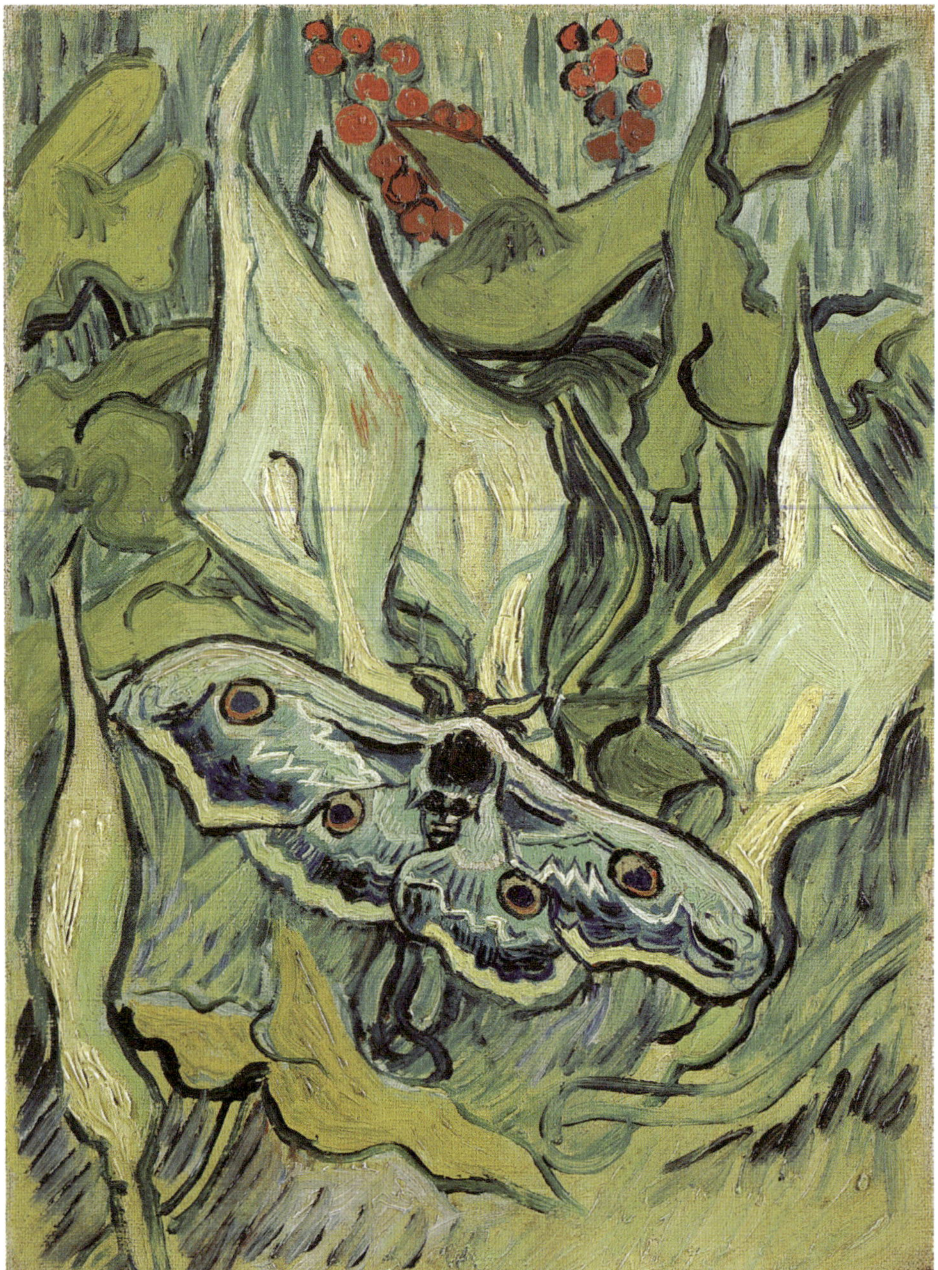

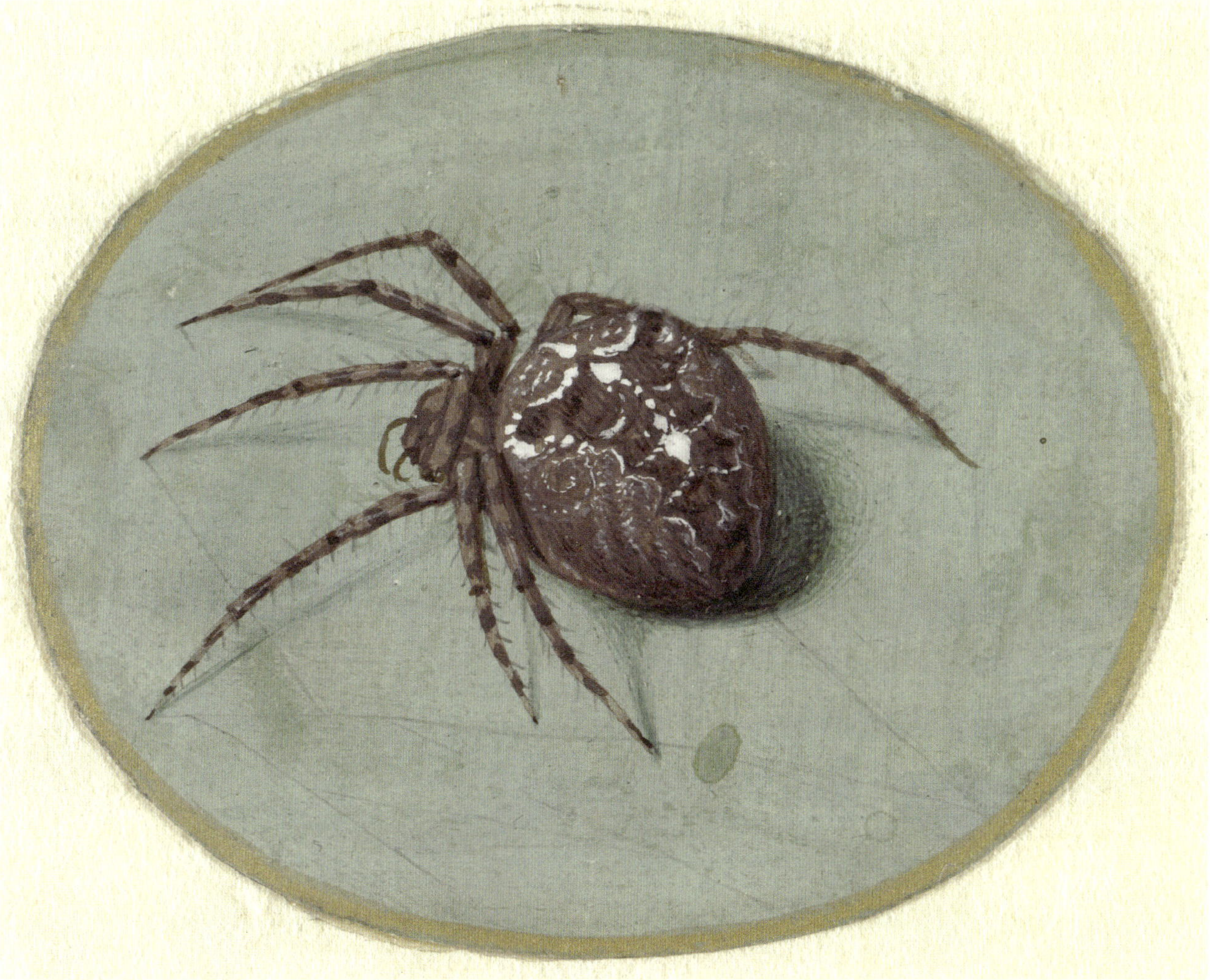

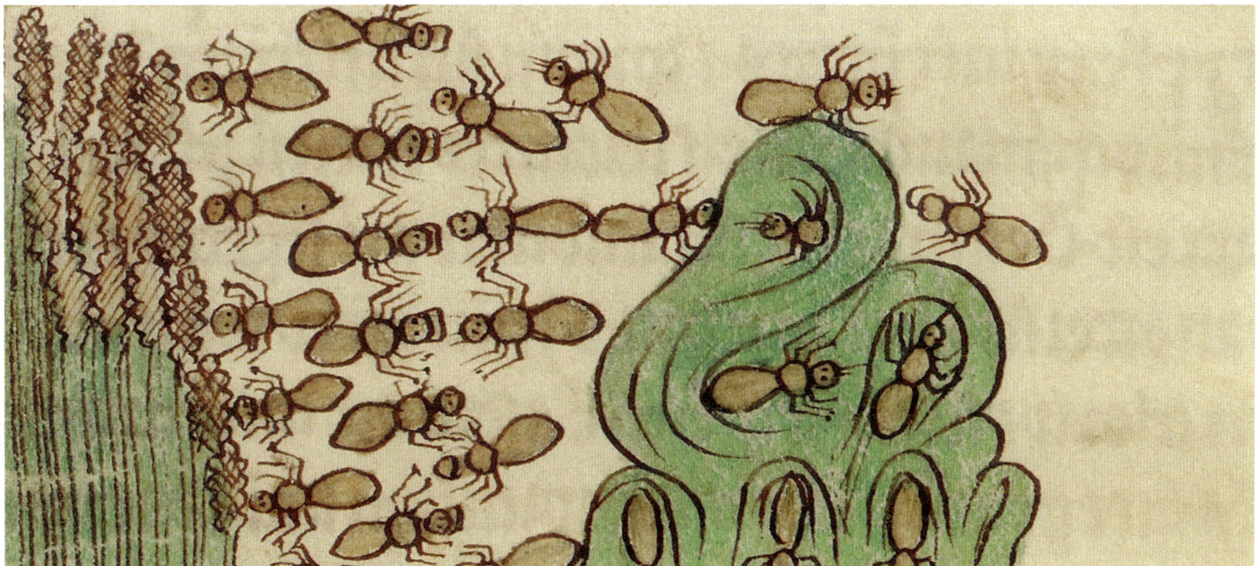

(111)

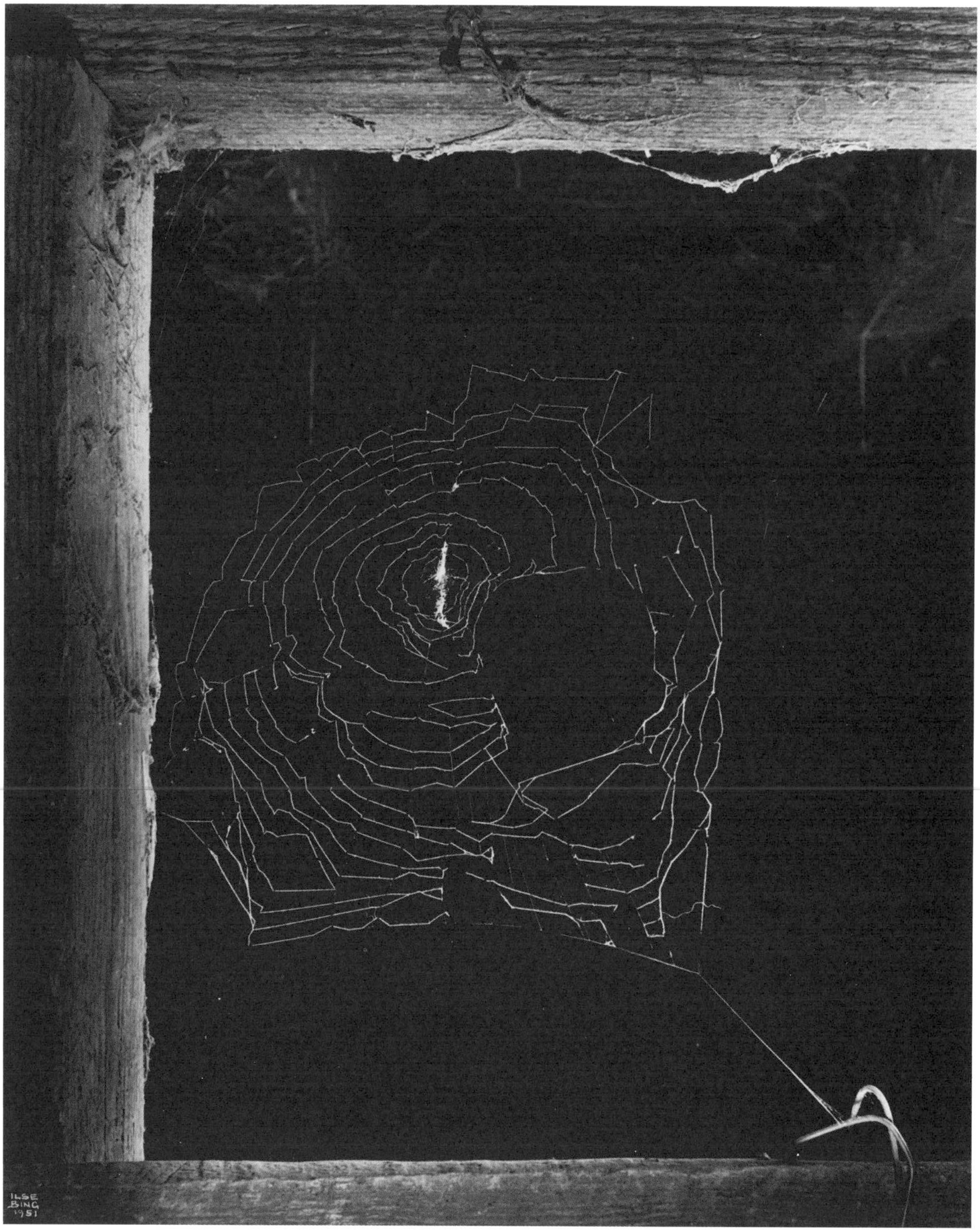

*'It launch'd forth filament, filament, filament out of itself;
Ever unreeling them – ever tirelessly speeding them.'*

WALT WHITMAN, 'WHISPERS OF HEAVENLY DEATH', 1868

←
German-American photographer Ilse Bing's *Spider Web and Stables* (1951) captures the beauty of these elegant traps. The radial threads provide the scaffolding, and the spiralling threads are sticky to catch prey.

→
'Ant trails observed by a scientist' is from *Ants, Bees and Wasps: A Record of Observations on the Habits of the Social Hymenoptera* (1882) by scientist Sir John Lubbock, in which he records his experiments into the behaviour of insects. Lubbock wanted 'to test their mental condition and powers of sense'.

↘
Snails are celebrated in haiku poetry for their perseverance. 'Snail and Poem by Buson', from a series of haiku postcards, reproduces an illustration by Saitō Shōshū (born 1870) and a poem by the 18th-century poet Yosa Buson.

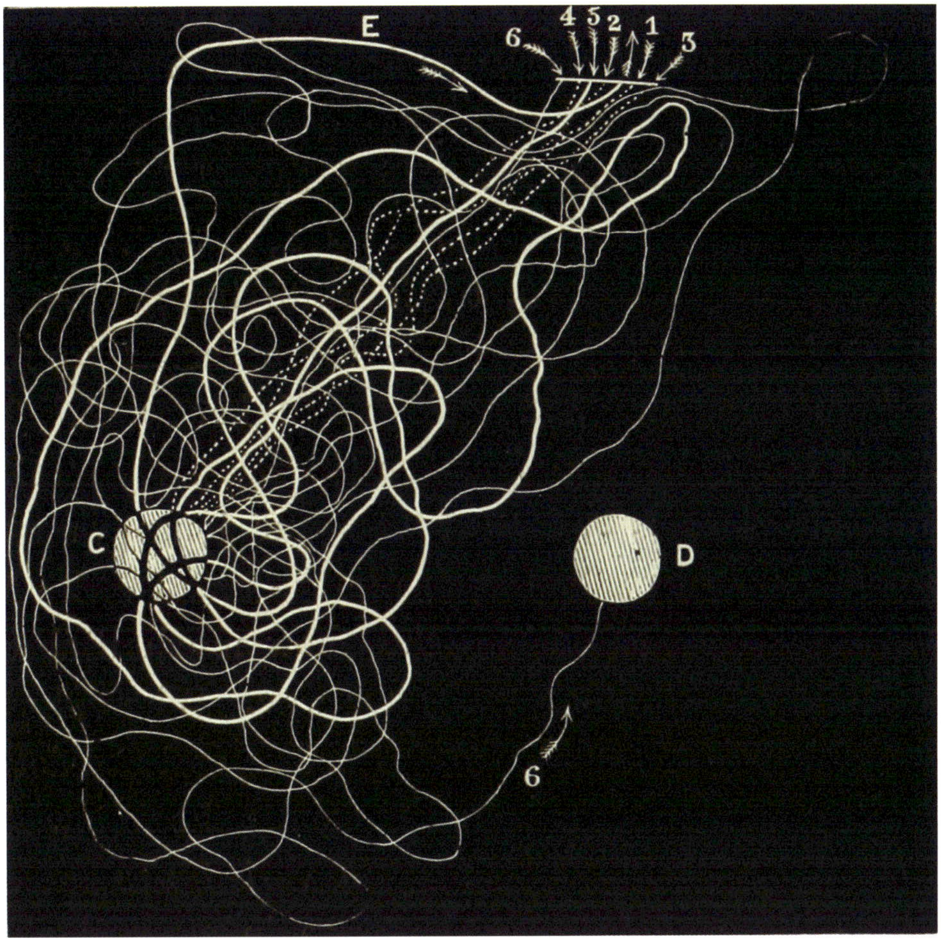

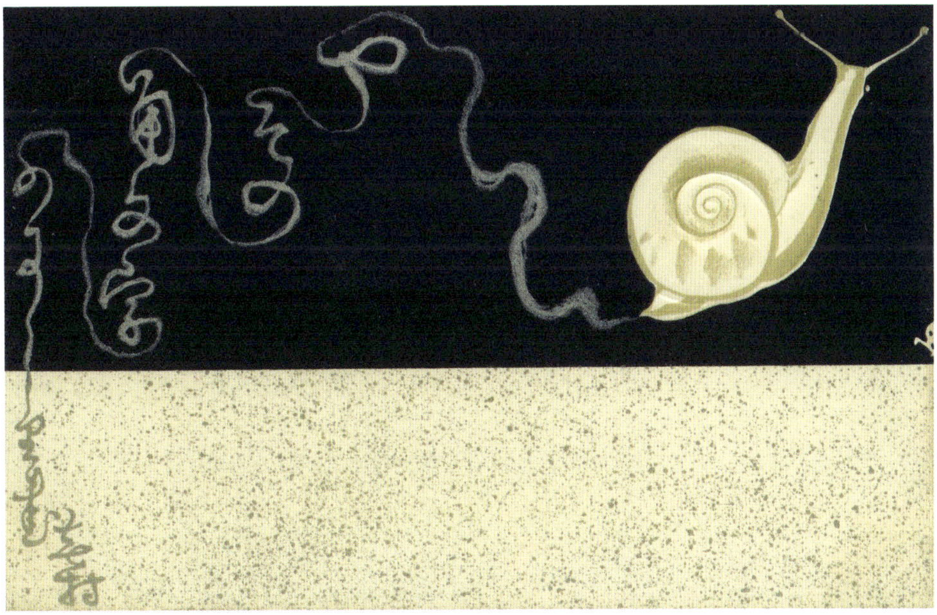

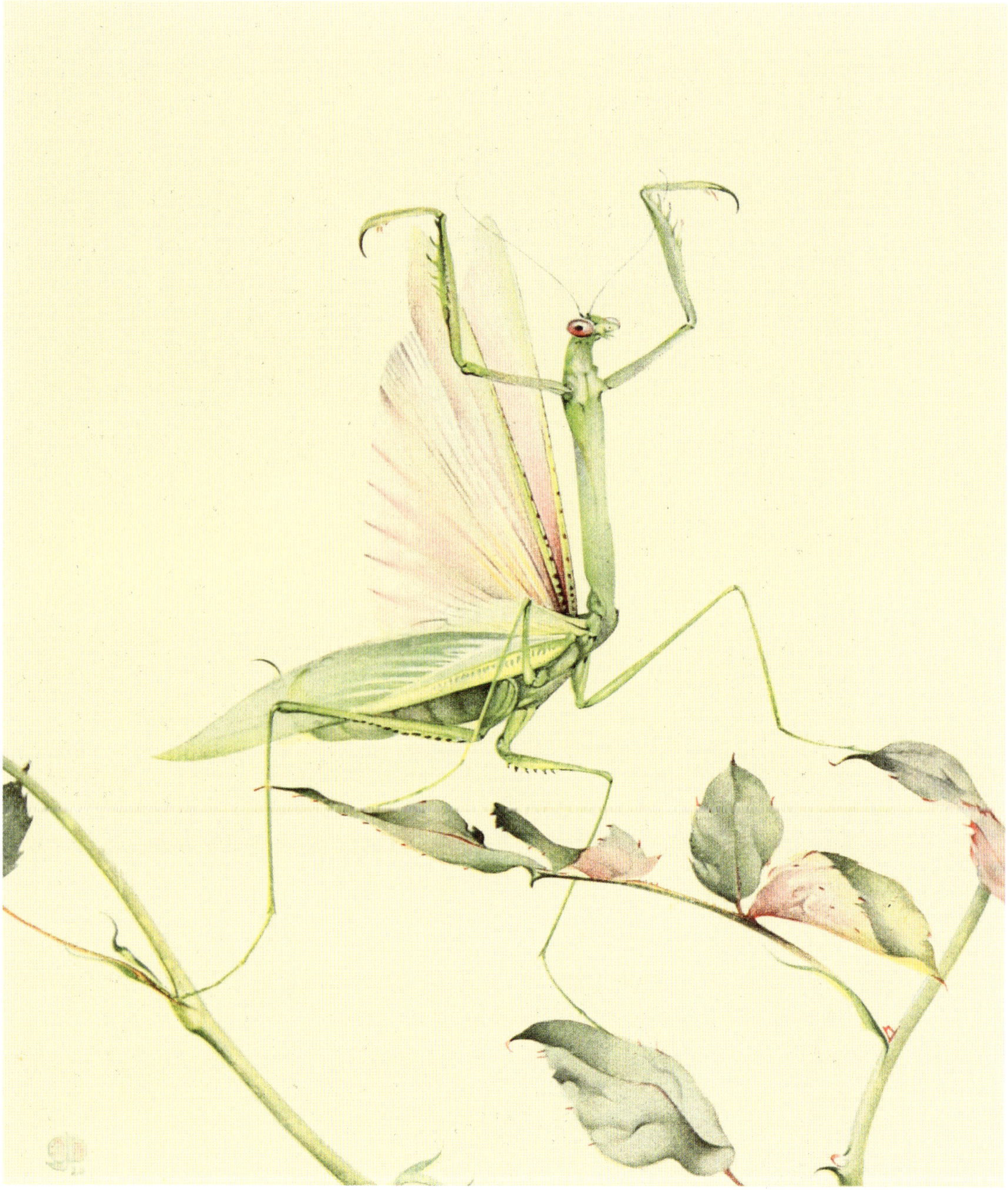

↑
Edward Julius Detmold's illustration in Fabre's *Book of Insects* (1921) shows the praying mantis performing its characteristic dance. This behaviour may be aimed at attracting or battling for a mate.

↗
Chinese artist Ju Chao's *Silkworm* (1859) is produced on a fan in ink and colour. Silkworm larvae feed on the leaves of mulberry trees and spin tough cocoons of silk. As producers of silk they are of considerable economic value.

→
An illustration from a mid-16th-century Persian manuscript of Al-Qazwini's *Ajā'ib al-Makhlūqāt va-Gharā'ib al-Mawjūdāt* ('The Wonders of Creation') shows earthworms gathered on the ground. They do this after rain not just to avoid being in the wet earth, but also because they breathe through their skin. Earthworms break down organic matter in the soil, making them essential to soil health.

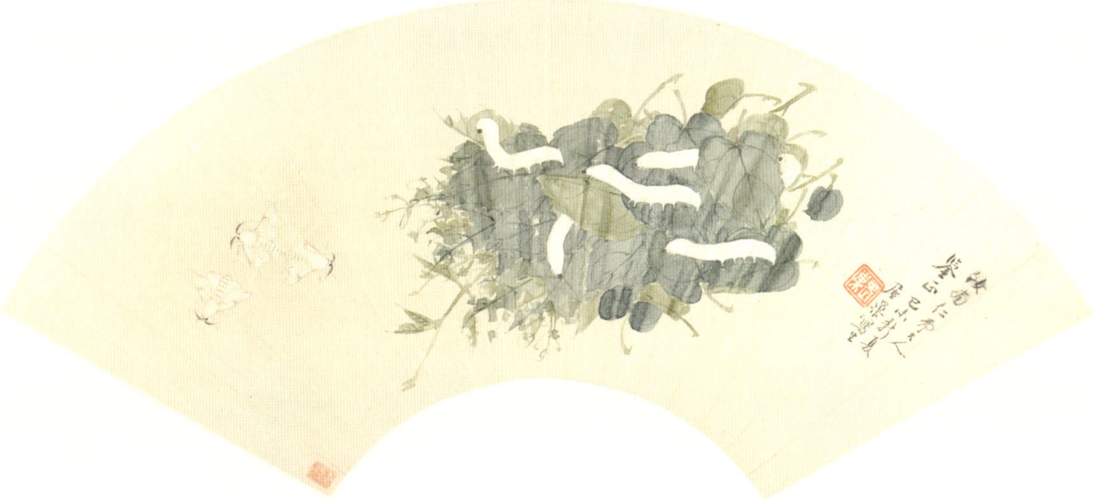

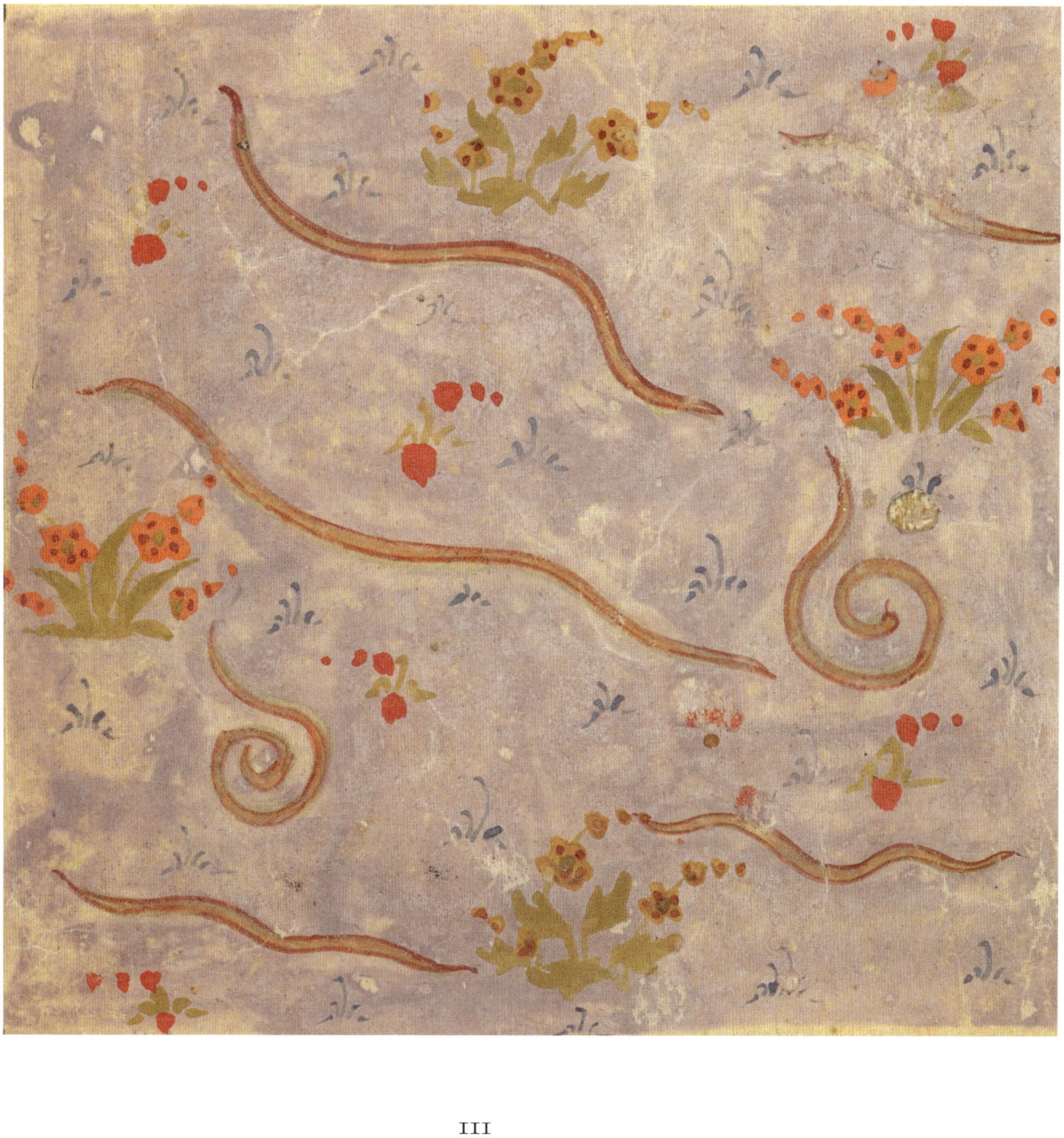

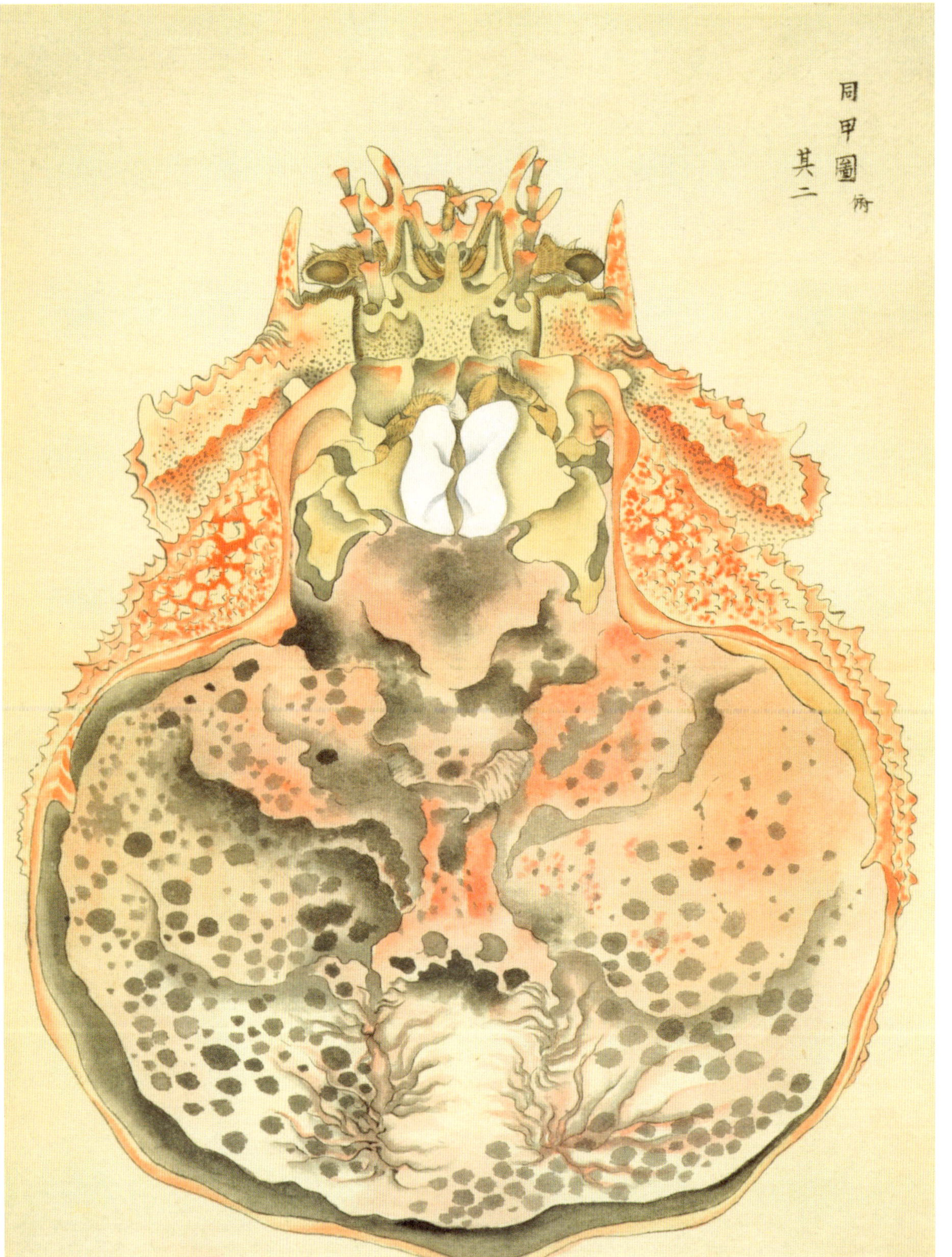

(III)

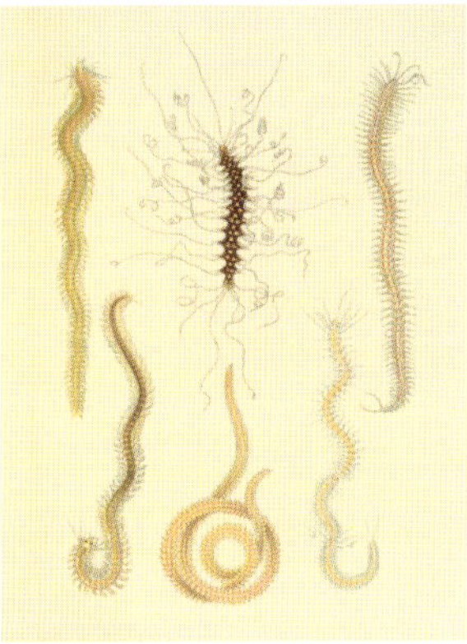
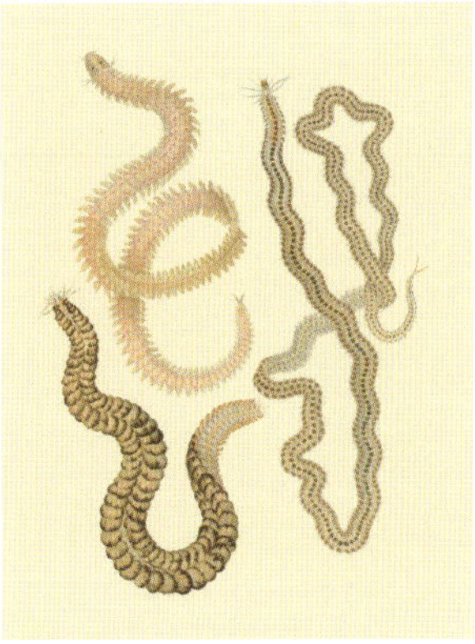
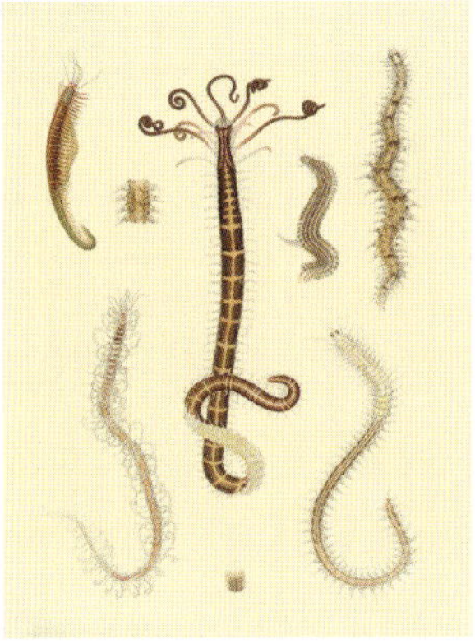
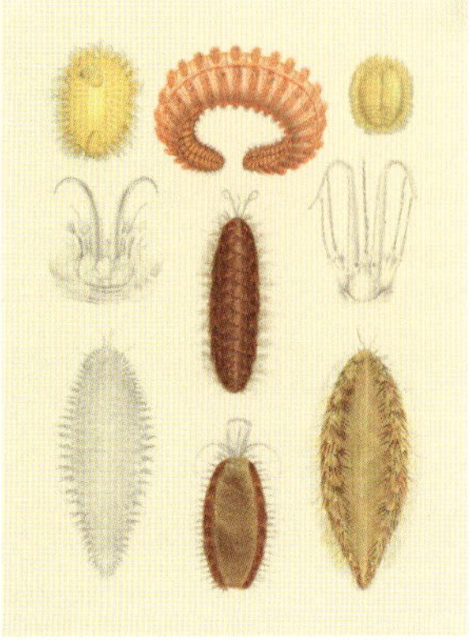
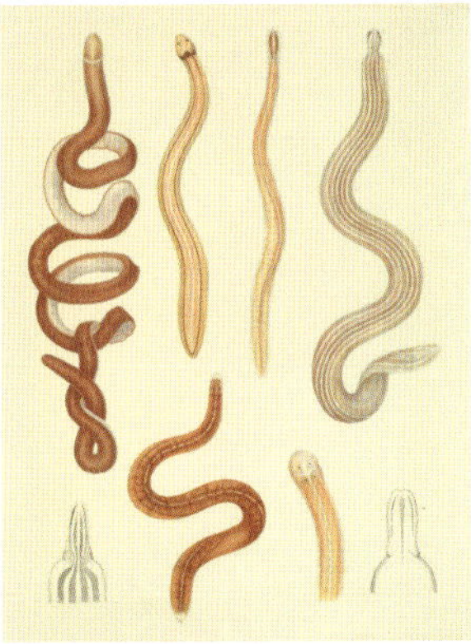
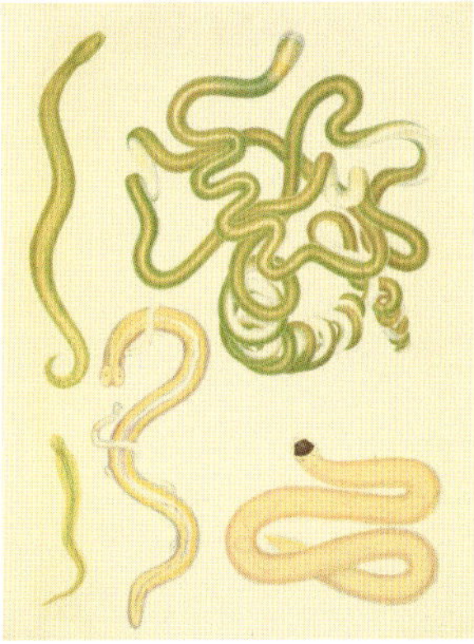
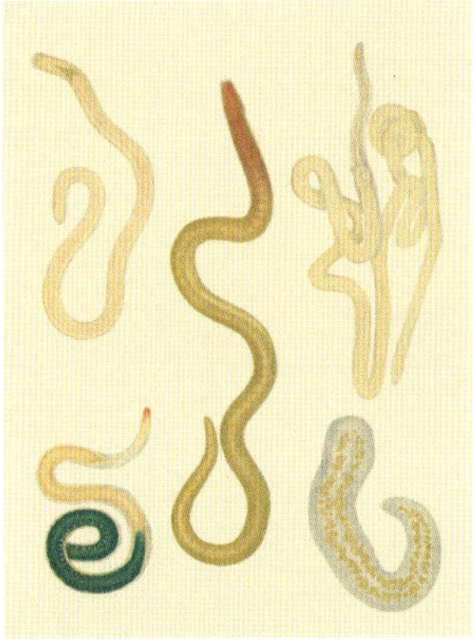
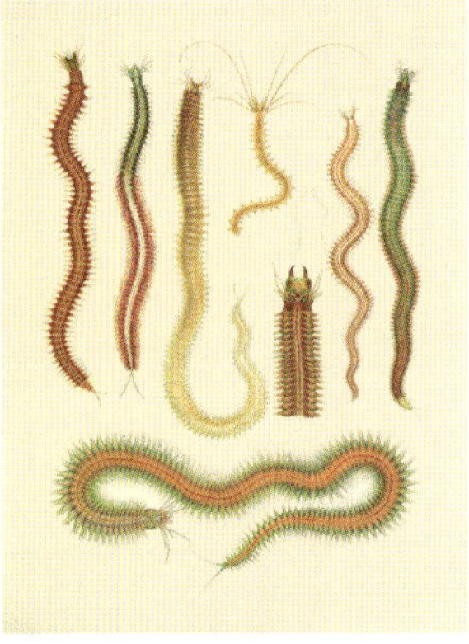
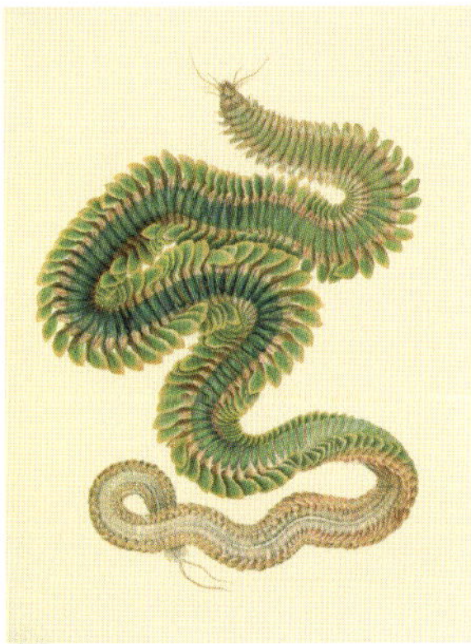
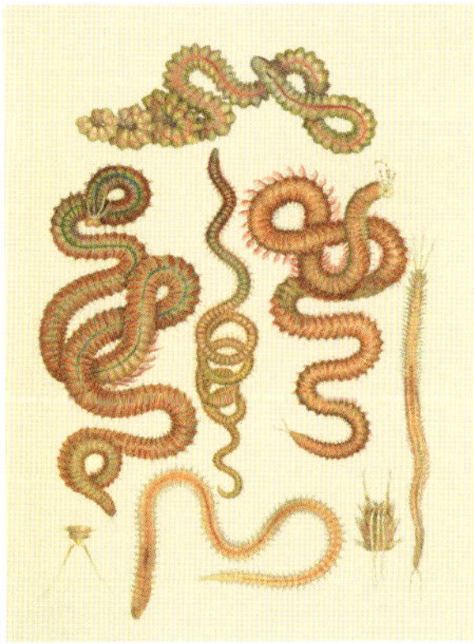
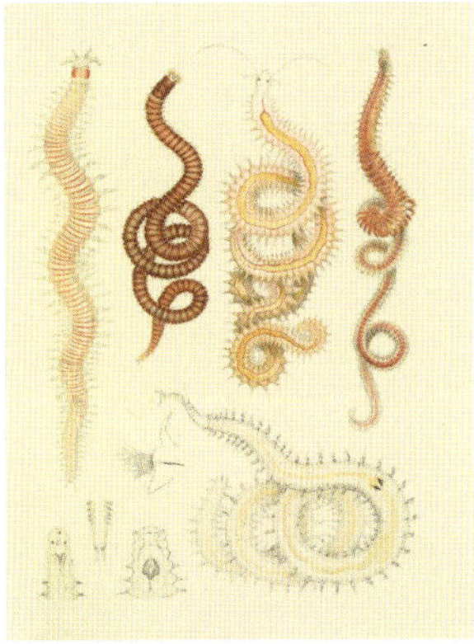
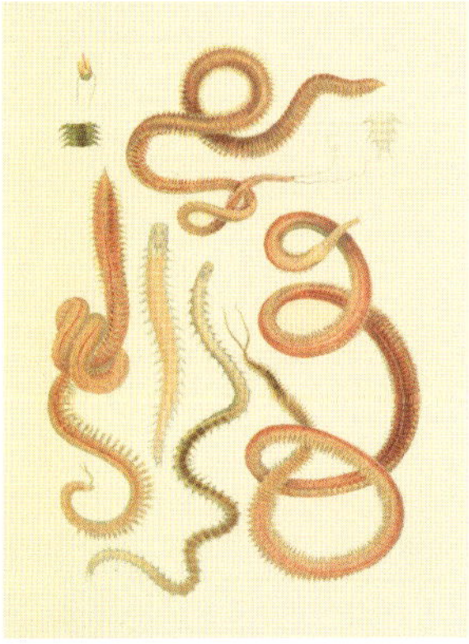

←
◎ The Japanese horseshoe crab (*Tachypleus tridentatus*), illustrated here by Mōri Baien, appears in a 19th-century atlas of marine invertebrates. Fossils of horseshoe crabs have been found dating back hundreds of millions of years, but this species is now at risk from overfishing and habitat decline.

→
A series of plates from William McIntosh's *A monograph of the British marine annelids* (1908).

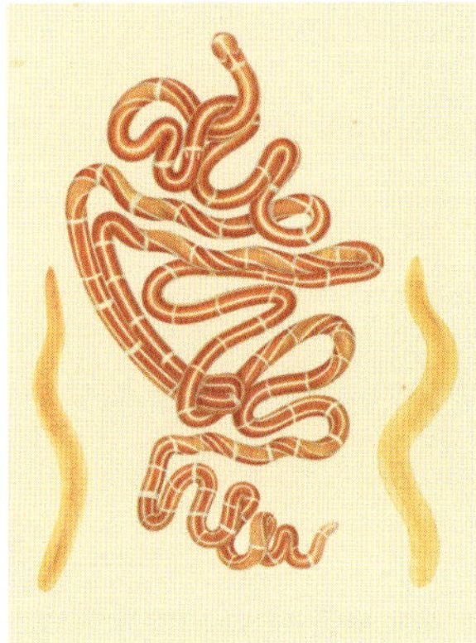
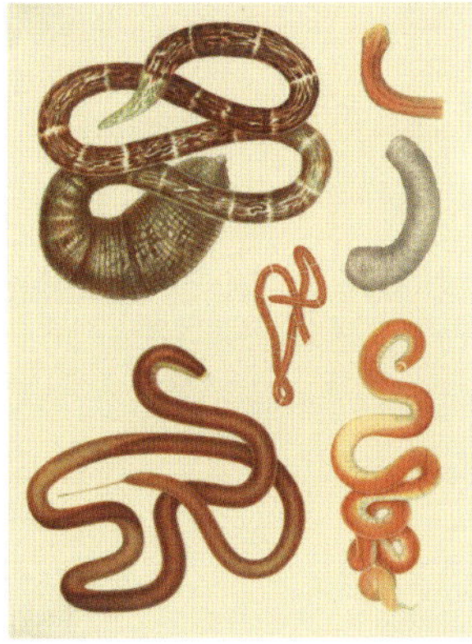
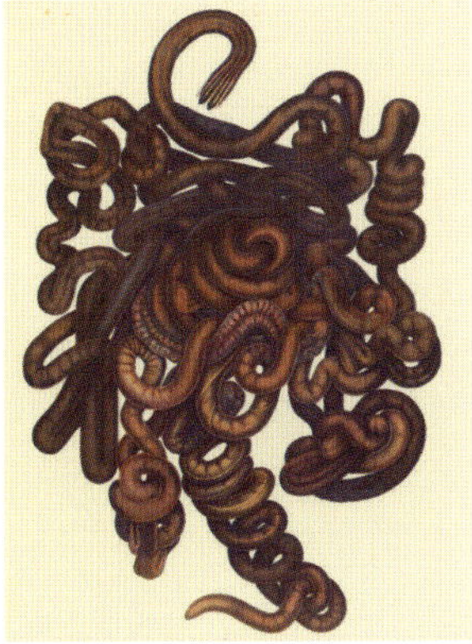

(111)

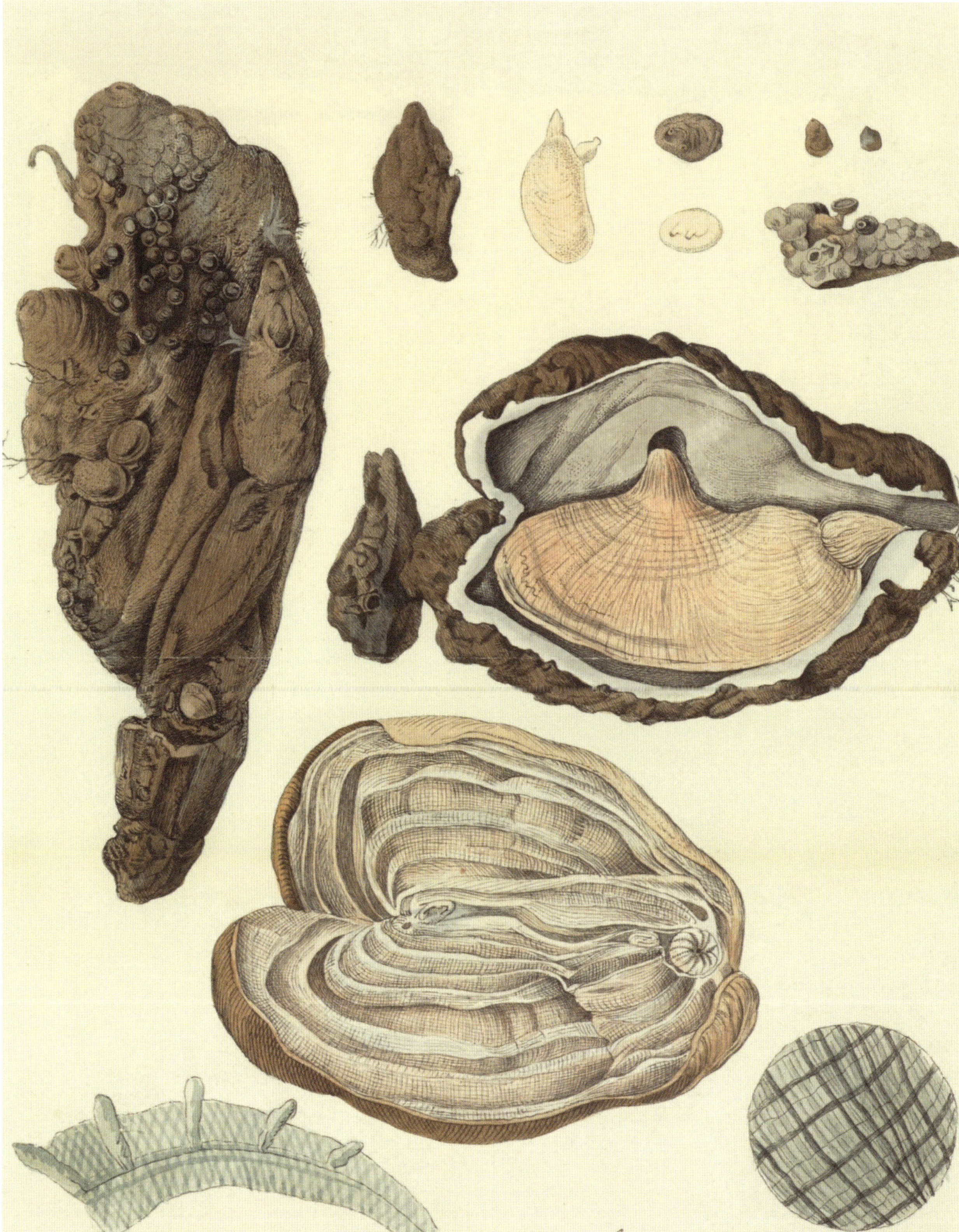

'From within were heard
Murmurings, whereby the monitor expressed
Mysterious union with its native sea.'

WILLIAM WORDSWORTH, 'DESPONDENCY CORRECTED', *THE EXCURSION: BEING A PORTION OF THE RECLUSE*, 1814

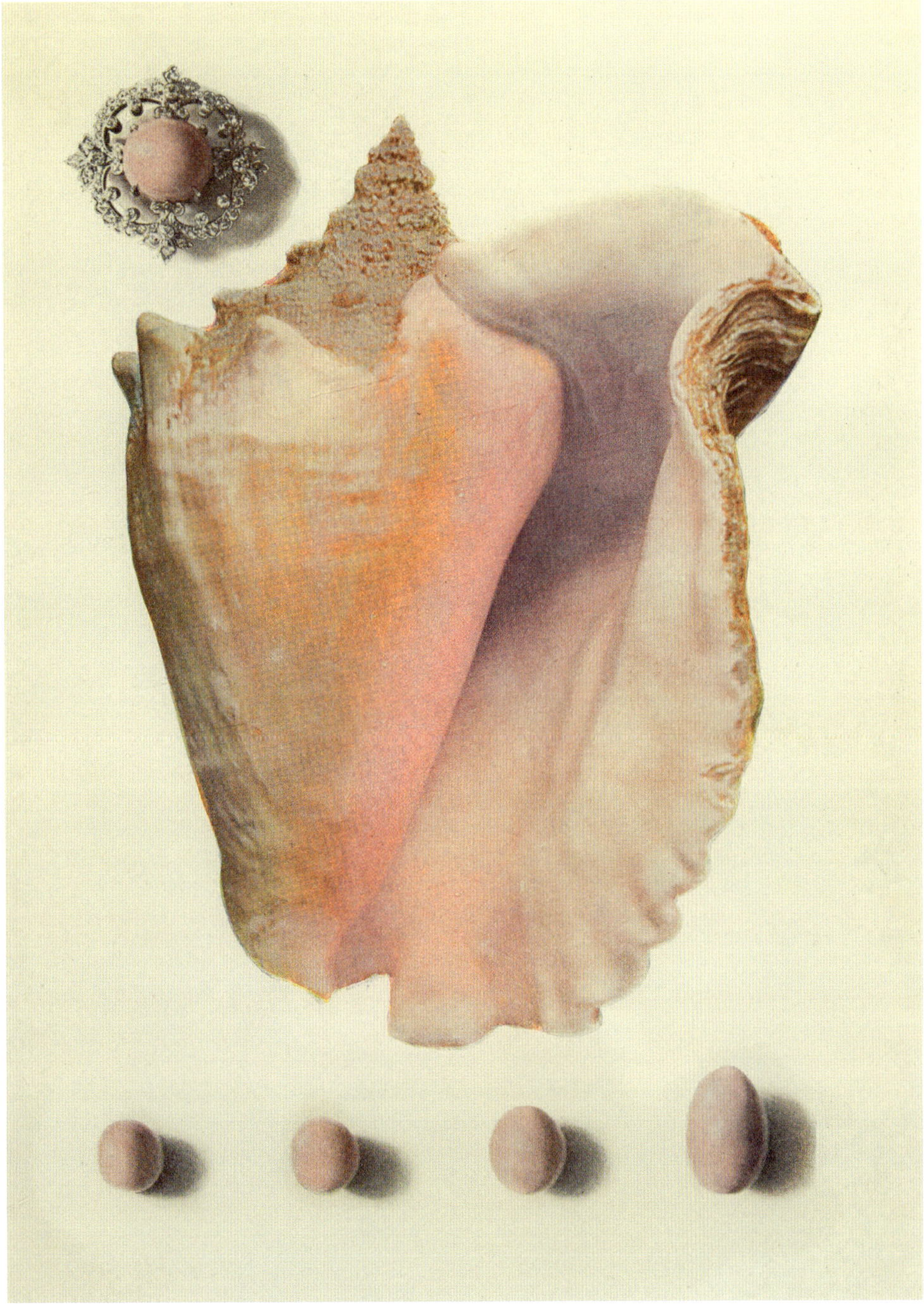

←
Originally identified as *Cynthia microcosmus* in 1816, this species of sea squirt has since been accepted as *Pyura microcosmus*. They are a form of marine invertebrates that attach to hard substrates, including shells (as is shown here), stones and man-made structures.

→
⊖ The queen conch (*Aliger gigas*) is the most prominent pearl-producer of the the West Indies and the Florida coast and can measure up to 35 centimetres (14 inches) in length. Its pearls are usually pink, as is shown here in George Frederick Kunz and Charles Hugh Stevenson's *The Book of the Pearl* (1908), but can also be yellow.

'Gather a shell from the strown beach
And listen at its lips: they sigh
The same desire and mystery,
The echo of the whole sea's speech.'

DANTE GABRIEL ROSSETTI, 'THE SEA LIMITS', 1849

(111)

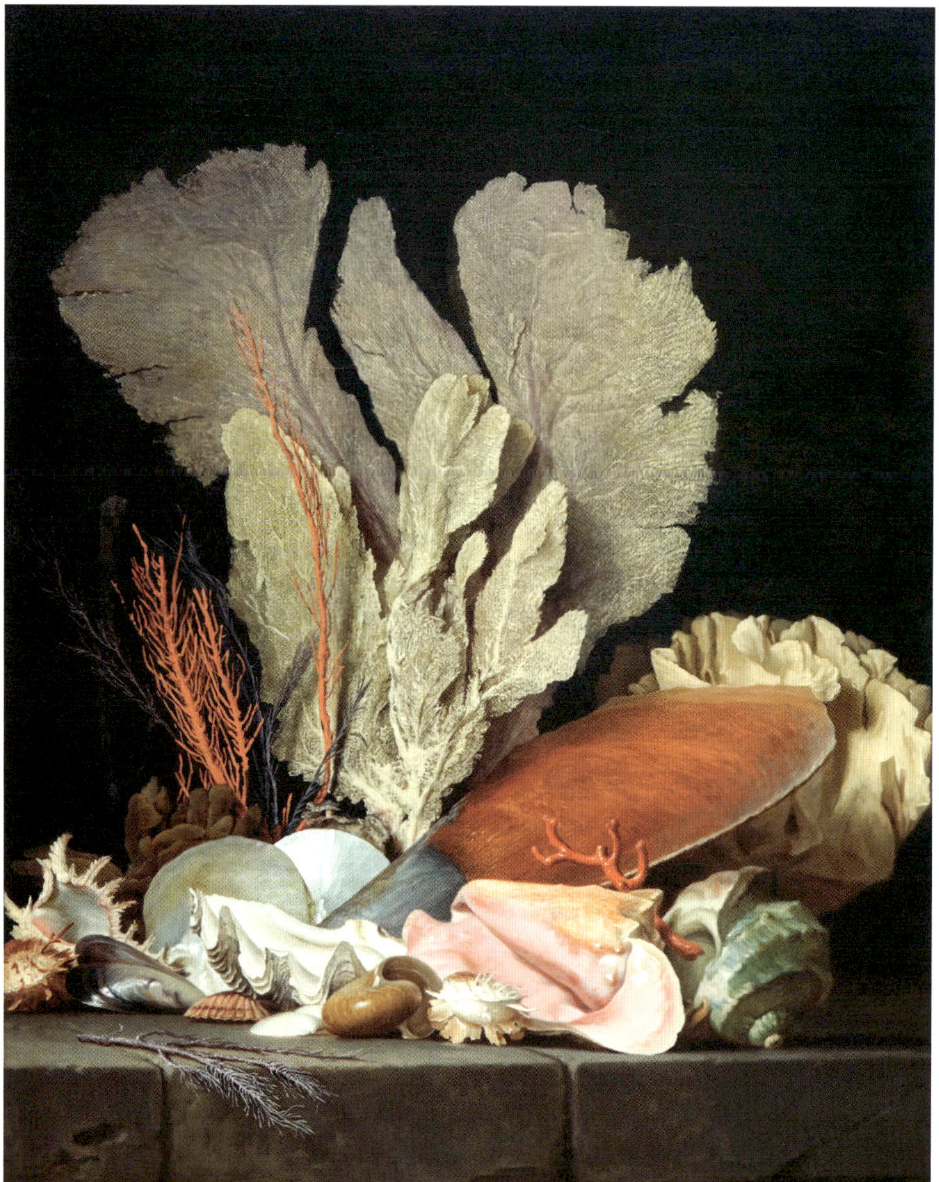

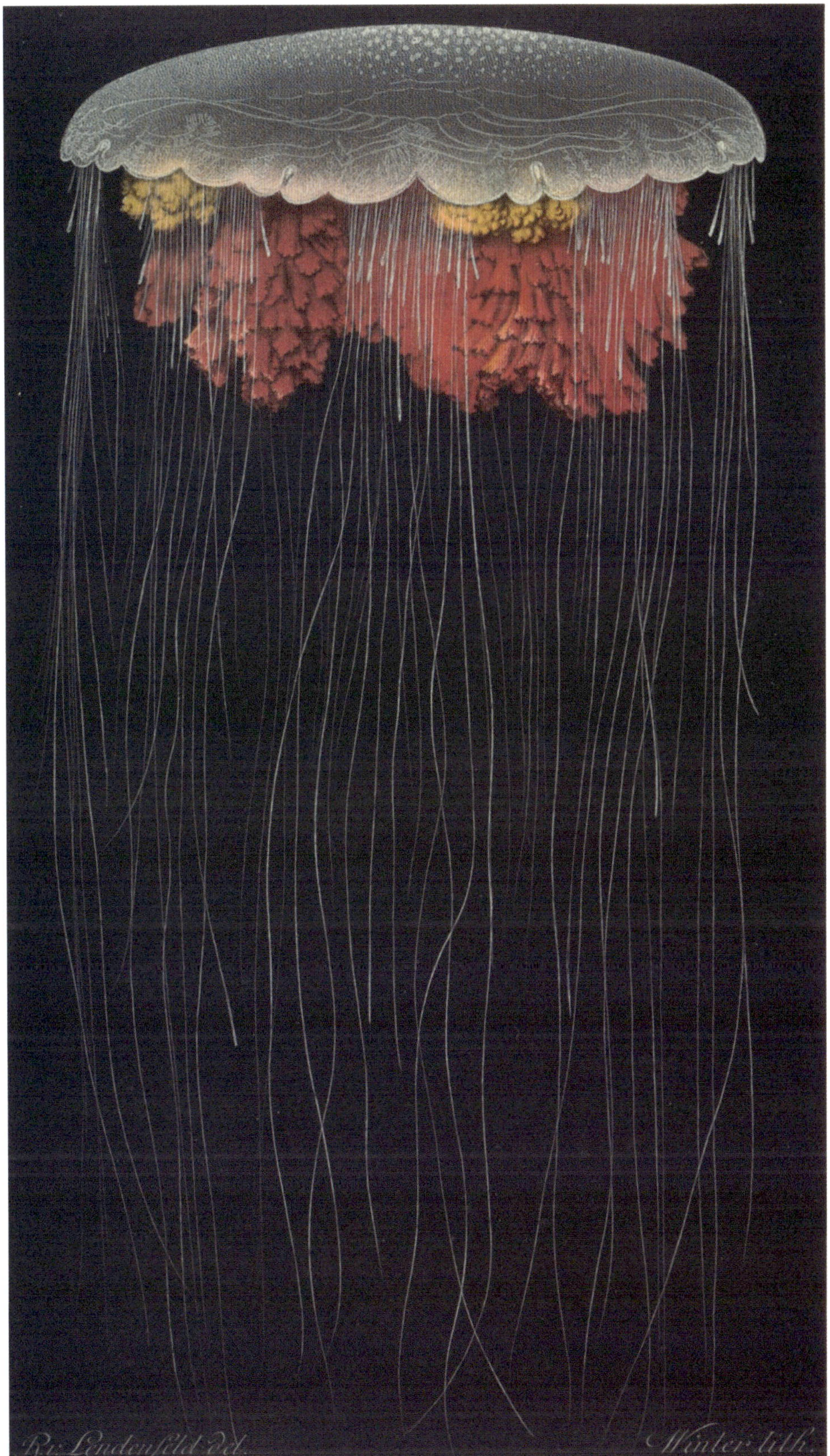

←
The 18th-century French painter Anne Vallayer-Coster specialized in still-life paintings. In *Panaches de mer, lithophytes et coquilles* ('Seaweed, Lithophytes and Seashells', 1769), she has rendered a wide selection of shells, together with corals and dried seaweed, in exquisite detail.

→
The *Cyanea annaskala* species of lion's mane jellyfish, illustrated here in a print from the vast, 19th-century Iconographia Zoologica collection, is endemic to the waters around southeastern Australia.

(III)

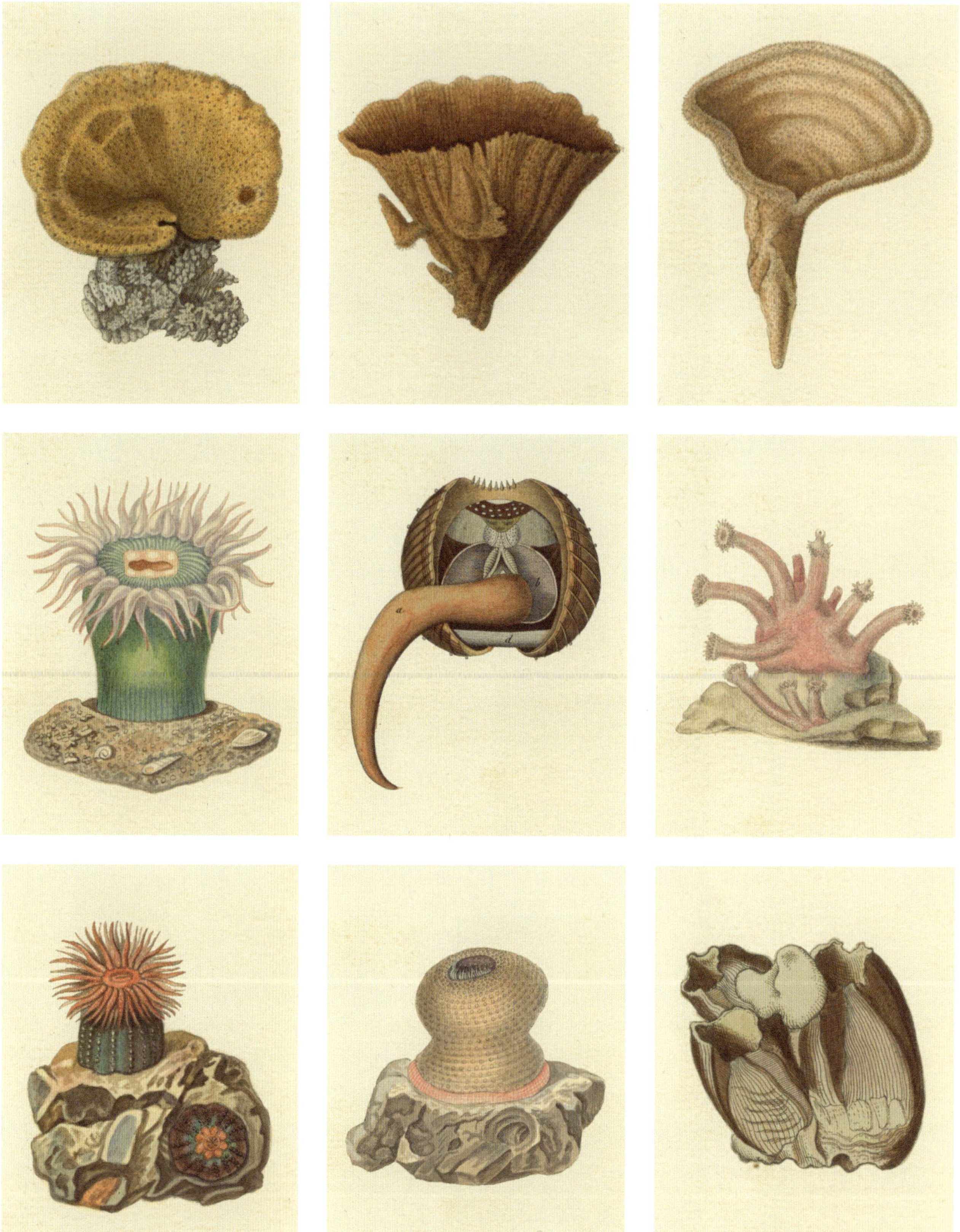

'The trawlers ... rake the seashores in their endeavours to catch the flat-fish tribe, and indeed whatever comes into their net: for a trawl is an omnium gatherum, and takes everything that is not strong enough to break its meshes.'

ARTHUR ARNOLD, 'ON OUR FISHERIES', *ONCE A WEEK*, 1863

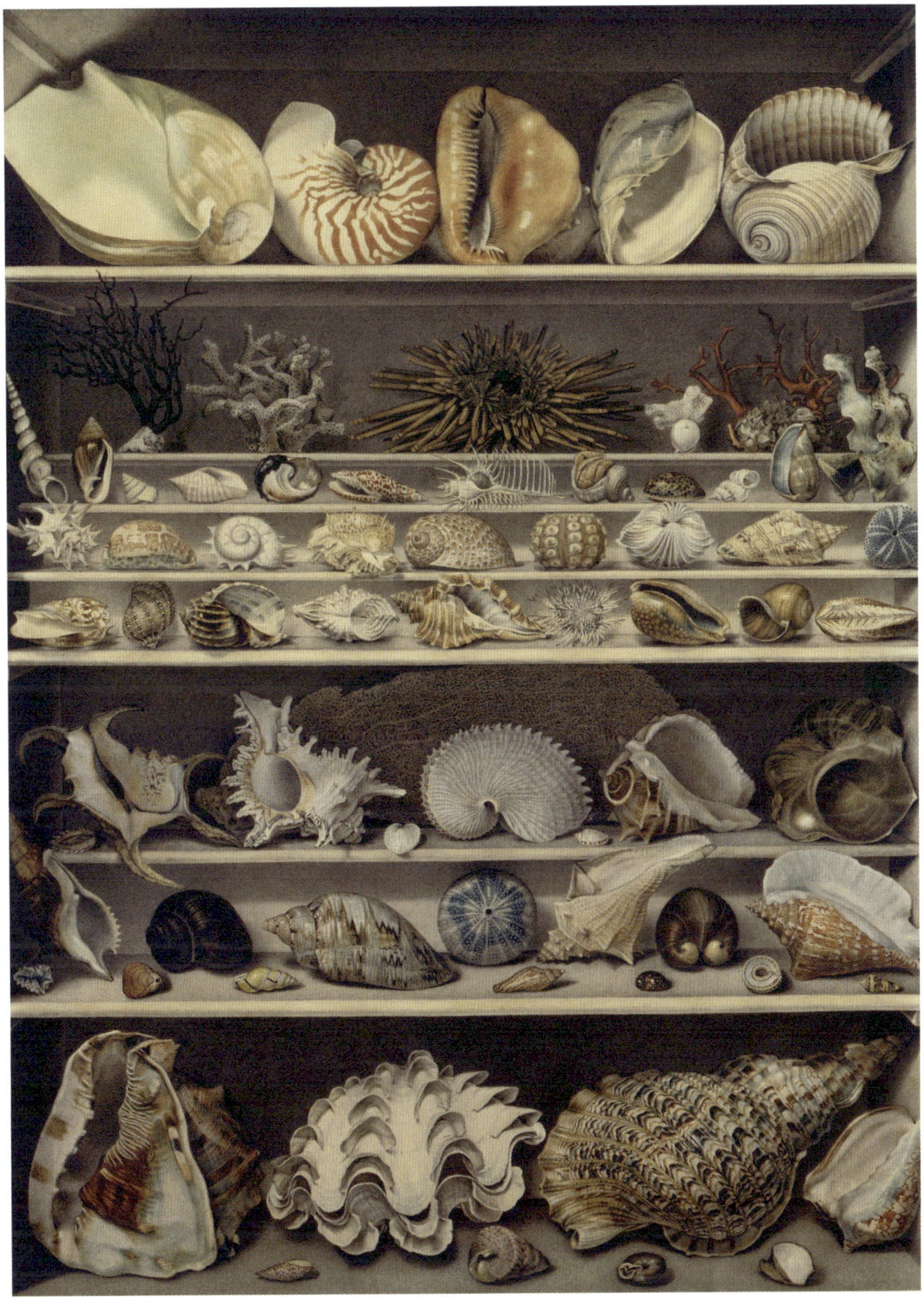

←
This collection of plates from the Iconographia Zoologica illustrates several sponges and sea anemones, including, bottom centre, the knobbly anemone (*Bunodosoma capense*), named for the columns of raised spots on its body; and, top left, the elephant ear sponge (*Spongia agaricina*).

→
Alexandre-Isidore Leroy de Barde painted *Choix de Coquillages* ('Selection of Shells') between around 1803 and 1810. Shell-collecting was a popular hobby at the time, and the painting was once in the collection of Louis XVIII of France.

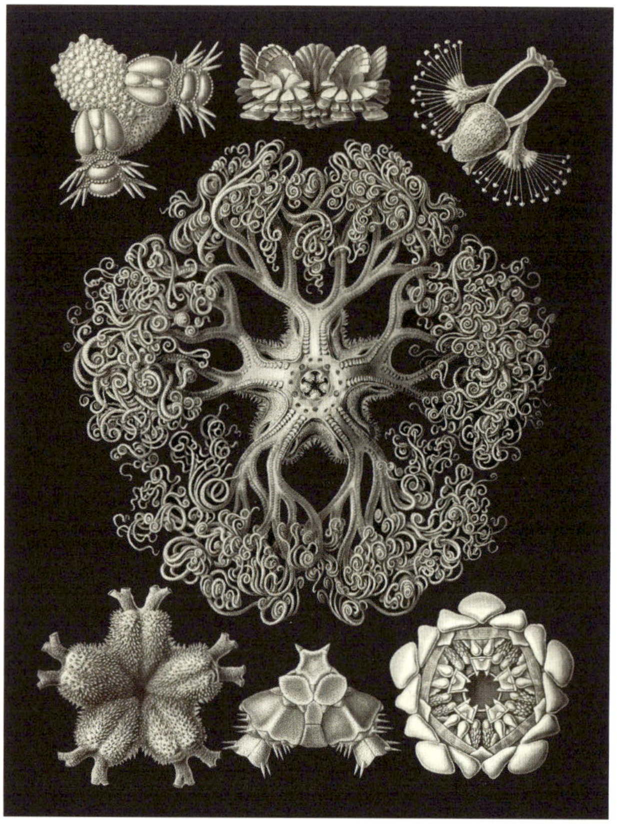

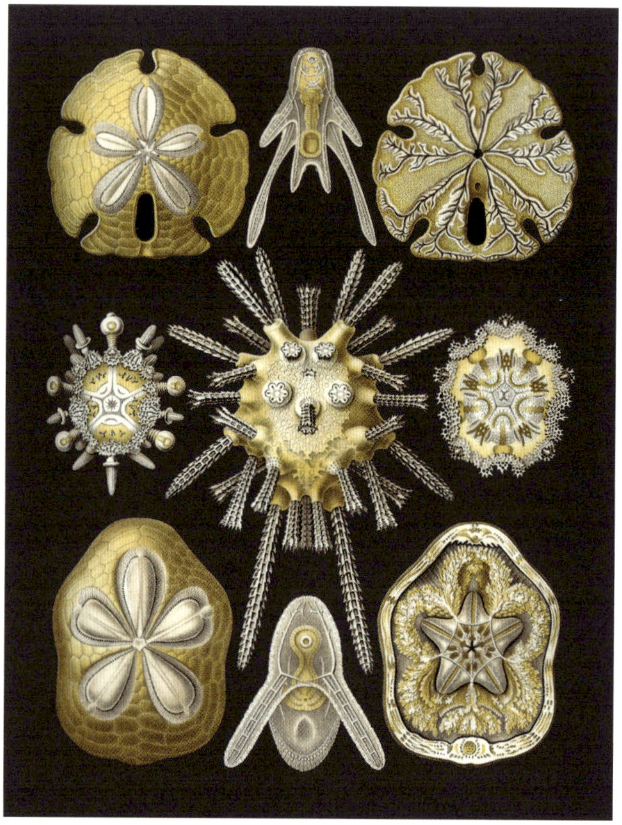

(111)

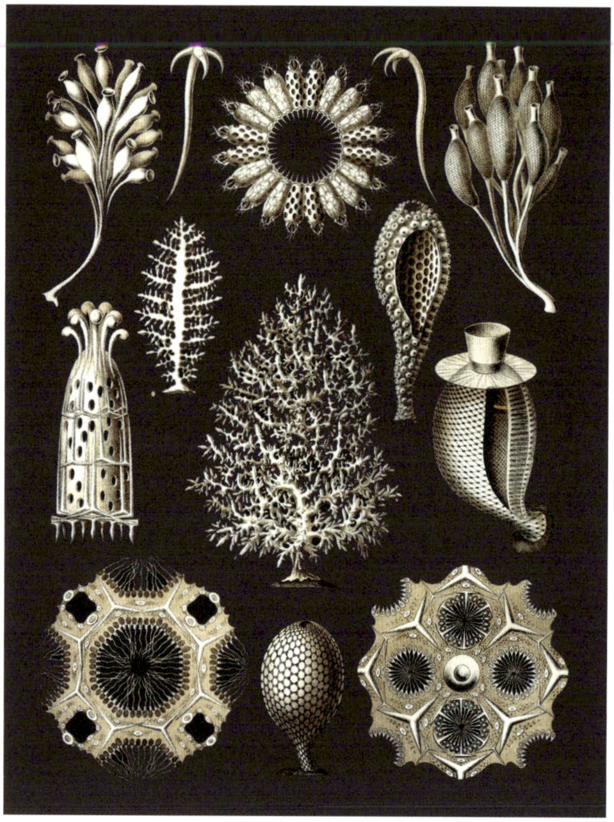

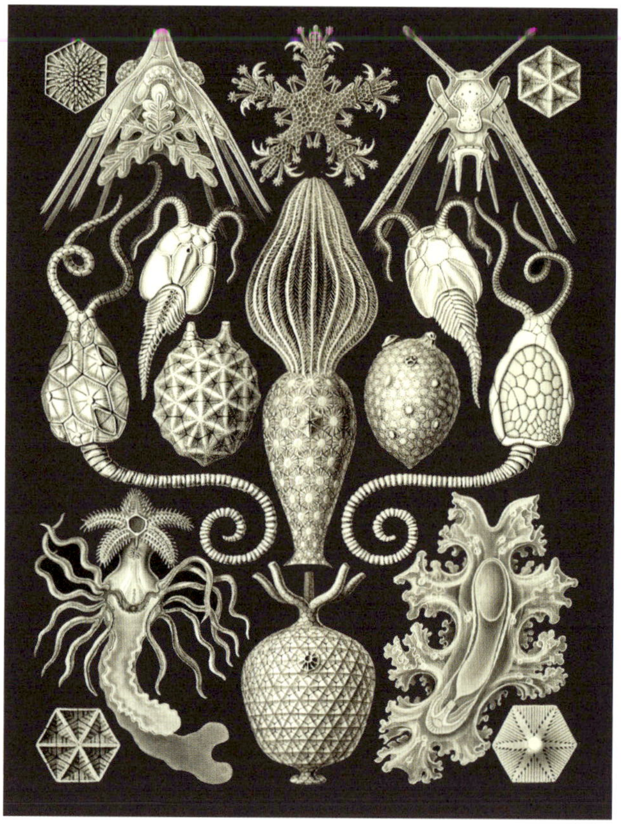

←
Ernst Haeckel's *Kunstformen der Natur* (*Art Forms in Nature*, 1899–1904) was a fantastical, aesthetic celebration of nature. These plates from the book feature, clockwise from top left: brittle stars including the giant basket star (*Astrophyton muricatum*); a selection of echinoderms including cake urchins; crustaceans from the *Amphoridea* genus; and examples of calcareous sponges.

↓
Photographer Richard Collins has captured a jellyfish of the *Aequoreidae* genus as it feeds on an unknown species of comb jelly. The jellyfish's tentacles are trapping the comb jelly so that it cannot escape.

'The finest crystal vase was clumsiness itself when compared with it. It was as fine as the transparent soap-bubble blown out of a pipe.'

EDWARD FORBES, *A MONOGRAPH OF THE BRITISH NAKED-EYED MEDUSAE*, 1848

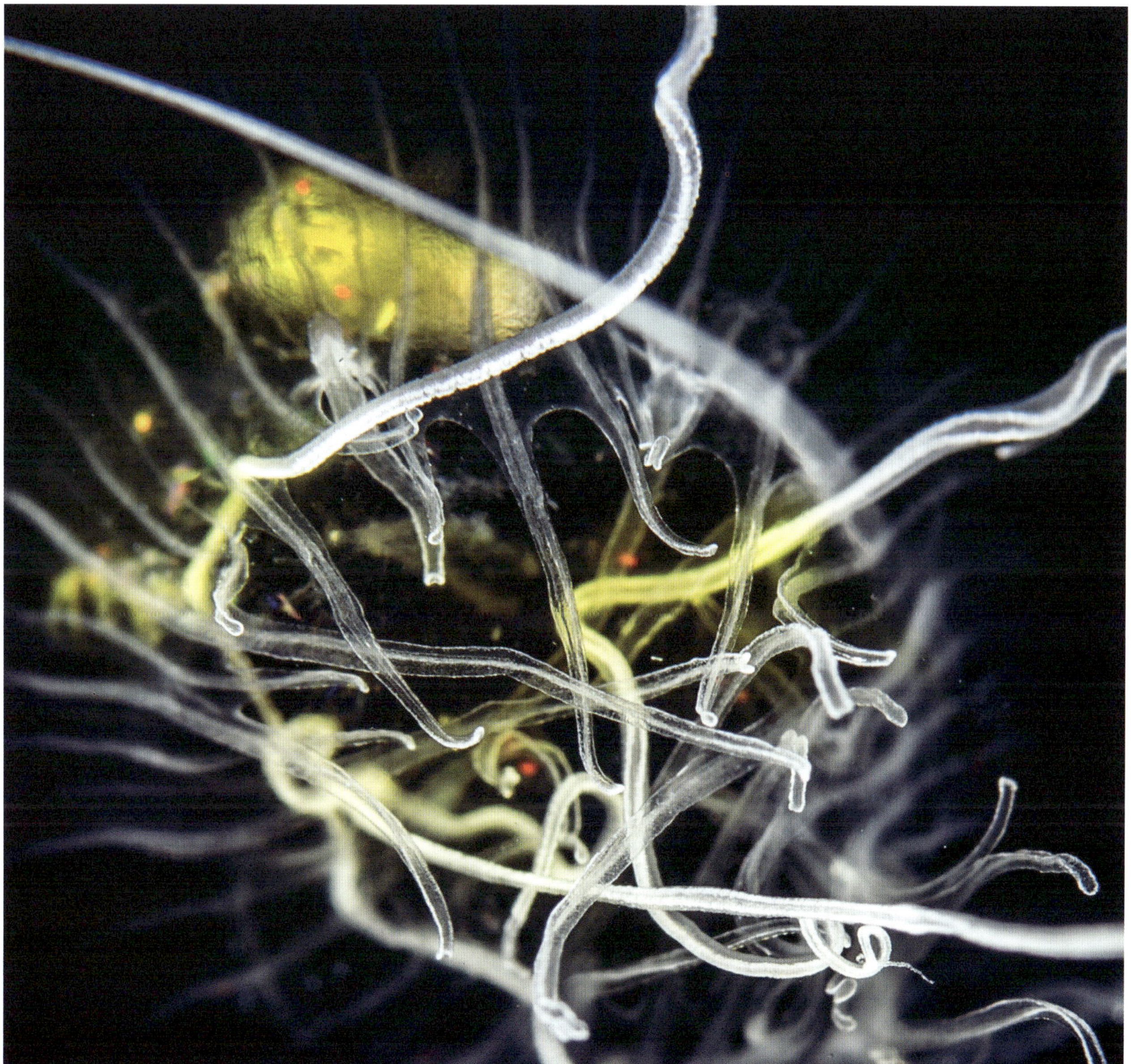

↓
Photographed in 2023 by free diver Helen Walne is a starfish in its natural habitat, sitting under the kelp forests off the coast of Cape Town, South Africa, among colourful anemones. Several starfish species can regenerate arms; for some, a severed arm can even regenerate the rest of the starfish body.

→
◎ In *The Great Barrier Reef of Australia* (1893), William Saville-Kent's underwater photographs, which were ground-breaking at the time of publication, are paired with watercolours, including these examples. Several of the corals depicted are now endangered, including *Madrepora pulchra* (top left, no. 12), a species of staghorn coral, and *Isis hippuris*, or sea bamboo (top right, no. 1).

(III)

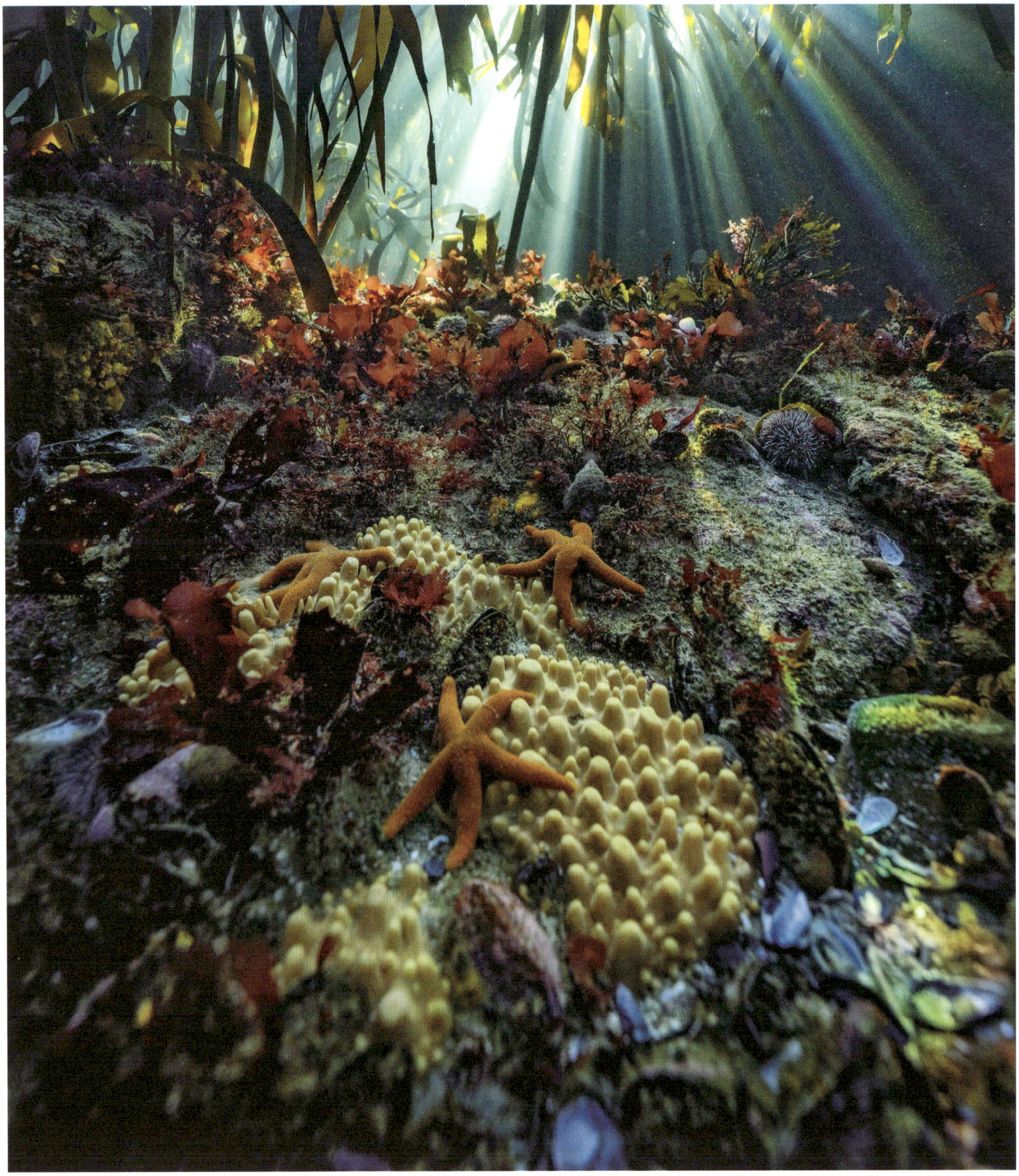

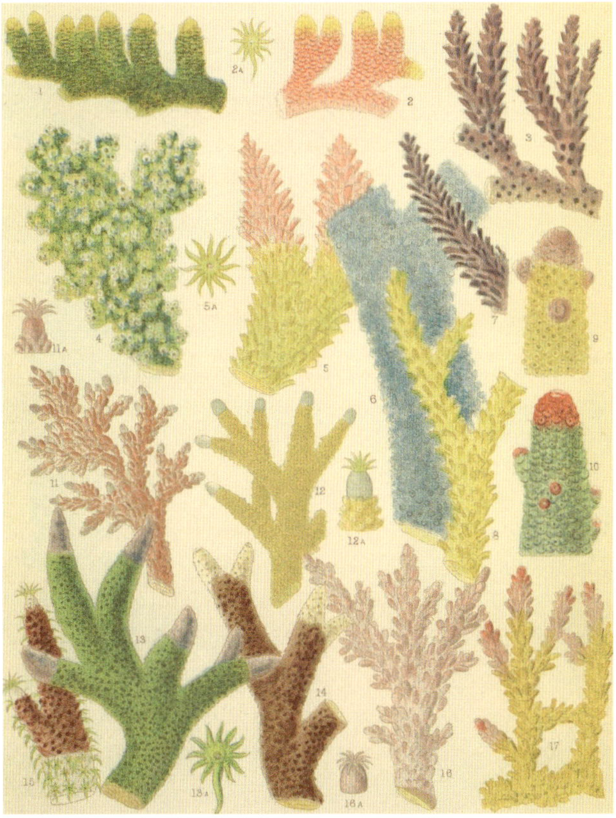

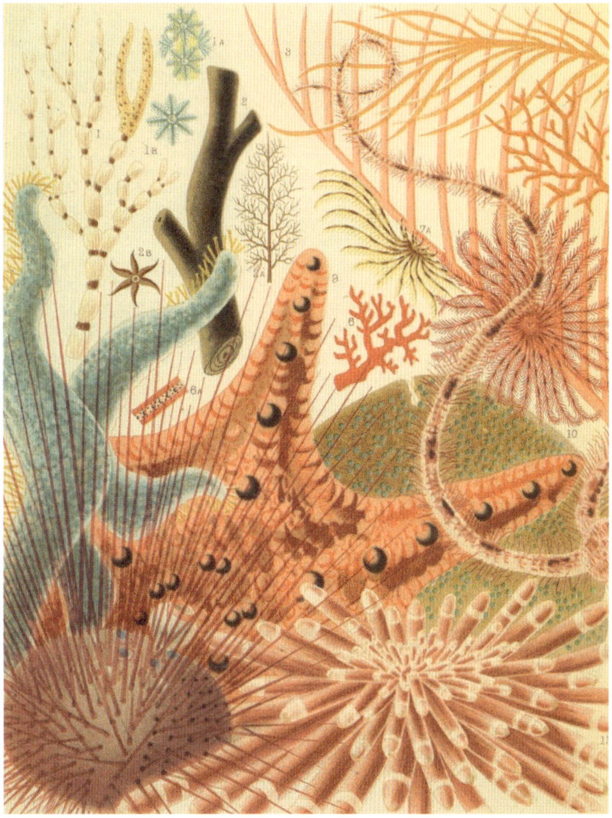

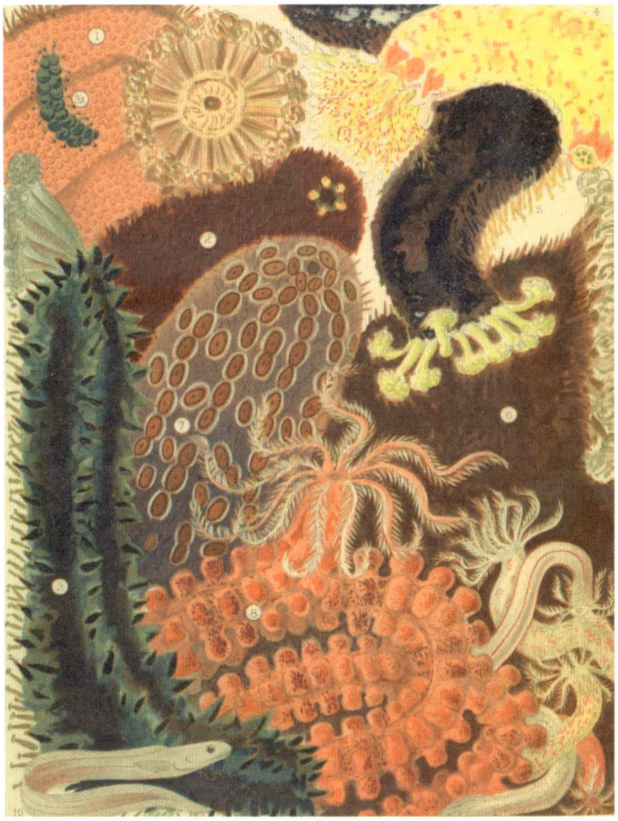

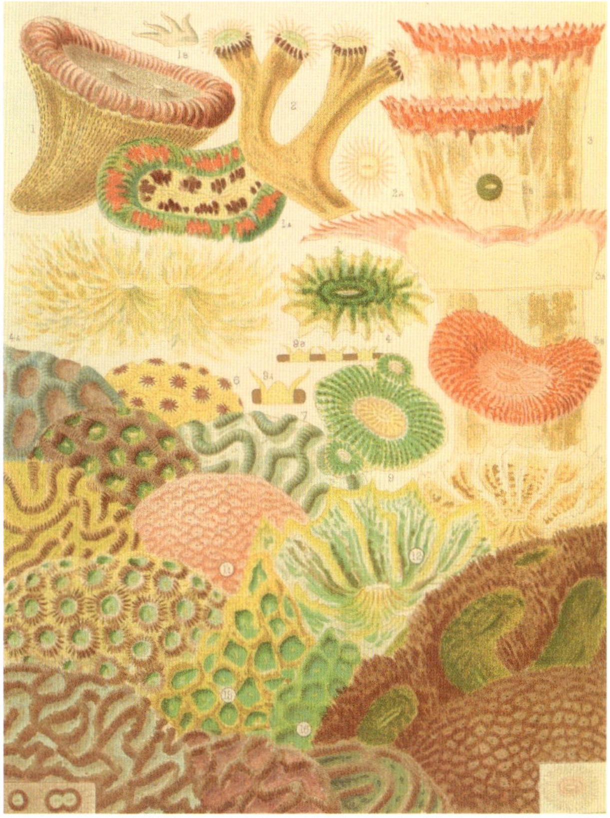

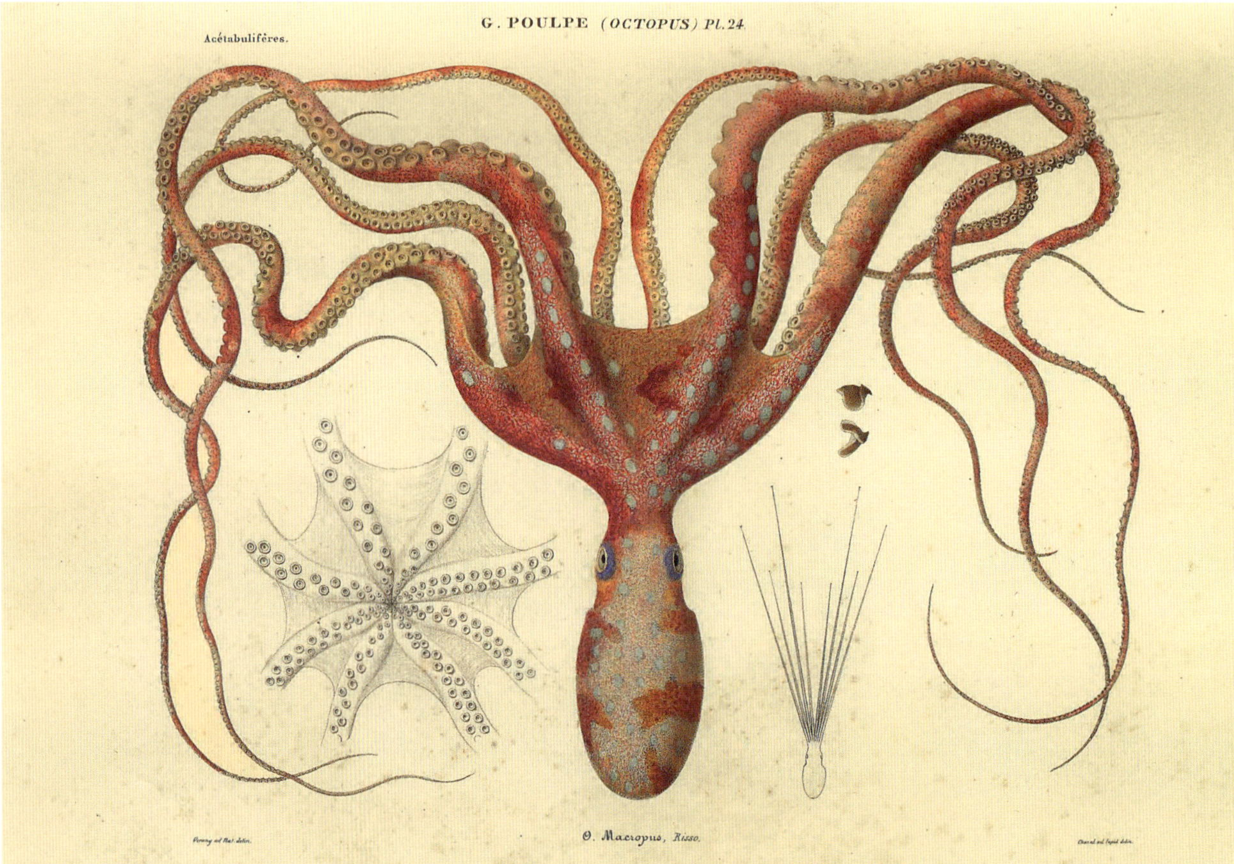

(111)

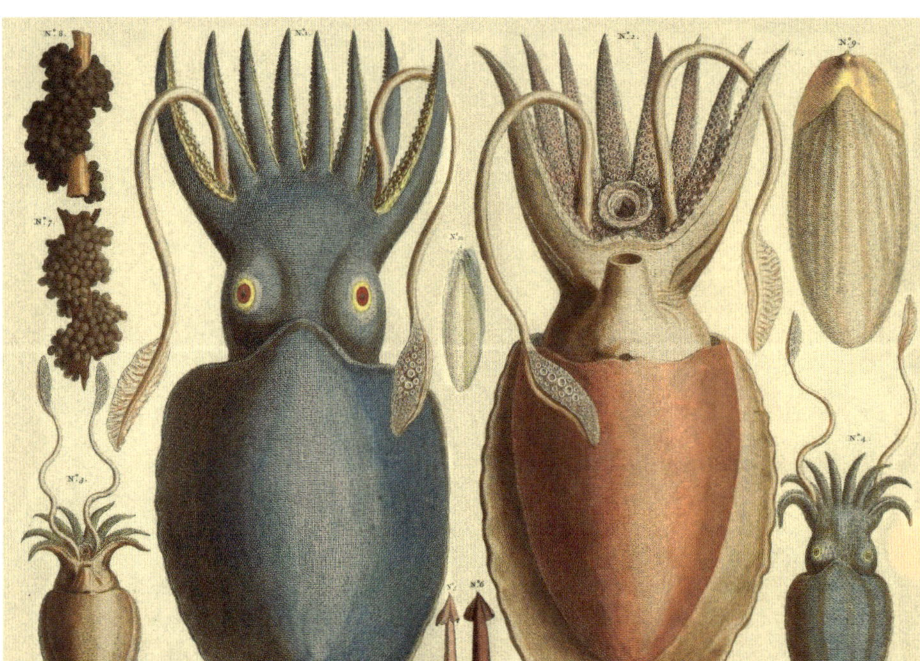

↑
Here illustrated by André Étienne d'Audebert de Férussac in 1835, the Atlantic white-spotted octopus (*Callistoctopus macropus*) is an ocean-dweller that feeds around coral reefs. Octopuses have a large number of neurons (sensory cells for transmitting messages) spread through their tentacles, giving them semi-autonomous minds of their own.

←
Albertus Seba's illustration of flying squids is from the *Cabinet of Natural Curiosities* (1734–65). These cephalopods have three hearts, and use their ten tentacles to catch and hold onto prey. The mobility of squid species makes assessment of their conservation status difficult.

→
In his *Crabs Near the Water's Edge* (*c.* 1830) Japanese artist Yashima Gakutei presents a trio of crabs. Crabs are threatened by overfishing, climate change and pollution.

'Who has known the ocean? Neither you nor I, with our earth-bound senses, know the foam and surge of the tide that beats over the crab hiding under the seaweed of his tide-pool home.'

RACHEL L. CARSON, 'UNDERSEA', *THE ATLANTIC*, SEPTEMBER 1937

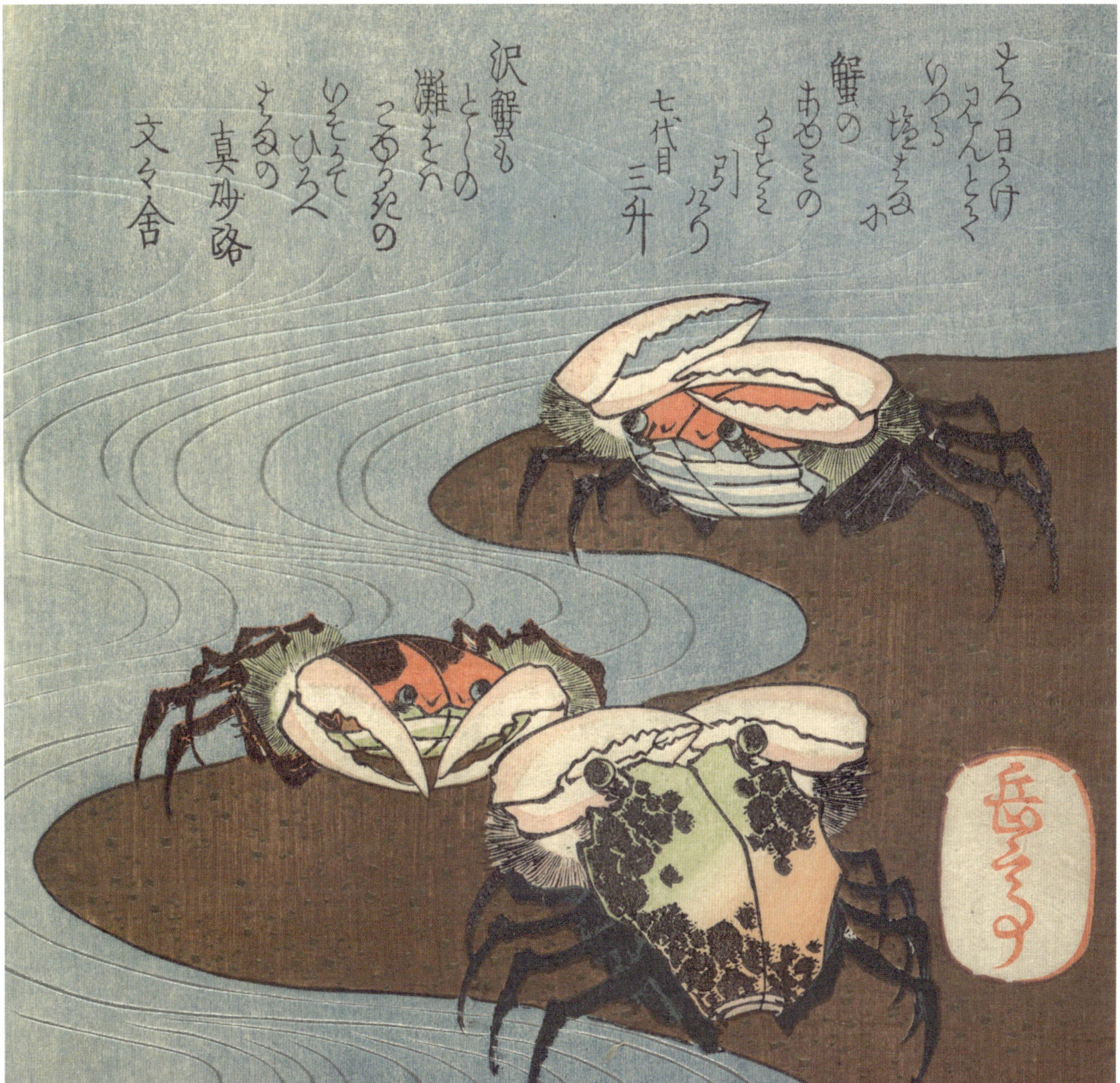

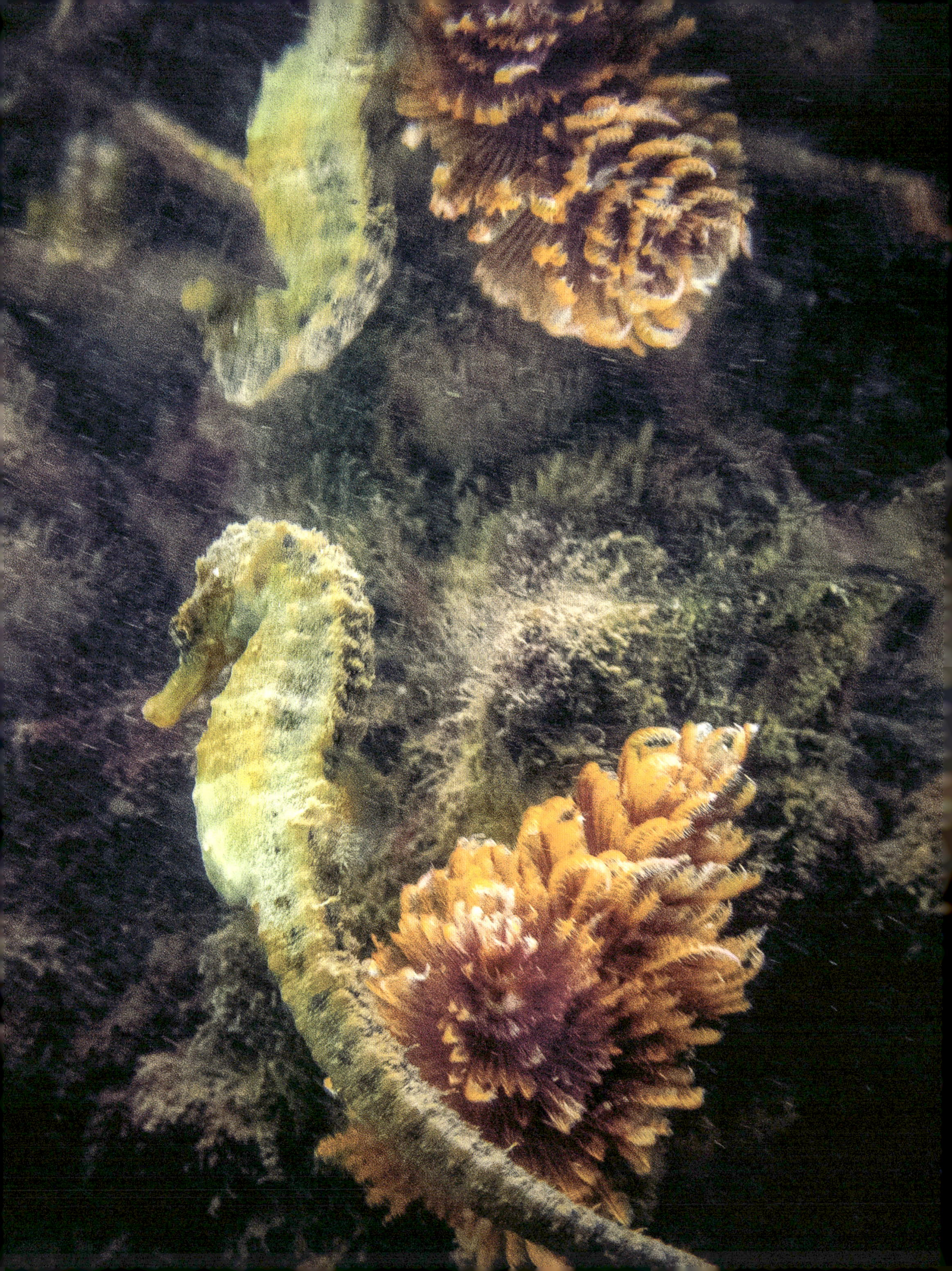

(IV)
FISH

KNYSNA SEAHORSE	GOLDFISH
SCORPIONFISH	RED HANDFISH
JOHN DORY	MAGPIE PERCH
SILVER FISH	BUTTERFLY FAN LOBSTER
HAMMERHEAD SHARK	TROUT
SKATE	OCEAN SUNFISH
TRIGGERFISH	SAILFIN TANG
EMPEROR ANGELFISH	EASTERN TRIANGLE BUTTERFLYFISH
GREY TRIGGERFISH	RIO SKATE
FLATHEAD FISH	FLYING FISH
STOPLIGHT LOOSEJAW	BOA DRAGONFISH
GREAT WHITE SHARK	GREAT HOGFISH
TUB GURNARD	DRAGONET
CARP	WHITE-SPOTTED PUFFER
POMPANO FISH	BARRACUDA
EEL	GARFISH
LONG-SNOUTED SEAHORSE	LEAFY SEADRAGON

FISH are aquatic, cold-blooded, finned vertebrates that have gills for extracting oxygen from water. The majority of fish – about 20,000 species, including goldfish and sunfish – are bony fish, with hard skeletons and swim bladders that help them move up and down in water. The majority also have scales. Cartilaginous fish – encompassing fewer than 600 species of sharks, rays, skates, sawfish and chimeras – are the oldest group of living jawed vertebrates. They have a skeleton made of cartilage and large noses filled with sensory cells.

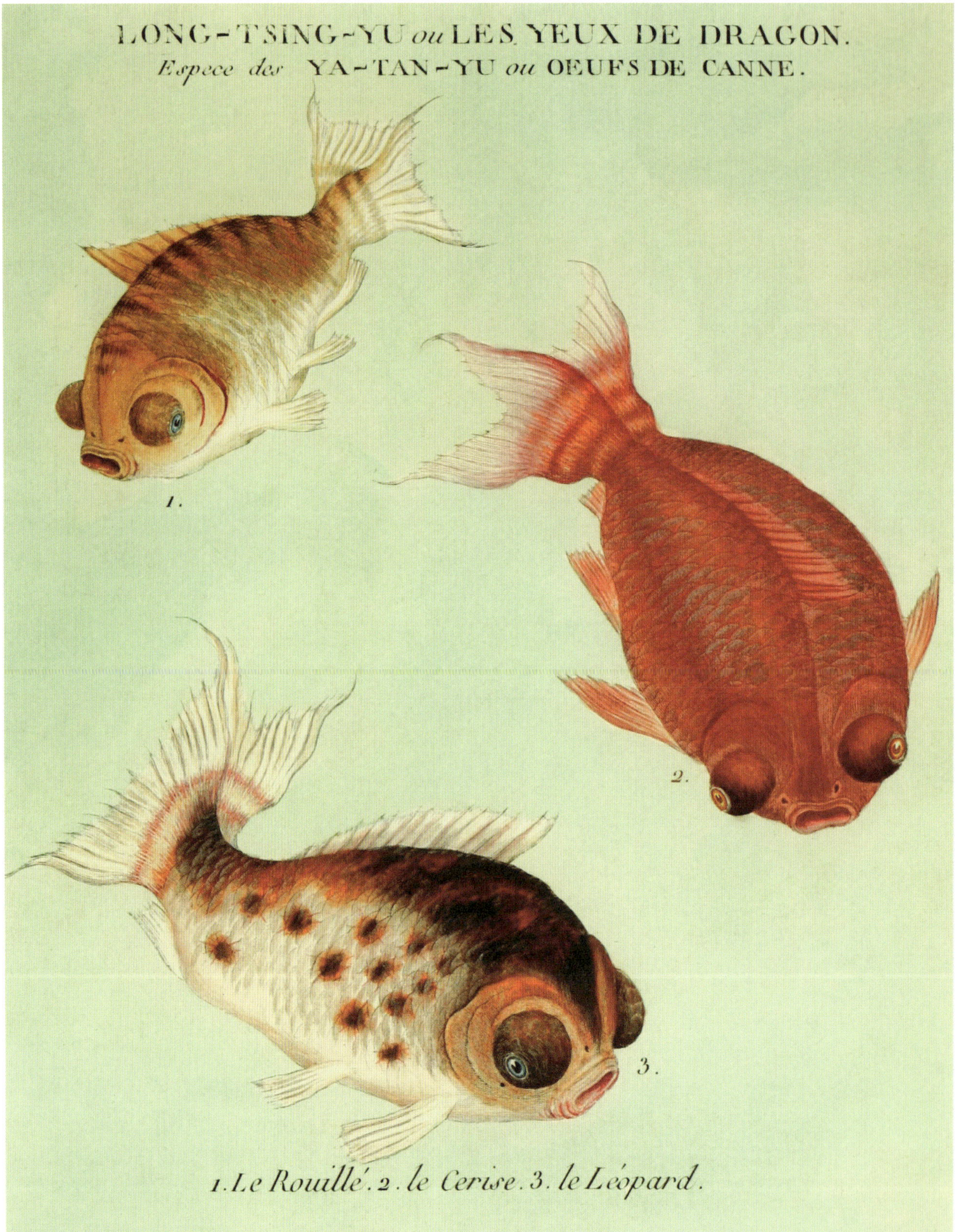
LONG-TSING-YU ou LES YEUX DE DRAGON.
Espece des YA-TAN-YU ou OEUFS DE CANNE.
1.
2.
3.
1. Le Rouillé. 2. le Cerise. 3. le Léopard.

(IV)

'The goldfish will do the darting about, and we can get our feeling of life and movement in watching them.'

HARRIET OSGOOD TAYLOR, *JAPANESE GARDENS*, 1912

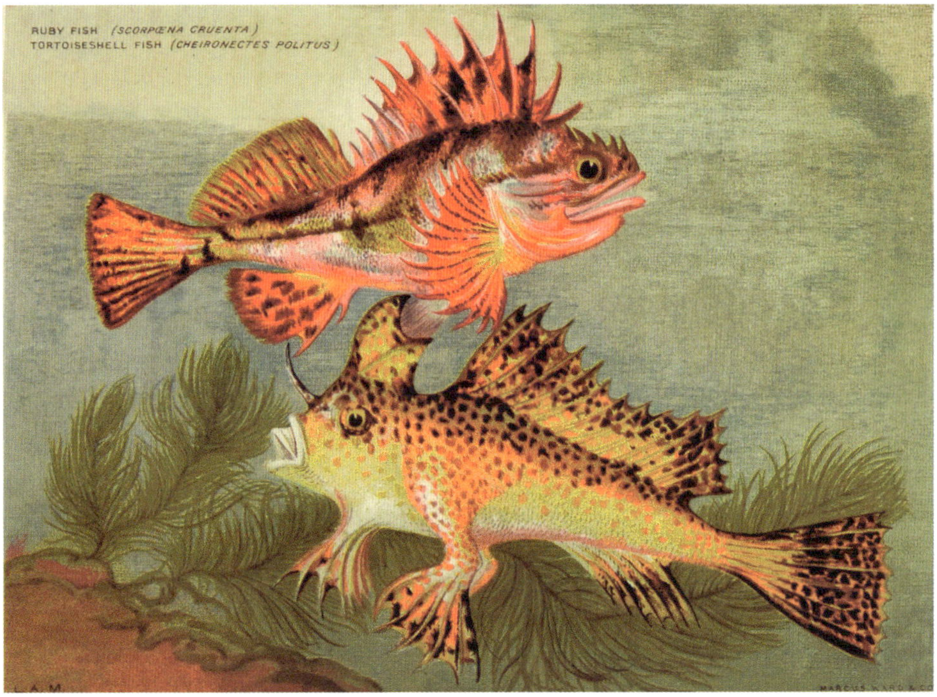

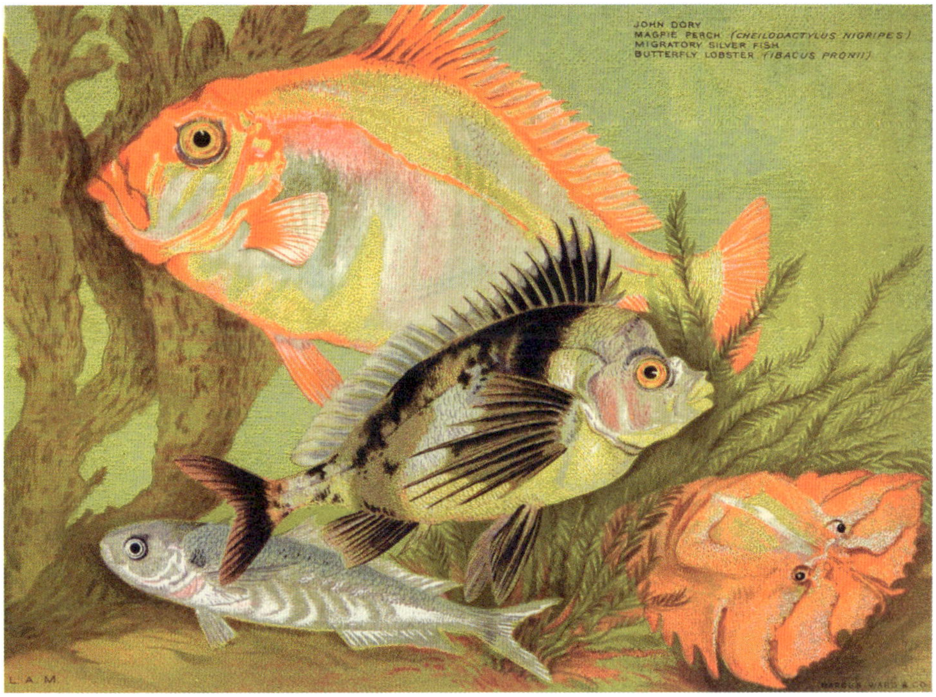

↲
◎ The Knysna, or Cape, seahorse (*Hippocampus capensis*), captured here in *Losing Time* (2021) by underwater photographer Helen Walne, lives among clumps of eelgrass in three river estuaries along the southern coast of South Africa, where its survival is threatened by urban development.

←
François-Nicolas Martinet's *Histoire naturelle des dorades de la Chine* ('Natural History of the Gilded Fish of China'), published in 1780, was the first monograph on goldfish published in Europe. Goldfish were domesticated from crucian carp (both are *Carassius auratus*) in ancient China, and the yellow variety was the symbol of the Imperial Family during the Song Dynasty (960–1279).

→
◍ Anglo-Australian writer and illustrator Louisa Anne Meredith included these two plates in her *Tasmanian Friends and Foes, Feathered, Furred and Finned* (1880). At the top of the top image is the 'ruby fish', now known as the scorpionfish or red rock cod (*Scorpaena papillosa*), and the 'tortoiseshell fish' below it is now the red handfish (*Thymichthys politus*), a critically endangered fish that walks on oversized 'hands' rather than swimming. The bottom plate illustrates, from top down: the John Dory (*Zeus faber*), the magpie perch (*Cheilodactylus nigripes*), a migratory species of silver fish (possibly silver warehou, *Seriolella punctata*), and the butterfly fan lobster (*Ibacus peronii*).

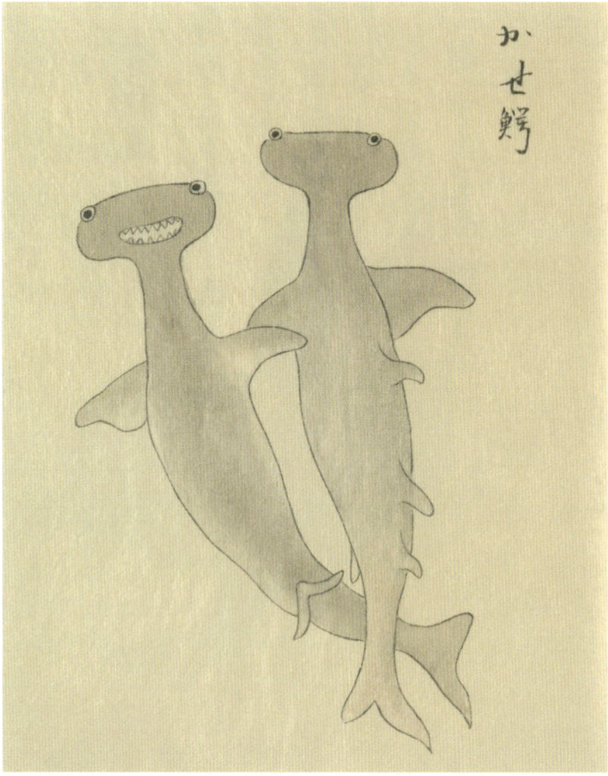

'No human being, however great, or powerful, was ever so free as a fish.'

JOHN RUSKIN, *THE TWO PATHS*, 1859

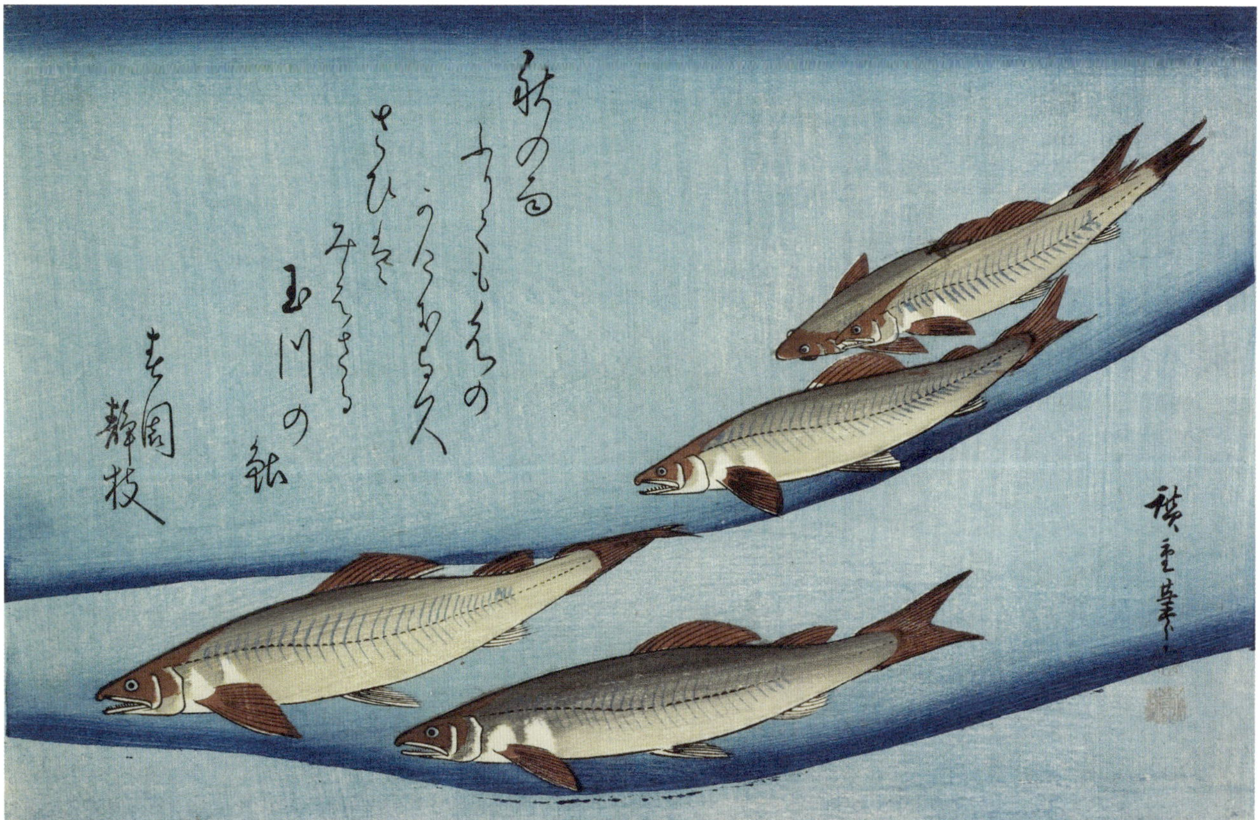

↖
Hammerhead sharks have wide heads and wide-set eyes, which give them binocular vision and good depth perception, although they cannot see directly in front of themselves. They also use their broad heads to ram and pin down prey. This Japanese illustration from 1735 borders on caricature.

←
Best known for his landscapes, the Japanese artist Utagawa Hiroshige also produced detailed studies of animals and plants. His *Trout* is from a series of woodblock prints known as *Uo-zukushi*, or 'Every Variety of Fish' (*c.* 1832–33).

→
The French biologist and filmmaker Étienne-Jules Marey sought to record the movement of all types of living creatures on film. To do this, in 1882 he invented a sequence camera, which he called a chronophotographic gun, that took rapid-fire images on a single piece of film. This sequence, *Undulations of the Fins of a Skate Viewed from the Side* (1891–92), is the watercolour and ink reproduction of a chronophotographic film that he made.

↳
Henri Matisse's *Polynésie, la mer* (*Polynesia: the Sea*, 1946) is filled with the shapes and forms of sea creatures, including fish, corals, starfish and a diving bird. It was inspired by the colours and the abundance of ocean life around the islands of the South Pacific, which he had visited in the 1930s.

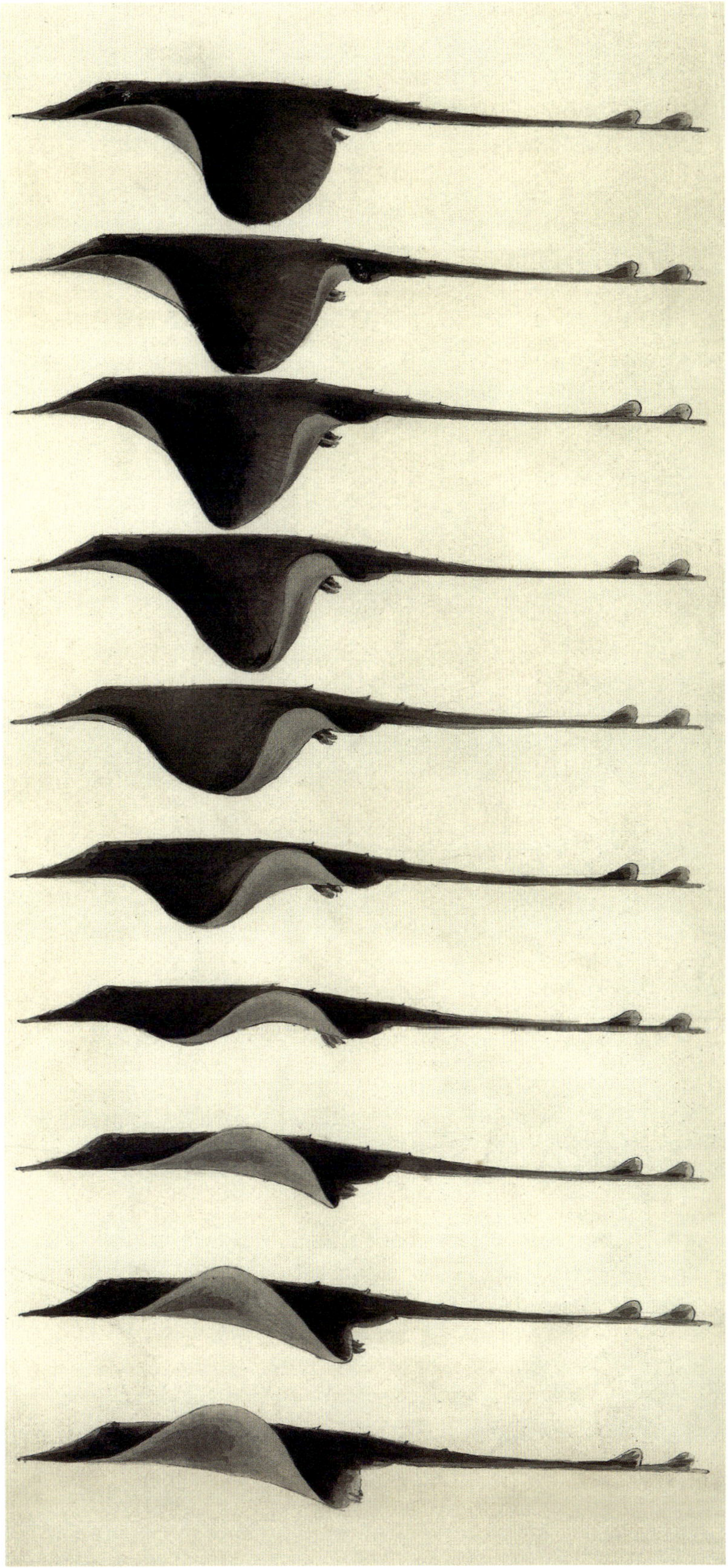

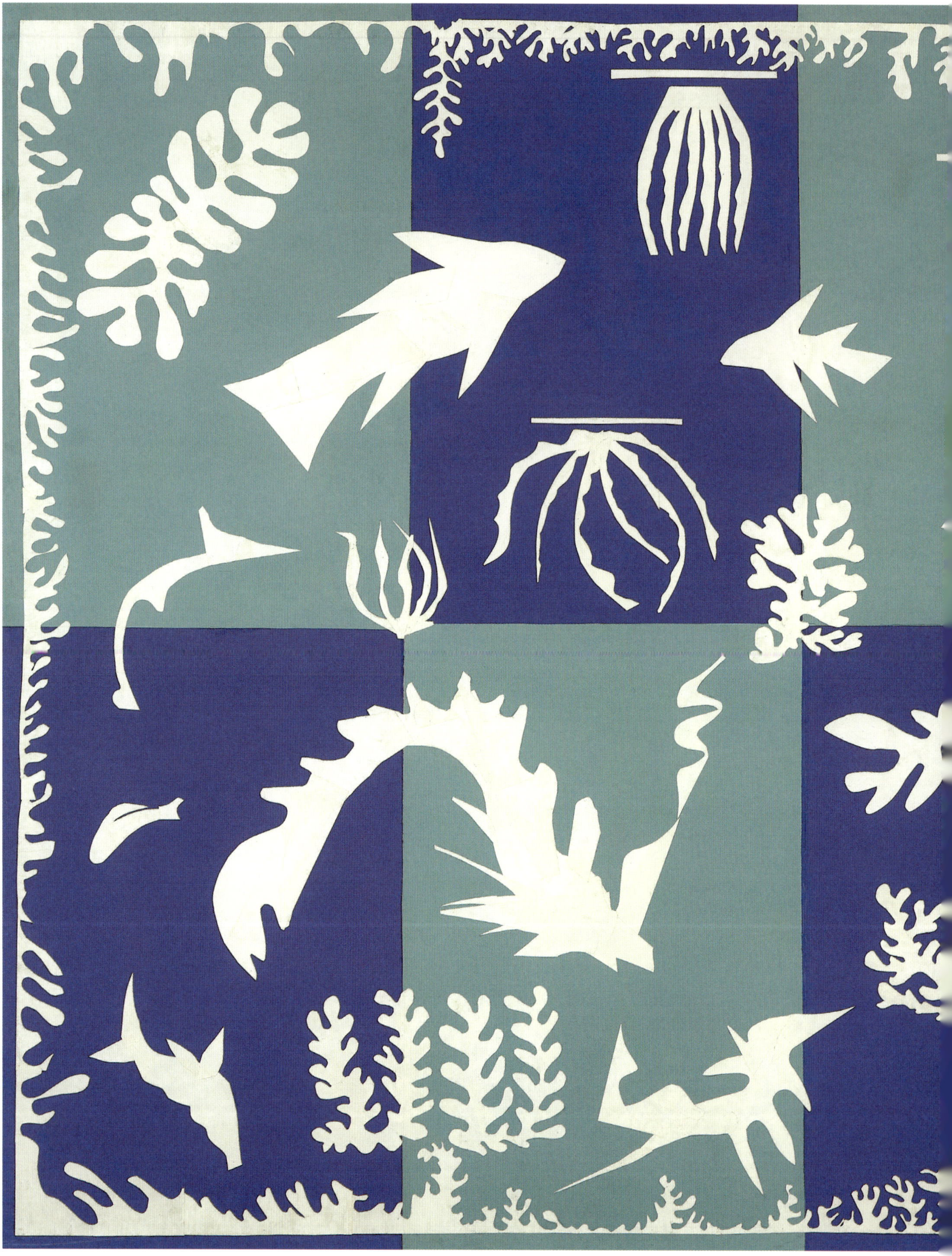

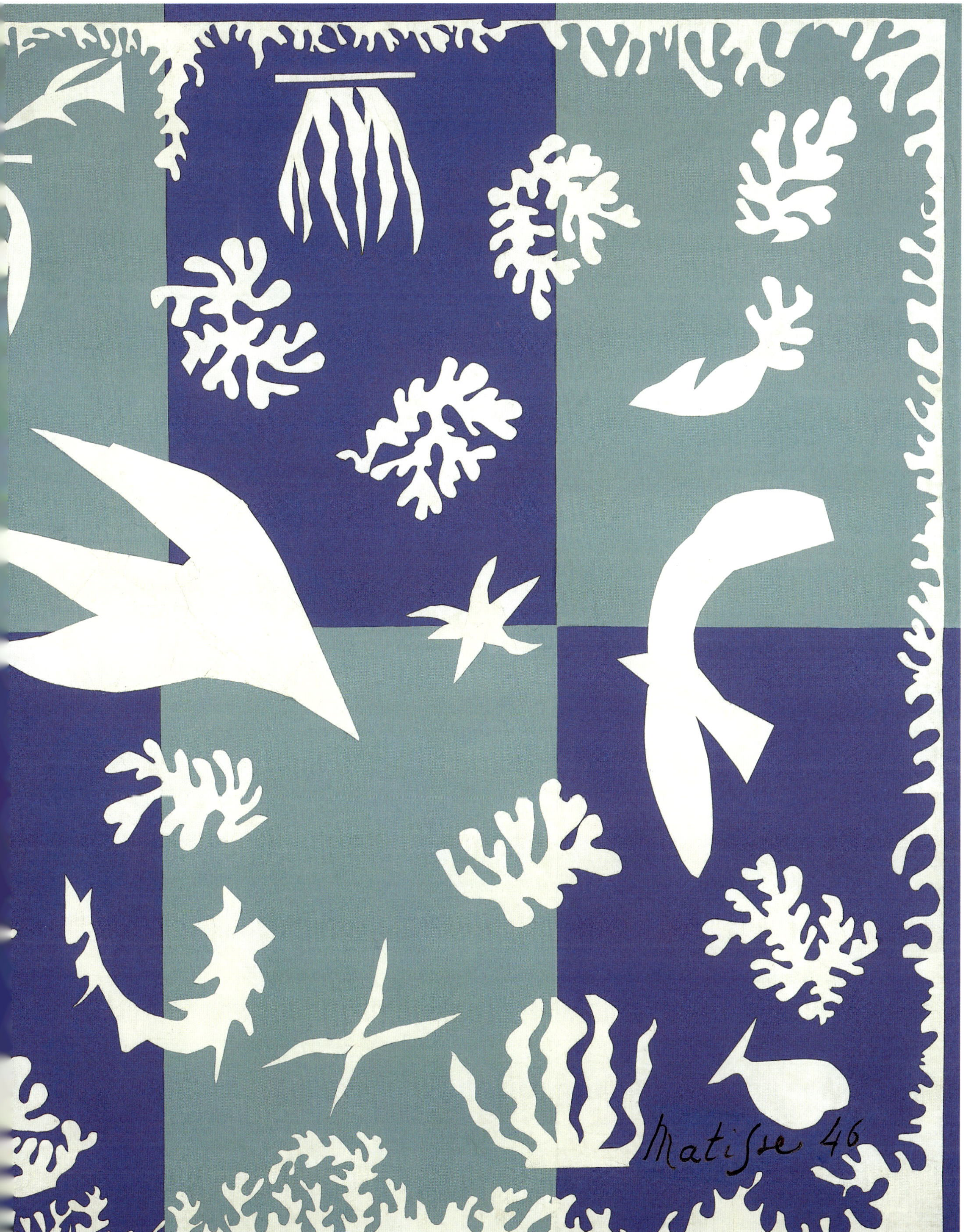
Matisse 46

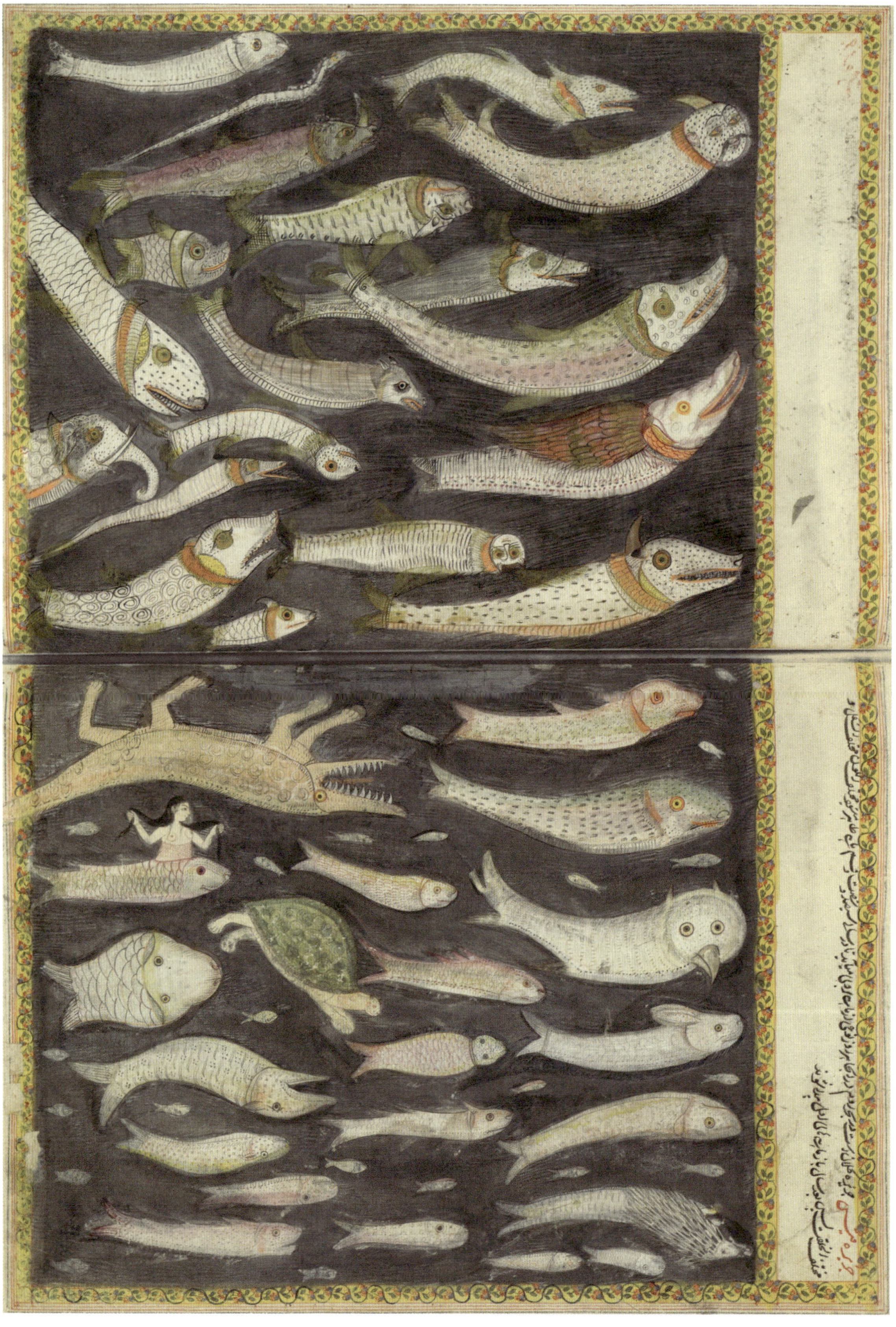

(IV)

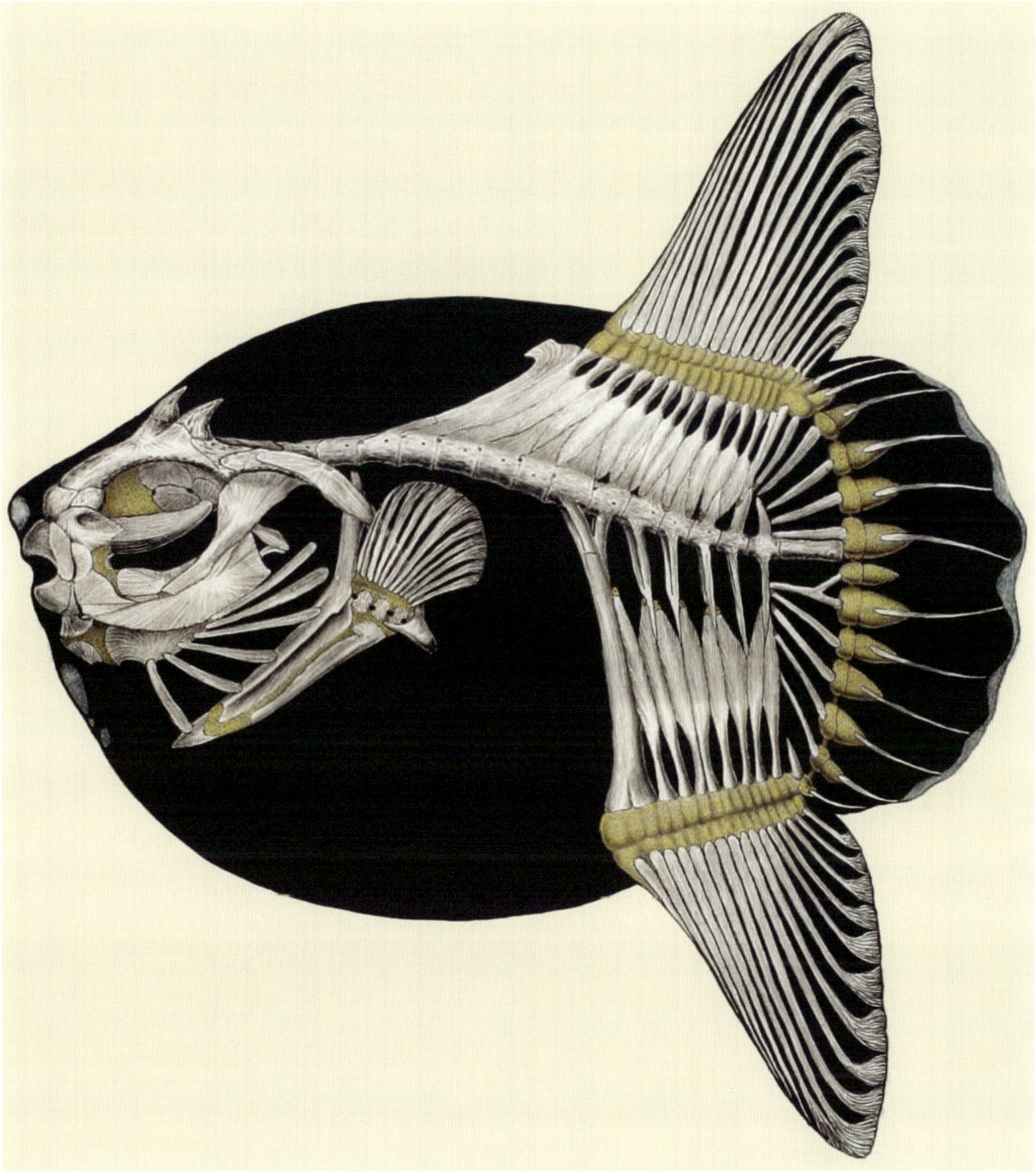

←
This array of fantastical and anthropomorphized fish appears in a 17th-century illustrated manuscript of Al-Qazwini's *Ajā'ib al-Makhlūqāt va-Gharā'ib al-Mawjūdāt* ('The Wonders of Creation'), a Persian work that gives an account of land and sea creatures, and their peculiarities.

↑
◎ Ocean sunfish (*Mola mola*) are the largest bony fish on earth, with a length of up to 3 metres (10 feet) and an average weight of 1,000 kg (220 lb). This illustration is from *Spolia Atlantica* ('The Spoils of the Atlantic', 1898) by Johannes Japetus Steenstrup and Christian Lütken. Sunfish numbers are in decline due to plastic pollution and accidental entanglement in fishing nets.

(IV)

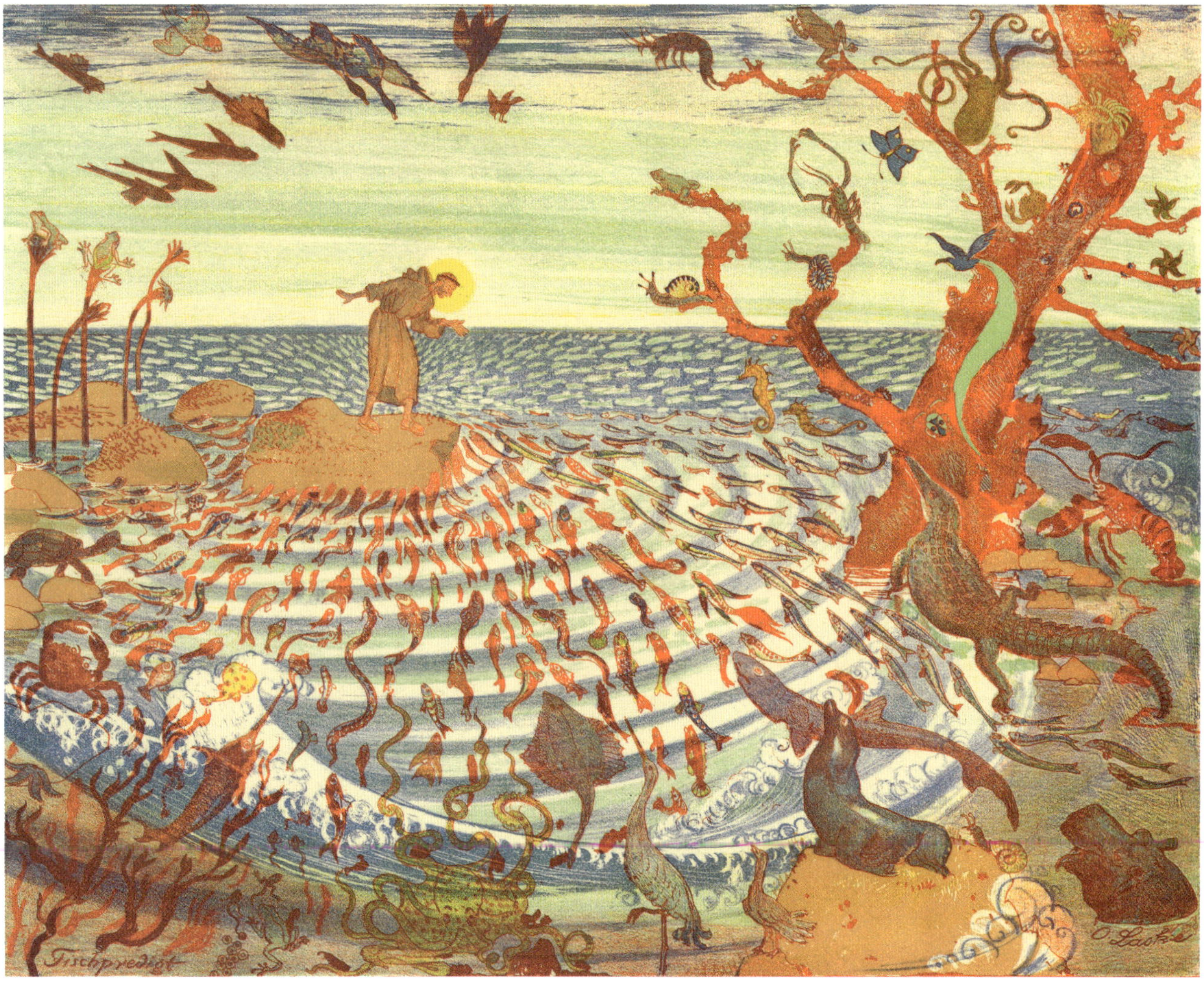

'My brothers the fishes, you are bound, as much as is in your power, to return your thanks to your Creator .... In the universal deluge, all other creatures perished; you alone did God preserve from all harm. He has given you fins to enable you to go where you will.'

THE LITTLE FLOWERS OF SAINT FRANCIS, LATE 14TH CENTURY, TRANSLATED BY REV. H. E. MANNING

←
When Saint Anthony of Padua could not engage the people of Rimini, Italy, to listen to his preaching, he turned instead to the mouth of the river and preached to the fish. As Oskar Laske captures in this 1914 artwork, *Fischpredigt* ('Fish Sermon'), the words of Saint Anthony drew a multitude of fish to the surface of the water, peeping their heads out to hear him preach.

↓
Louis Renard had not been an eyewitness to any of the tropical fish or crustaceans – or, for that matter, the two stick insects, mermaid and dugong – that he depicted in his 1719 book *Poissons, Ecrevisses et Crabes* ('Fishes, Crayfishes and Crabs') so his illustrations are fantastical and sometimes entirely fictional. This selection, however, bears similarities to some real species: clockwise from top left, triggerfish (possibly titan triggerfish, *Balistoides viridescens*), sailfin tang (possibly *Zebrasoma veliferum*), emperor angelfish (*Pomacanthus imperator*), and Eastern triangle butterflyfish (*Chaetodon baronessa*).

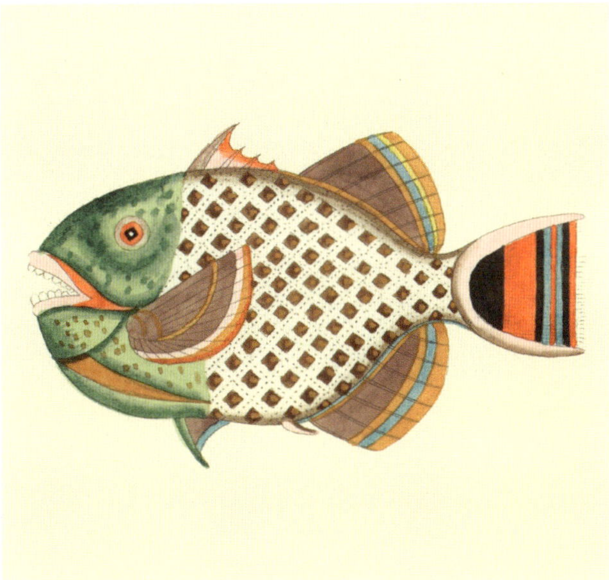

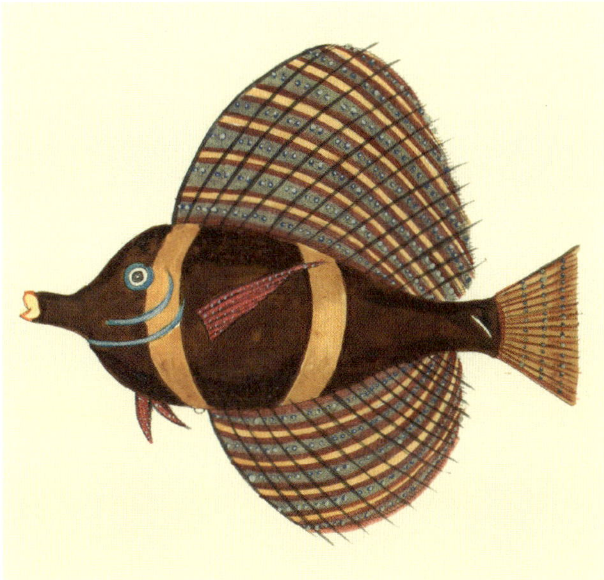

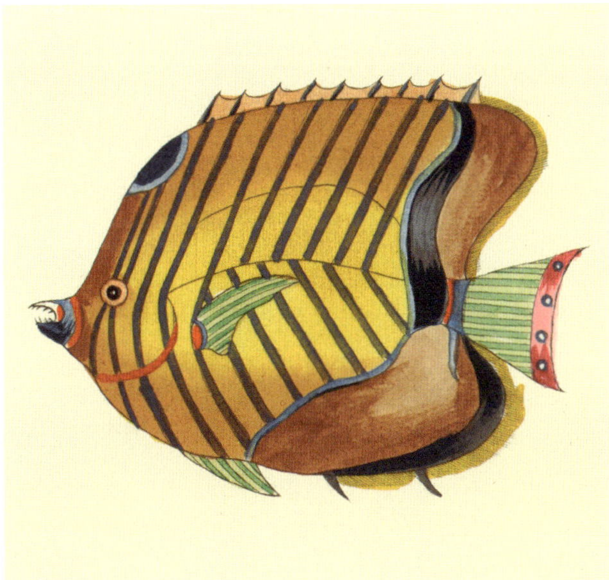

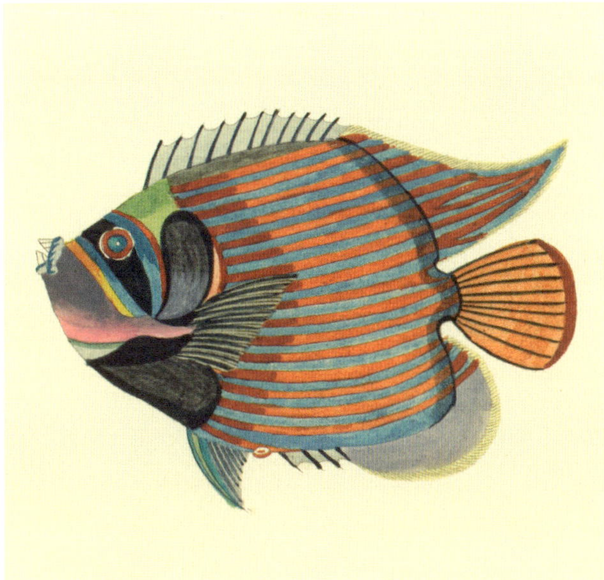

(IV)

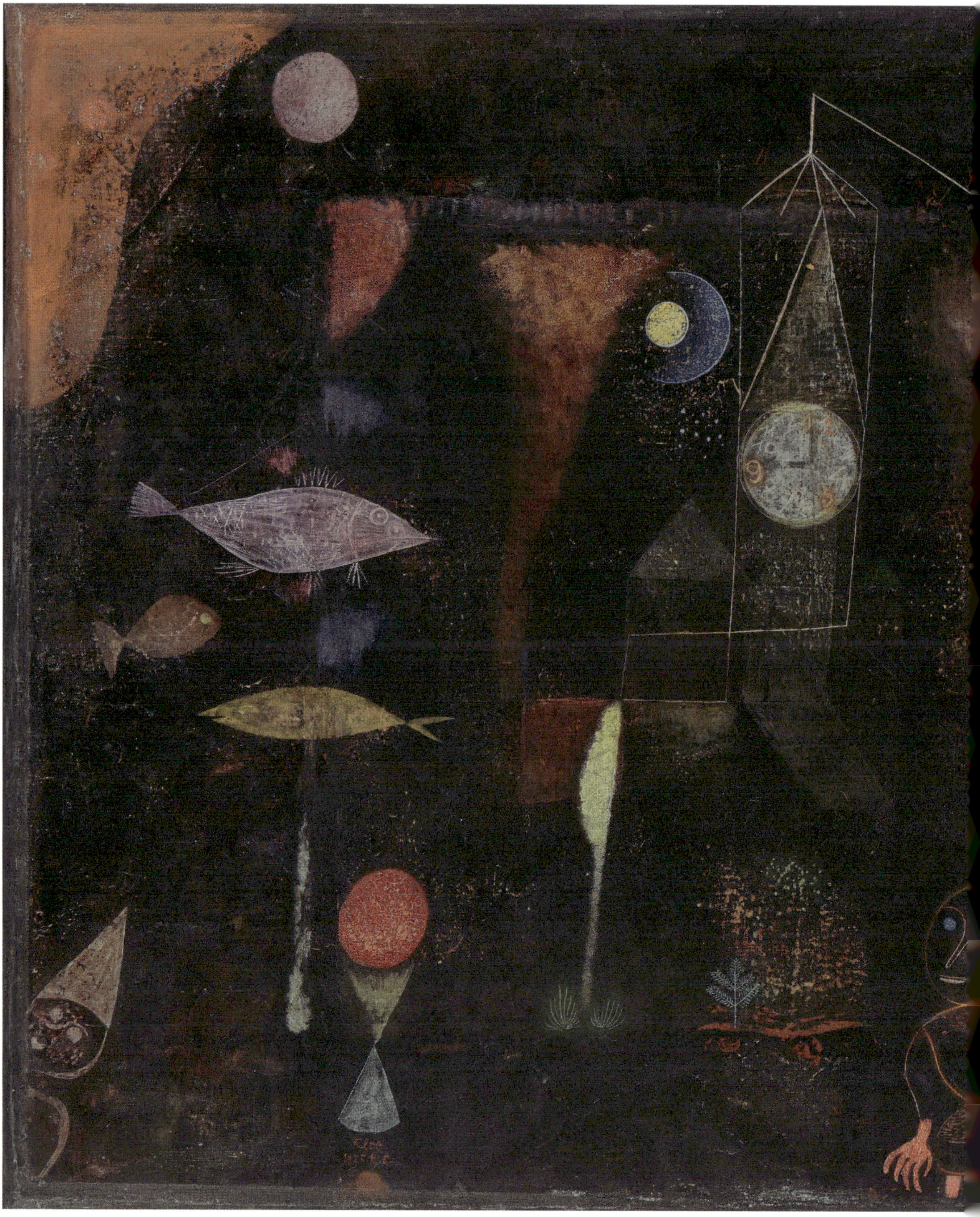

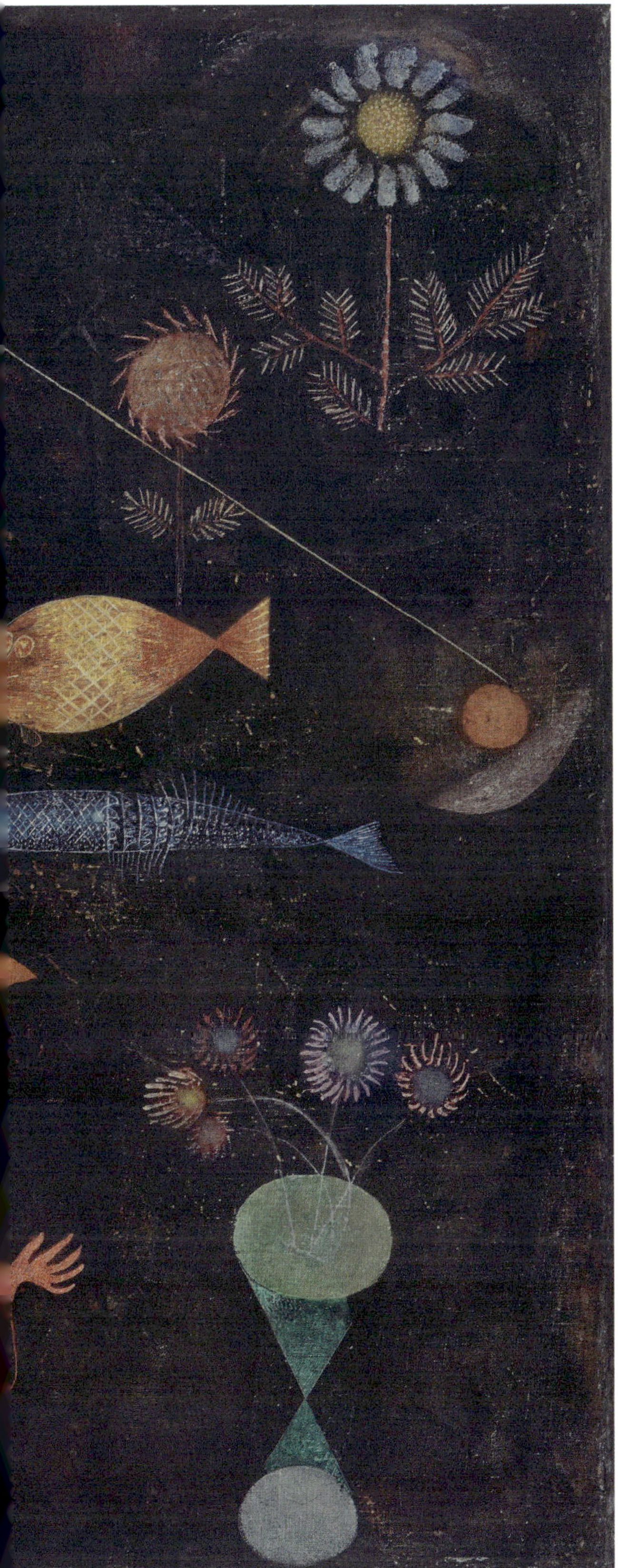

←
A fantastical collection of fish, plants, people and celestial bodies float through the darkness in Swiss artist Paul Klee's *Fish Magic* (1925), which seems to combine aquatic, earthly and celestial realms.

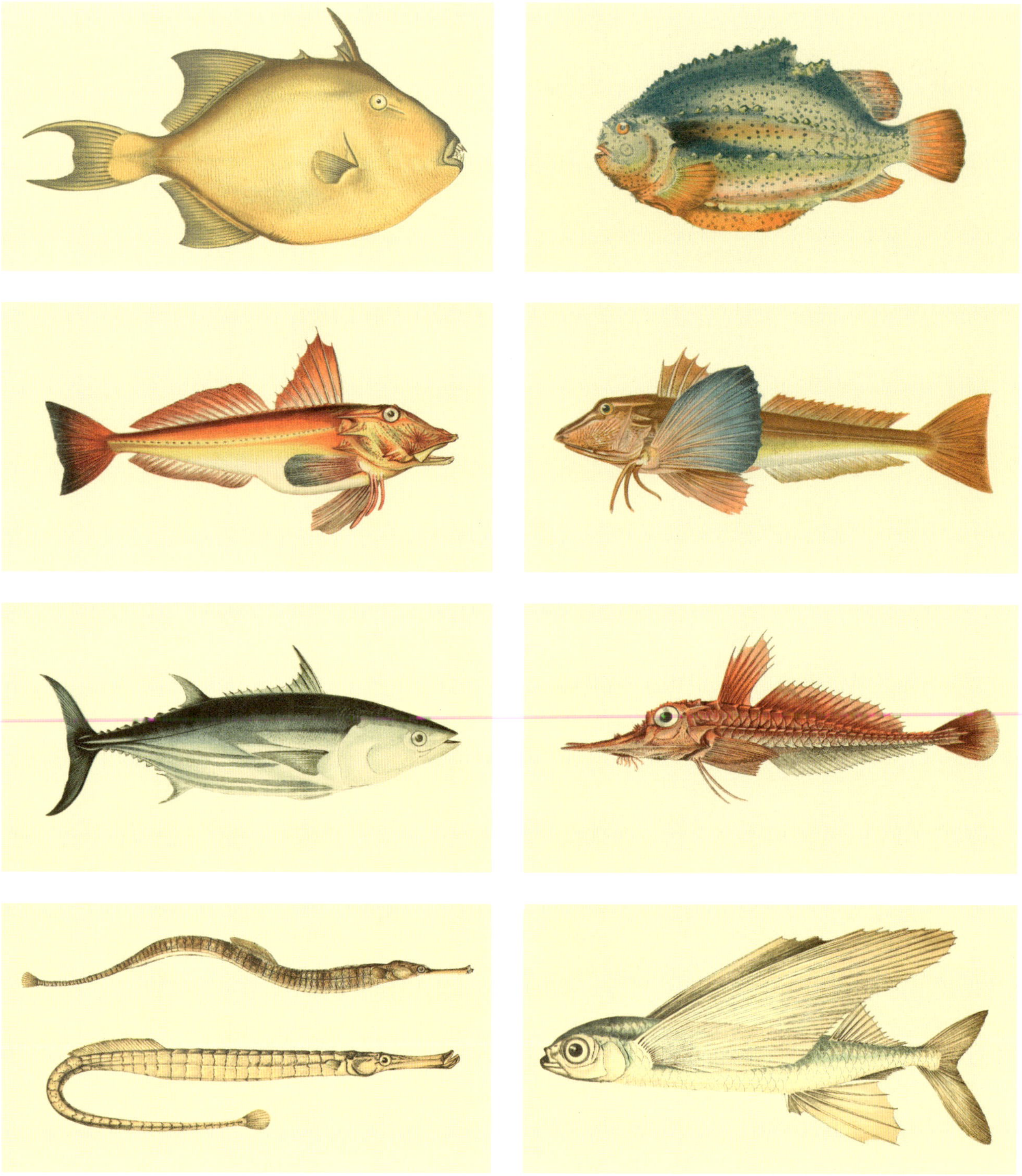

↑
◎ These hand-coloured illustrations from *A History of the Fishes of the British Islands* (1868–69) by Jonathan Couch include, at top left, the grey triggerfish (*Balistes capriscus*), a species that is sometimes observed chewing on hard materials to wear down their teeth, which grow continually.

→
The photographic series *The Mouth of Krishna* by Albarrán Cabrera was inspired by the myth in which Yashoda, stepmother of the child-god Krishna, looks inside his mouth and sees the whole universe within him.

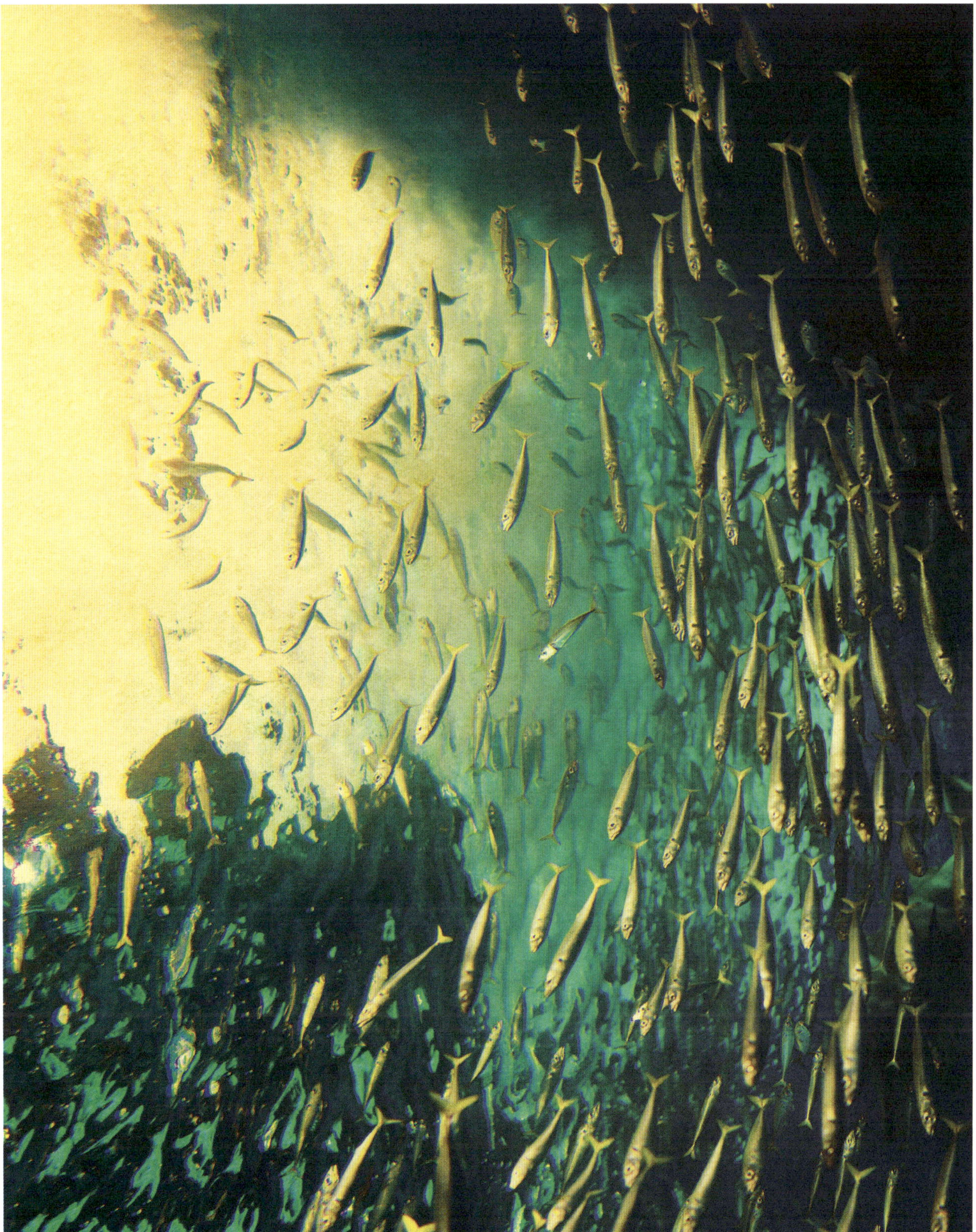

Fish

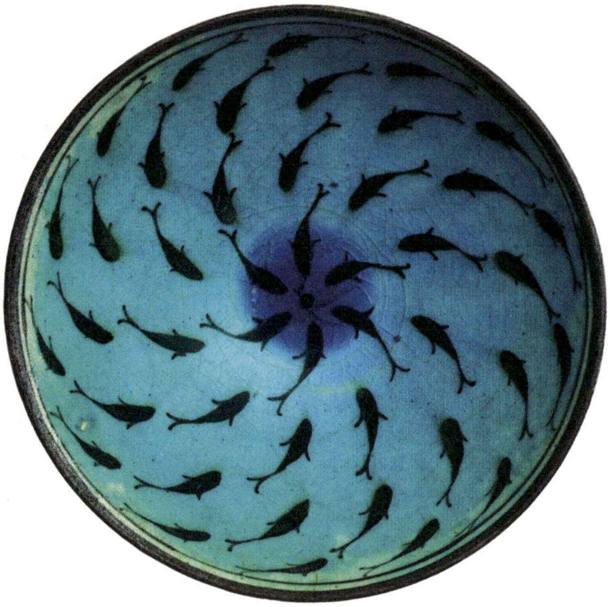

'Flying-fish ... dart out of the water when pursued by their enemies, or frightened by an approaching vessel ... they rise without regard to the direction of the wind or waves.'

ALBERT CARL LUDWIG GOTTHILF GÜNTHER,
AN INTRODUCTION TO THE STUDY OF FISHES, 1880

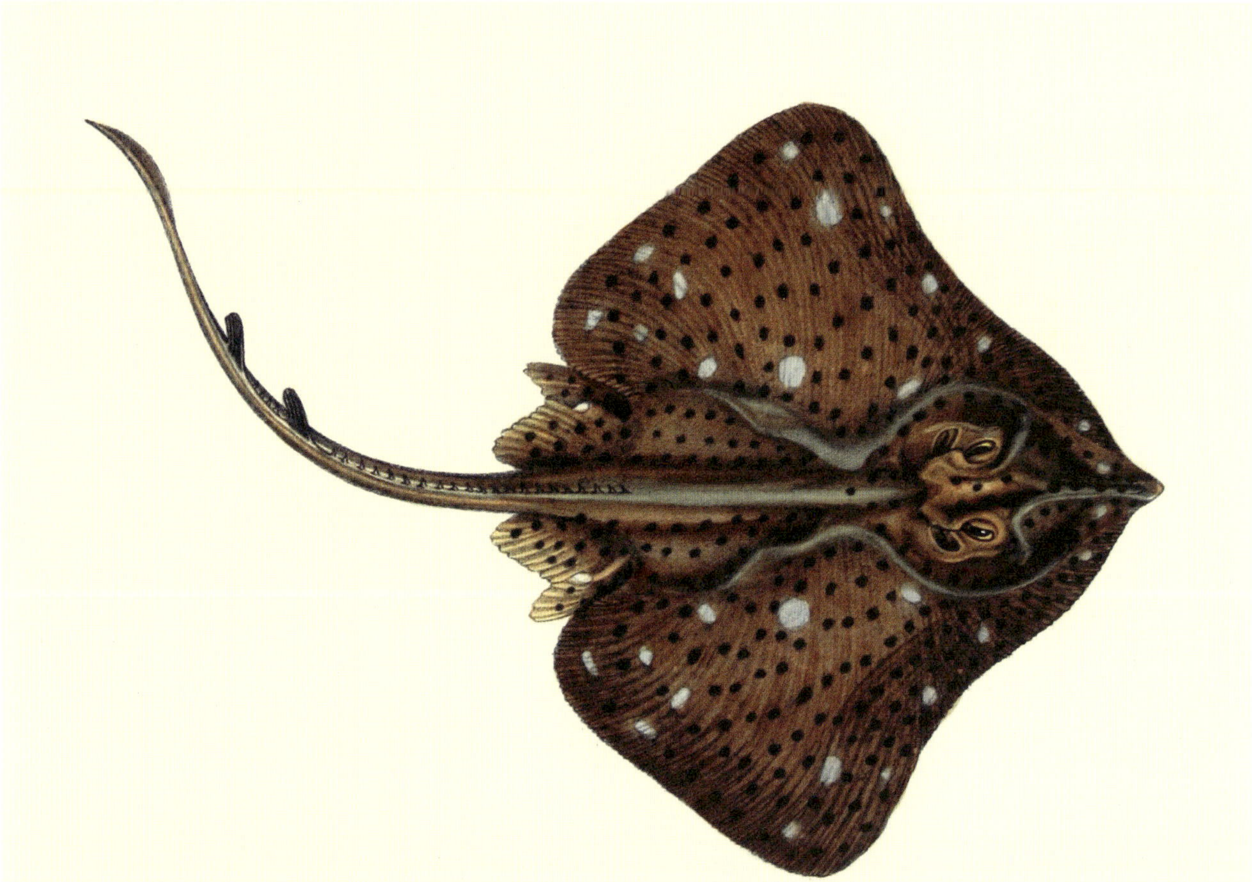

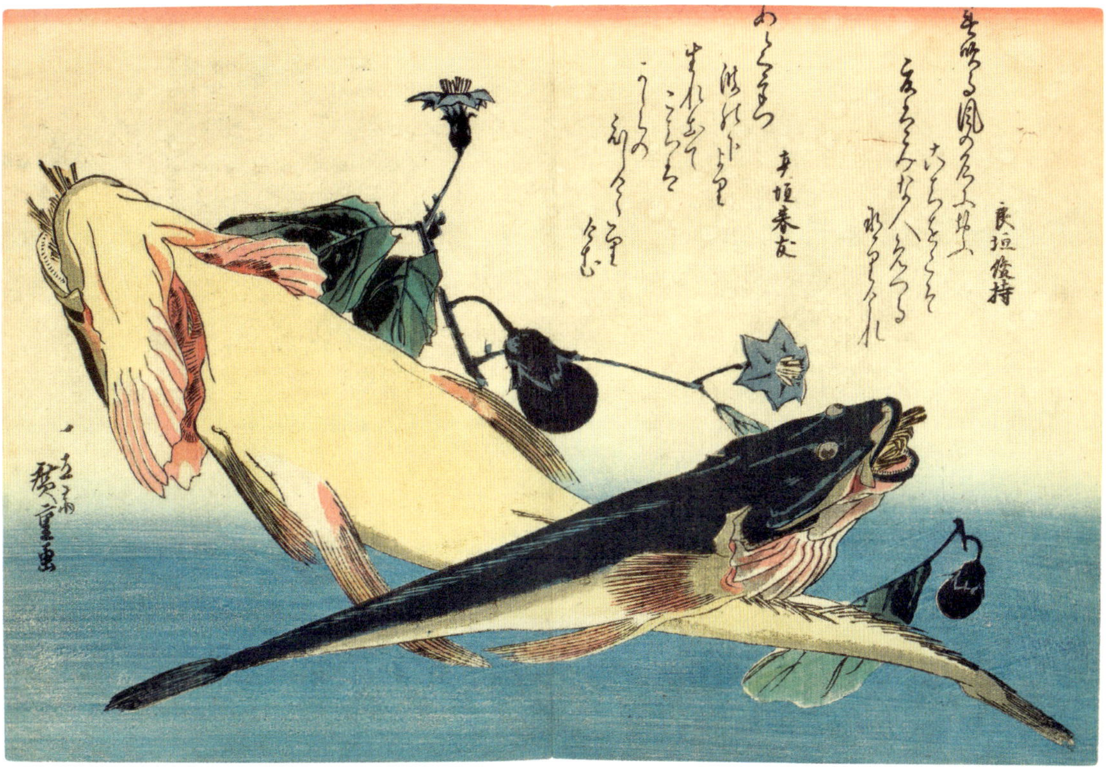

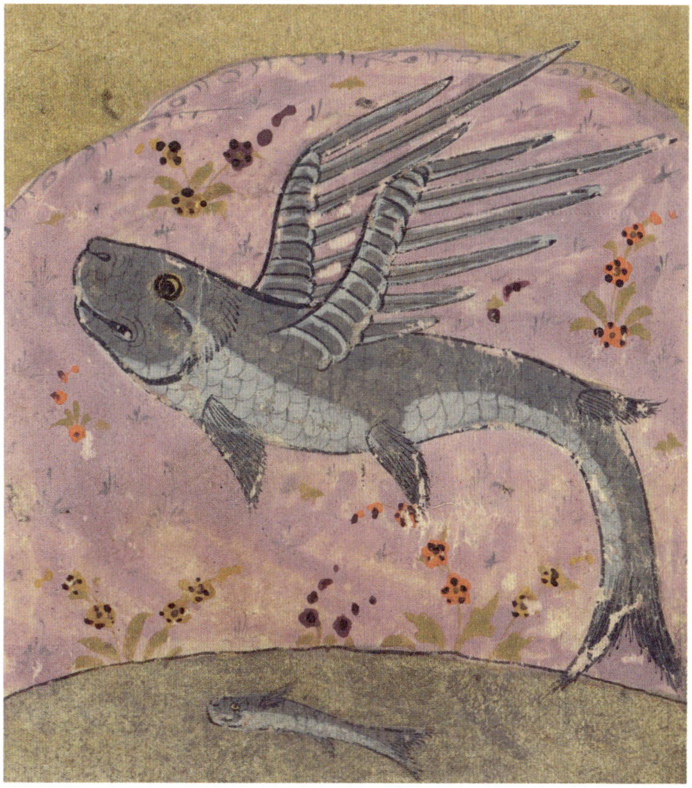

↖
A shoal of fish swims in formation around a turquoise glazed bowl from Iran, dating from between 1275 and 1350 CE.

←
◎ The Rio skate (*Rioraja agassizii*) inhabits shallow waters in the southwestern Atlantic. It is illustrated in this plate from *Expédition dans les parties centrales de l'Amérique du Sud* ('Expedition to the central parts of South America', 1850–59) by Francis de Castelnau.

↗
Flatheaded fish generally live on the ocean floor, burying themselves in sand, and pounce on passing fish and crustaceans. Utagawa Hiroshige's woodblock print of 'Kochi and nasu' ('two flatheads and flowering eggplant') is from his series *Uo-zukushi*, or ('Every Variety of Fish'), published in the early 1830s.

→
It may be hard to believe from this mid-16th-century Persian manuscript illustration, 'A Fish With Wings Flying Above the Sea', but flying fish can glide above the water at speeds of up to 56 kilometres (35 miles) per hour.

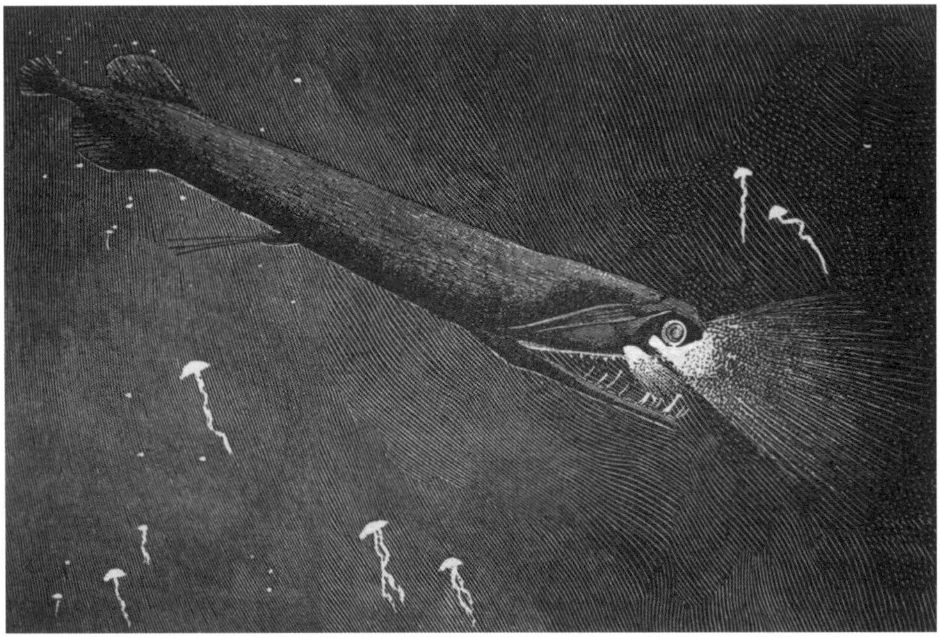

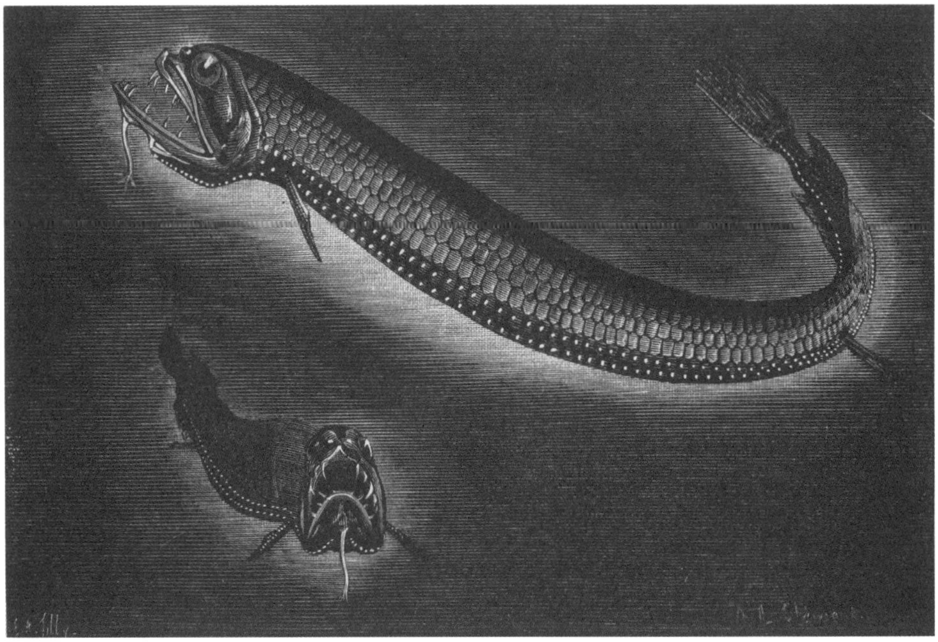

←
The stoplight loosejaw (*Malacosteus niger*) is a dragonfish that exists in the deep sea below 500 metres (1,600 feet) depth. It is one of the only bioluminescent fish that produces a red light – a colour that is undetectable by most of their prey, meaning they can illuminate their food while hunting without being seen. The illustrations of this fish (top) and the boa dragonfish (*Stomias boa*) (bottom) are both by Armand-Lucien Clément, published in *Living Lights* (1887) by Charles Frederick Holder. Uniquely, the boa dragonfish is able to detect the amount of light in its surroundings and adjust its own emission of light accordingly.

→
◎ Robert Longo's imposing, large-scale charcoal drawing of a great white shark (*Carcharodon carcharias*), dated 2008, is part of his series *Perfect Gods*. The dynamic depiction of the shark's bite is made possible by Longo's use of photographic sources, which allows for a precise level of anatomical detail while depicting a living creature in motion.

'It is advisable to look from the tide pool to the stars, and then back to the tide pool again.'

JOHN STEINBECK, *THE LOG FROM THE SEA OF CORTEZ*, 1951

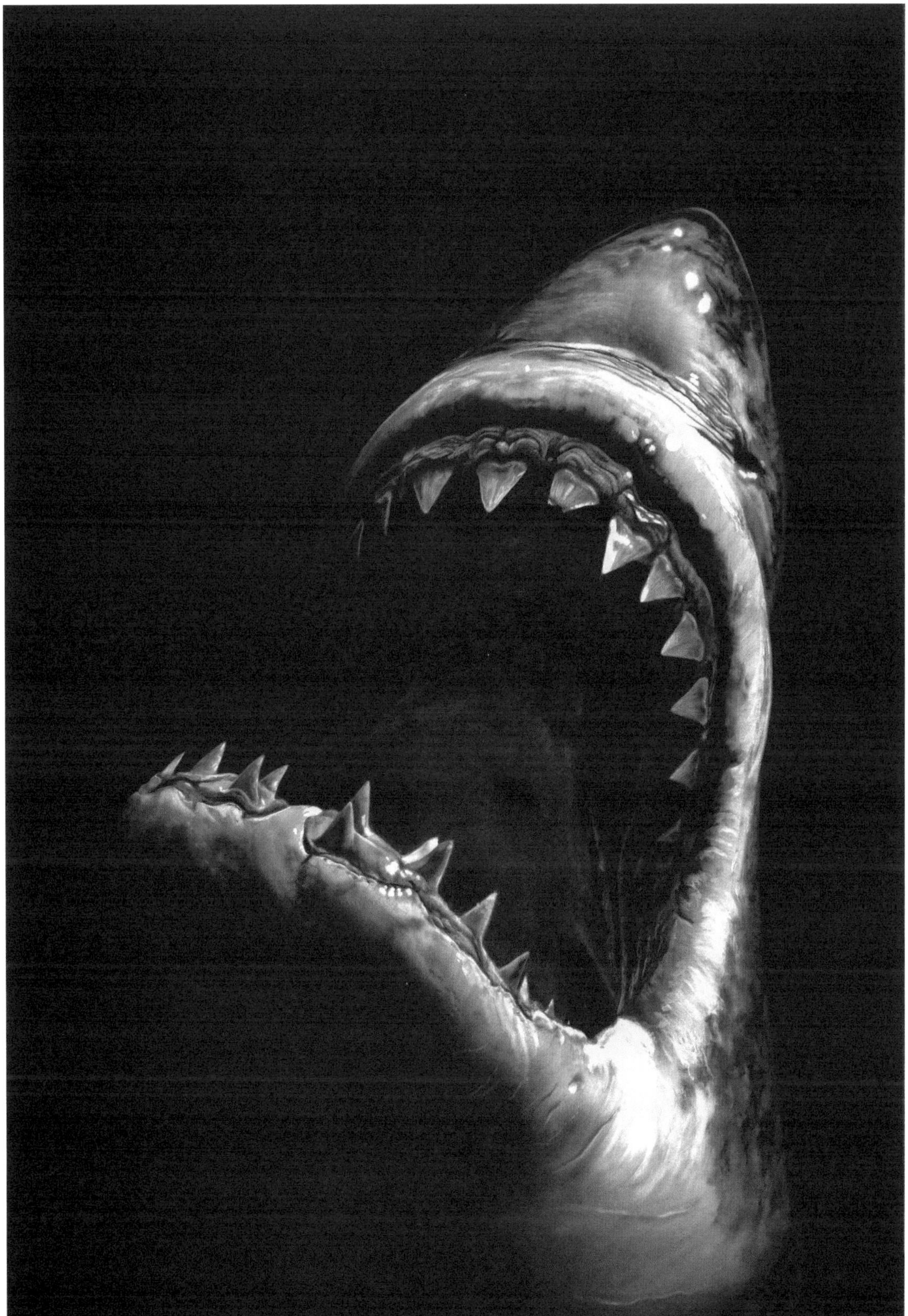

Fish

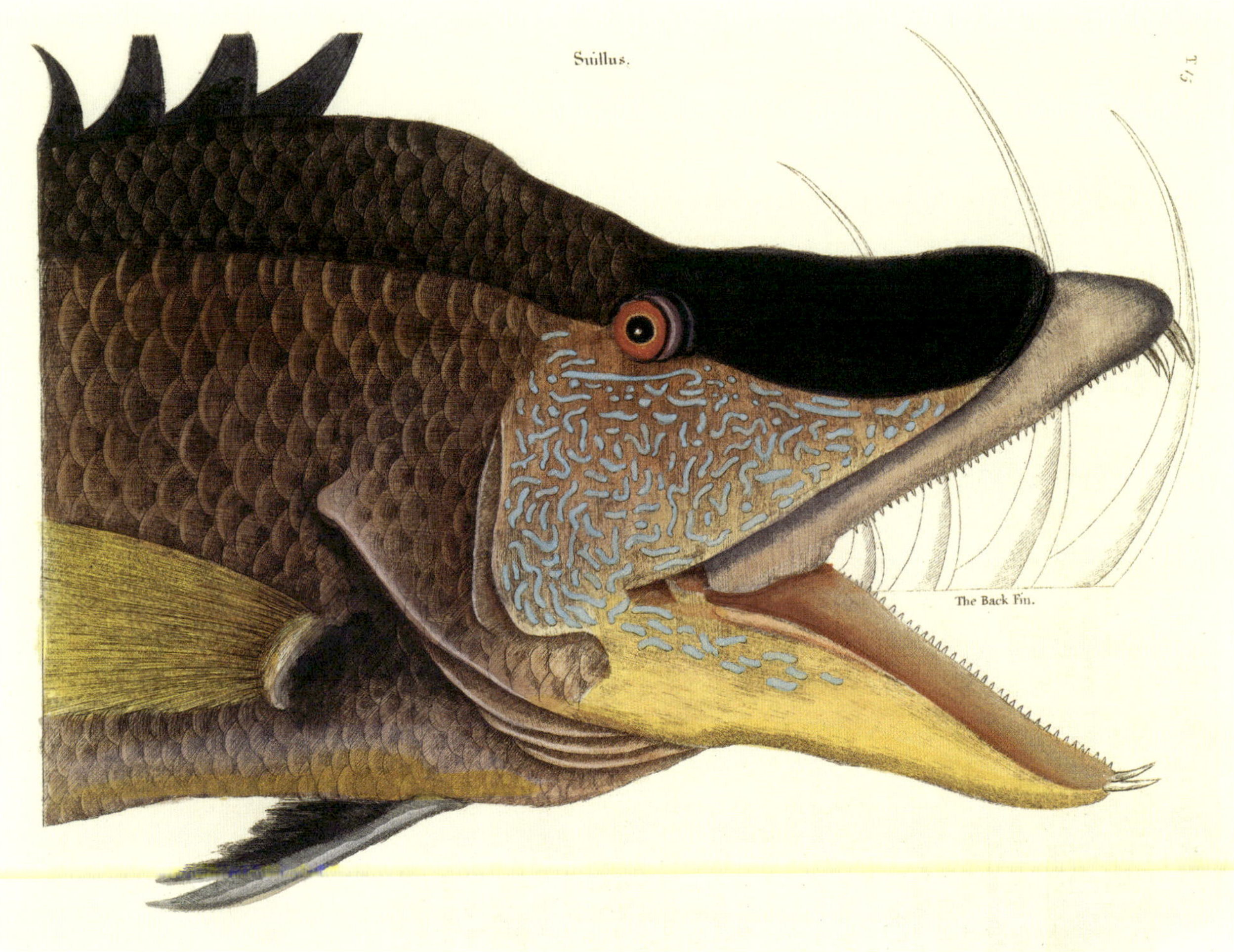

(IV)

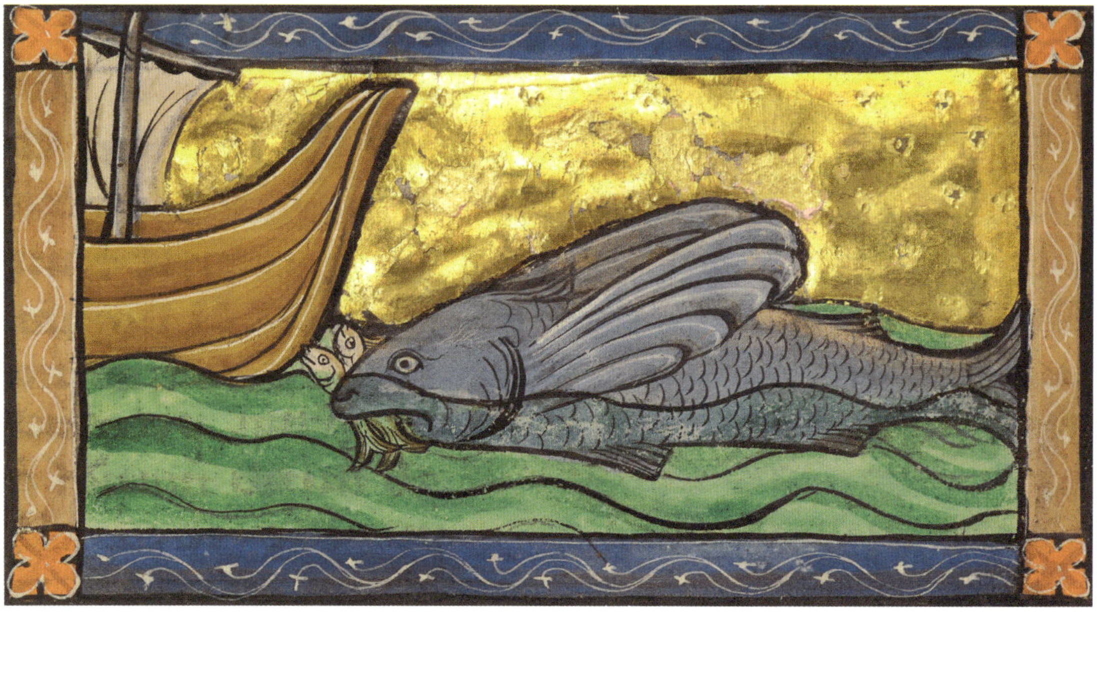

'Scientific studies ... show that if you close a place to fishing, the number of species increases twenty per cent, the average size of a fish increases by a third, and the total weight of fish per hectare increases almost five times – in less than a decade.'

ENRIC SALA, QUOTED IN 'CAN WE EAT OUR FISH AND PROTECT THEM TOO?', *FORBES*, 21 FEBRUARY 2012

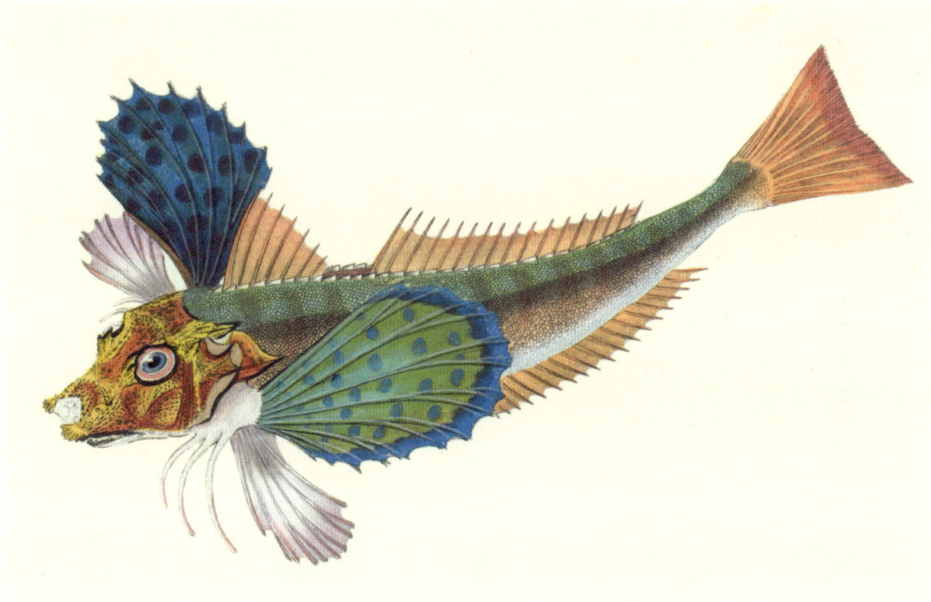

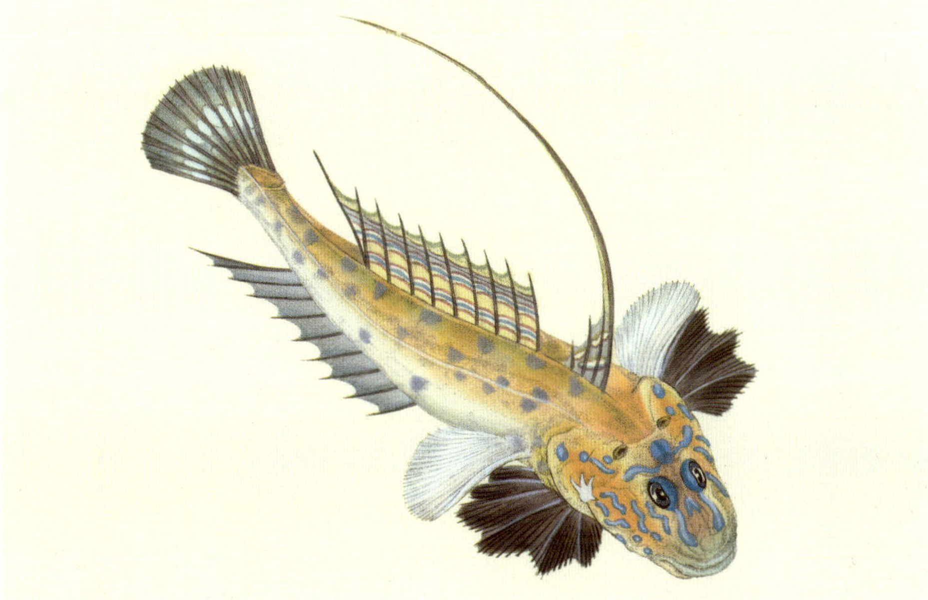

↖
◎ The detailed painting by Mark Catesby of the head and jaws of a great hogfish (*Lachnolaimus maximus*) dates from 1731. Hogfish are protogynous hermaphrodites, meaning they begin life as females and can switch from female to male when they mature.

←
This illustration is from a bestiary made in Flanders around 1270 and known as *De avibus* ('The Aviary'), though it curiously includes this fish as well as other animals that aren't birds. Medieval bestiaries contained descriptions and illustrations – sometimes realistic and sometimes humorous – of real and mythical animals intended to convey a moral message.

→
Two illustrations from *The Natural History of British Fishes* (1802) by Edward Donovan show two species that live on gravel seabeds, burying themselves and emerging to eat: at the top is the the tub gurnard (*Chelidonichthys lucerna*); and, right, the dragonet (*Callionymus lyra*).

↓
The motif of carp in rough water originated in China, where the expression 'carp jumping over the Dragon Gate' refers to attaining success in life. This Japanese embroidered silk gift cover (*fukusa*) dating from the Meiji period (1868–1912) would likewise have celebrated the recipient's achievements.

→
To make themselves difficult for predators to eat, pufferfish swell themselves up by taking water into their stomachs, as shown in this 18th-century illustration of the white-spotted puffer (*Arothron hispidus*) by Marcus Elieser Bloch.

↘
American artist H. L. Bridwell's illustration of a pompano fish was published in 1913 in *Keramic Studio*, a magazine for potters.

(IV)

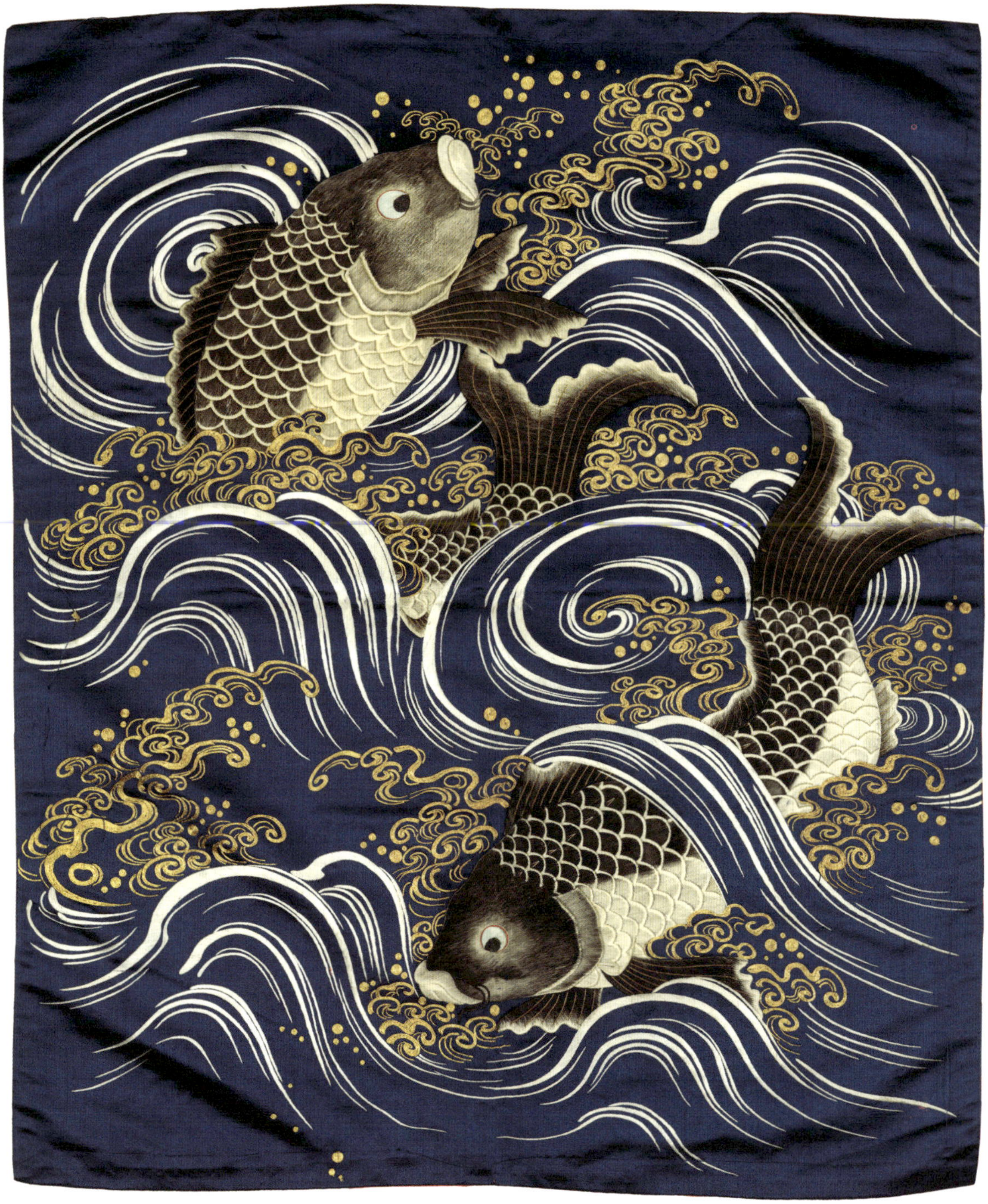

'Every year at the end of spring, yellow carp fight their way upstream. Those which reach [Dragon Gate] turn into dragons.'

CARP LEAP OVER DRAGON GATE, FROM *T'AI-P'ING KUANG-CHI* (*A GENERAL RECORD OF THE WRITINGS OF THE T'AI-P'ING ERA*), COMPILED 978, TRANSLATED BY ANNE BIRRELL

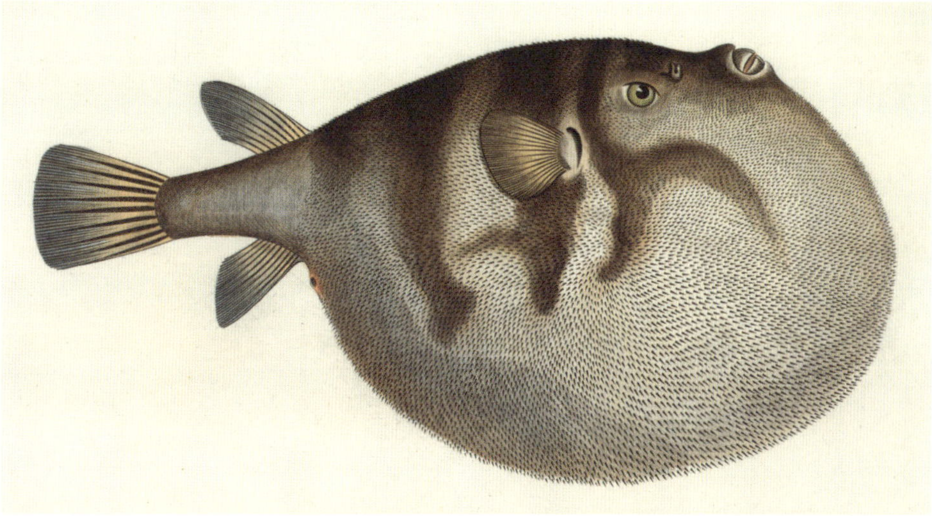

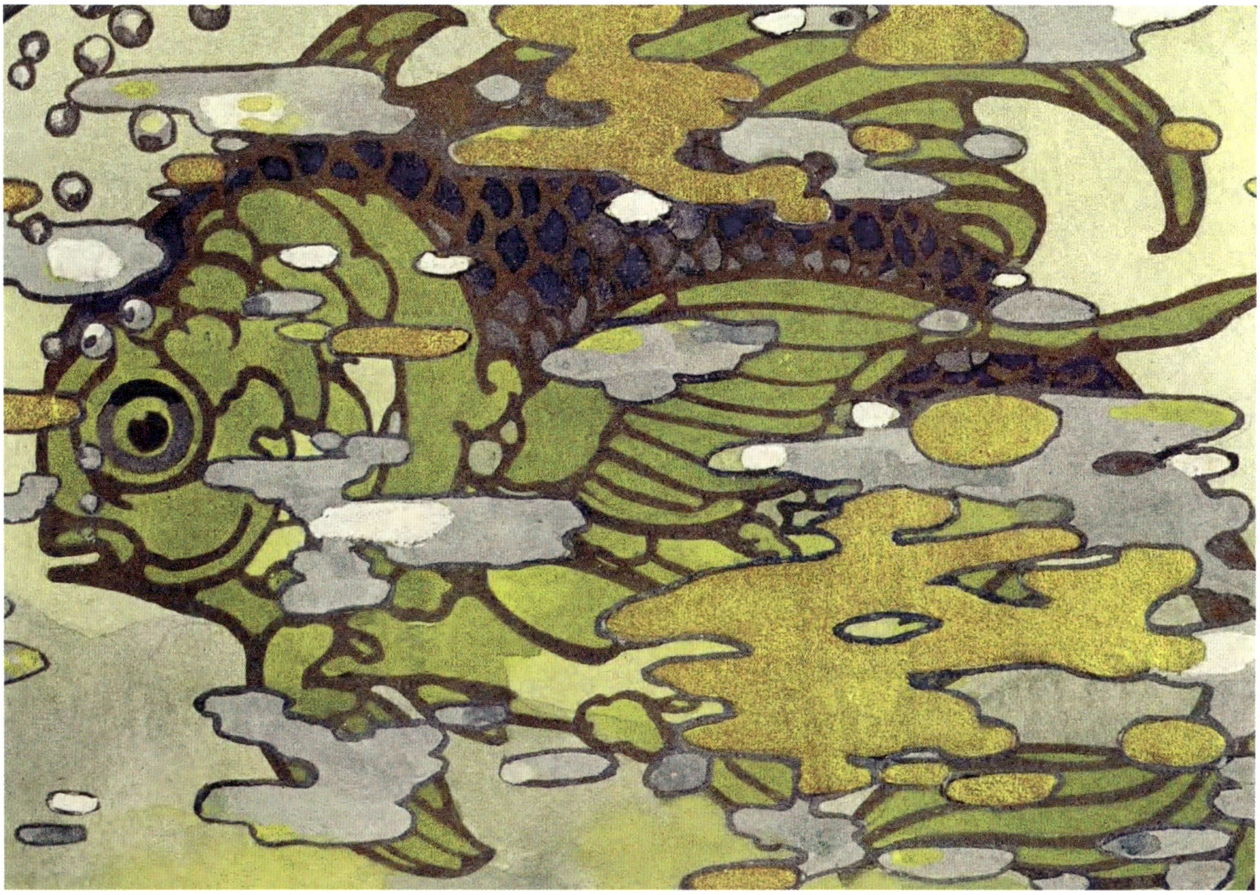

↓
In *School VI: Barracuda under Skipjack Tuna* (1978), British artist Michael Andrews shows the predatorial barracuda (*Sphyraena*) all facing to the left, the tuna to the right. Andrews used the motif of shoals of fish several times in his work as an analogy for human behaviour within complex social groups.

→
The top plate from Louis Renard's 1719 book *Poissons, Ecrevisses et Crabes* ('Fishes, Crayfishes and Crabs') shows a heavily embellished crayfish and an eel. The bottom plate illustrates a garfish (top) and two unknown species that may owe more to Renard's imagination than to reality.

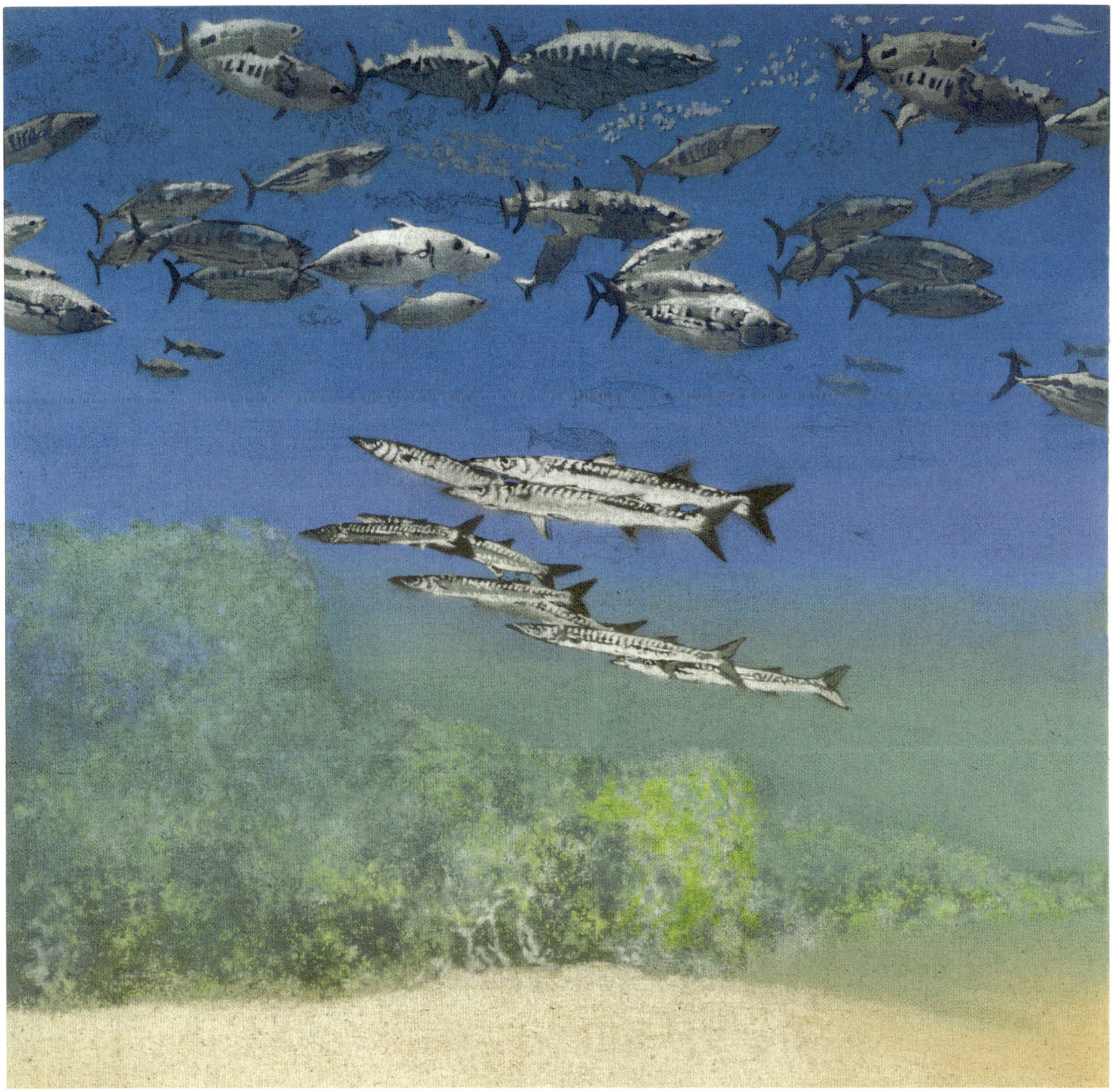

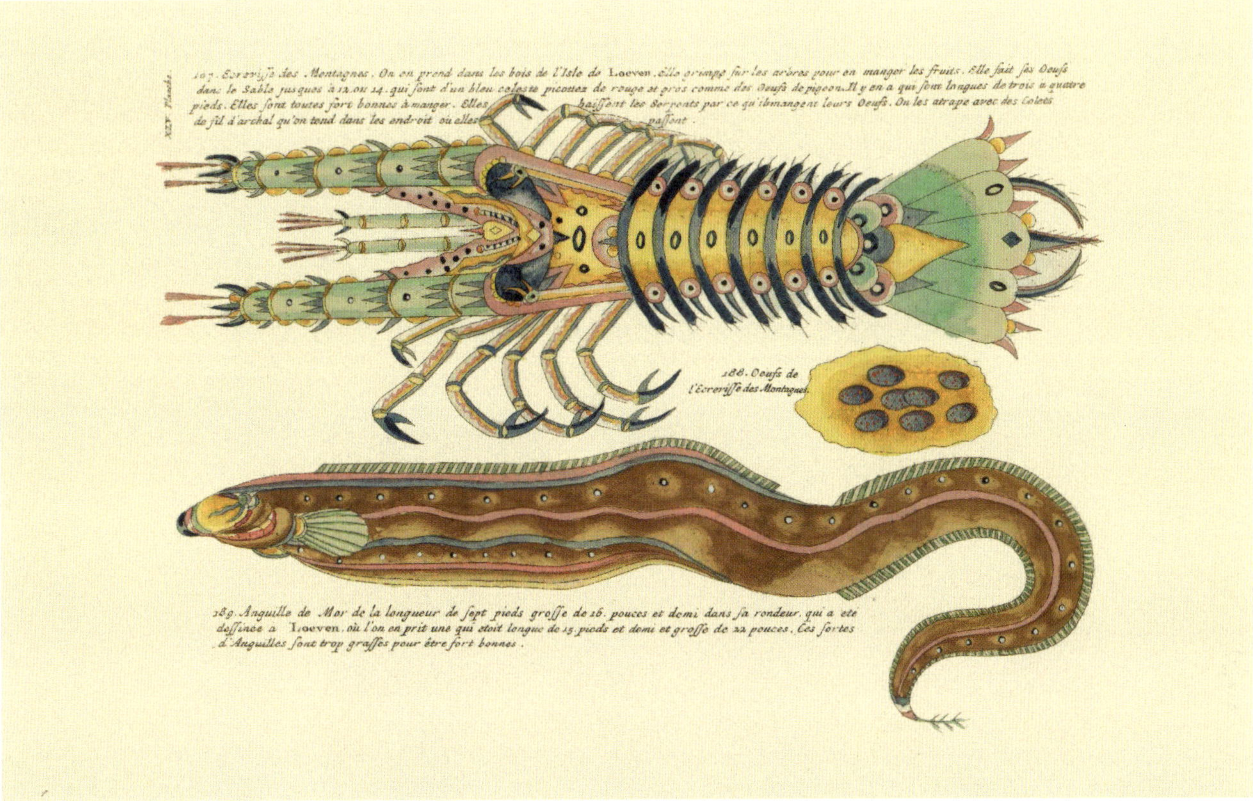
XLV. Planche.
167. Ecrevisse des Montagnes. On en prend dans les bois de l'Isle de Loeven. elle grimpe sur les arbres pour en manger les fruits. Elle fait ses Oeufs dans le Sable jusques à 12. ou 14. qui sont d'un bleu celeste picottez de rouge et gros comme des Oeufs de pigeon. Il y en a qui sont longues de trois a quatre pieds. Elles sont toutes fort bonnes à manger. Elles haissent les Serpents par ce qu'ils mangent leurs Oeufs. On les atrape avec des Colets de fil d'archal qu'on tend dans les endroit où elles passent.
168. Oeufs de l'Ecrevisse des Montagnes.
169. Anguille de Mer de la longueur de sept pieds grosse de 16. pouces et demi dans sa rondeur. qui a été dessinée a Loeven. où l'on en prit une qui etoit longue de 15. pieds et demi et grosse de 22 pouces. Ces sortes d'Anguilles sont trop grasses pour être fort bonnes.

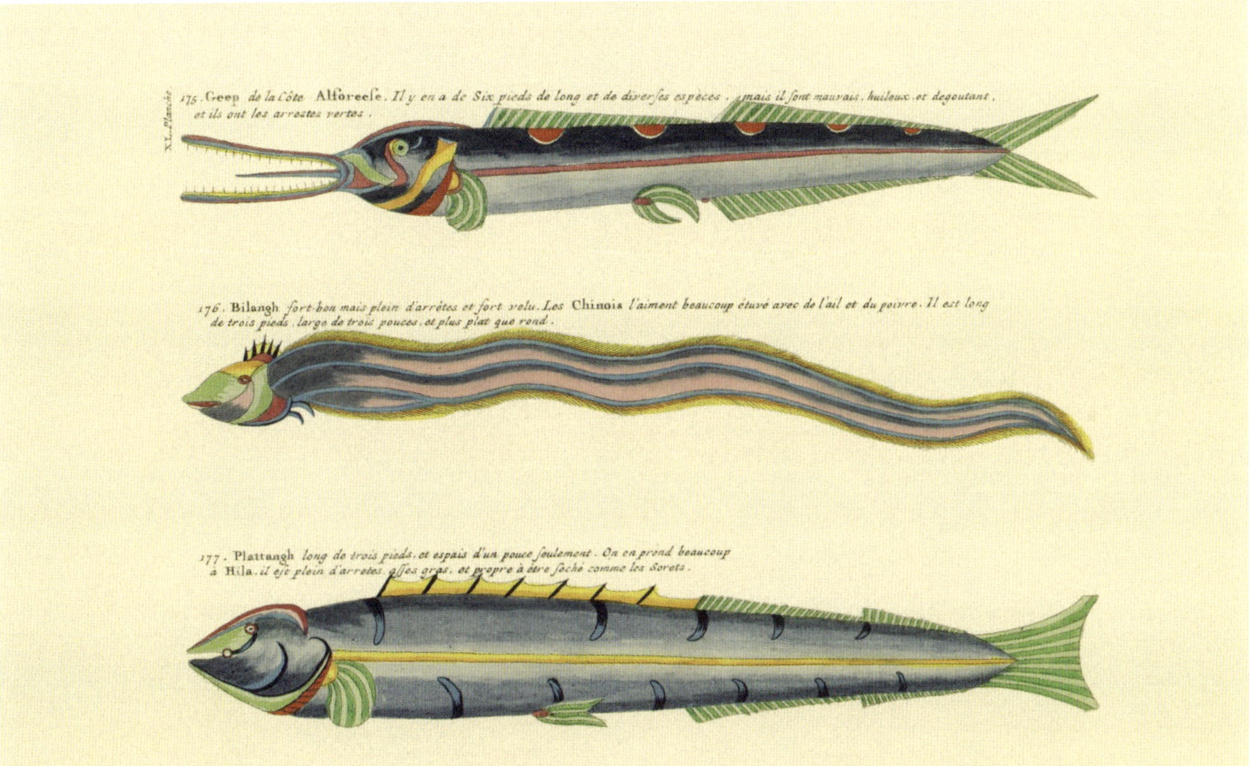
XL. Planche.
175. Geep de la Côte Alforeese. Il y en a de Six pieds de long et de diverses espèces, mais il sont mauvais, huileux et degoutant, et ils ont les arrestes vertes.
176. Bilangh fort-bon mais plein d'arrêtes et fort velu. Les Chinois l'aiment beaucoup étuvé avec de l'ail et du poivre. Il est long de trois pieds, large de trois pouces, et plus plat que rond.
177. Plattangh long de trois pieds, et espais d'un pouce seulement. On en prend beaucoup à Hila. il est plein d'arretes, asses gras, et propre à être seché comme les Sorets.

'A host of young sea-colts are ... so helpless that many of them – sometimes the entire brood – fall a prey to other fishes. They are, however, provided with a means of protection by their resemblance to plants.'

C. F. HOLDER, 'FISHES AND THEIR YOUNG', *ST. NICHOLAS*, 1886

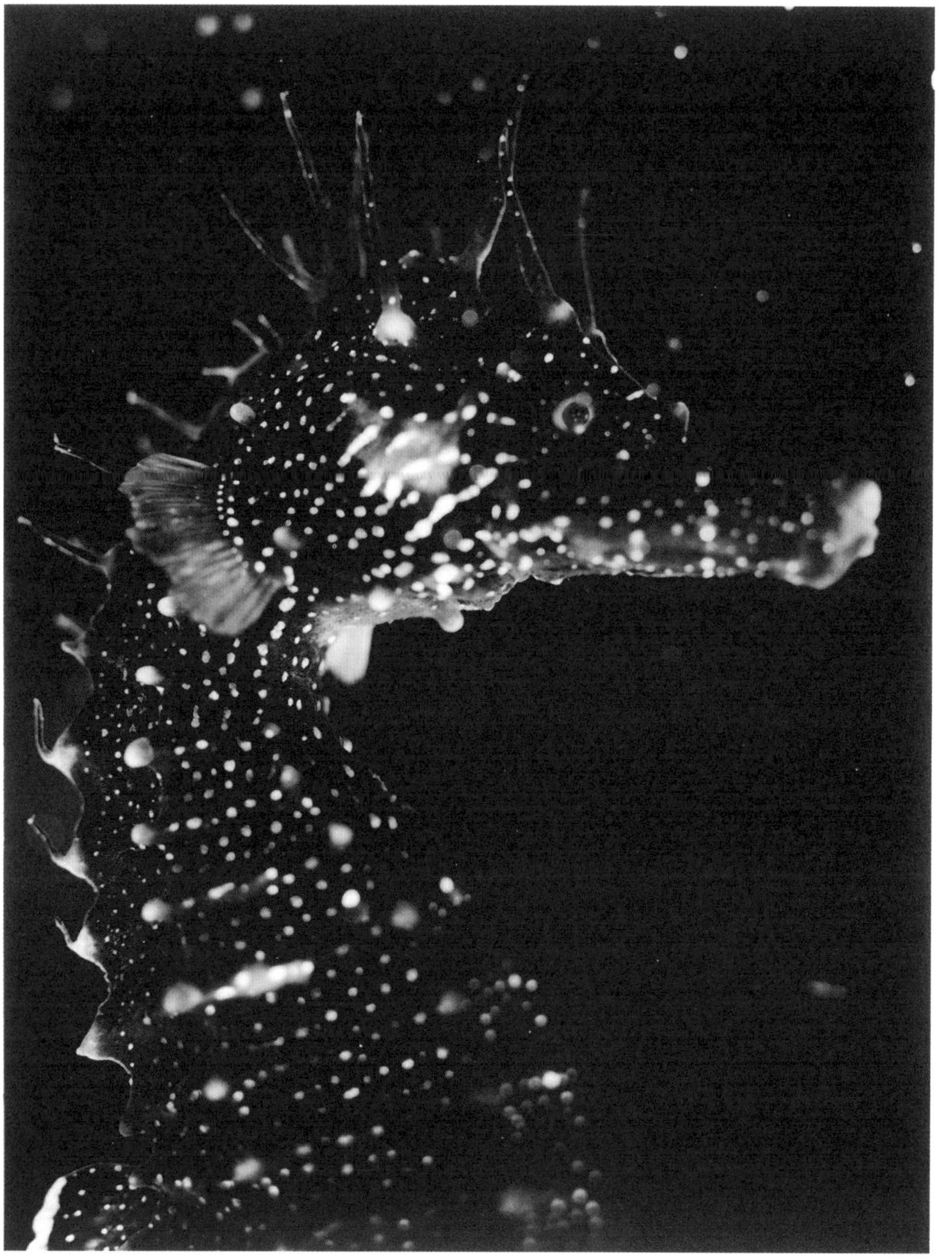

←
The long-snouted, or spiny, seahorse (*Hippocampus guttulatus*) is the subject of French filmmaker and photographer Jean Painlevé's *Buste d'Hippocampe* ('Seahorse Bust', 1933). It inhabits the northeast Atlantic Ocean and the Mediterranean and is an indicator species, meaning that its presence or absence signals a change in conditions in a particular environment.

↗
The leaf-like appendages of the leafy seadragon (*Phycodurus eques*) provide the perfect camouflage as it feeds in seagrass fields among the rocky reefs of South and Western Australia. Instead of having scales, the seadragon is covered in bony plates for protection.

→
Aquarium (2008) captures a sense of the vastness of the natural world and the fascination that it holds.

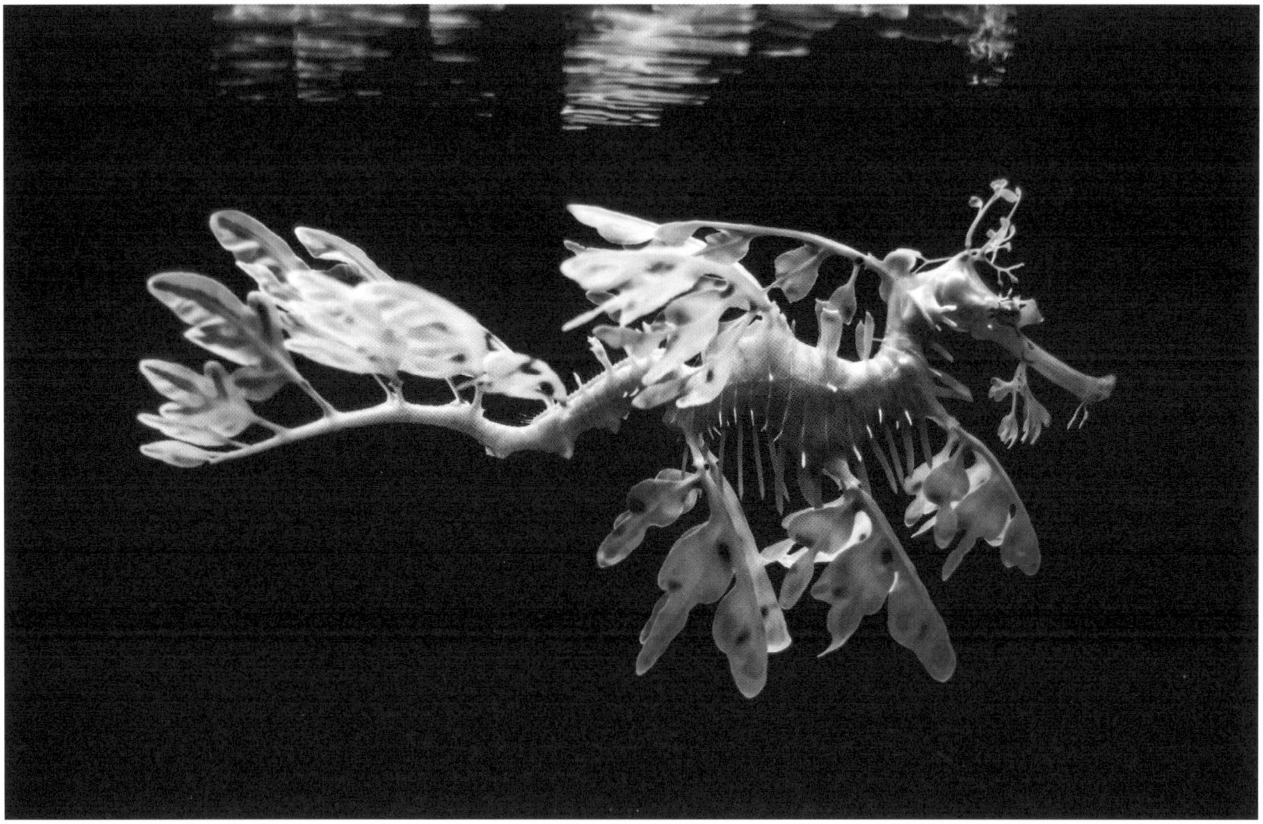

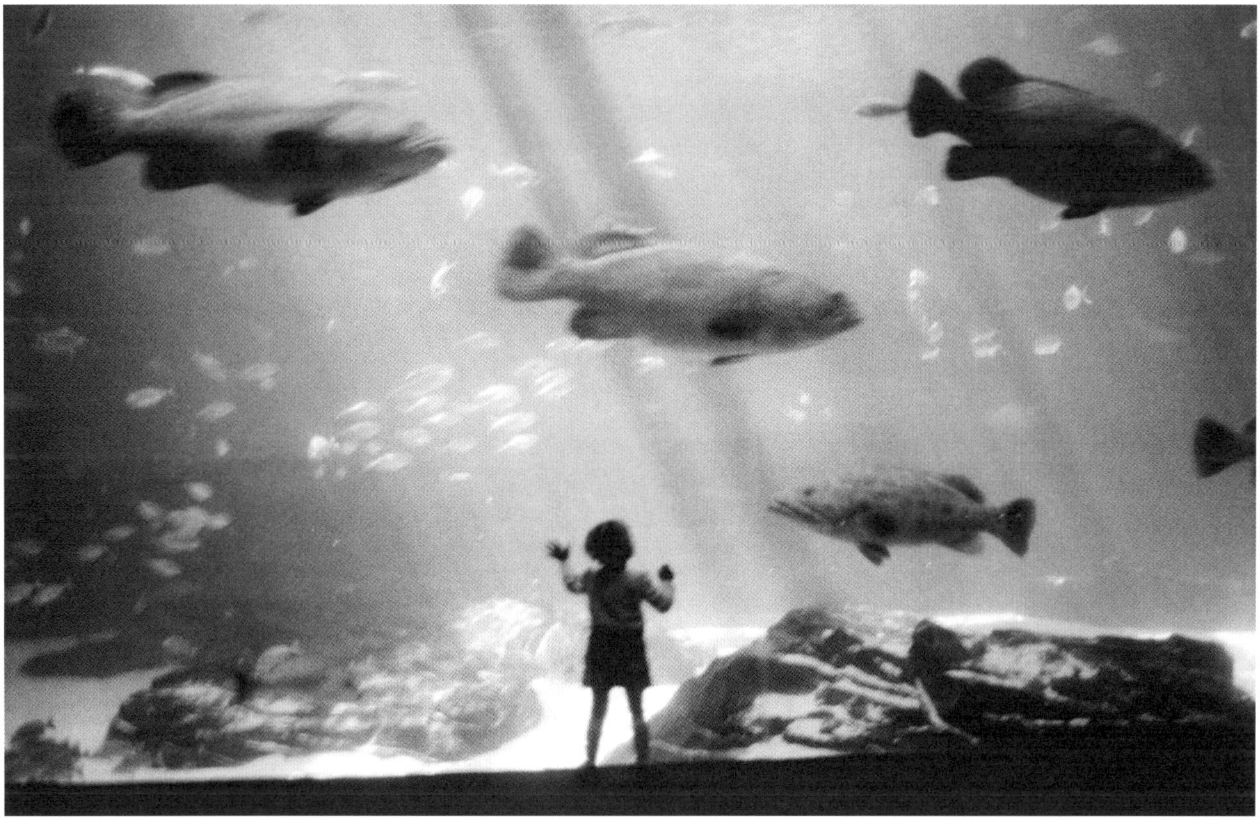

KEEPING THE BEAUTIFUL SEA ALIVE

Helen Scales

A moment in the ocean that's lodged in my mind is a close encounter I had with a sea turtle. I was scuba diving in Australia, visiting a shipwreck that, after almost a century of lying on the seabed and being protected from fishing, had become encrusted in vibrant sea life. Sea snakes were winding their way through pink and purple corals. Silvery fish were flitting in agitated clouds. I hovered above the ship's deck, totally immersed in the scene, when I felt a gentle tug on one of my dive fins: a sign my dive buddy wanted to get my attention. He gestured for me to turn around, and only then I noticed an enormous sea turtle right next to me, so close I could have reached out and touched it. I watched in wonderment as this huge, old animal gracefully passed by with sweeps of their long fore flippers, flying not swimming.

People have always been drawn to the sea and its spellbinding inhabitants and, since time immemorial, have filled in the gaps of knowledge with stories, myths and legends of fabulous sea beasts. As French oceanographer and film-maker Jacques-Yves Cousteau (1910–1997) put it in *Life and Death in the Coral Sea* (1971), 'The sea, once it casts its spell, holds one in its net of wonder forever.'[1] Now, with ever-advancing technologies that look deeper and further from shore, the ocean is giving up more of its secrets and showing that real-life beasts are far stranger and more wonderful than anything people have imagined: scintillating swimming worms; corals that live for millennia; sponges that look like wide-eyed aliens peering through the dark deep.

The closer we look, however, the more human-made problems in the ocean are also falling into sharp focus.

More than a third of all fished populations are being overfished;[2] not enough animals are left in the wild to reproduce and replace those taken. Pelagic sharks, magnificent animals that roam open seas, are especially vulnerable to being snagged on thousands of miles of longlines that bristle with deadly hooks. Three out of four pelagic shark species are endangered, among them oceanic whitetip sharks, which not so long ago were considered among the most abundant large animals on earth but are now vanishing. Many of the most damaging fisheries don't catch food for people who depend on the ocean for nutrition. Supertrawlers, longer than a football field, vie with whales and penguins to catch Antarctic krill. The krill fisheries mash down these ecologically vital animals to turn them into pet food.

At the same time that the ocean is being stripped of living species, the seawater itself is filling up with more trash and pollutants than at any point in earth's history. Fragments of microplastics are everywhere, from the deepest ocean trenches to polar sea ice.[3] Toxic, industrial chemicals have been accumulating in the ocean for decades, poisoning animals and food webs. Long-lived predators, such as orca and seals, are becoming so contaminated with artificial chemicals that they're losing their fertility and succumbing to disease.[4] Sewage and farm run-off add to the deluge of pollutants that leach into rivers and flow into the ocean, upsetting the balance of ecosystems.

Mingling with the threats of overexploitation and pollution, the climate crisis is a menacing threat to ocean life. The vast size of the ocean is warding off catastrophic climate breakdown by absorbing around 90% of the excess heat that greenhouses gases trap in the atmosphere. But all that heat sinking into the sea comes at a huge cost. In human history the ocean has never been hotter. Since the late 1980s the rate of ocean warming has more than quadrupled, rising from 0.06 to 0.27 °C (0.11 to 0.49 °F) per decade. That might not sound like much warming, but the energy required to heat so much water is truly colossal, equivalent to at least seven Hiroshima atomic bombs detonating in the sea every second.[5] Fish, crabs and other mobile animals are moving north, south or deeper underwater in search of cooler seas, thereby transforming ecosystems. Narwhals, emperor penguins and a host of other polar animals are losing their essential habitats as the earth's frozen seas melt. As well as the relentless rise in average temperatures, the shorter spikes in heat, known as ocean heat waves, have become more intense and frequent, devastating coral reefs and kelp forests and killing off swathes of sea life.

The ocean also absorbs huge amounts of carbon dioxide and, in the process, seawater is becoming more acidic. This spells disaster for animals such as the minuscule swimming snails, known as sea butterflies, which have delicate, transparent shells that could soon begin dissolving in the corrosive seas. Sea butterflies

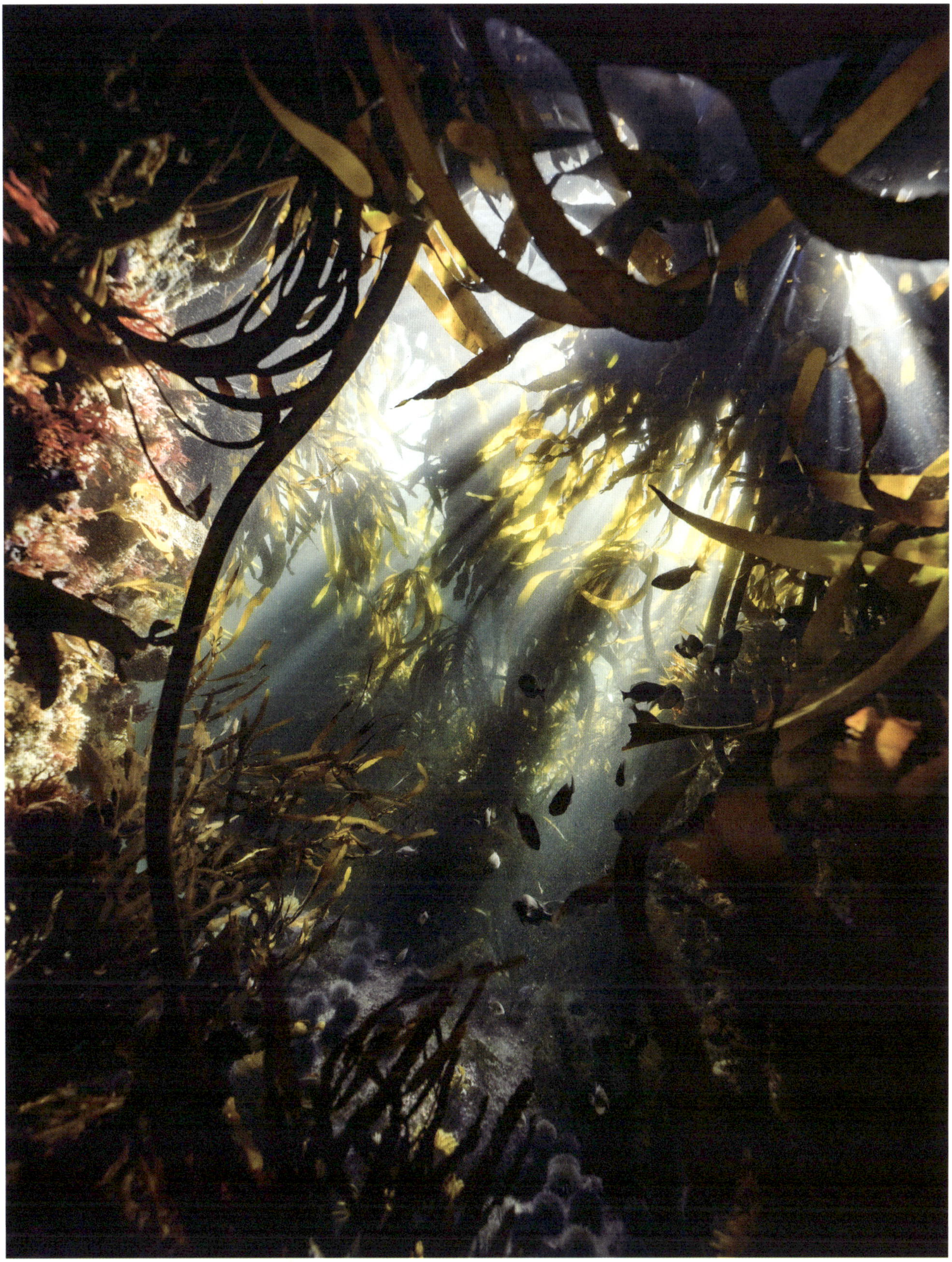

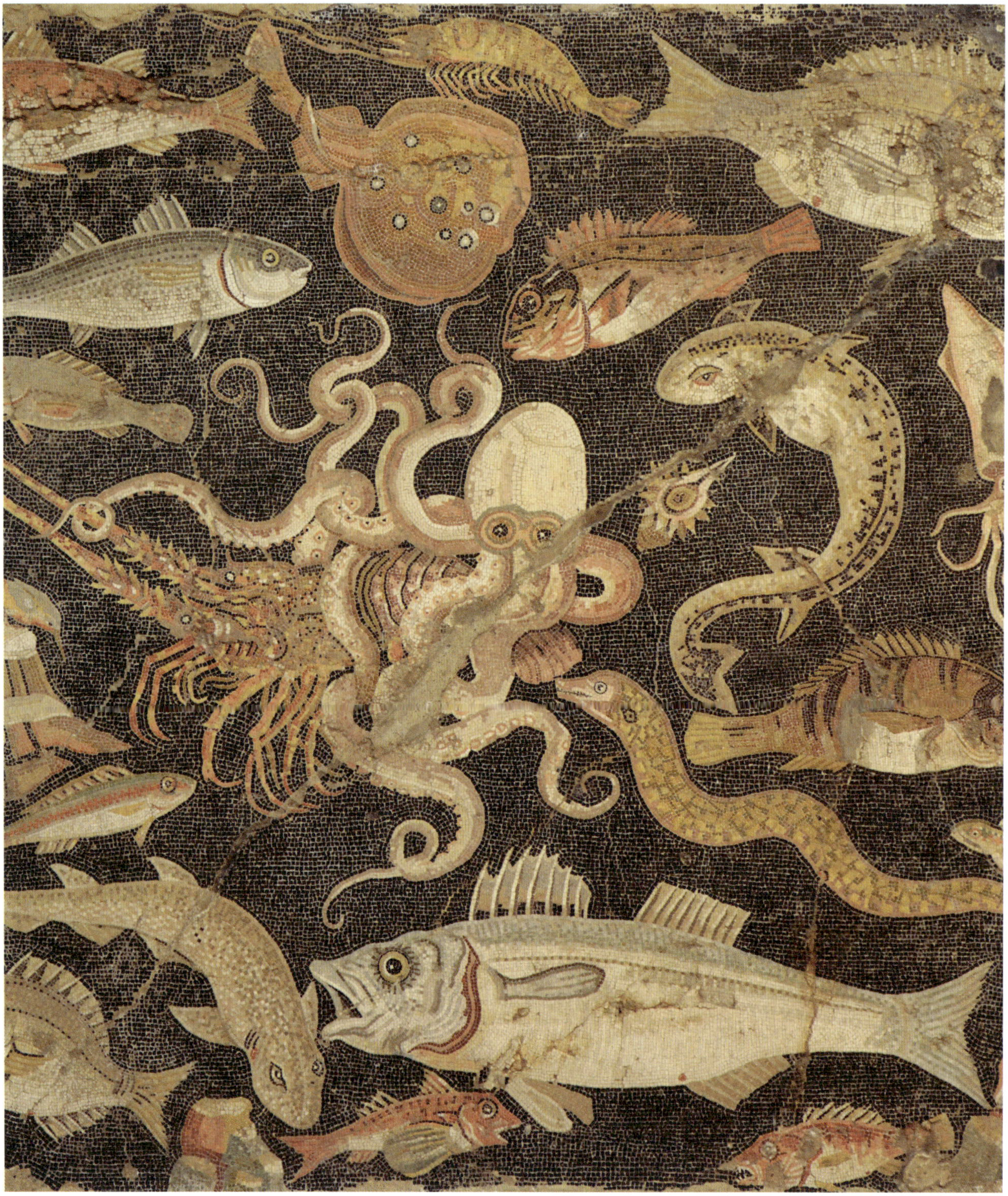

↲
Helen Walne captures sunlight penetrating deep into a kelp forest off the South African coast in *Golden Curl* (2024). As well as providing a home for a diverse array of seaweeds and creatures, kelp forests are vital carbon sinks.

↑
In ancient Roman times the Mediterranean teemed with fish species, portrayed in this mosaic (*c.* 100 BCE), from the House of the Geometric Mosaics, Pompeii. Biodiversity in the Mediterranean is now threatened due to overfishing.

aren't just beautiful creatures flitting around on tiny wings, they're a cornerstone in many oceanic food webs.

All these threats act together, painting a bleak outlook for the future of the ocean. However, ocean life has immense power to recover and rebuild, given a chance. Elephant seals and sea otters are among the marine species once hunted to the brink of extinction but were protected before it was too late and made a comeback. Fishing can be carried out more sustainably, by targeting fast-growing species and by keeping catch quotas within ecological limits.

Protecting areas of sea from fishing and other damaging activities is another important aspect of restoring health and abundance to the ocean. Even when relatively small areas of sea are declared off limits, the results can be astonishing. Within marine protected areas, delicate seabed habitats such as coral reefs, seagrasses and seaweed gardens remain intact and provide food and shelter for animals that grow bigger and older; their offspring then drift and swim away, replenishing the surrounding seas.

The ocean faces global, systemic problems and to overcome them requires solutions on a similarly large scale. But that doesn't mean that individuals are powerless. Anyone can be part of the growing collective movement that's pushing for a better future for the ocean.

You can make different choices. For example, if you eat seafood, you can choose to eat more resilient species, caught using less harmful fishing gear. You can support fisheries that are working hard to produce healthier, more sustainable seafood and be part of building a better industry.

There's also a direct link between the ocean and what you throw away, especially the things you flush down the toilet. Many water companies in high income nations such as the United Kingdom dump untreated sewage in rivers, which ultimately ends up in the ocean. You can support campaigns pushing against these environmental crimes, and you can flush only things that are fully biodegradable – no wet wipes, ear buds or dental floss.

Cutting down on single-use plastics is an essential way to limit plastic pollution in the ocean and help to protect the health and well-being of marine species. The amount of plastic manufactured each year is skyrocketing.[6] More than half of the plastic ever made has been produced since 2002, and in the past two decades annual production has more than doubled, now exceeding 400 million tonnes per year. Amid all this, the rising tide of pointless plastics can feel overwhelming. Nevertheless, consumer pressure can make change happen. In France, for instance, pressure from shoppers led to regulations in 2022 banning supermarkets from wrapping most fresh fruit and vegetables in plastic.

The fight for the ocean must also be a fight against the climate crisis. Each one of us can decide to take responsibility for reducing our carbon emissions. You can take a train instead of flying, purchase a smaller car, move your money to a bank that won't invest in fossil fuel corporations.

You can join forces with like-minded people to fight together for change on a bigger scale. Some ocean communities are fighting for their neighbourhood seas. A decade of campaigning by Chumash people, whose ancestors have lived in the region of the Central Californian coast for more than 10,000 years, led to the 2024 designation of the Chumash Heritage National Marine Sanctuary to protect 11,766 square kilometres (4,543 square miles) of wildlife-rich ocean.

Mining corporations are in a frenzy to open the world's first deep-sea mines. They want to exploit seabed rocks, miles beneath the sea surface, that are full of valuable minerals but are also encrusted in priceless living things. People around the world – scientists, conservationists, youth-led organizations, Indigenous peoples, governments, corporations – are collectively pushing back against the race to recklessly exploit the deep sea. In 2025, forty countries were in support of a proposed temporary prohibition on deep-sea mining, including France, Germany, the United Kingdom, Canada, Mexico and Brazil.[7] Deep-sea scientists forecast that the industry would cause widespread and irreversible ecological damage, wiping out species that live nowhere else. Damage to the deep wouldn't stay in the deep. Millions of organisms dive up and down through the ocean, connecting the surface and the deep, which means deep-sea mining could contaminate far and wide, including fish that end up in the human food chain. If it were to go ahead, deep-sea mining would bring industrial-scale destruction to some of the most unspoiled parts of the planet.

My hope for the future of the ocean is strengthened knowing that so many people feel the same way as I do. More people than ever are outraged at how the oceans are being mistreated and they're transforming their anger into action. More people than ever are curious about sea life and find joy wherever they can, from the breath-taking glimpse of a leaping whale to the simple elegance of a seashell. The most powerful thing we can all do is to hold tightly to those feelings of awe and wonder, and keep the beautiful sea in our minds.

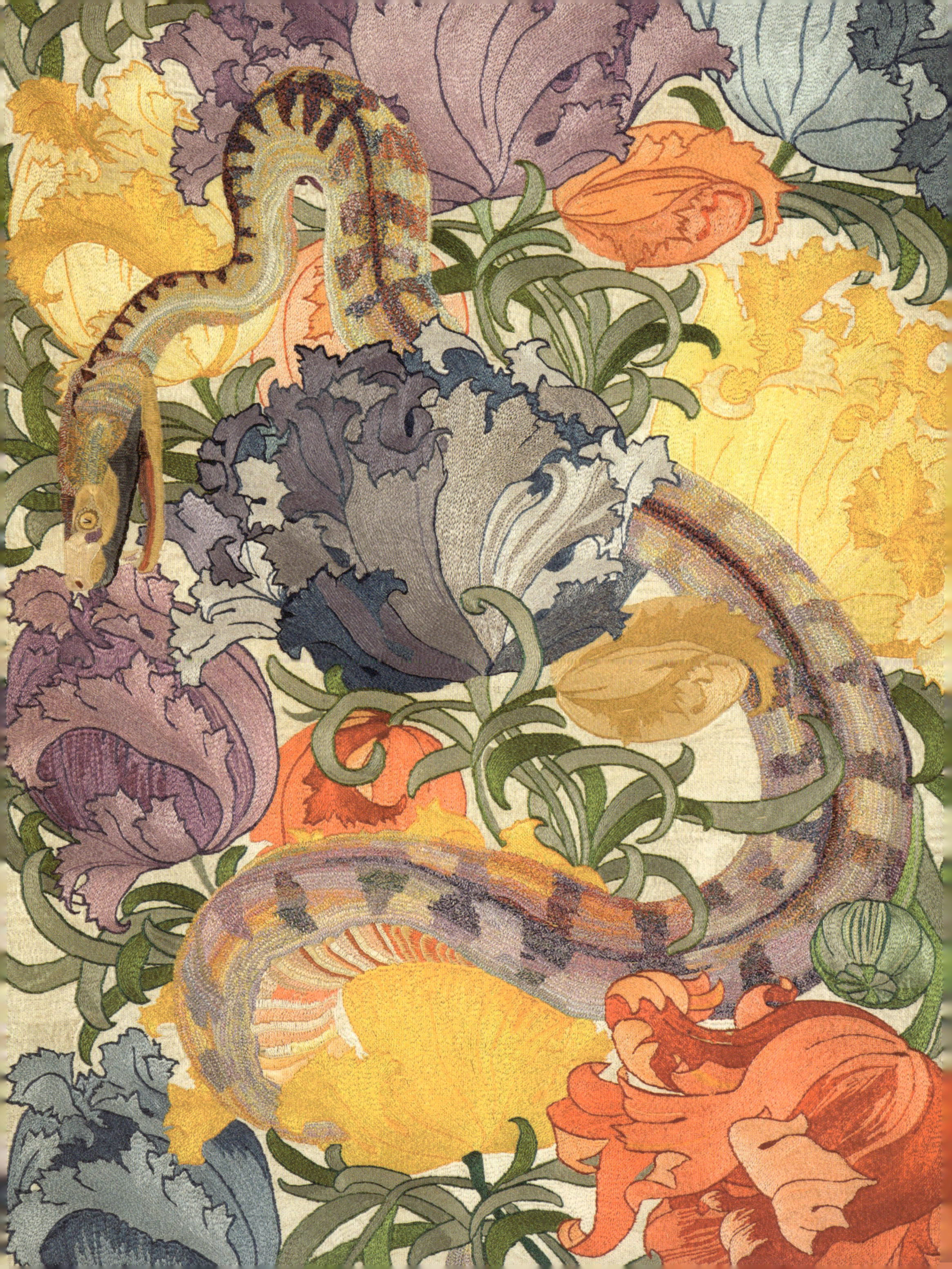

(V)

AMPHIBIANS & REPTILES

AMPHIBIANS & REPTILES are cold-blooded vertebrates. Amphibians, such as frogs and toads, have soft skin that needs to be kept moist. They spend the larval stage in water and move onto land when they are mature, returning to water to breed. Unlike amphibians, reptiles have dry scales that cover some or all of their body. They lay eggs on land, though some species give birth to live young. Reptiles first evolved more than 300 million years ago; today, they include tortoises and turtles, iguanas, lizards and snakes.

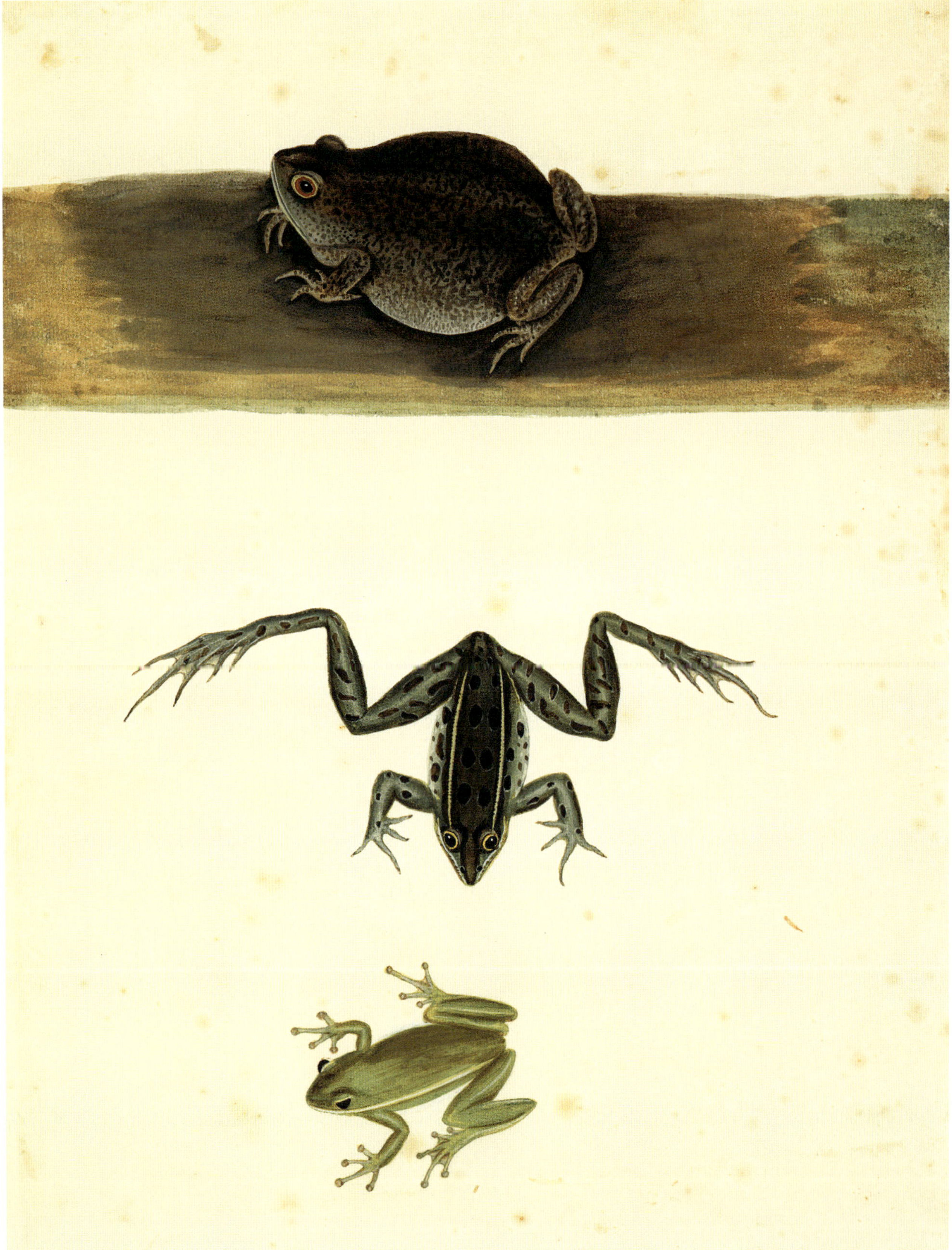

(v)

↲
Though the embroiderer of this piece, *Snake Amongst Flowers* (*c.* 1895), is unknown, it follows a design by architect and pattern maker Charles Francis Annesley Voysey.

←
Born in Suffolk, United Kingdom, Mark Catesby spent years on the East Coast of America, creating watercolours such as this one (*c.* 1722–26). This page depicts, from top to bottom, the Southern toad (*Anaxyrus terrestris*), Southern leopard frog (*Lithobates sphenocephalus*) and American green treefrog (*Hyla cinerea*).

→
Produced between 1575 and 1585, this Hans Hoffmann watercolour likely depicts the green frog (*Rana esculenta*), otherwise known as the edible frog. It is often the species of choice for frogs legs, which is commonly served in France, but also in Italy and Germany, where Hoffmann was born.

↓
Mink Pond, in the Adirondacks Mountains of New York state, was a frequent source of inspiration for Winslow Homer, who travelled there eighteen times between 1870 and 1910. A trip in 1891 resulted in this watercolour of a redbreast sunfish (*Lepomis auritus*) and frog, titled *Mink Pond*.

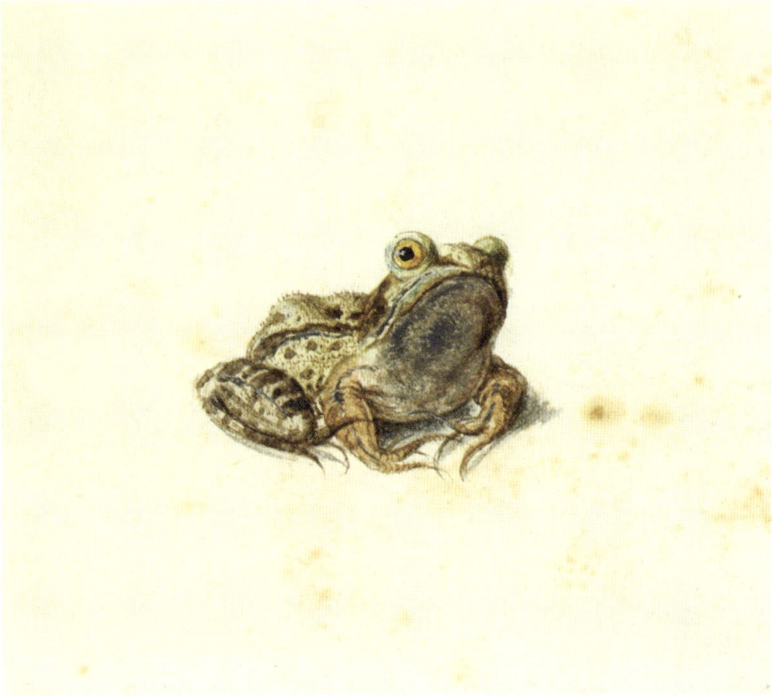

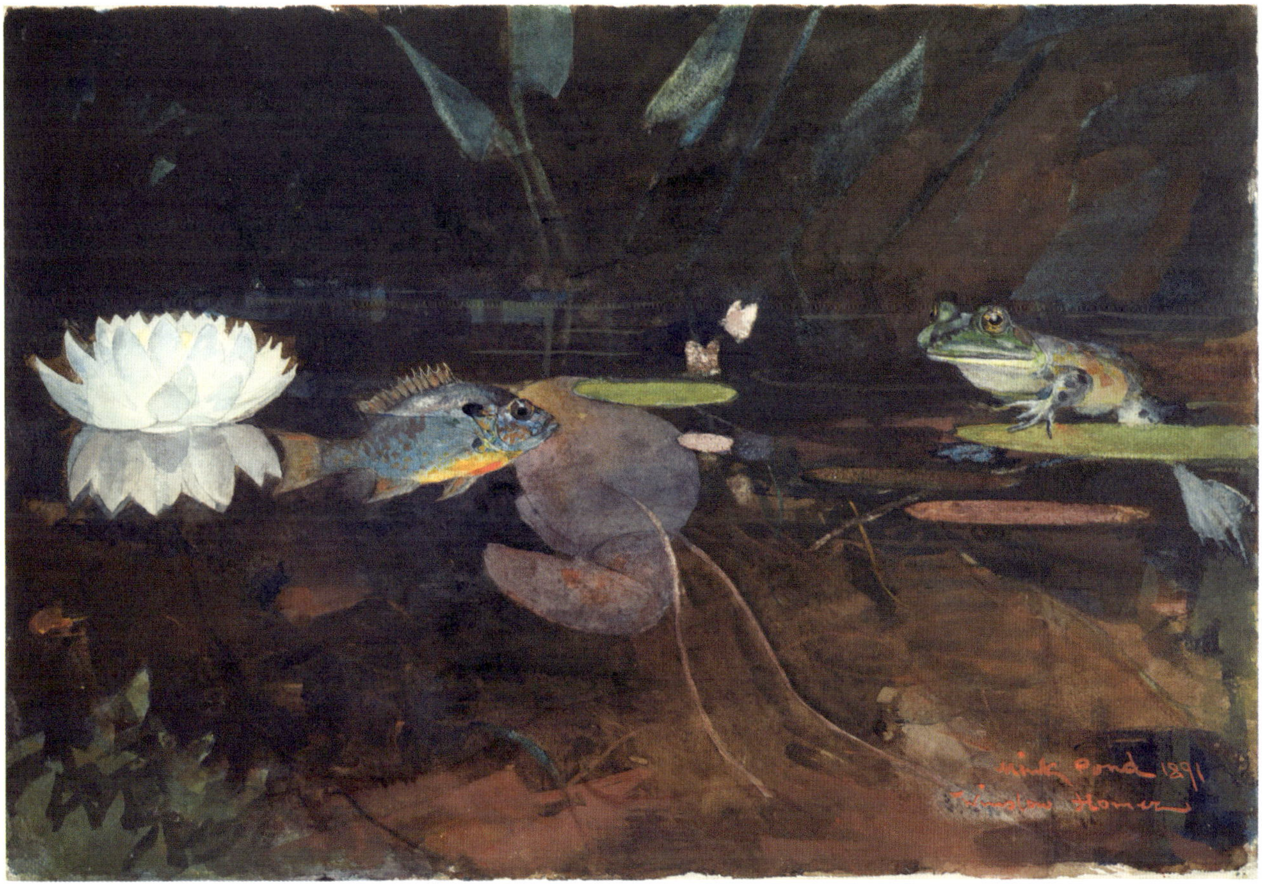

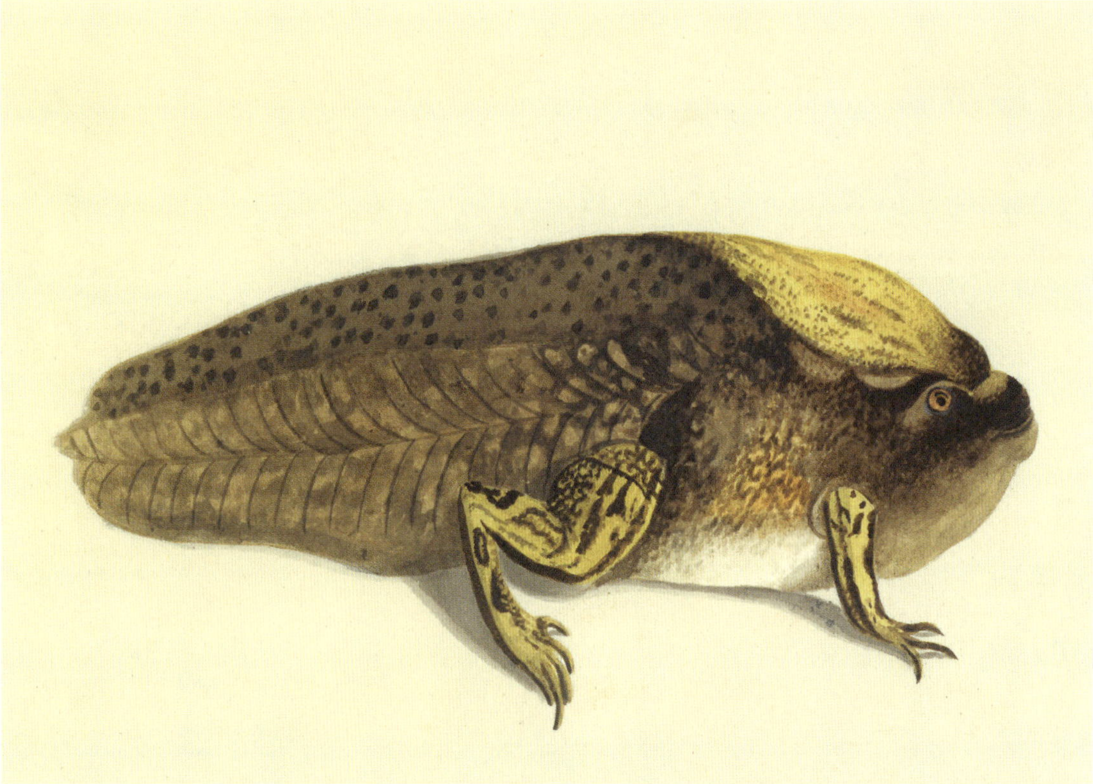

(v)

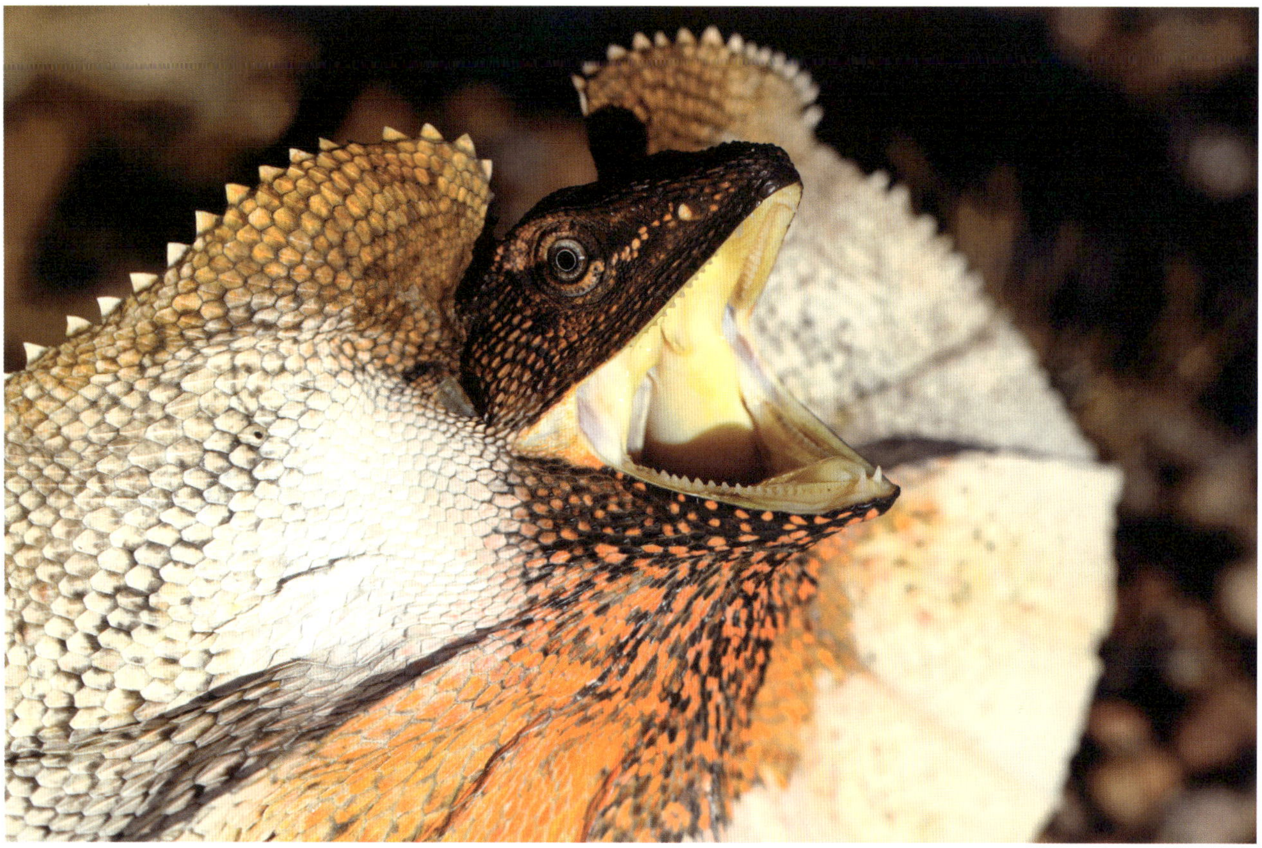

←
The South American paradoxical frog (*Pseudis paradoxa*), illustrated here by Gerrit Schouten in 1818, has the largest tadpoles in the world, reaching up to 27 centimetres (11 inches) in length. They shrink to around a third of that size as they metamorphose into adult frogs.

↙
The frill-necked lizard (*Chlamydosaurus kingii*) puffs out the skin flaps around its neck to scare off predators such as birds of prey, larger lizards, snakes, dingoes and feral cats. It can also run at speed on its hind legs to escape.

→
Contrary to the downcast expression of Taigadō Giryō's simple ink drawing, *Frog* (1830–65), in Japanese culture frogs are considered a symbol of good fortune and prosperity.

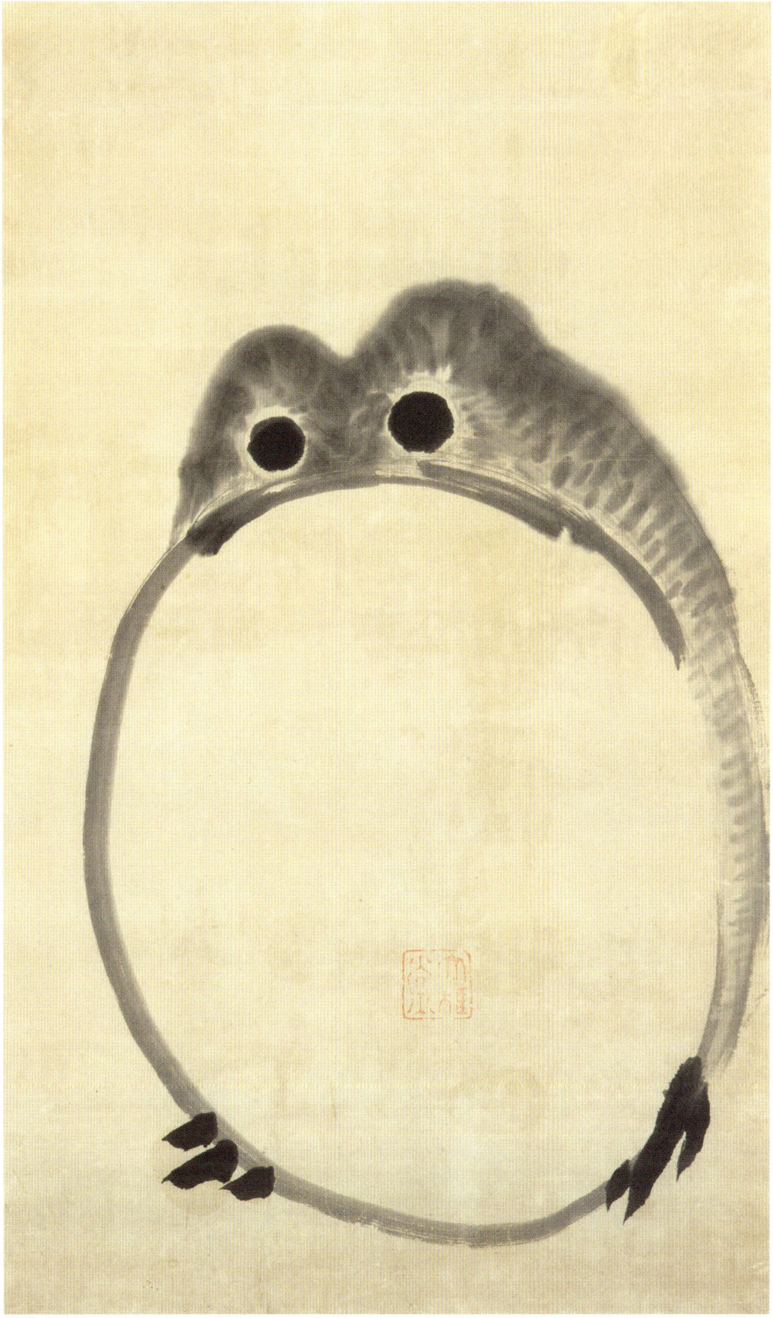

'An ancient pond!
With a sound from the water
Of the frog as it plunges in.'

MATSUO BASHŌ (1644–1694), TRANSLATED BY WILLIAM GEORGE ASTON

'Truly demonic are fantastic horned frogs of Brazil which devour other amphibians and small mammals. The largest of them do not hesitate to defy a human being .... They are six inches long or longer and as broad as long. ... All have enormous mouths, so that a mouse can be swallowed quite easily.'

THOMAS R. HENRY, *THE STRANGEST THINGS IN THE WORLD: A BOOK ABOUT EXTRAORDINARY MANIFESTATIONS OF NATURE*, 1958

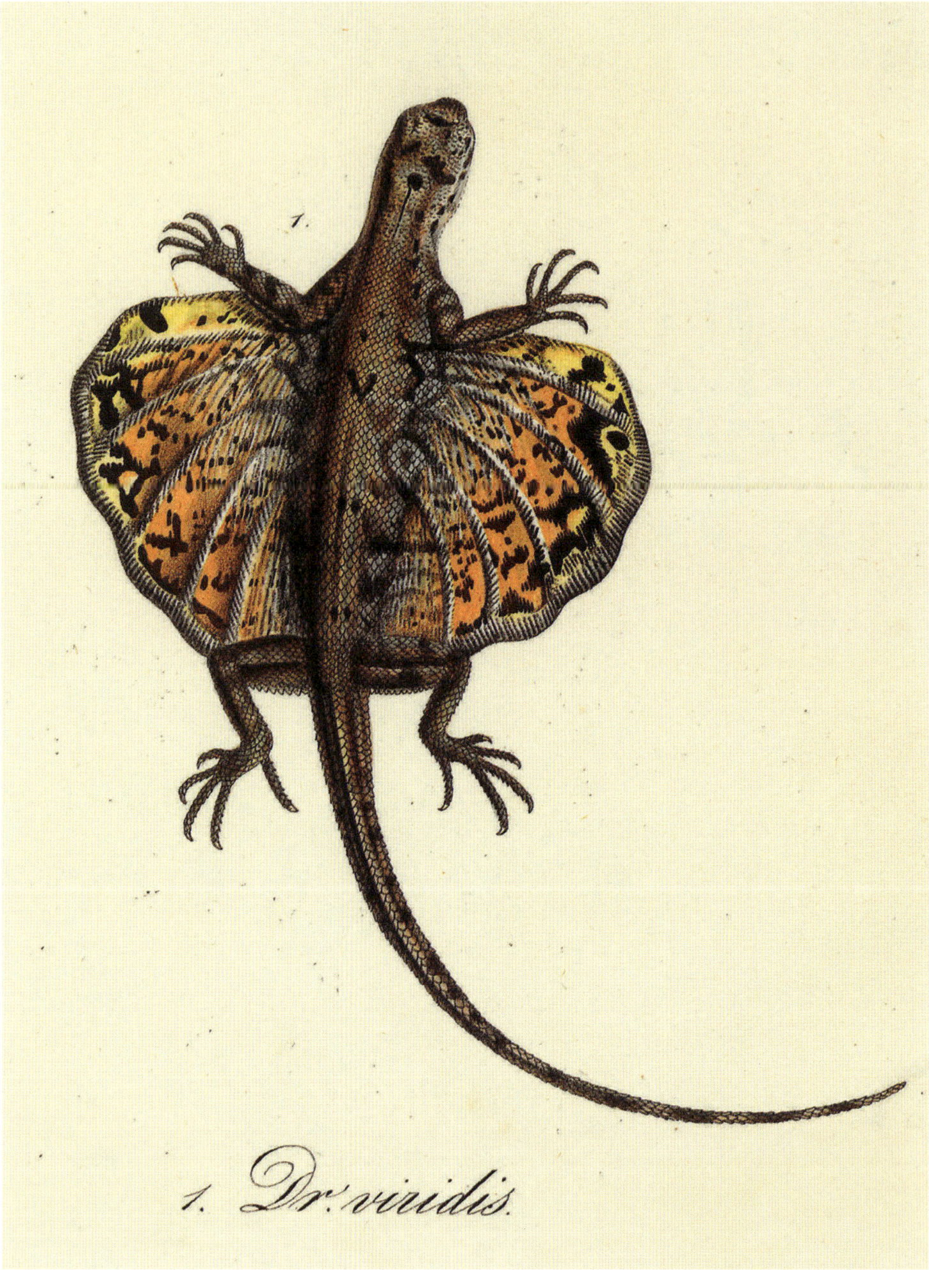

←
The flying dragon (*Draco volans*) is a native of the rainforests of Southeast Asia, where it glides through the trees to move about. After jumping off from a branch, it can glide for up to 8 metres (26 feet).

↗
Surinam horned frogs (*Ceratophrys cornuta*) from northern South America have very wide mouths, as can be seen in this illustration from *Brehms Tierleben* (*Brehm's Life of Animals*, 1890–93). They also have voracious appetites and sometimes try to eat prey larger than themselves.

→
The great crested newt (*Triturus cristatus*), seen here in an illustration from *Die Reptilien und Amphibien mitteleuropas* ('Reptiles and Amphibians of Central Europe', 1912), is threatened by habitat loss and is protected across Europe. The pattern of black spots on their bellies is unique to each individual.

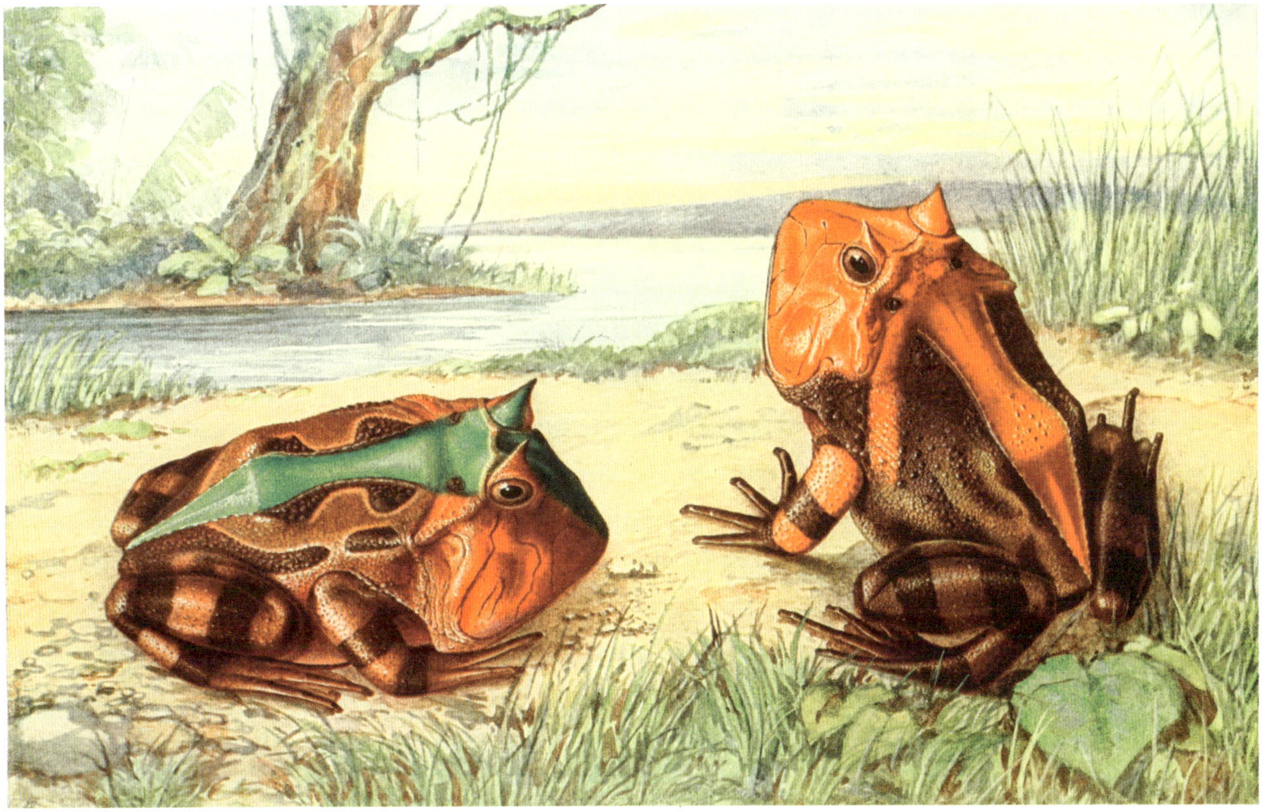

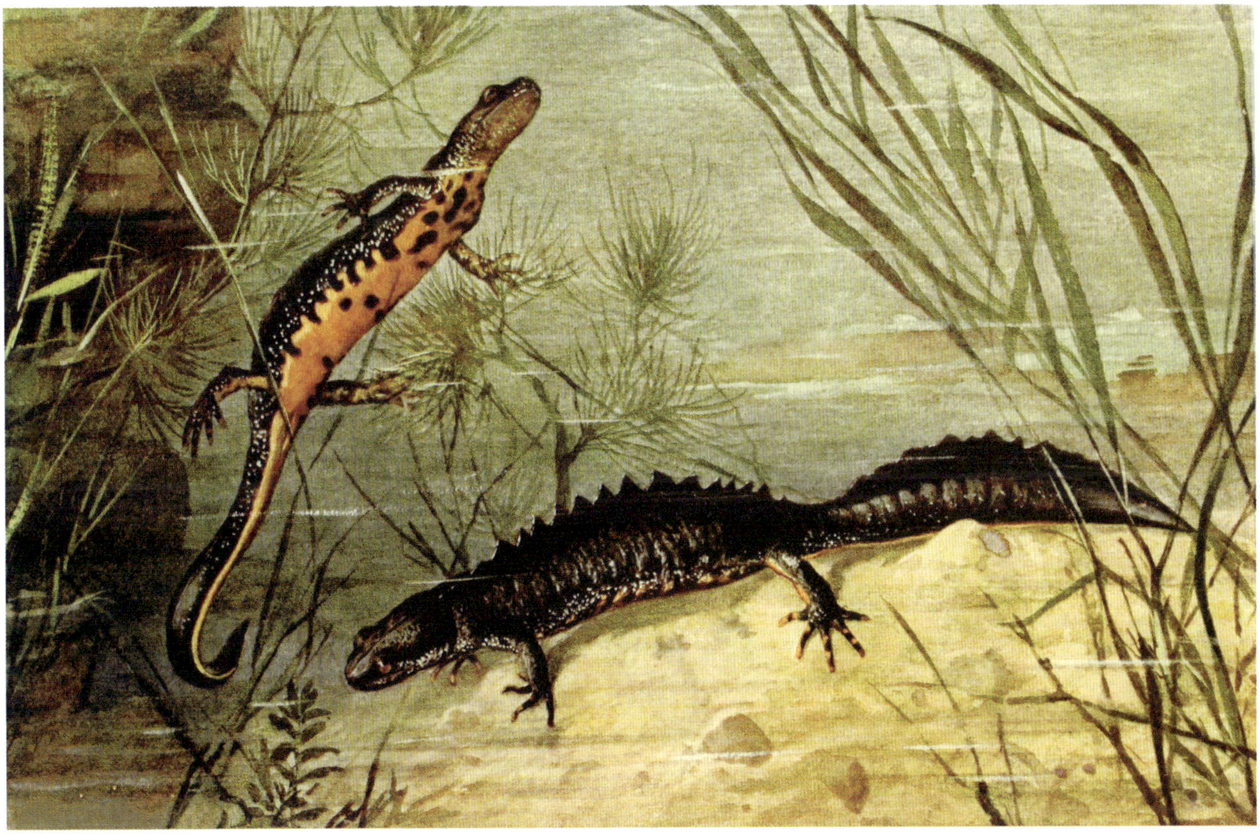

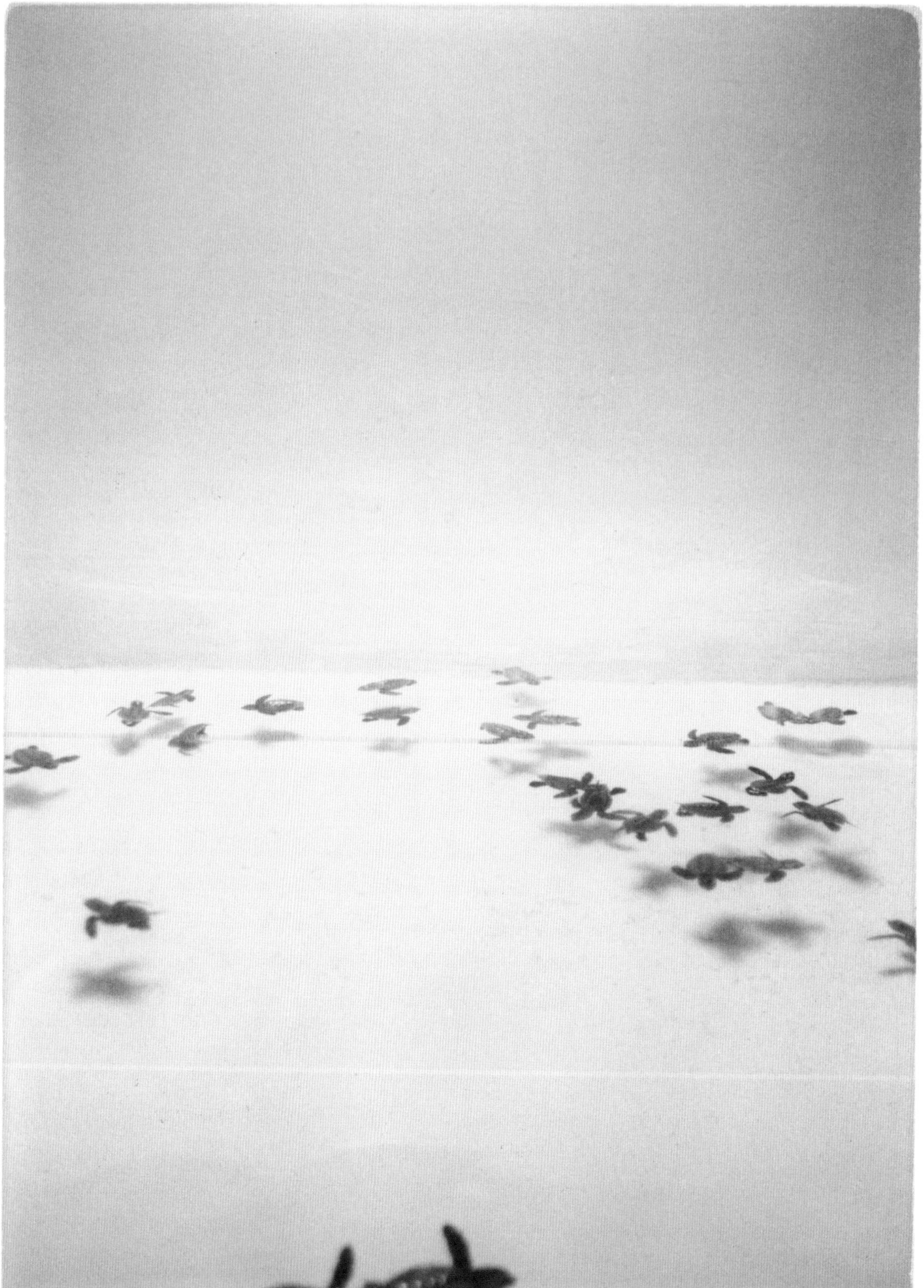

(v)

'all tangled up
in heat shimmers...
the little turtle'

KOBAYASHI ISSA, 1813, TRANSLATED
BY DAVID G. LANOUE

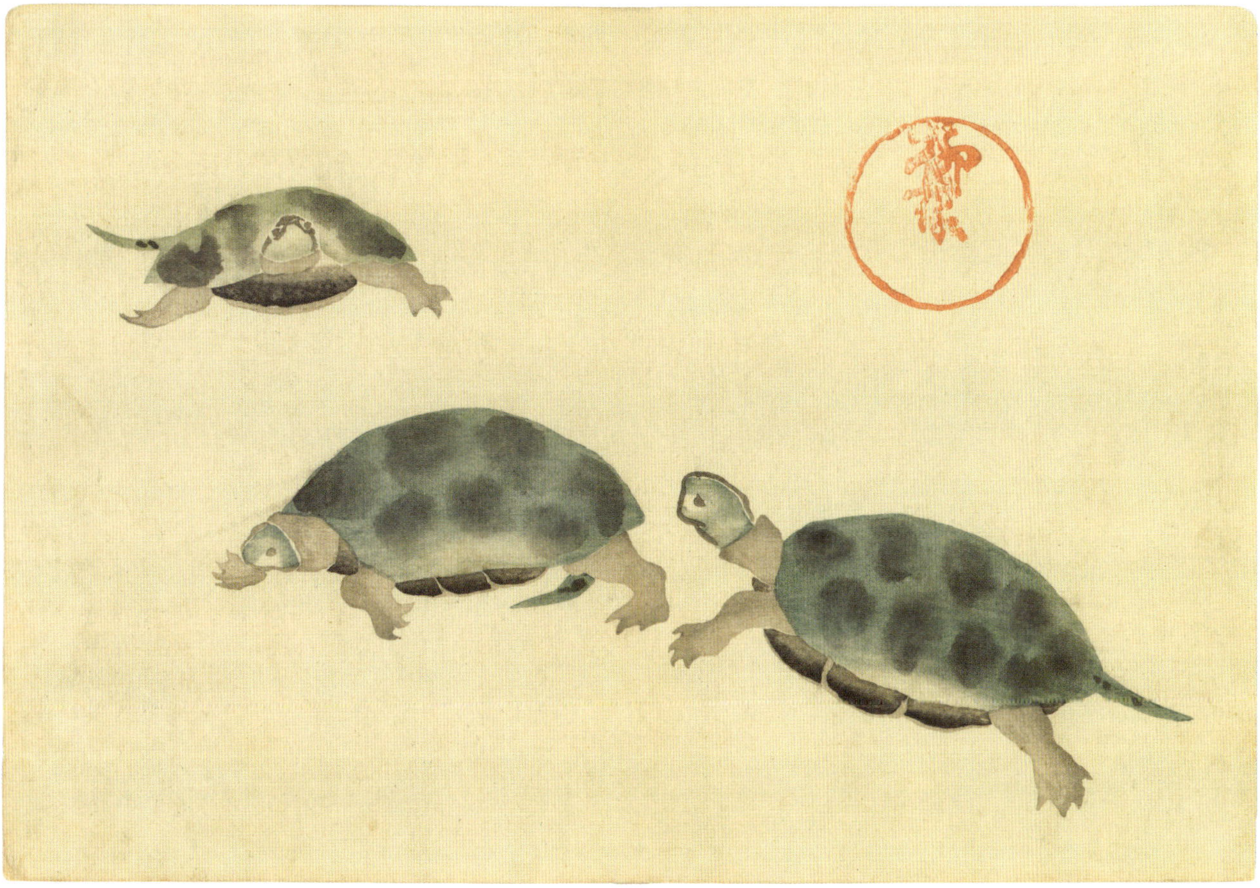

←
Japanese photographer Yamamoto Masao draws on Zen philosophy in producing images that encourage the viewer to reflect on nature. He manipulates his silver-gelatin prints by adding ink by hand. This is '#853' from his *Nakazora* series (1999).

↑
Nakamura Hōchū's woodblock print *Three Turtles* is from the *Kōrin Gafu* (1802), an album of pictures based on the style of revered artist Ogata Kōrin. In many cultures, including Japan, turtles are symbols of longevity.

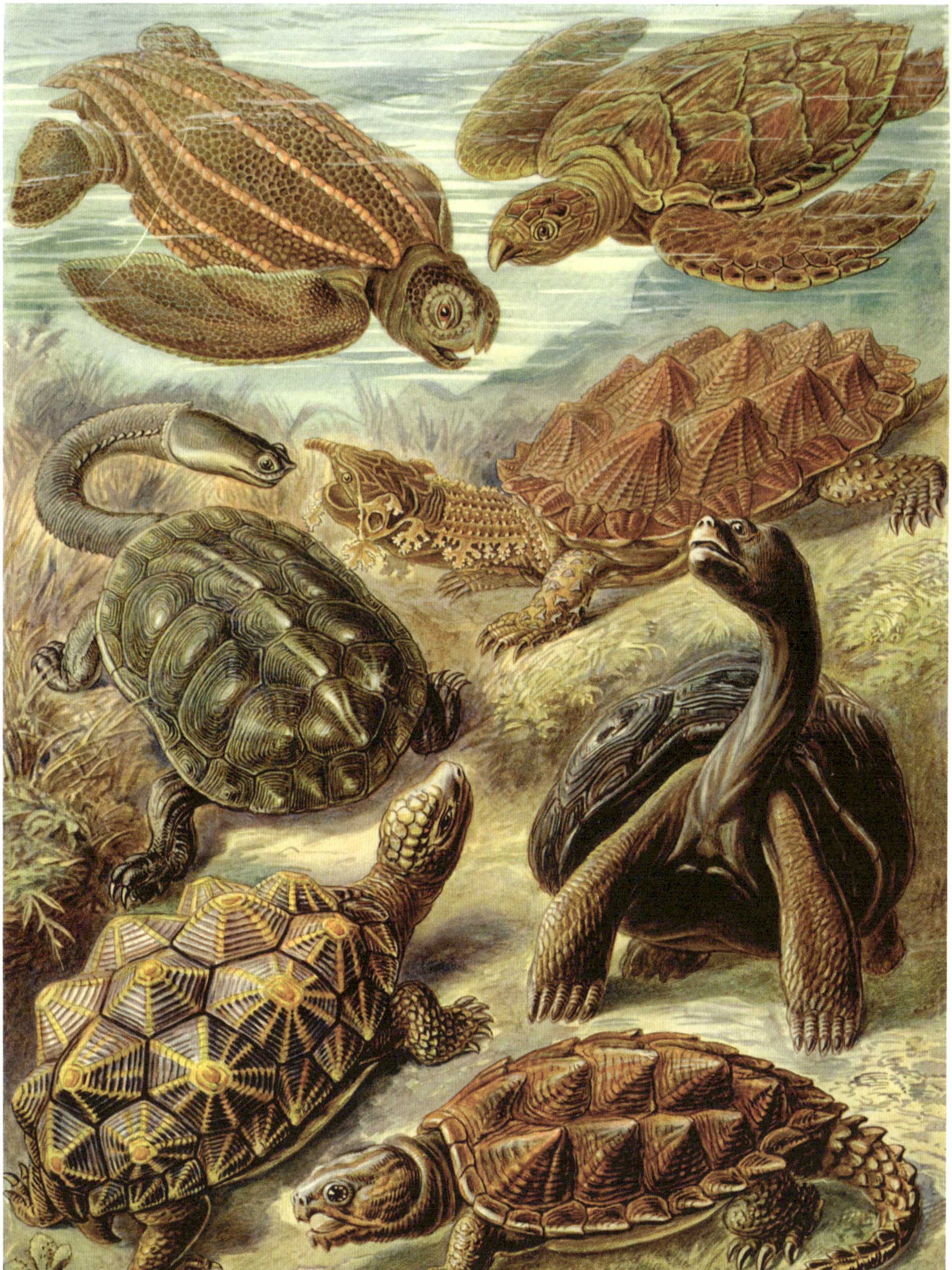

(v)

'We find long-necked turtles in wide open country, where there are flood plains. Women of our people would go out onto the floodplain looking for the small holes the turtles breathe through as they lie buried under the mud.'

DEFINITION OF 'MIM': THE BREATHING HOLES OF LONG-NECKED TURTLES, CRABS ETC., DICTIONARY OF THE BININJ KUNWOK LANGUAGE OF THE WEST ARNHEM LAND, AUSTRALIA

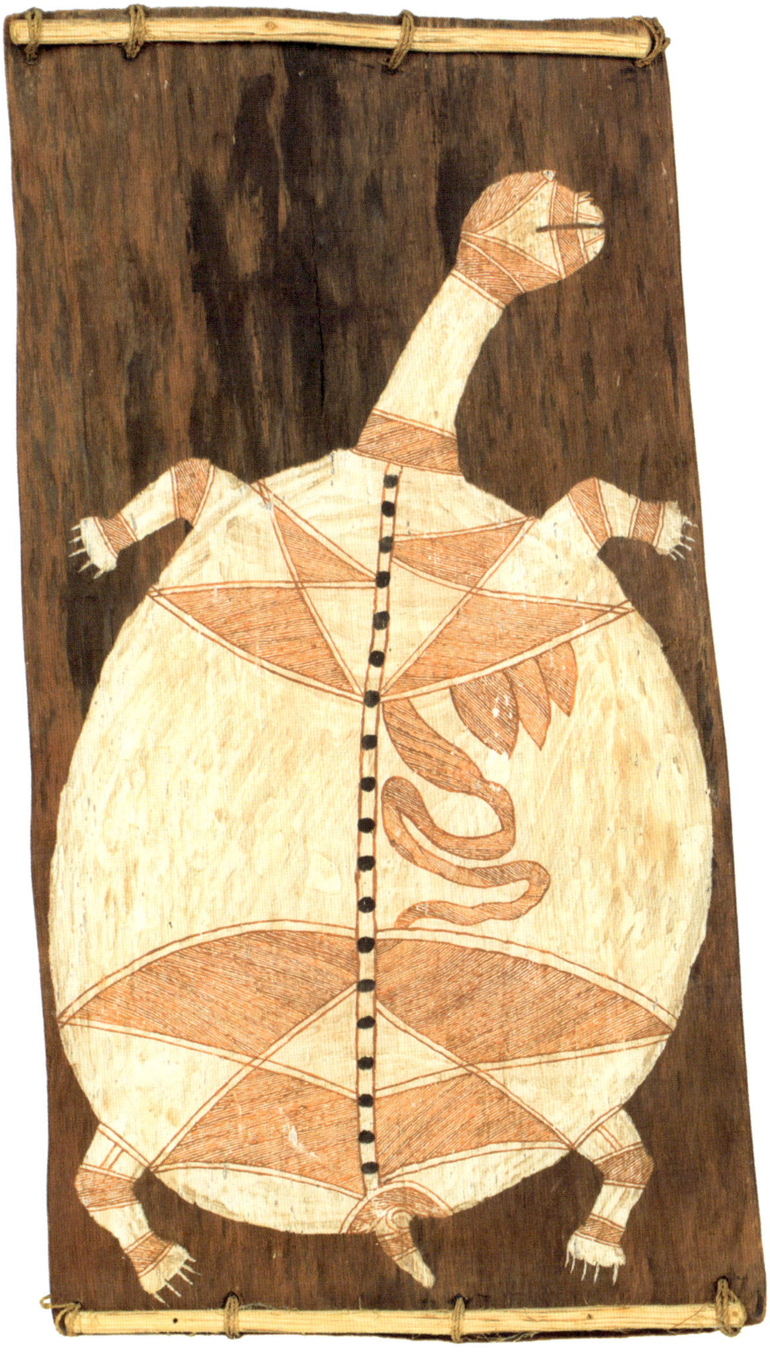

←
Ⓓ This illustration from Ernst Haeckel's *Kunstformen der Natur* (*Art Forms in Nature*, 1904) shows species including, at bottom left, the geometric tortoise (*Psammobates geometricus*), which inhabits the Fynbos region of South Africa, a critically endangered ecosystem; above it, the Argentine snake-necked turtle (*Hydromedusa tectifera*), which uses its long neck as a snorkel when searching for food; and, facing that species, the mata mata turtle (*Chelus fimbriata*), a bottom-feeder that hoovers up prey with its long snout. Both the hawksbill sea turtle (*Eretmochelys imbricata*) (top right) and the geometric tortoise are critically endangered.

→
Aboriginal Australian artist Dick Nguleingulei Murrumurru (*c.* 1920–88) used natural earth pigments to make this bark painting of a freshwater turtle. Turtles are important in Aboriginal culture both as a food source and for their spiritual significance as a mystical ancestor.

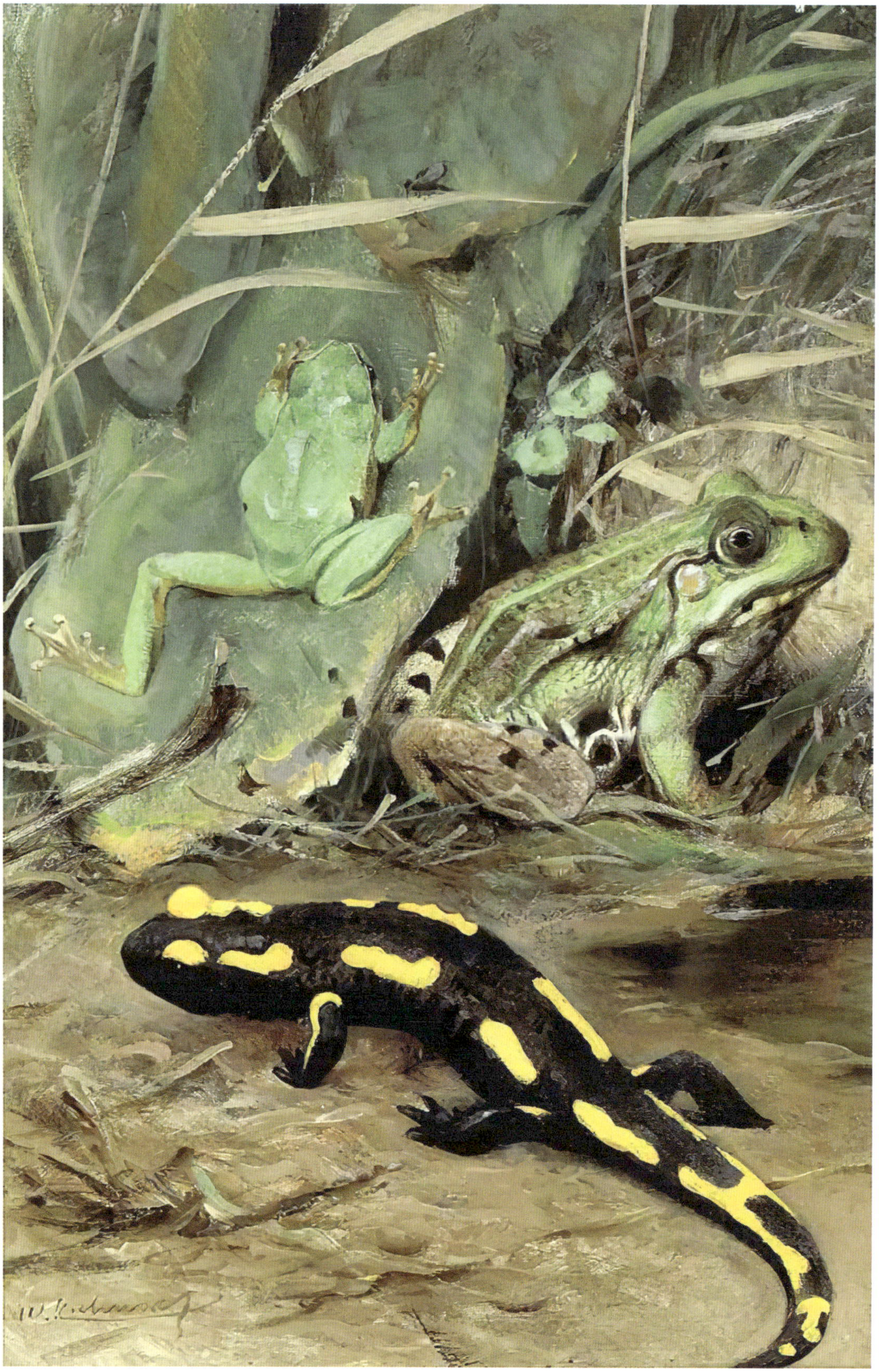

(v)

←
◎ The fire salamander (*Salamandra salamandra*), pictured in Wilhelm Kuhnert's *A Tree Frog, Water Frog and Fire Salamander* (19th century), is threatened by a deadly fungus, the *Batrachochytrium salamandrivorans* (or *Bsal*), literally meaning salamander-eating virus. The fungus is believed to have arrived in Europe with newts imported as pets from Asia.

↓
◎ Isabel Cooper's illustration of a giant land iguana (*Conolophus subcristatus*) appears in William Beebe's *Galapagos: World's End* (1924). By the beginning of the 20th century, land iguanas had become extinct on Santiago island due to the presence of invasive species. In 2019, having rid the island of these, conservationists introduced around 2,500 land iguanas. Three years later, the iguanas were breeding successfully, though they remain a vulnerable species.

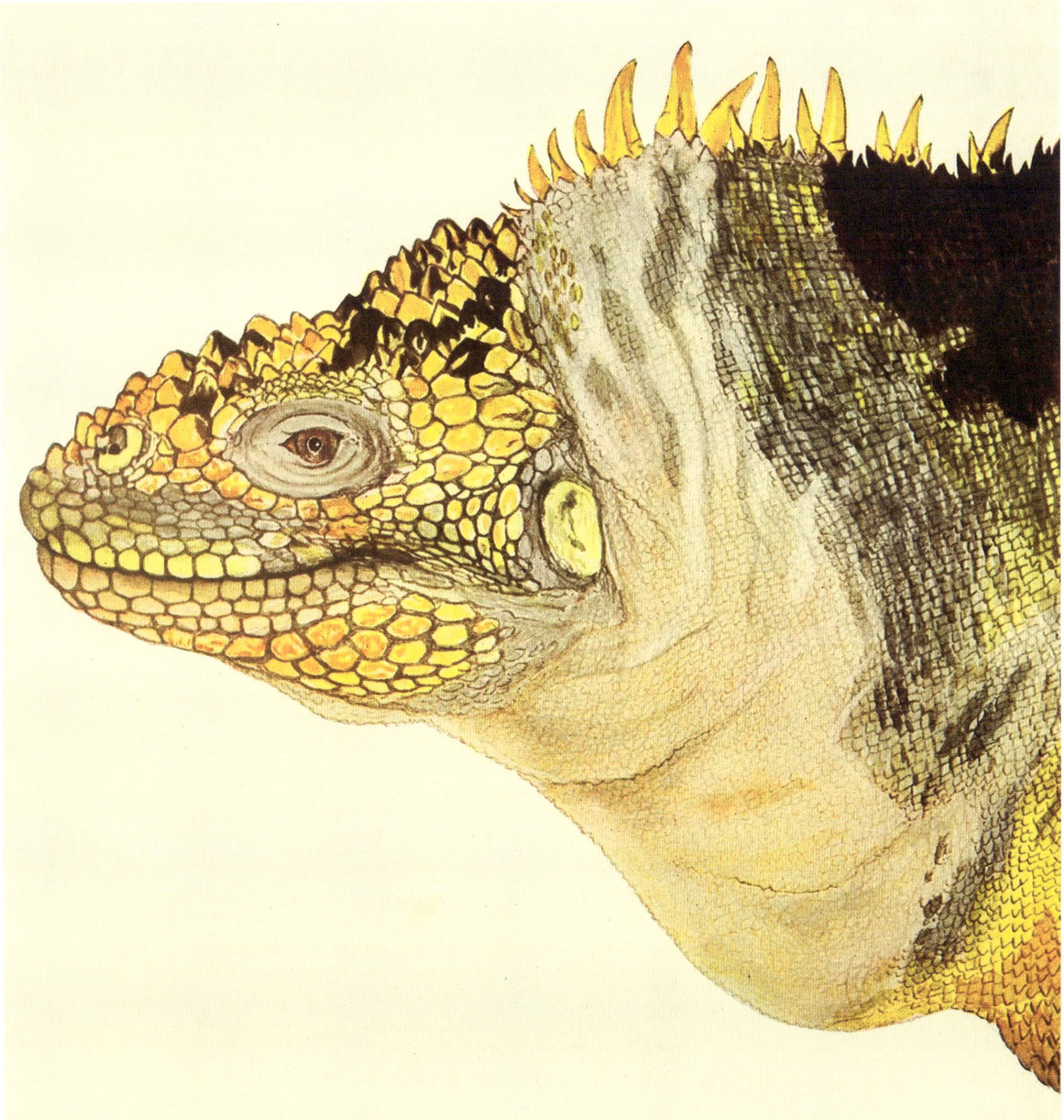

'Lizard ran out on a rock and looked up, listening no doubt to the sound of the spheres. And what a dandy fellow! the right toss of a chin for you and swirl of a tail!'

D. H. LAWRENCE, 'LIZARD', *PANSIES*, 1929

→
◍ The Mexican axolotl (*Ambystoma mexicanum f. leucistic*) typically retains its 'tadpole' appearance into maturity, although individuals can metamorphose into adult salamanders under certain conditions. The axolotl is critically endangered as urbanization and pollution are destroying the lakes and wetlands where it lives.

↓
Around 1700, the German naturalist and botanical artist Maria Sibylla Merian spent two years in Suriname, South America, studying the life cycle of insects. The resulting *Metamorphosis Insectorum Surinamensium* ('The Metamorphosis of the Insects of Suriname', 1705) contained her hand-coloured engravings of the insects, plants and animals of the region, including this golden tegu lizard (*Tupinambis teguixin*).

↘
A fly catches the attention of a blue lizard in this detail from Rachel Ruysch's *Fruit Still Life with Stag Beetle and Nest* (1717). Ruysch, a Dutch still-life painter who studied her subjects closely, combined native and non-native plants and animals in her paintings.

(V)

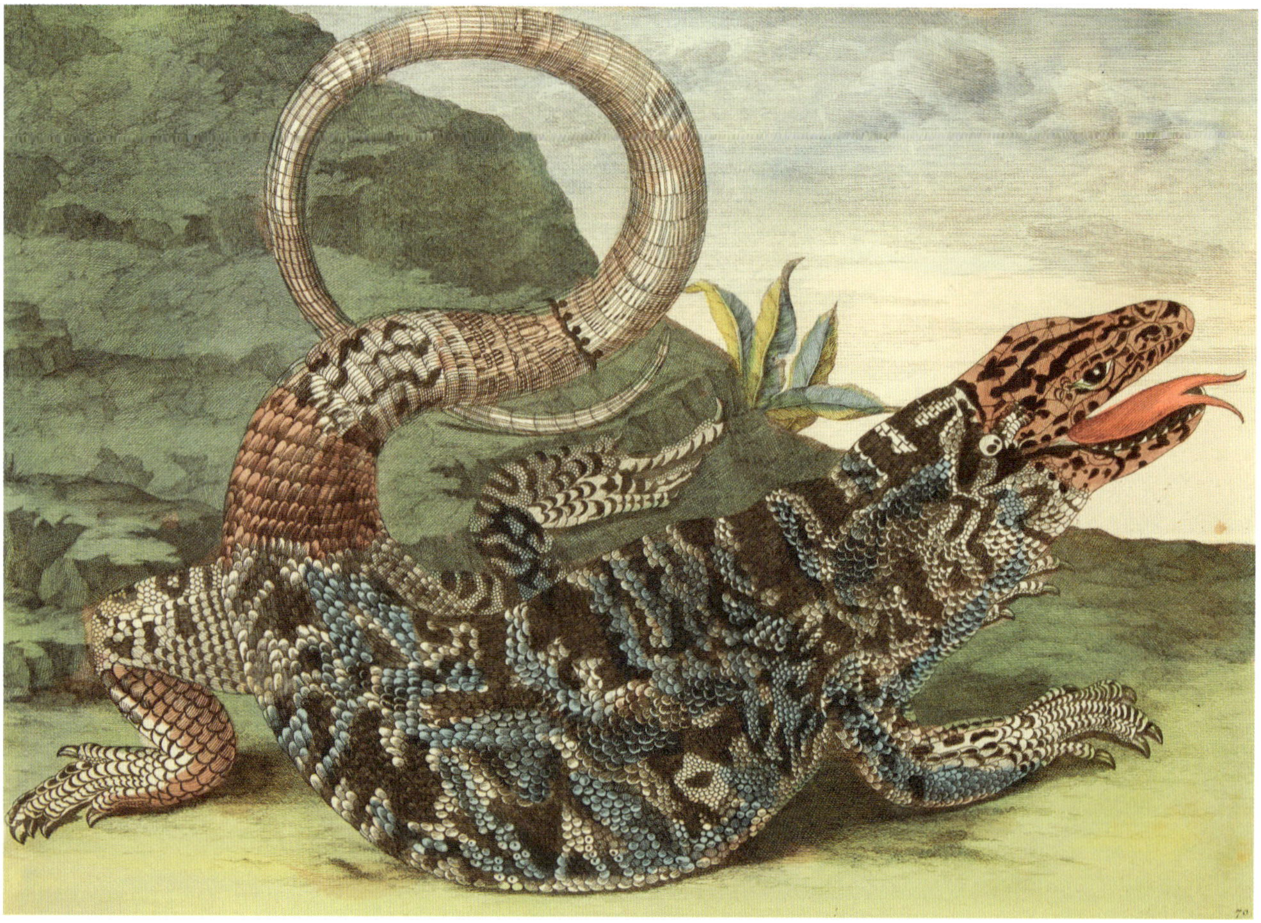

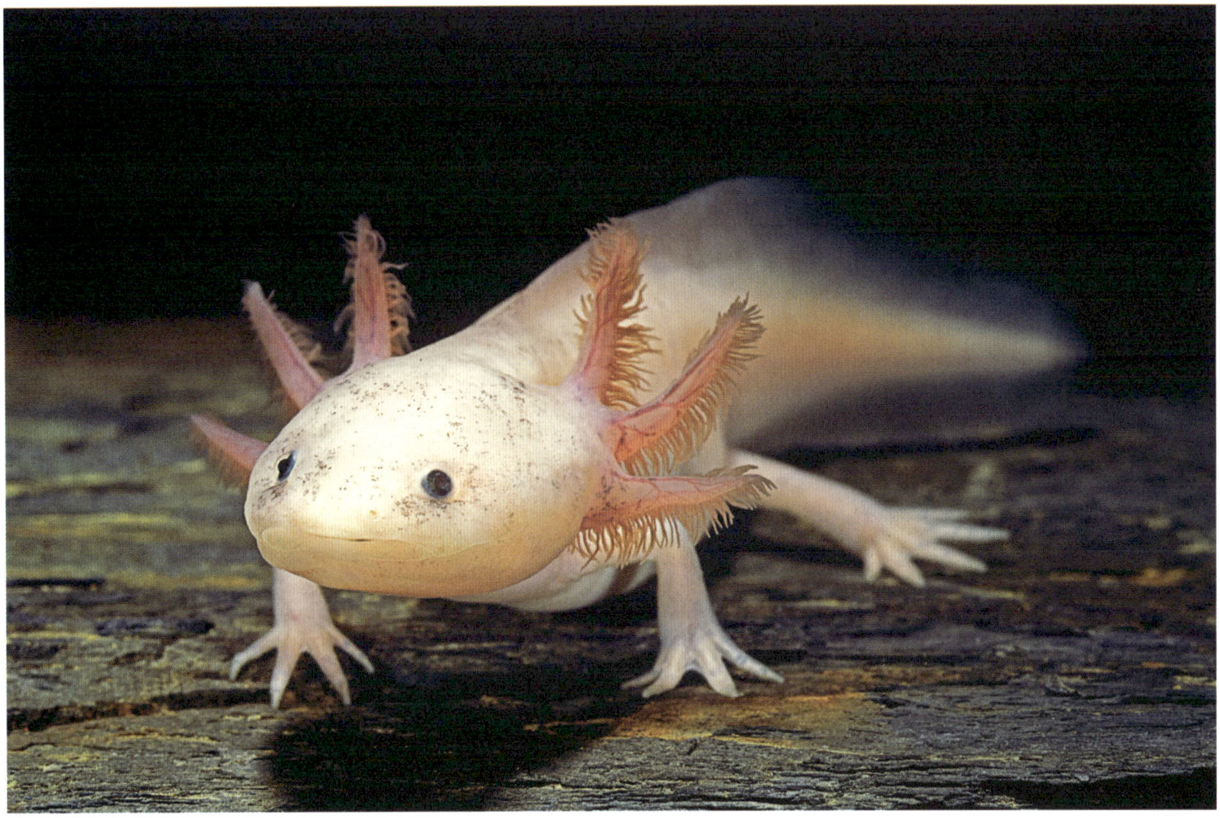

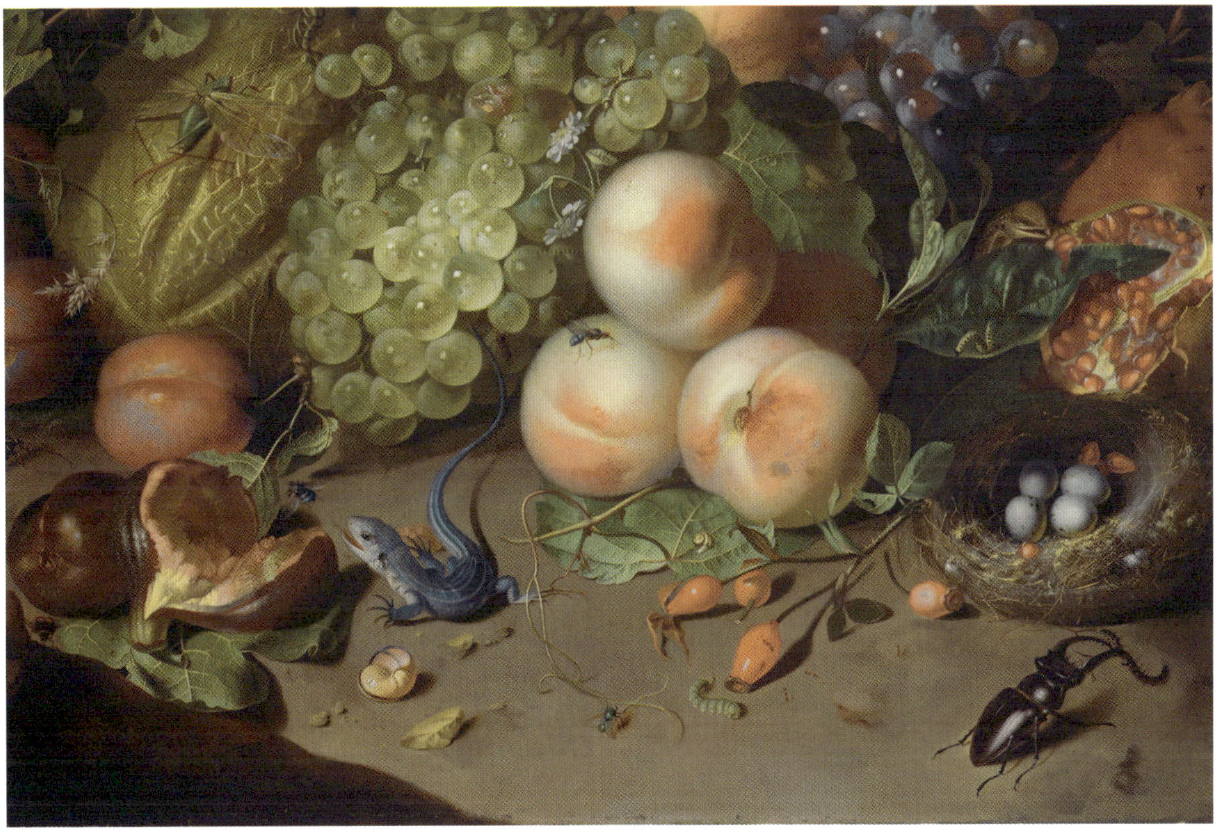

↓
At the time that the *Rochester Bestiary* was created, around 1230, European depictions of crocodiles were based entirely on descriptions of the animal that had been passed around the world. With this in mind, this illumination is a suprisingly accurate portrayal of the species.

→
In 2018 the Union of Concerned Scientists found that, by 2100, one million homes in Florida could be at risk of flooding. This threat is the focus of Anastasia Samoylova's photographic series *FloodZone*. Though the population of American alligators (*Alligator mississippiensis*, the species depicted in this work, 'Gator', 2017) in Florida recovered after illegal hide hunters pushed them onto the endangered species list, their freshwater wetland habitats remain threatened by saltwater intrusion from rising sea levels.

↘
Kaballa, an artist from the Democratic Republic of Congo, depicts the two crocodiles in this mid-20th-century oil painting as fiercely protective parents guarding their eggs.

(V)

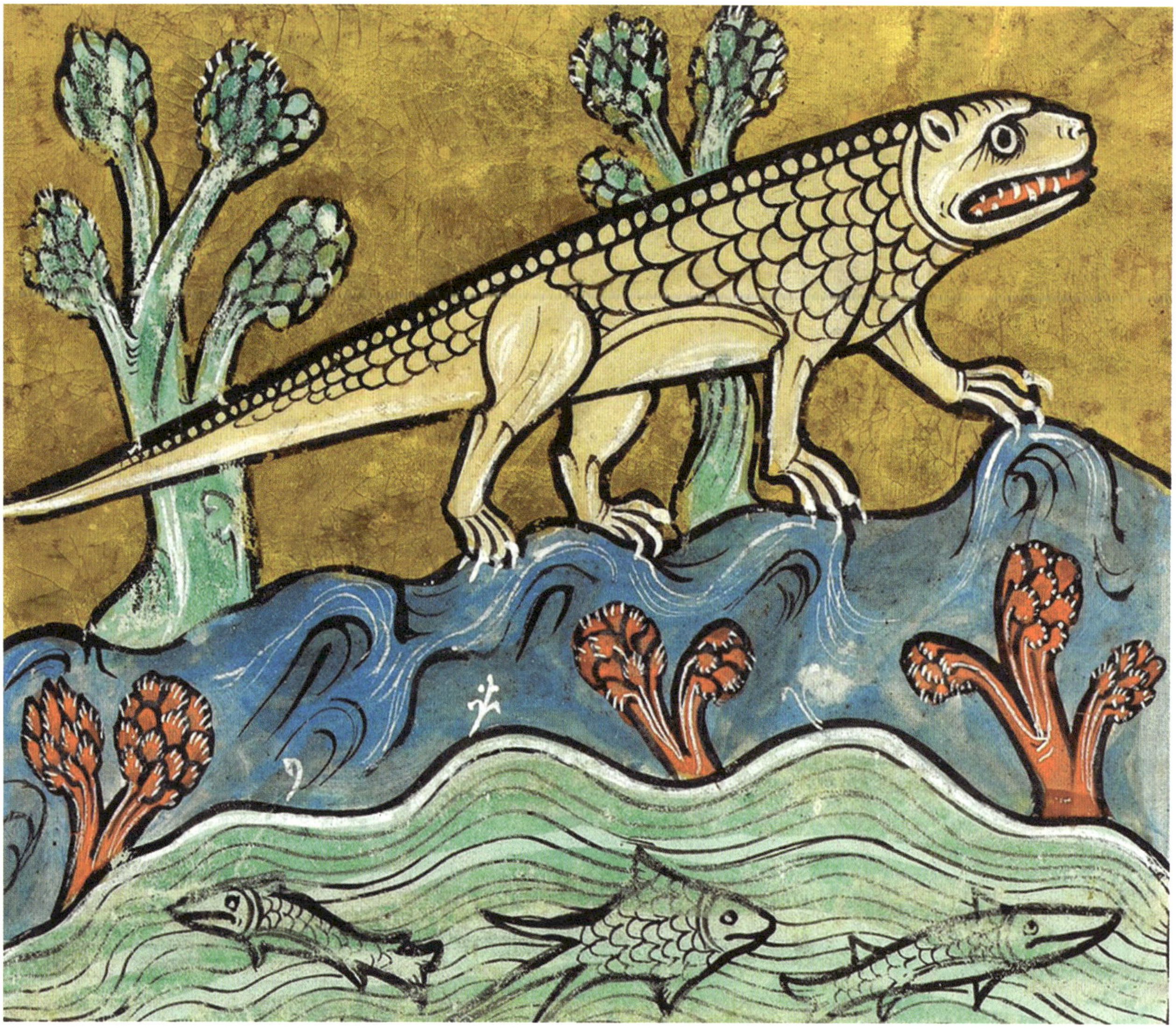

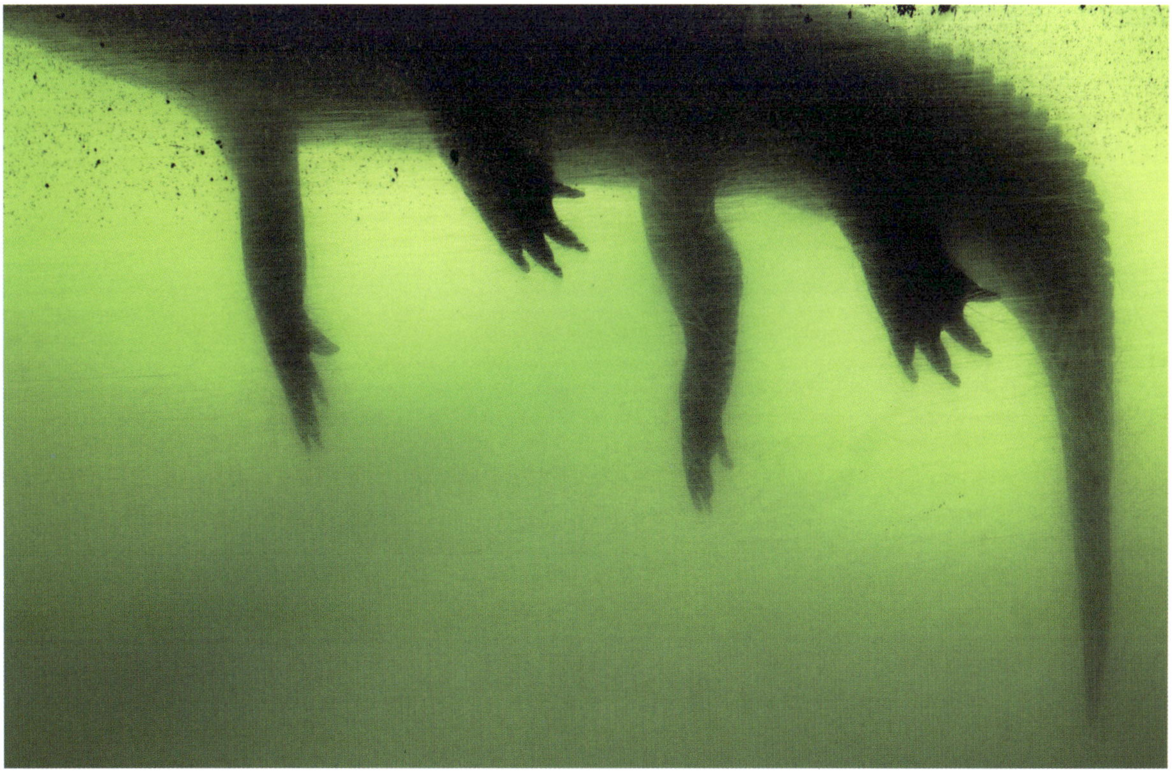

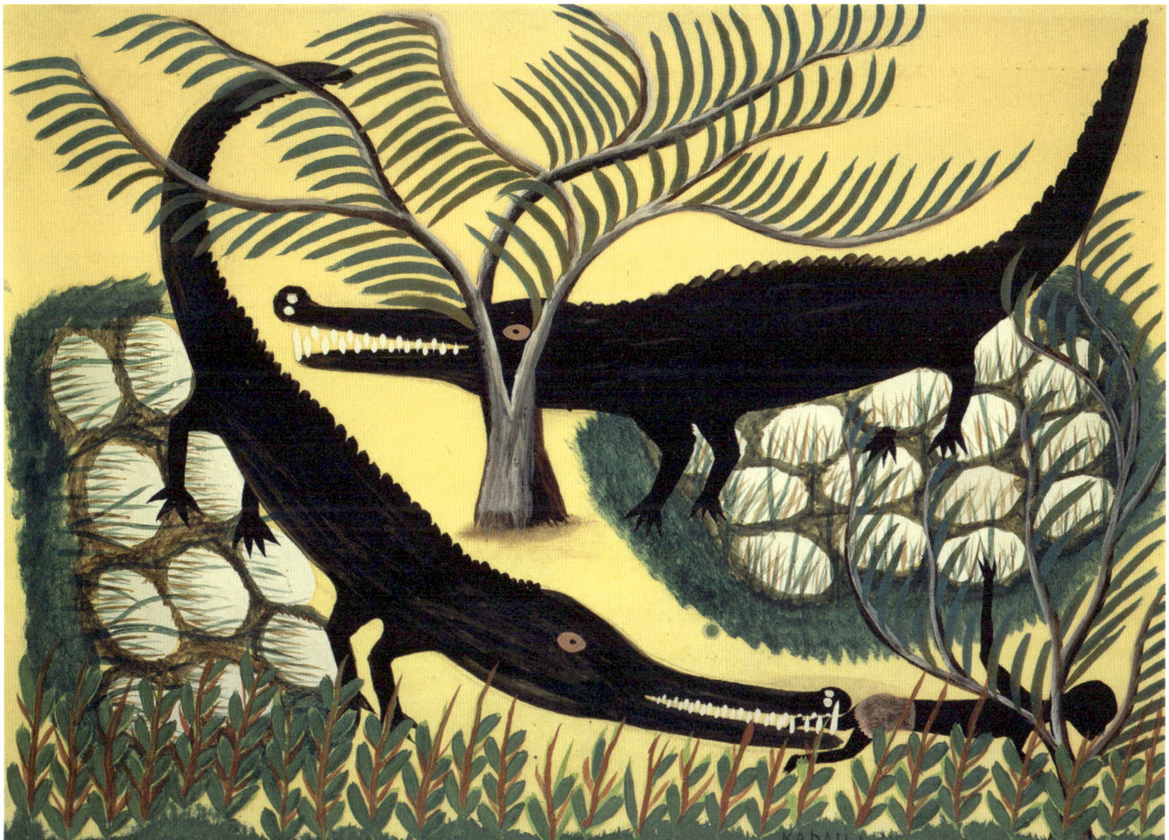
KABAILA

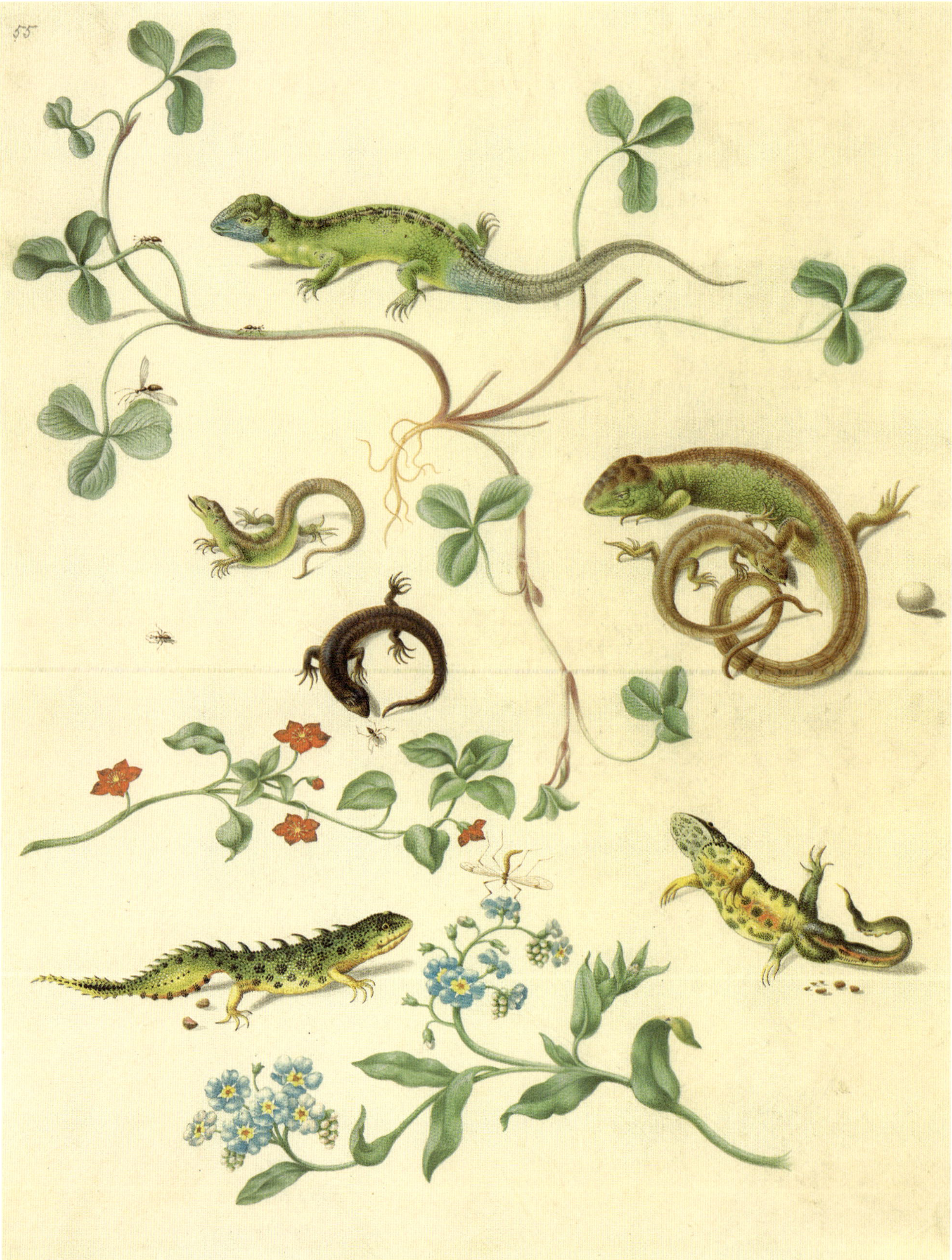

(V)

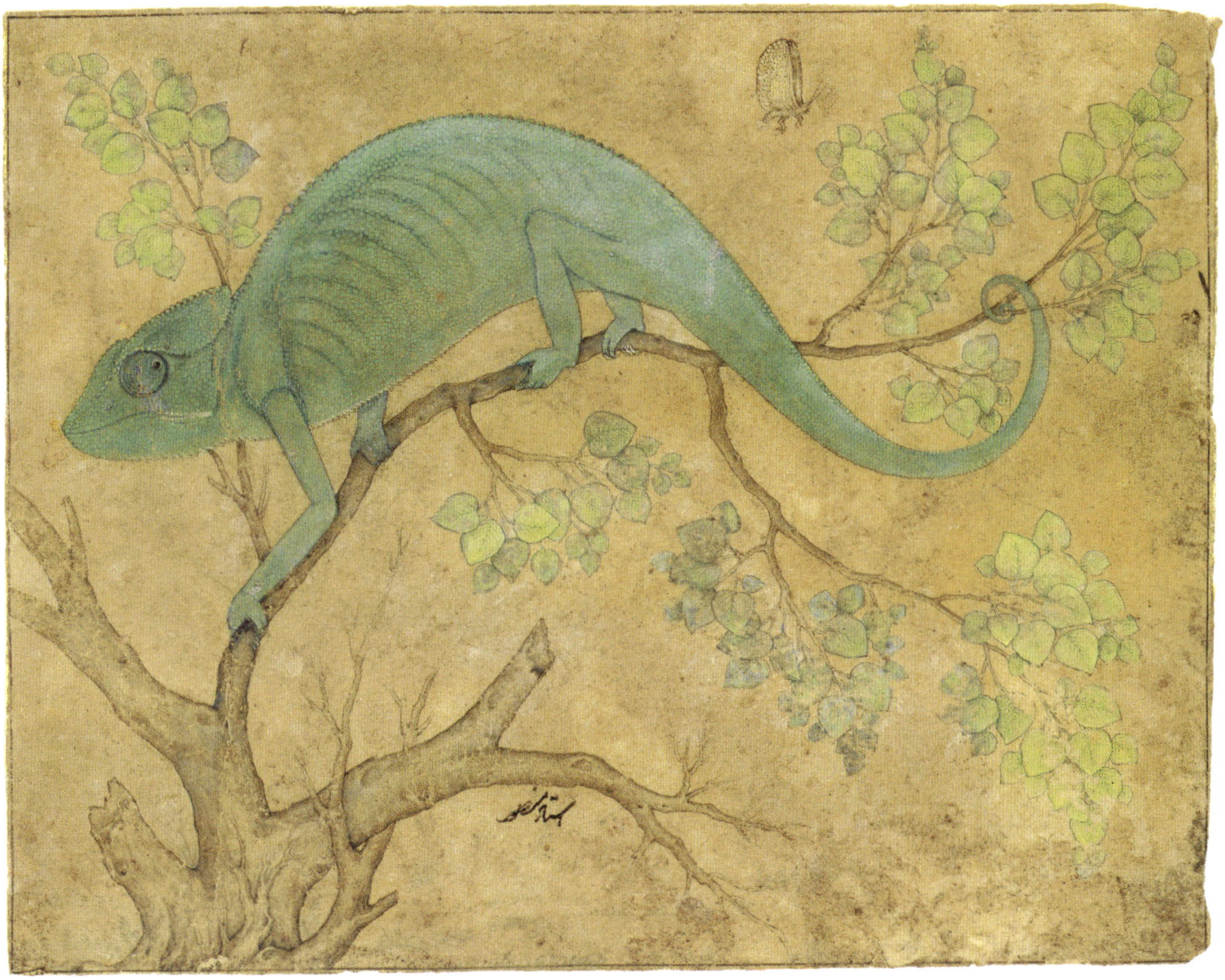

'Dull and weary, he scarce raises from the ground, but drags, his footstep amazedly, and moves forward .... His colour is properly one, yet, whenever anything has approached him, then he blushes. To the chameleon alone has been granted – as our common saying has it – to sport with his own hide.'

TERTULLIAN (*C.* 155–*C.* 220 CE), 'ON THE PALLIUM', TRANSLATED BY REV. S. THELWALL

←
Lizards, Insects and Plants, Including Clover and Forget-Me-Nots is attributed to the German botanical artist Johanna Helena Herolt (1668–1723), daughter of the artist Maria Sibylla Merian.

↑
Ustad Mansur, a leading animal painter at the court of Mughal Emperor Jahangir, has accurately recorded the attitude and markings of a chameleon that is eyeing a butterfly behind it in this miniature, dated *c.* 1612.

'Far away in the twilight time
Of every people, in every clime,
Dragons and griffins and
monsters dire,
Born of water, and air, and fire,
Or nursed, like the Python,
in the mud
And ooze of the old Deucalion
flood,
Crawl and wriggle and foam
with rage,
Through dusk tradition and
ballad age.'

JOHN GREENLEAF WHITTIER, 'THE DOUBLE-HEADED SNAKE OF NEWBURY', 1859

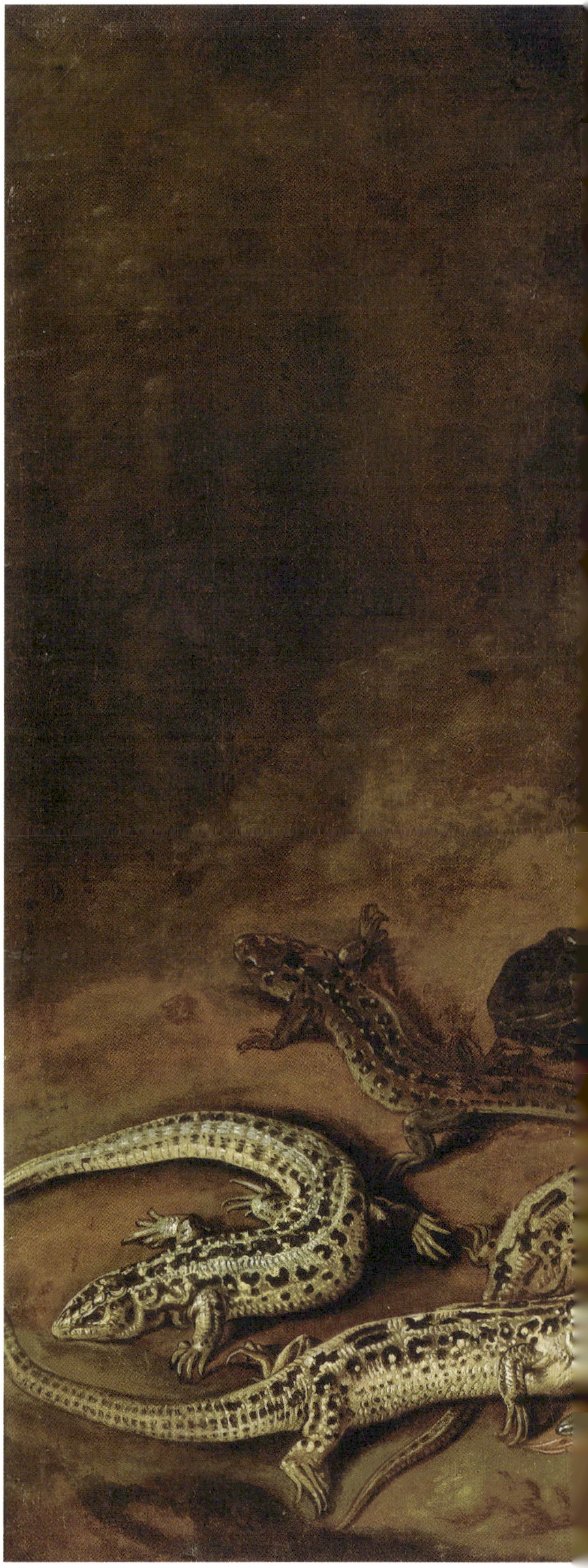

→
Forest with Snakes and Frogs is attributed to the 17th-century Flemish artist Carstian Luyckx, who specialized in all types of still-life paintings and animal scenes.

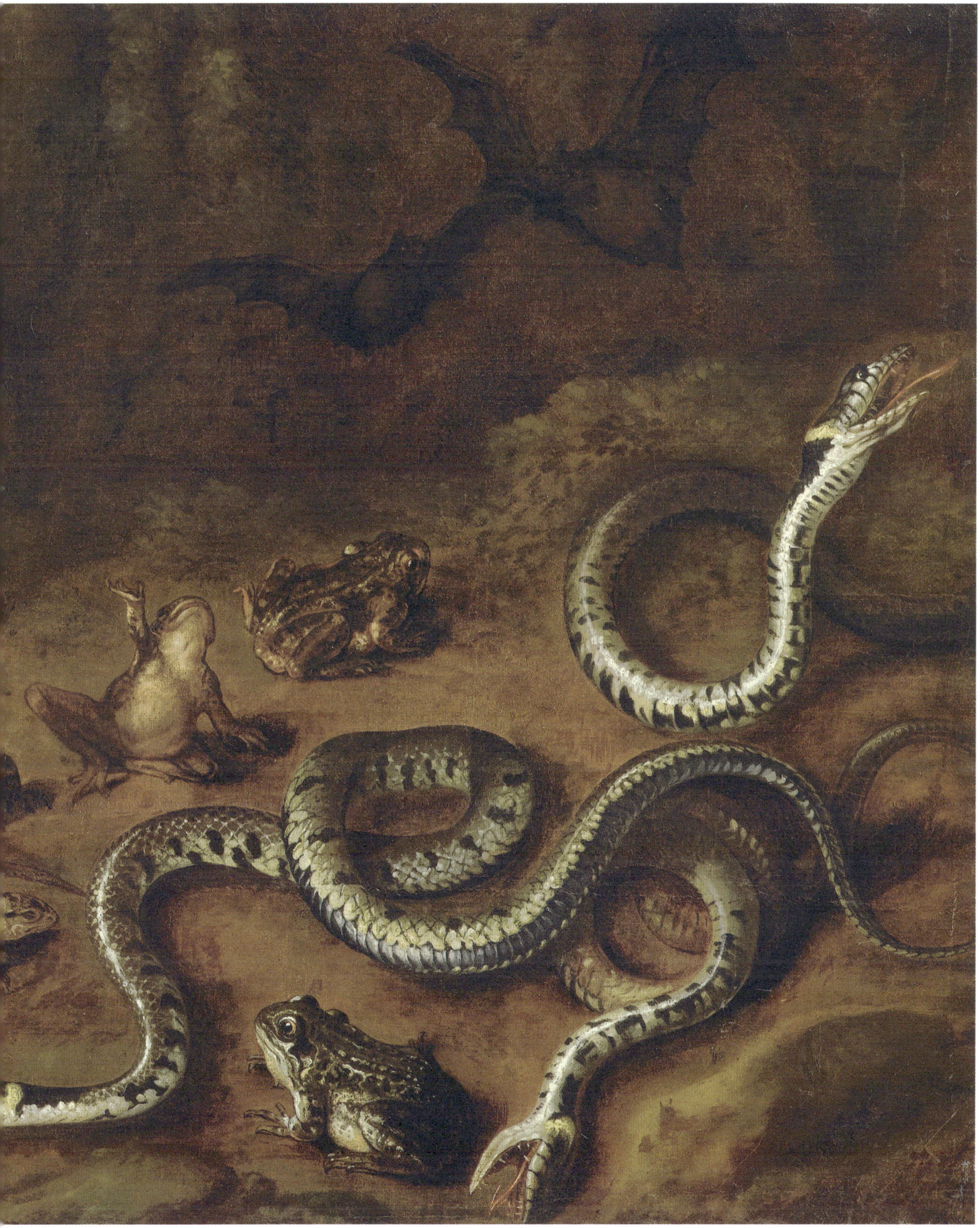

'When I made that picture of the hand of the iguana, I realized it's exactly the hand of a warrior from the Middle Ages. It's exactly the same! And in that moment, looking through my lens, I understood that the iguana was my relative.'

SEBASTIÃO SALGADO, QUOTED IN 'I WAS TRANSFORMED INTO AN ENVIRONMENTALIST', *THE GUARDIAN*, 2022

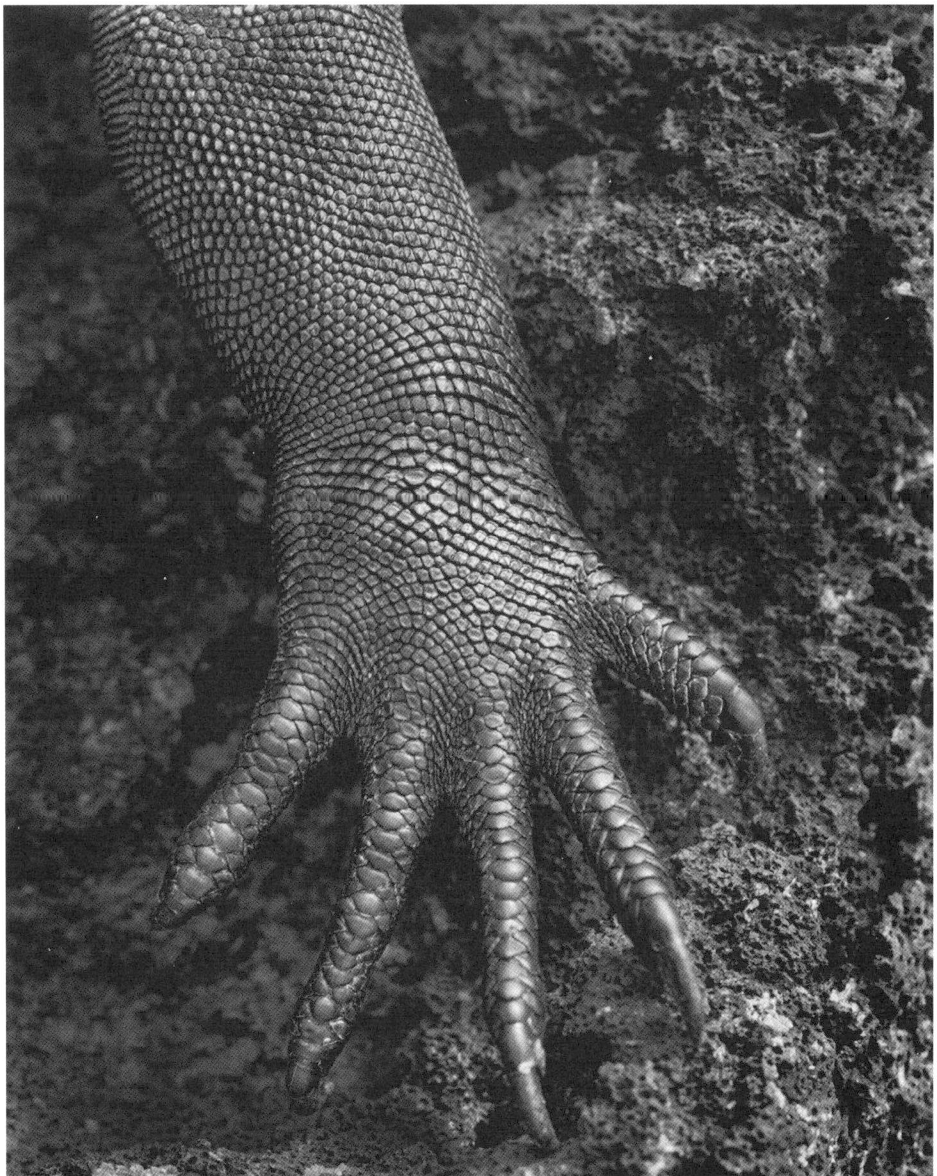

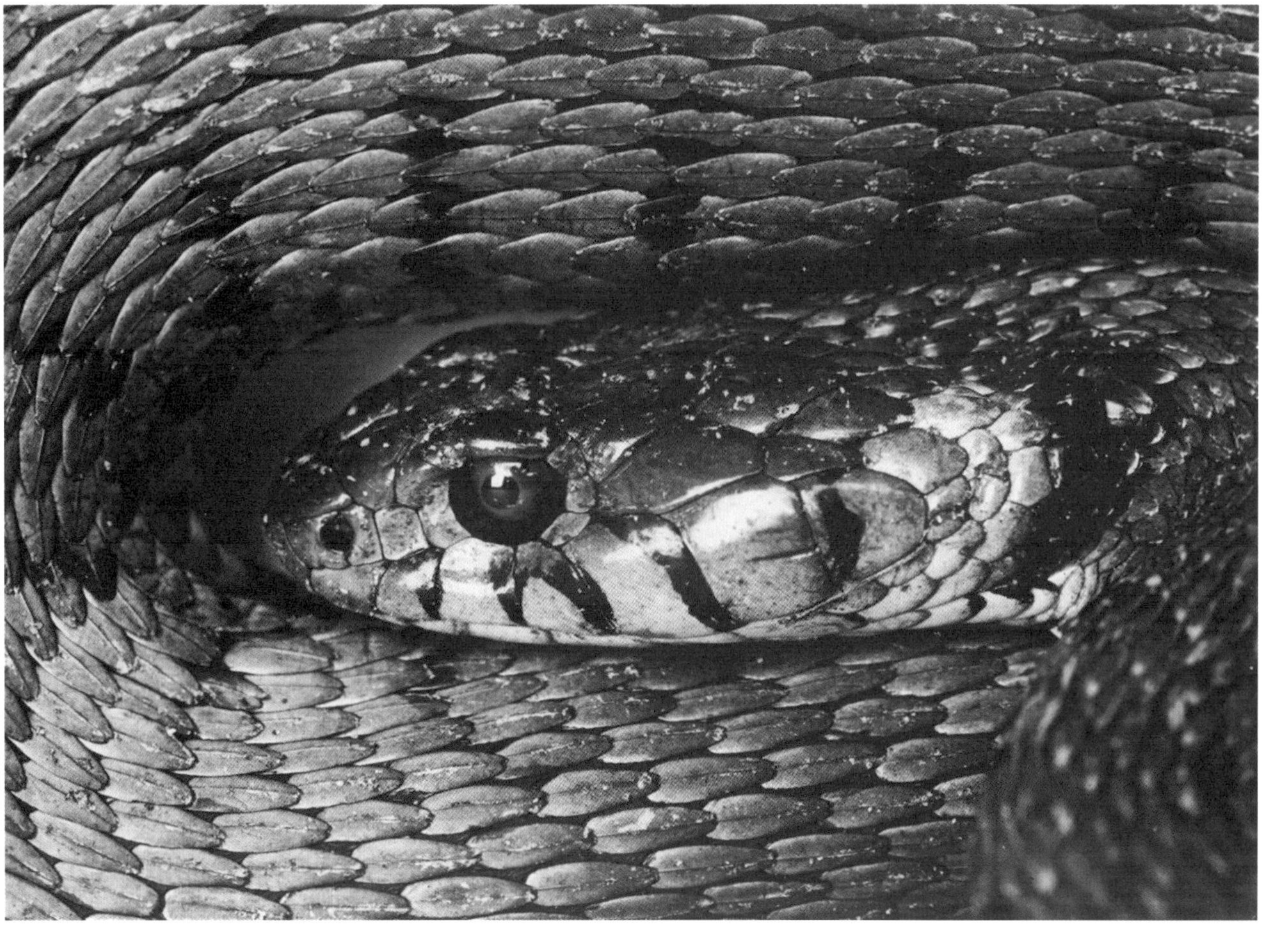

←
Alongside images of conflict and exploitation, the Brazilian documentary photographer Sebastião Salgado's work also captures the beauty of the natural world, including *Marine Iguana* (*Amblyrhynchus cristatus*), taken in 2004 on Rábida Island in the Galápagos, Ecuador.

↑
Albert Renger-Patzsch's *Snake Head*, a photograph of a grass snake (*Natrix natrix*), was included in his collection *The World is Beautiful*, published in 1927. Renger-Patzsch sought to reveal the beauty of the patterns and textures in his subjects in order to encourage people to see the world anew.

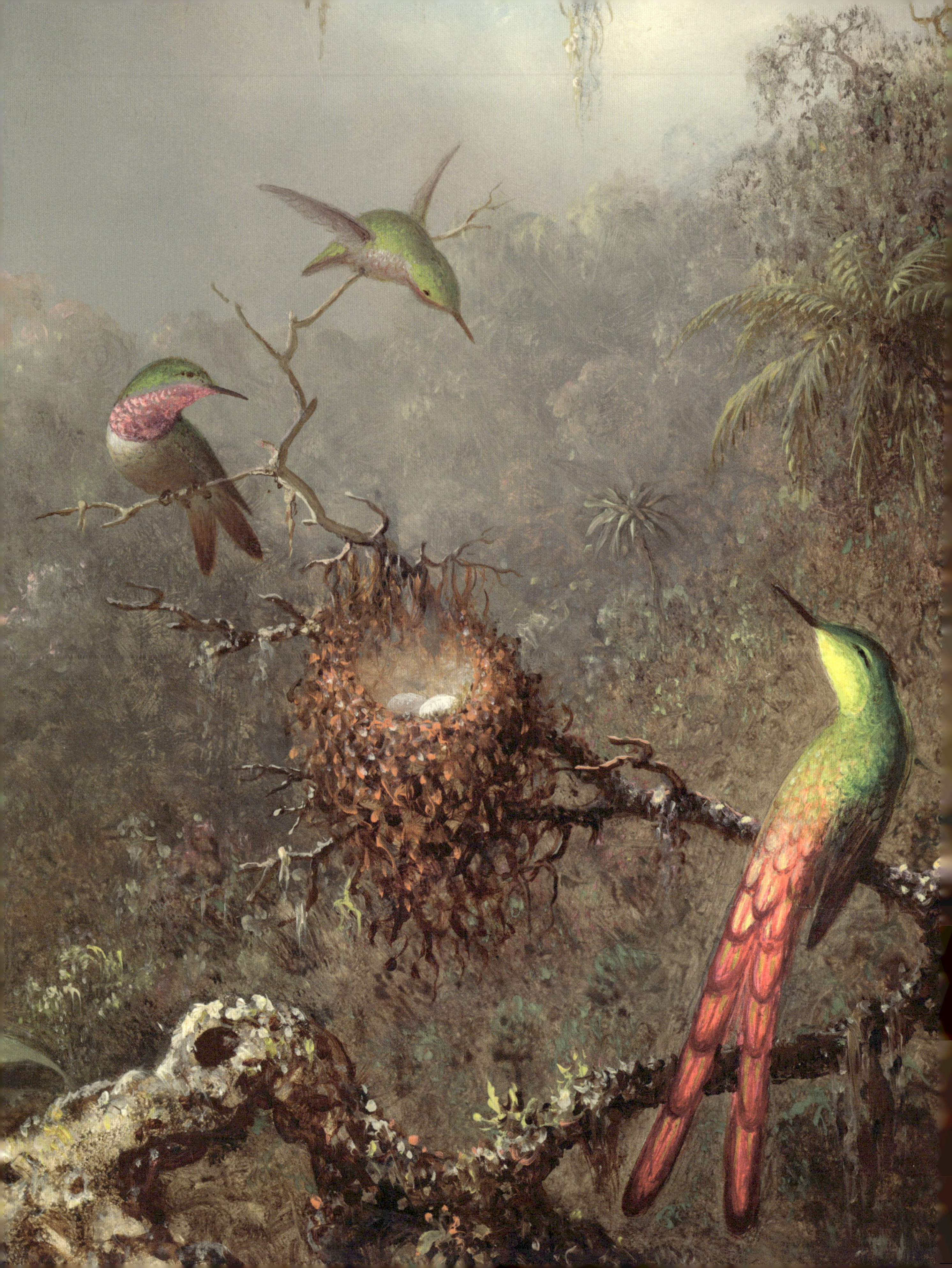

(VI) BIRDS

REDSTART CIRL BUNTING
GOLDFINCH EURASIAN SKYLARK
PURPLE MARTIN THRUSH
OSTRICH ADÉLIE PENGUIN
BROWN KIWI TURKEY
MAGPIE CHICKEN
BLUE POUTER COCK PIGEON JACOBINS PIGEON
BLACK CARRIER COCK PIGEON ENGLISH FANTAIL PIGEON
BULLFINCH GOLDEN EAGLE
ANDEAN CONDOR RÜPPELL'S VULTURE
BALD EAGLE HAWK
FALCON DUCK
IBIS CALIFORNIAN CONDOR
CHESTNUT-MANDIBLED TOUCAN NEW GUINEA ECLECTUS PARROT
AMERICAN FLAMINGO BLACK SWAN
SUPERB LYREBIRD BLACK-LEGGED KITTIWAKE
SILVERY-CHEEKED HORNBILL BLACK-CAPPED KINGFISHER
CRANE MALLARD
PLOVER EURASIAN HOOPOE
CROW BARN OWL
RED BIRD-OF-PARADISE GUIANIAN COCK-OF-THE-ROCK
WESTERN PAROTIA CASSOWARY
PEACOCK HERON
PELICAN DOVE

BIRDS are warm-blooded vertebrate animals that have feathers, wings, toothless beaked jaws, four-chambered hearts and are typically – although not always – able to fly. Instead of carrying the weight of their young inside their bodies, they lay hard-shelled eggs. The wings that have given these animals the ability to fly evolved from the forelimbs of theropod dinosaurs, but further evolution has led to the loss of flight in some species, including kiwis and penguins. There are more than 11,000 species of birds.

(VI)

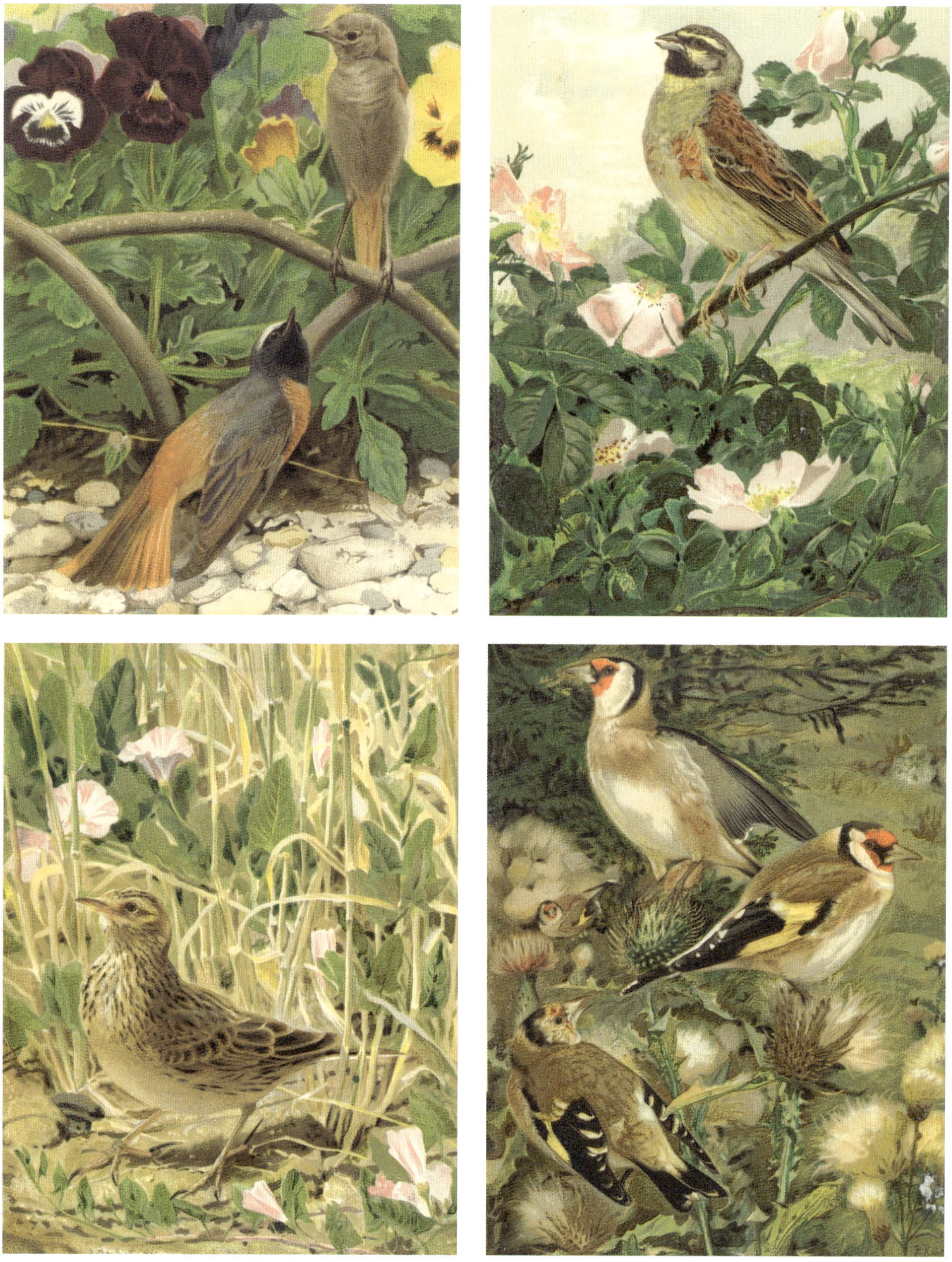

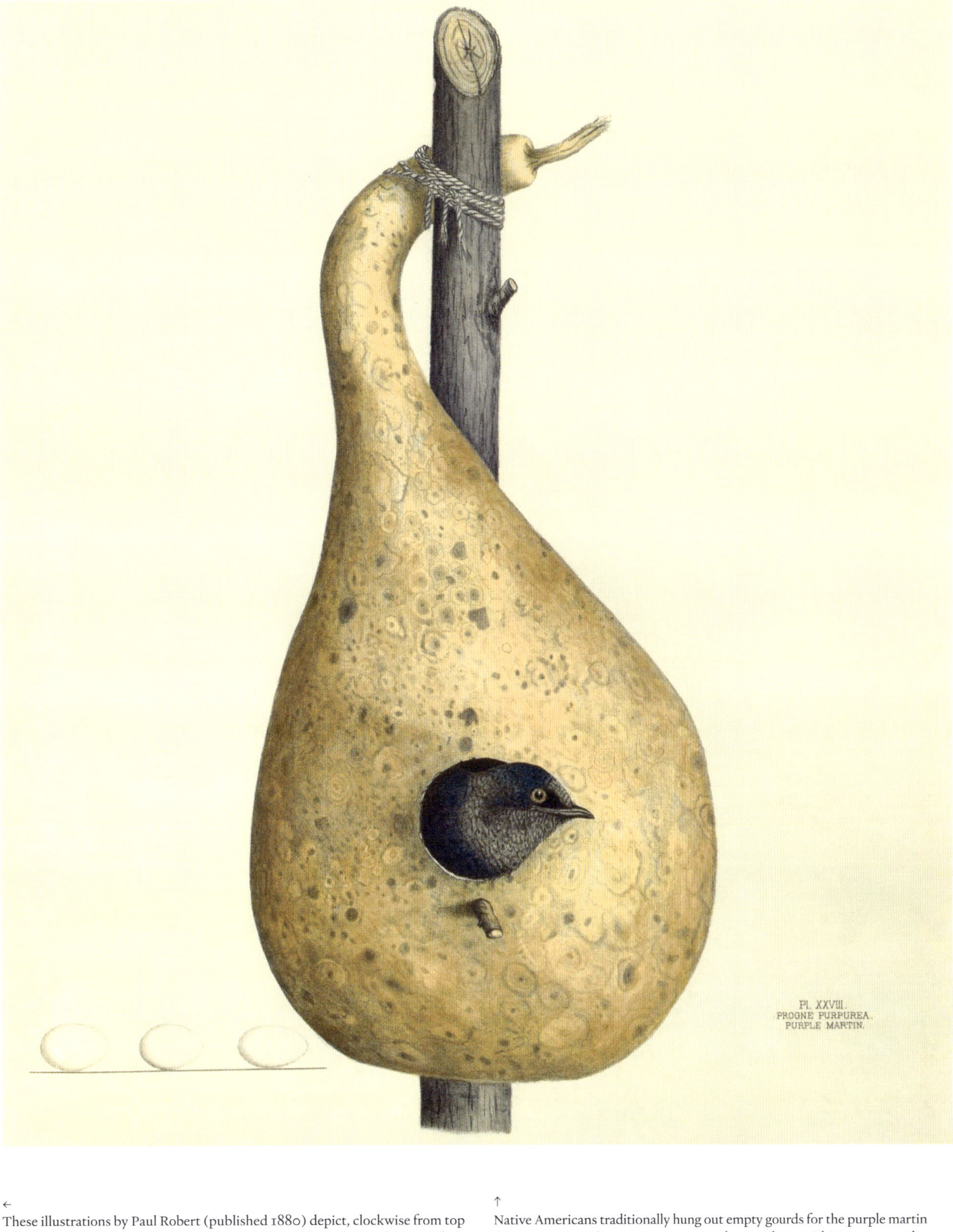

←
These illustrations by Paul Robert (published 1880) depict, clockwise from top left: the redstart (*Phoenicurus phoenicurus*), the cirl bunting (*Emberiza cirlus*), the goldfinch (*Carduelis carduelis*) and the Eurasian skylark (*Alauda arvensis*).

↑
Native Americans traditionally hung out empty gourds for the purple martin (*Progne purpurea*) to nest in, a practice that caught on with European settlers in North America. This 1886 illustration is by Howard and Genevieve Jones.

(VI)

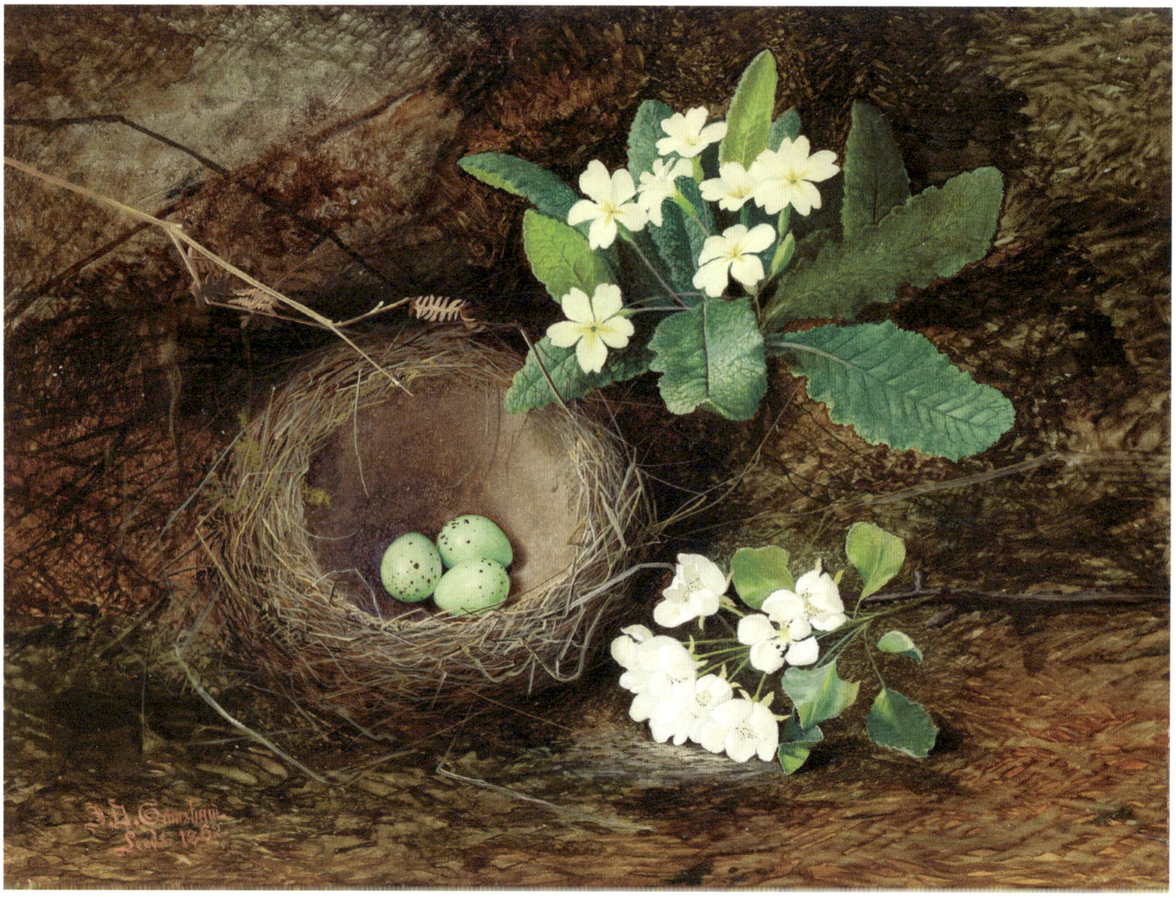

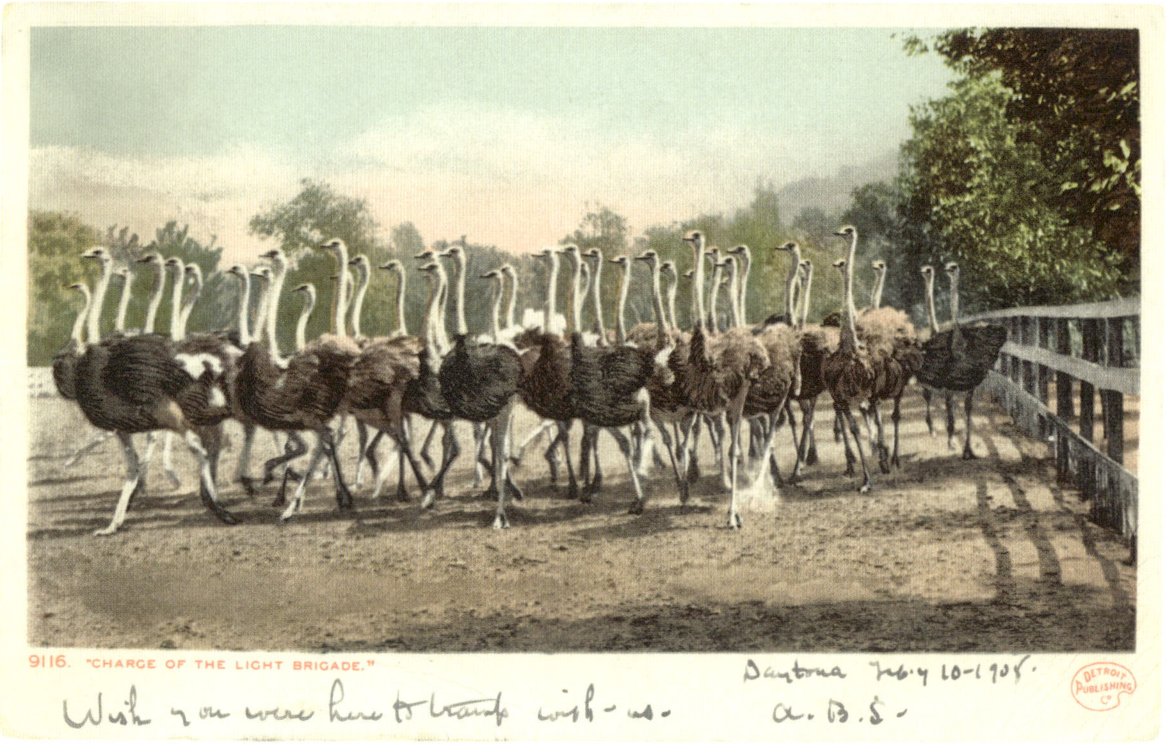
9116. "CHARGE OF THE LIGHT BRIGADE."
Daytona Feb'y 10-1905.
Wish you were here to tramp with us.
A. B. S.
DETROIT PUBLISHING Co

'Like alien gods who shun the world,
The flown owls wait with feathers furled'

CHARLES BAUDELAIRE, 'THE OWLS', 1857, TRANSLATED BY CLARK ASHTON SMITH

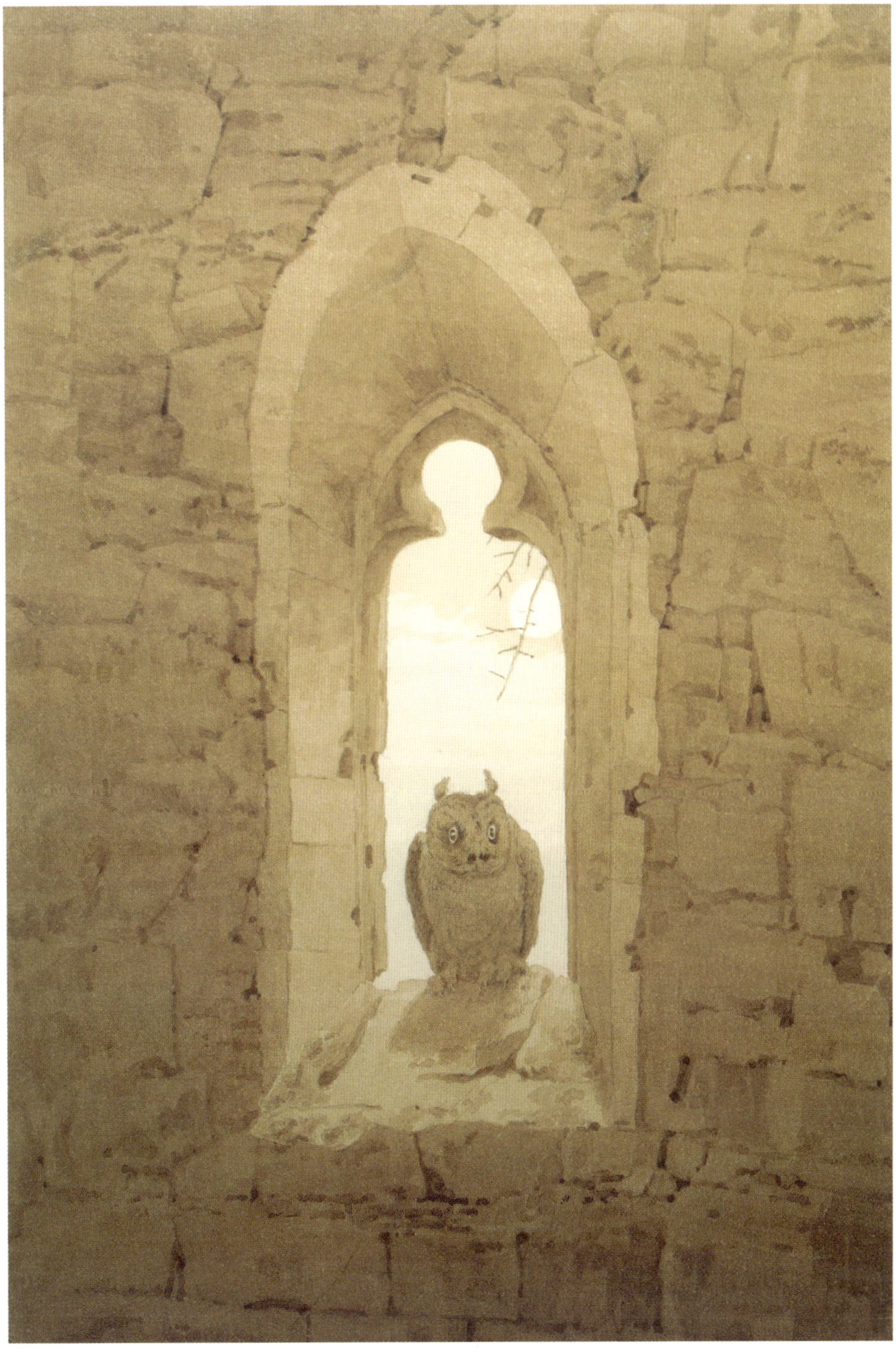

↖
British artist John Atkinson Grimshaw painted *Thrush's Nest, Primroses, Pear Blossom* in 1862. The detailed style that is characteristic of his earlier work is based on close observation and chimed with British art critic John Ruskin's insistence that art should be true to nature.

←
In around 1905, the Detroit Publishing Company issued this postcard, titled *Charge of the Light Brigade*, from a photograph taken at the Cawston Ostrich Farm, South Pasadena, California, USA.

→
Owls have sometimes been considered bad omens, particularly in medieval Europe, but they also have associations with wisdom and far-sightedness, owing in part to their connection to Athena/Minerva. The latter symbolism can be seen to be reflected in this depiction, *Owl in a Gothic Window* (*c.* 1837), by German Romantic painter Caspar David Friedrich.

(VI)

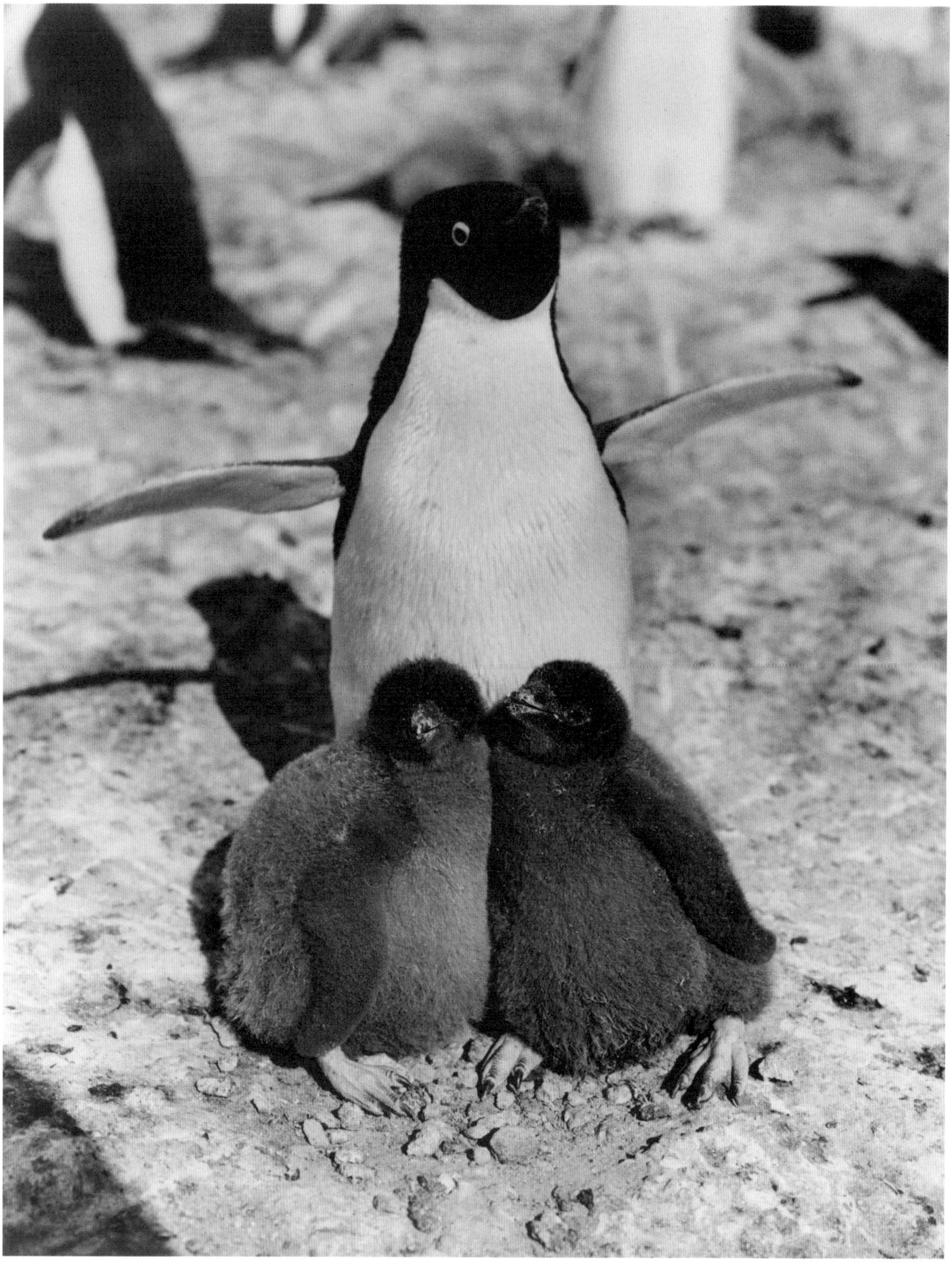

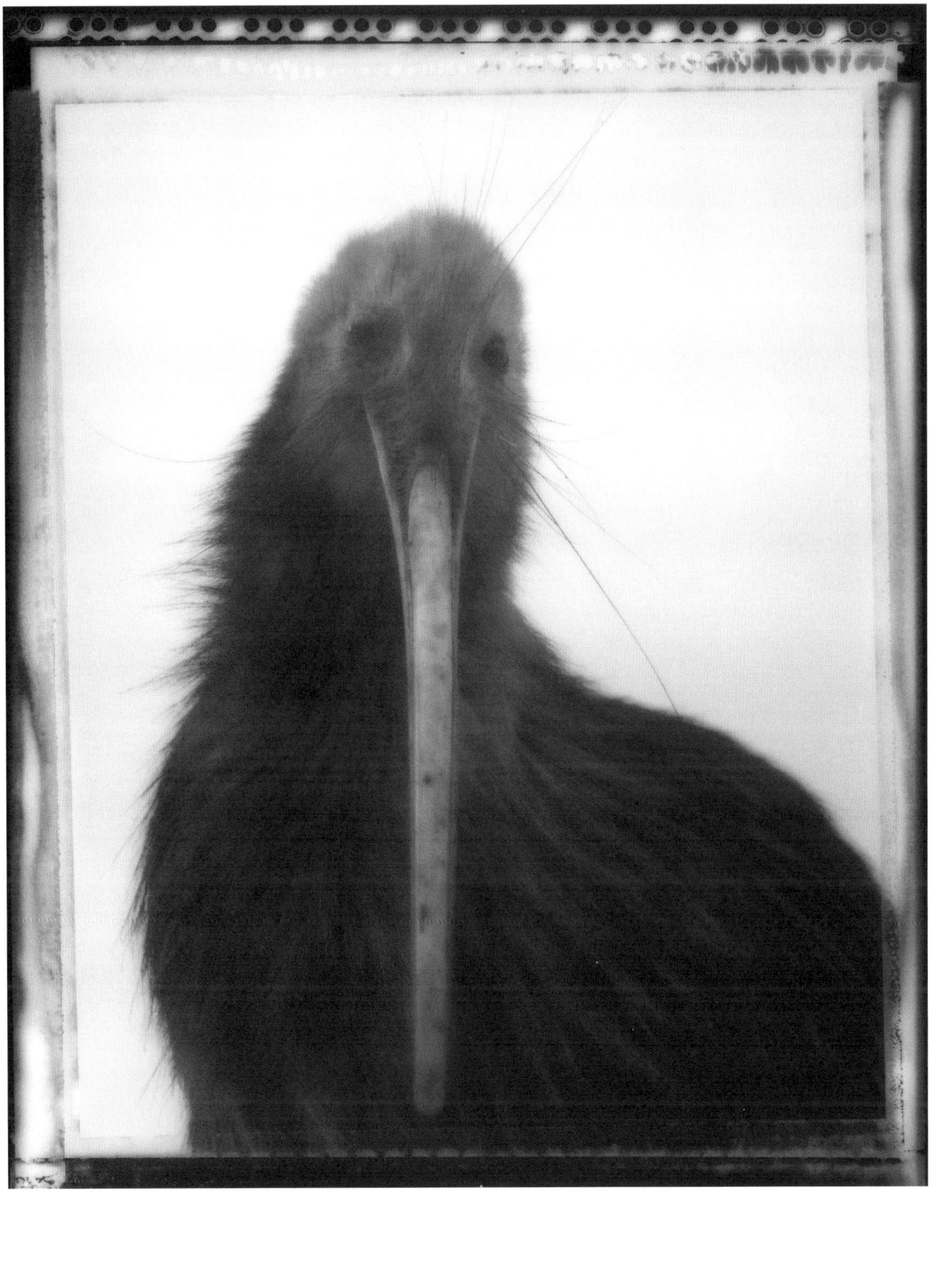

←
Herbert Ponting shot *Adélie Penguin and Chicks, Antarctica* (*c.* 1911) while acting as photographer for Robert Falcon Scott's Antarctic expedition of 1910–13. The Adélie penguin (*Pygoscelis adeliae*) is a vital link in the food chain, eating krill and small fish and being eaten by leopard seals and orcas.

↑
◎ In a challenge to established ways of documenting nature, New Zealand photographer Fiona Pardington's *Sweet Kiwi, Whaganui Museum* (2008) features a North Island brown kiwi, or kiwi-nui (*Apteryx mantelli*) from the museum's collection of taxidermied native birds.

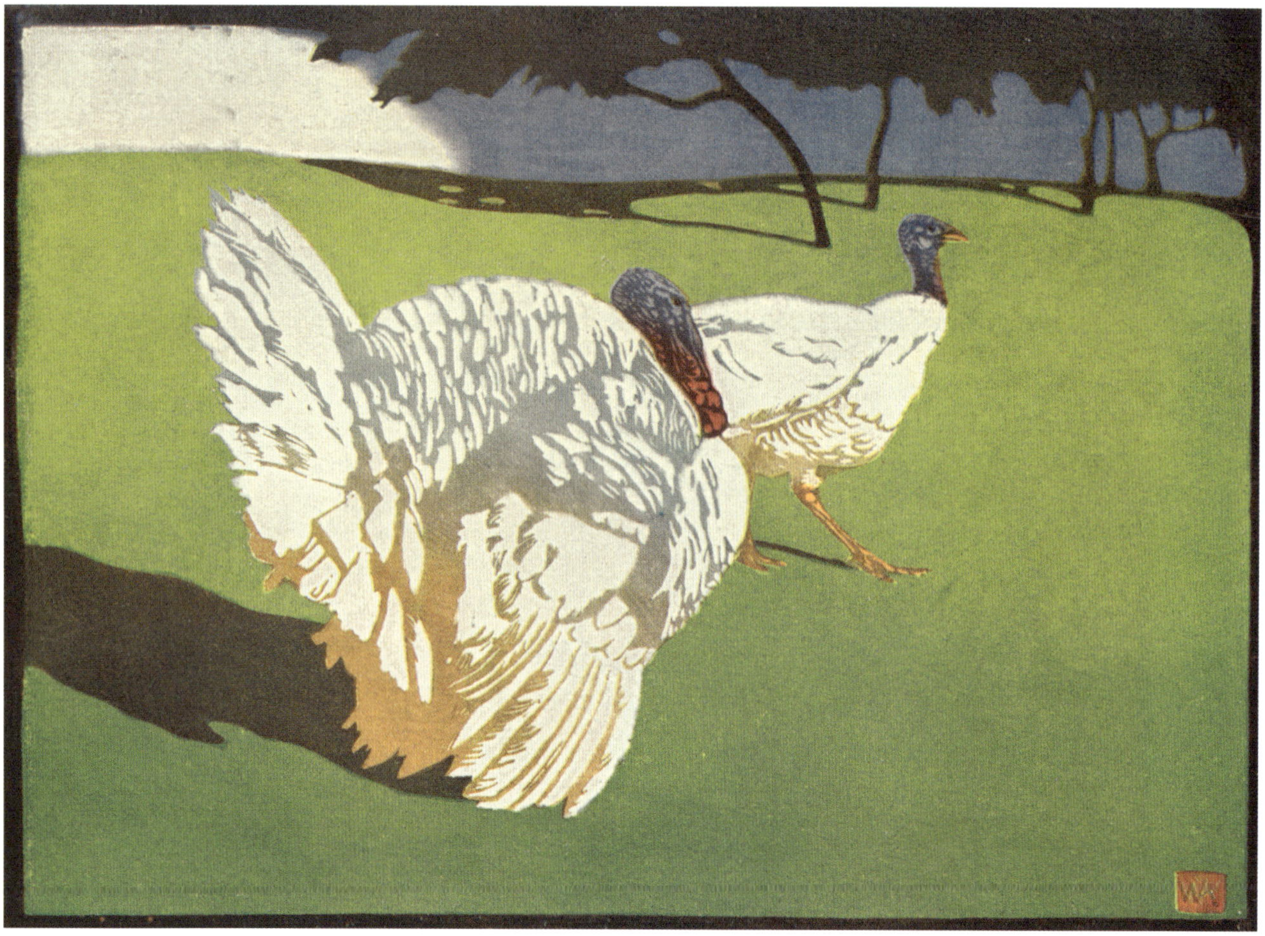

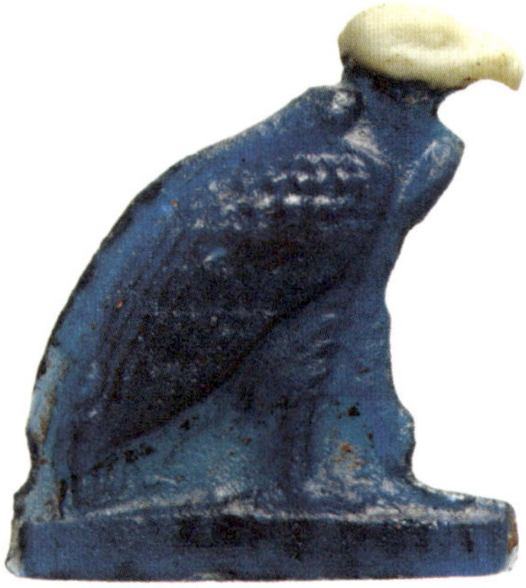

↑
German artist Walter Klemm was inspired by the bold, simplified shapes of Japanese printmakers. His *Turkeys* (1907), a coloured woodcut, was published in the magazine *Deutsche Kunst und Dekoration* ('German Art and Decoration').

←
This Egyptian glass vulture amulet dates from the Ptolemaic Period (400–30 BCE). In ancient Egypt, the vulture was a symbol of divine protection, and the amulet would have been tucked into the bandages wrapped around the body during mummification.

→
'The Concourse of the Birds' (*c.* 1600) by the Iranian artist Habiballah of Sava illustrates a scene in the *Mantiq al-Tayr* (*Language of the Birds*), an allegorical poem by the Sufi poet Farīd al-Dīn ʿAṭṭār about the human quest for the Divine. The birds, which each symbolize a soul, are about to begin a pilgrimage under the leadership of a Eurasian hoopoe (*Upupa epops*) (centre right, stood on a rock).

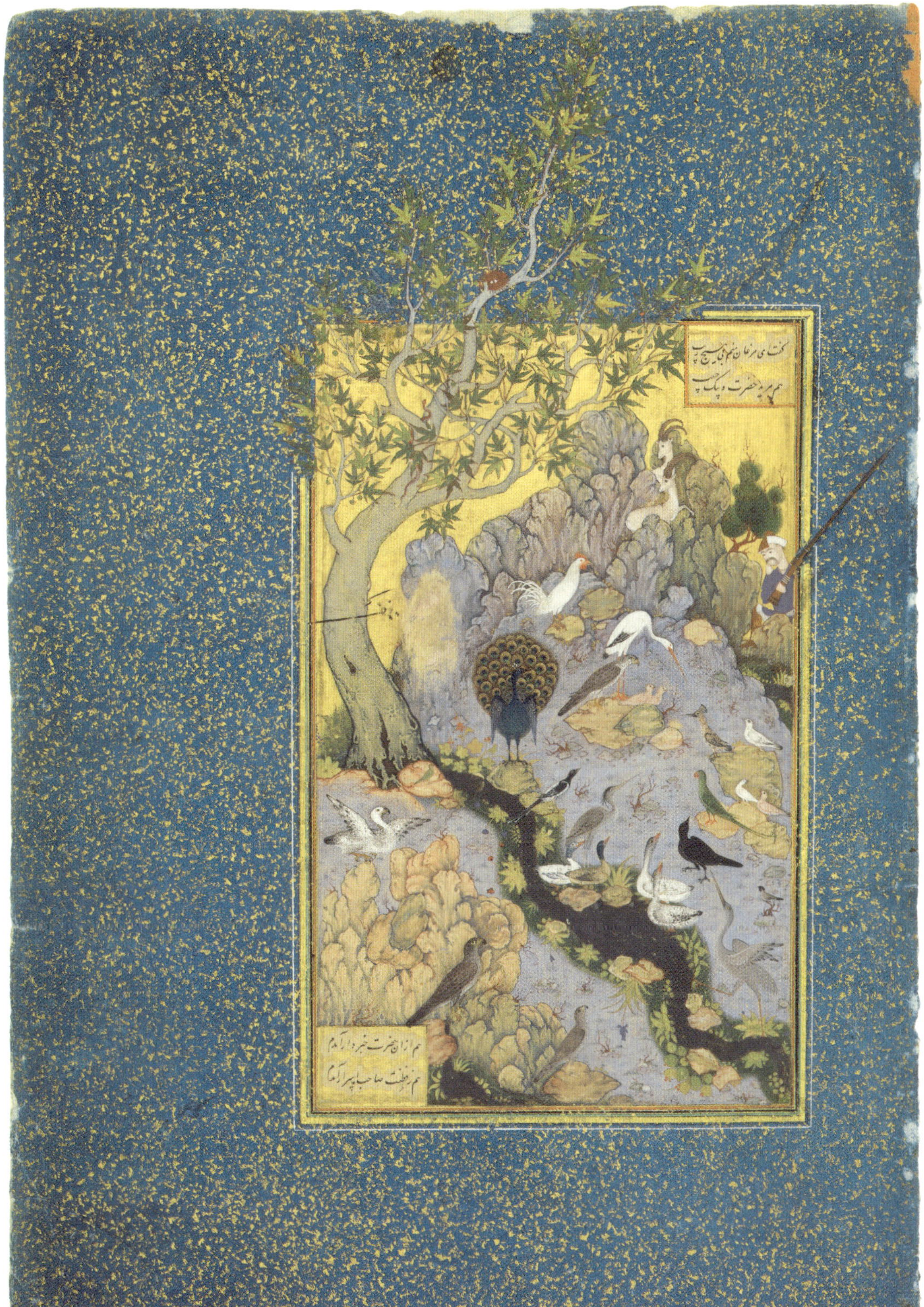
گفت ای مرغان منم بی هیچ ریب
هم برید حضرت و پیک غیب
هم ازان حضرت خبردار آمدم
هم ز فطنت صاحب اسرار آمدم

↓
John Constable's *The Skylark, Dedham* (*c*. 1830) captures a skylark (*Alauda arvensis*) hovering high over the Suffolk landscape. Males have been known to reach a height of 300 metres (1,000 feet) while delivering their melodious, unbroken songs, and their flights can last up to an hour. Skylark numbers are in decline due to habitat loss.

→
In *The Magpie* (1868–69), Claude Monet depicts a lone Eurasian magpie (*Pica pica*) perched on a gate in the winter sun. These intelligent and often raucous members of the crow family are wreathed in folklore. In China and Korea they symbolize joy, but in English-speaking countries they have long been seen as a harbinger of bad luck.

↘
American artist Lois Dodd almost always paints *en plein air*, whether in the city or, in the case of *Chickens* (1957–58), in rural Maine, USA.

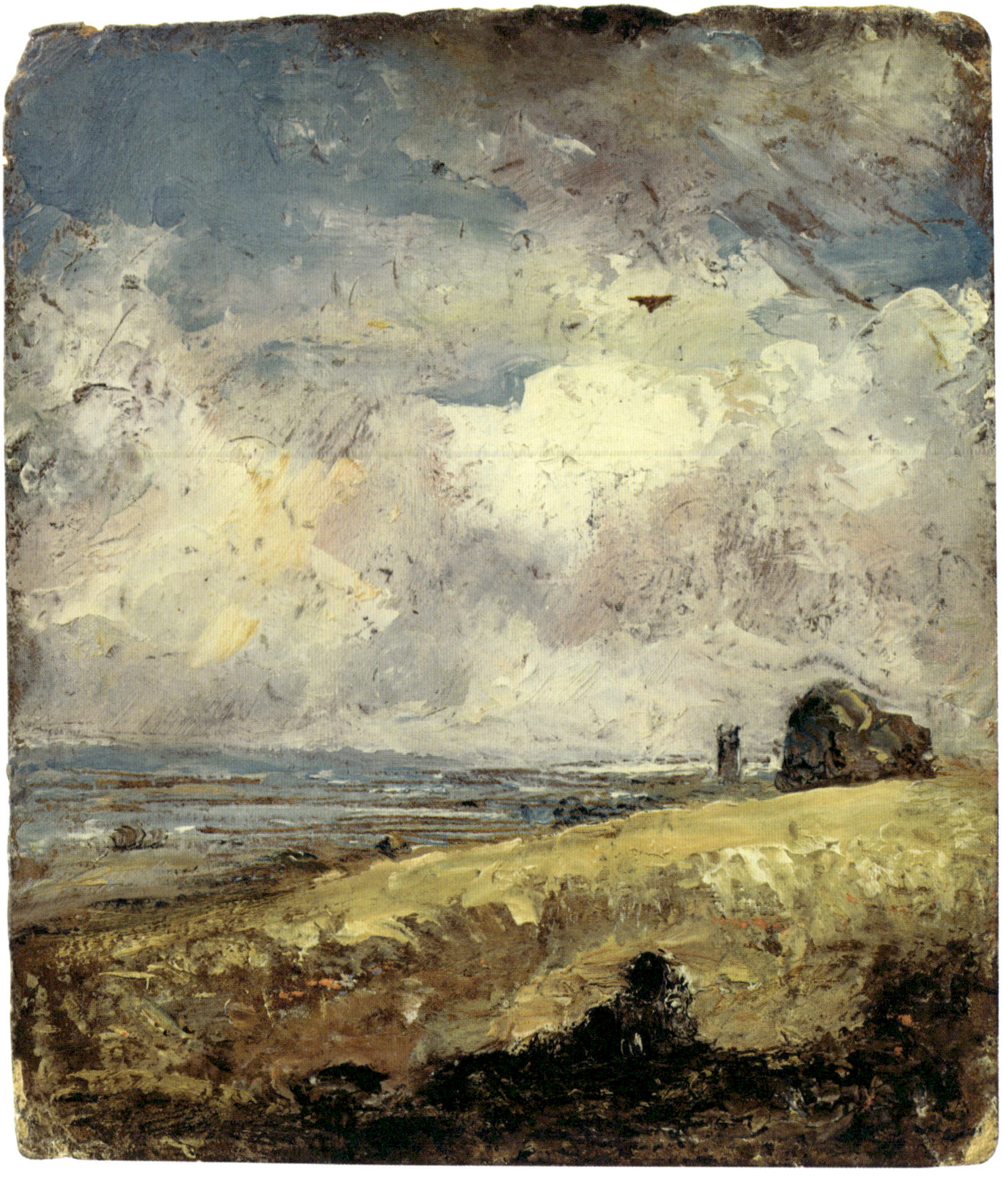

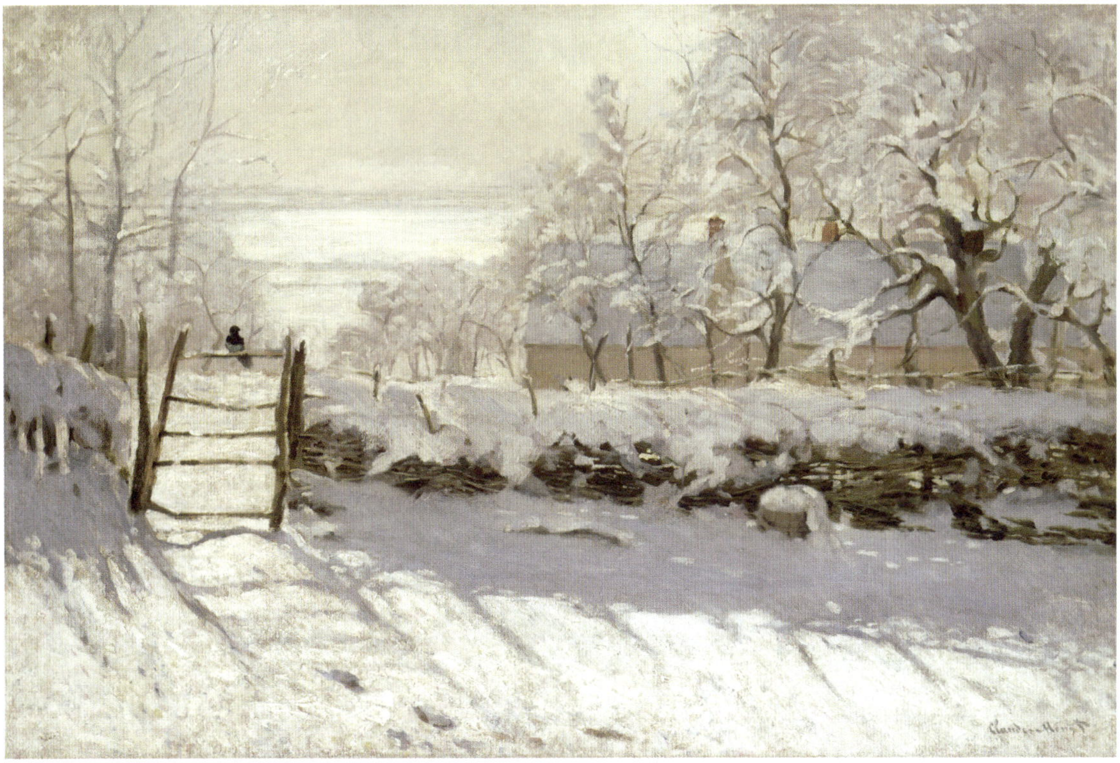

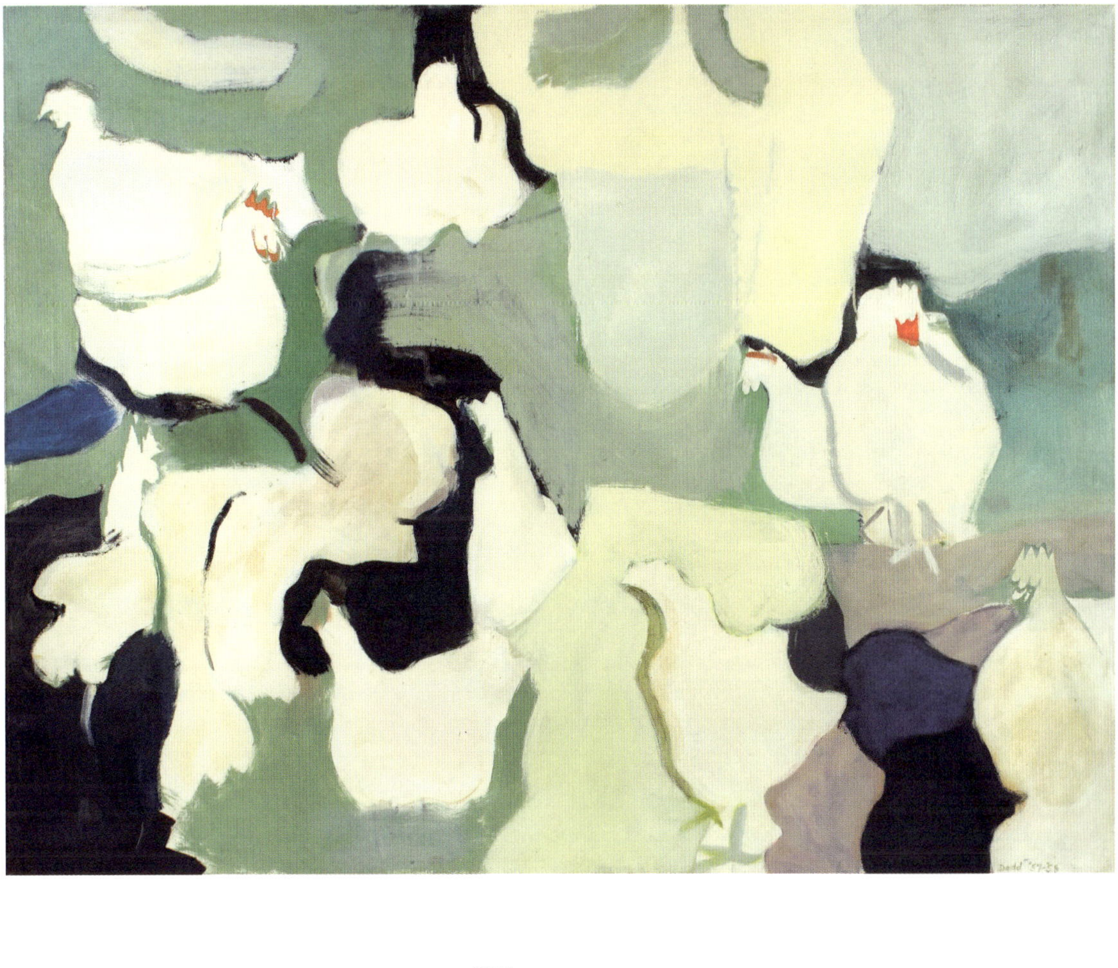

(VI)

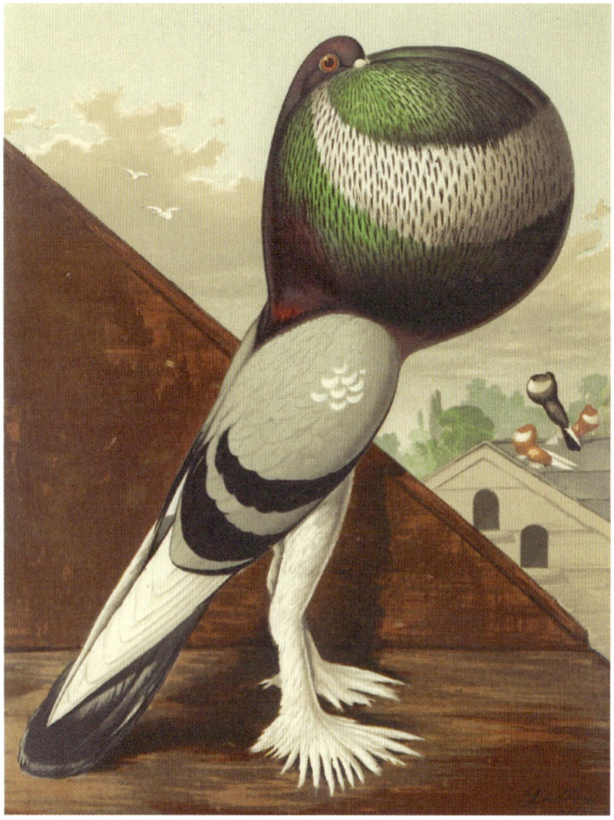

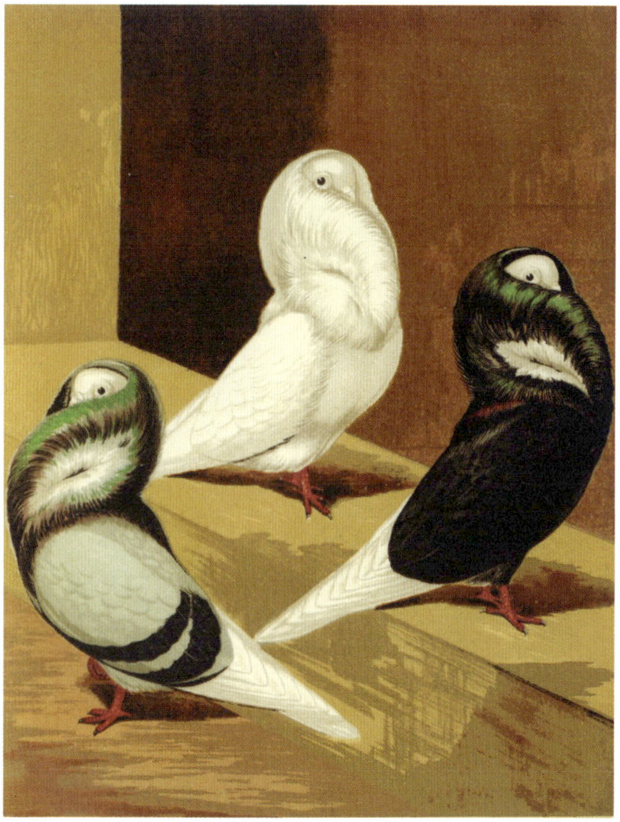

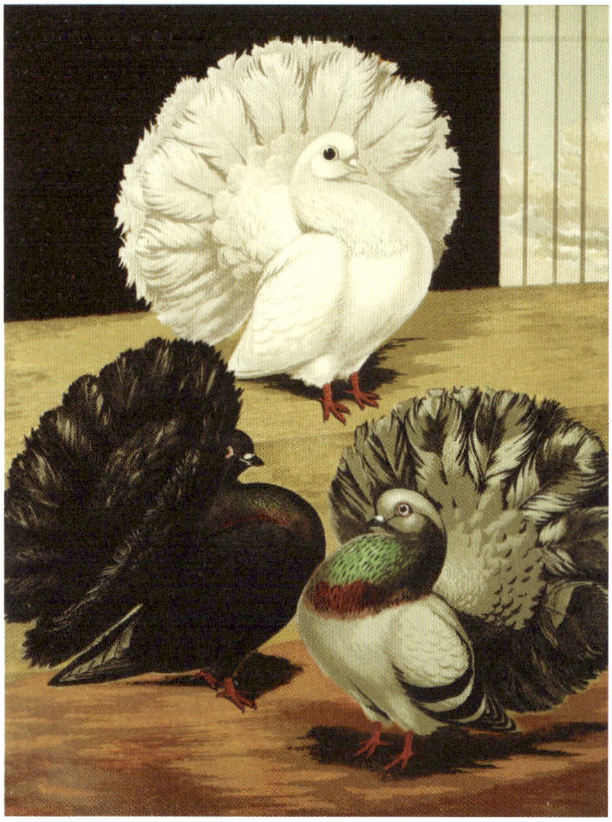

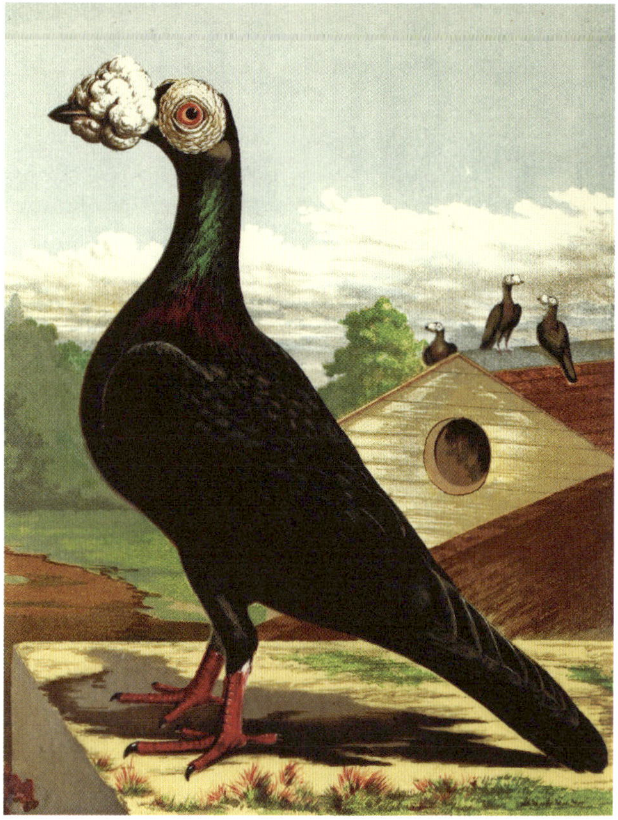

←
These four breeds of domesticated show pigeons are from *The Illustrated Book of Pigeons: With Standards for Judging*, published in 1876. Clockwise from top left: the blue pouter cock showing off its inflatable crop, Jacobins, the black carrier cock, and English fantails.

→
Albrecht Dürer's watercolour of *Three Studies of Tree Bullfinches* (1543) shows a male Eurasian bullfinch (*Pyrrhula pyrrhula*) from three directions. Bullfinches were often kept as cage birds in the past, and they were regarded as pests as they eat the buds on fruit trees. Their numbers are in decline due to the loss of woodland, orchards and hedgerow trees.

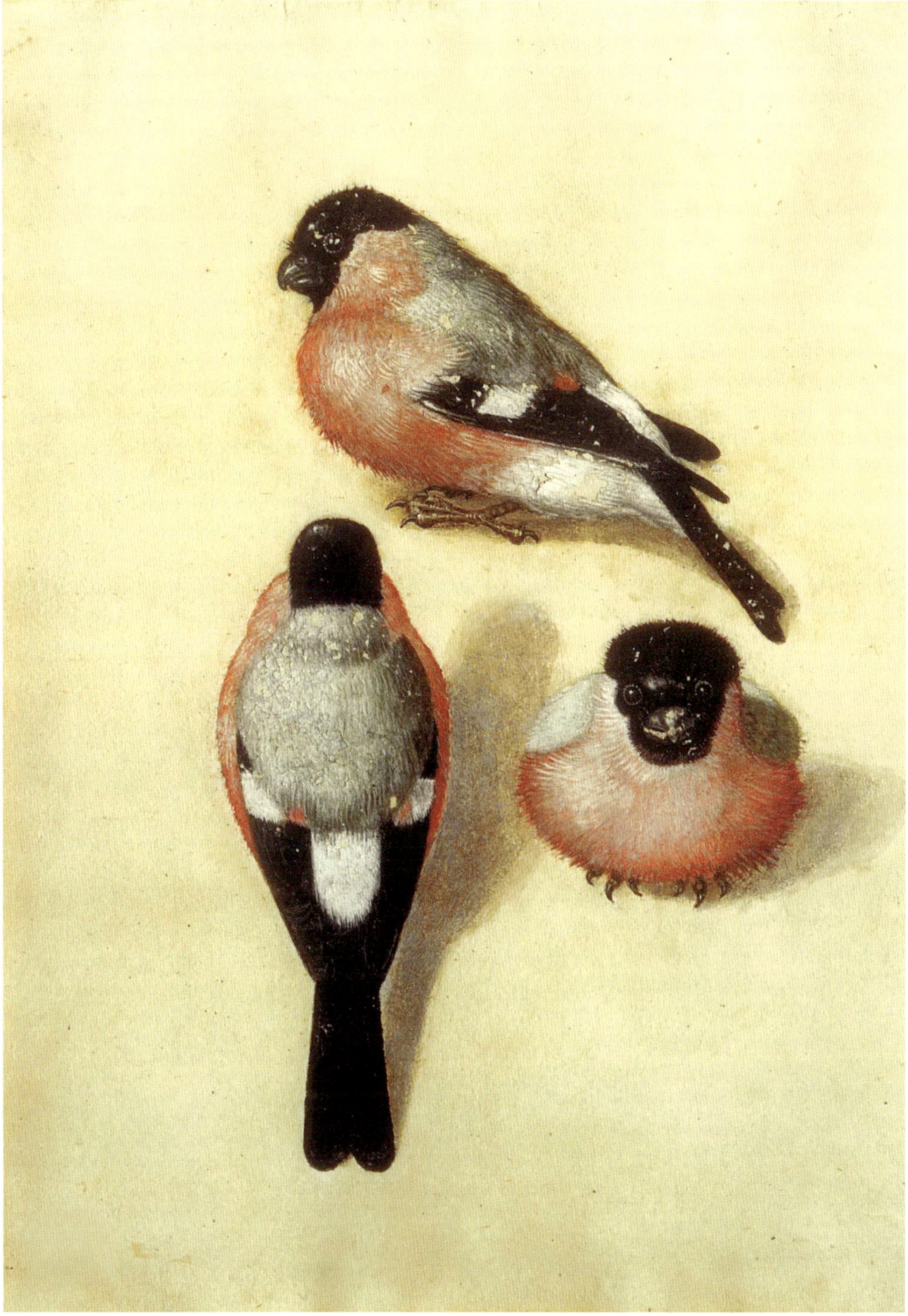

'A pair of pigeons will commonly sell at Rome for two hundred nummi, if they are well made, of good colour, without blemish, and of good breed.'

LUCIUS CORNELIUS MERULA (CONSUL IN 193 BCE), QUOTED IN *ROMAN FARM MANAGEMENT: THE TREATISES OF CATO AND VARO*, TRANSLATED BY FAIRFAX HARRISON

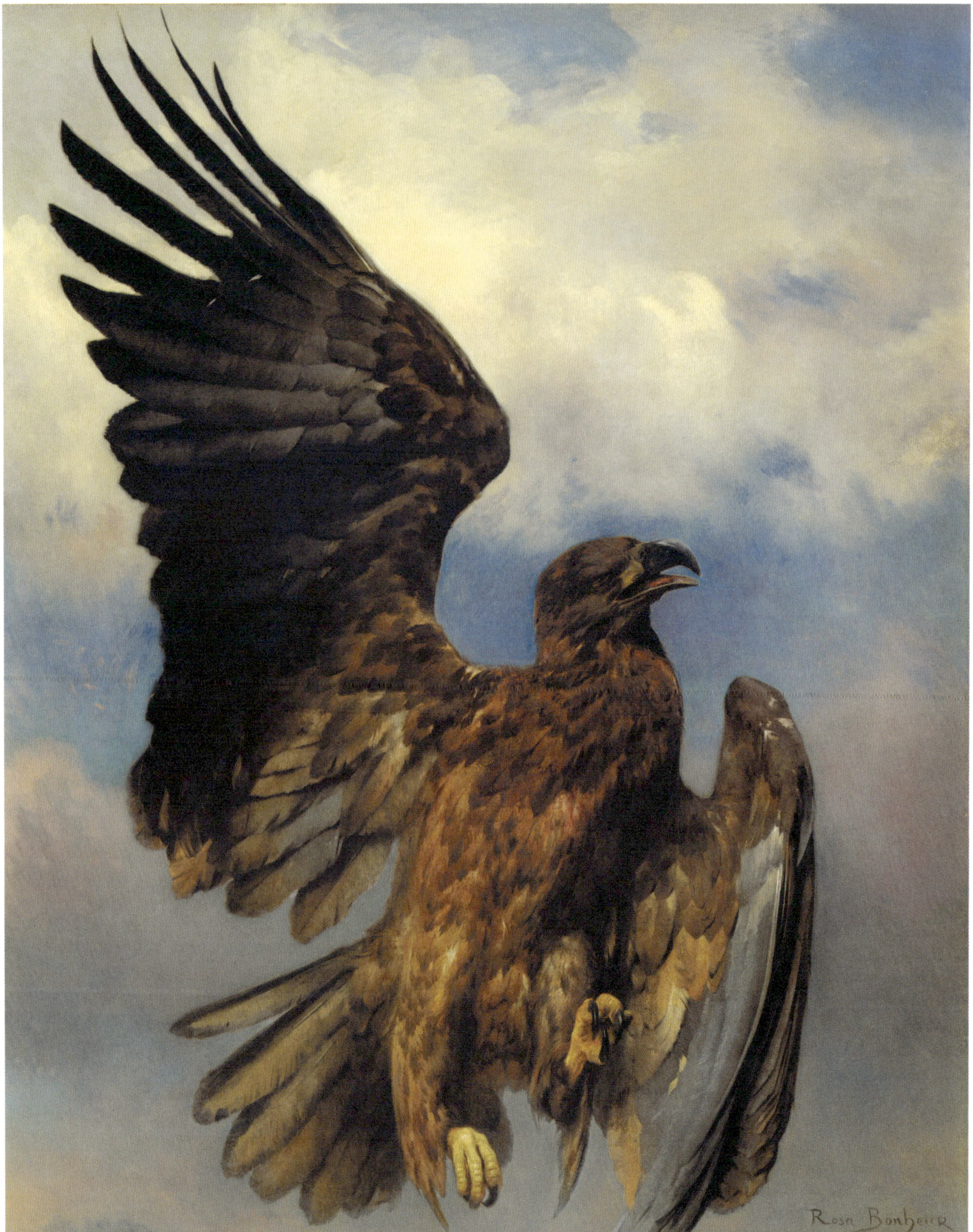

(VI)

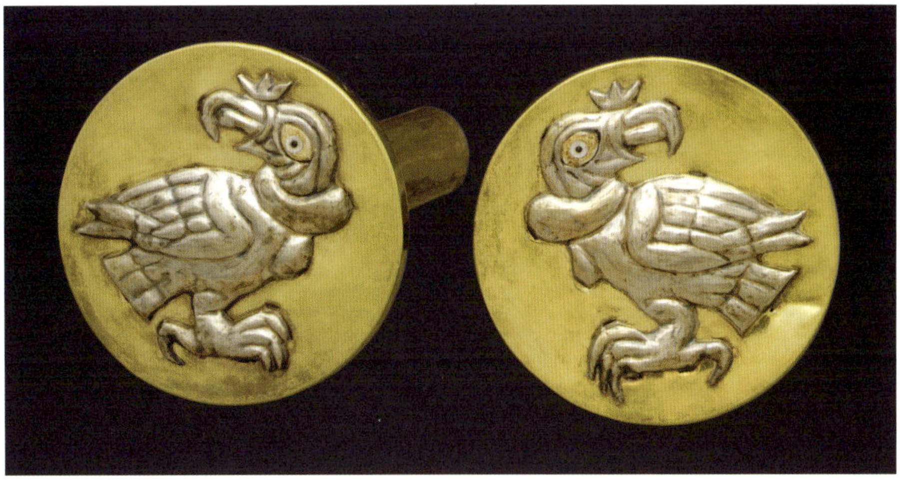

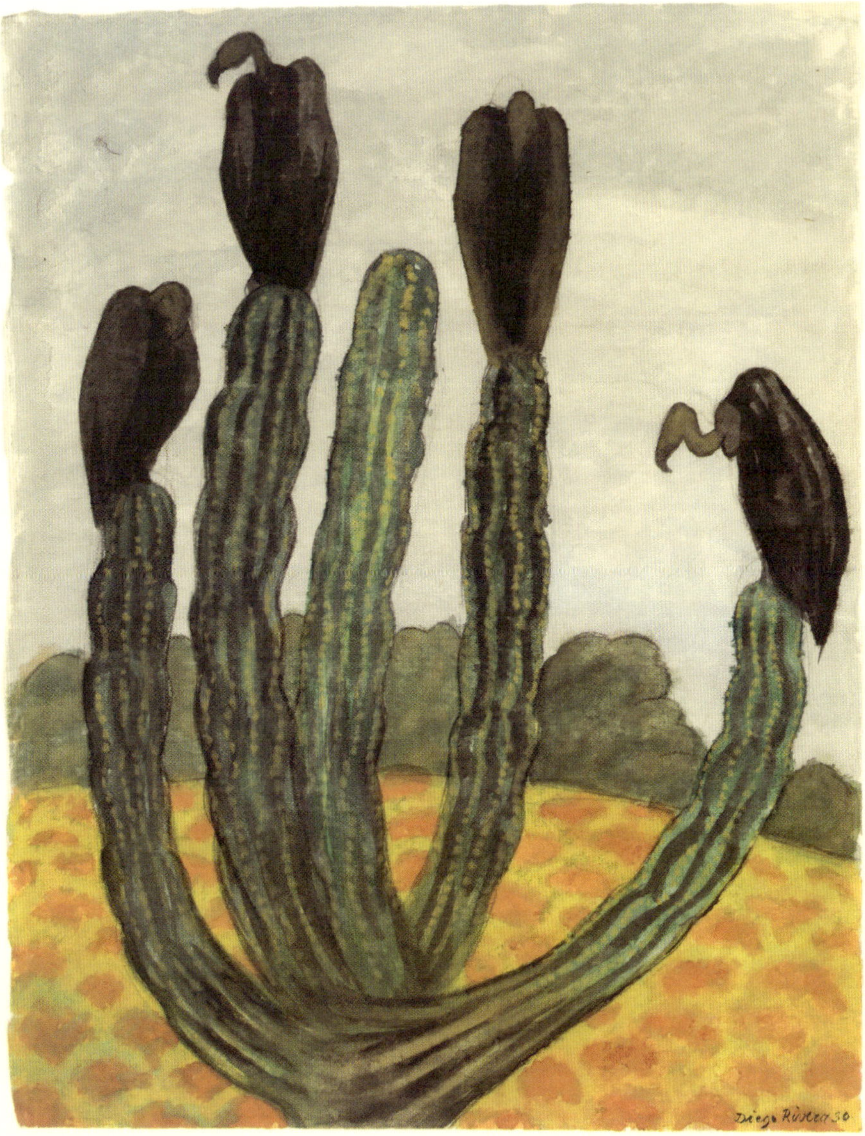

←
In *The Wounded Eagle* (*c.* 1870), the French painter Rosa Bonheur has captured the golden eagle (*Aquila chrysaetos*) in mid-flight. This massive bird of prey has a wingspan of up to 210 centimetres (7 feet).

↗
◎ Andean condors (*Vultur gryphus*) adorn this 6th–7th century pair of earflares, measuring 10 centimetres (4 inches) in diameter, from the Moche culture of Peru, whose people associated the condor with death and sacrifice. The condor feeds on large mammals, including livestock, and is persecuted by humans.

→
In his watercolour *Vultures on Cactus* (1930), Diego Rivera has captured the characteristic pose of vultures, which perch in groups on desert cactus.

'We need to witness our own limits transgressed, and some life pasturing freely where we never wander. We are cheered when we observe the vulture feeding on the carrion which disgusts and disheartens us, and deriving health and strength from the repast.'

HENRY DAVID THOREAU, *WALDEN; OR, LIFE IN THE WOODS*, 1854

(VI)

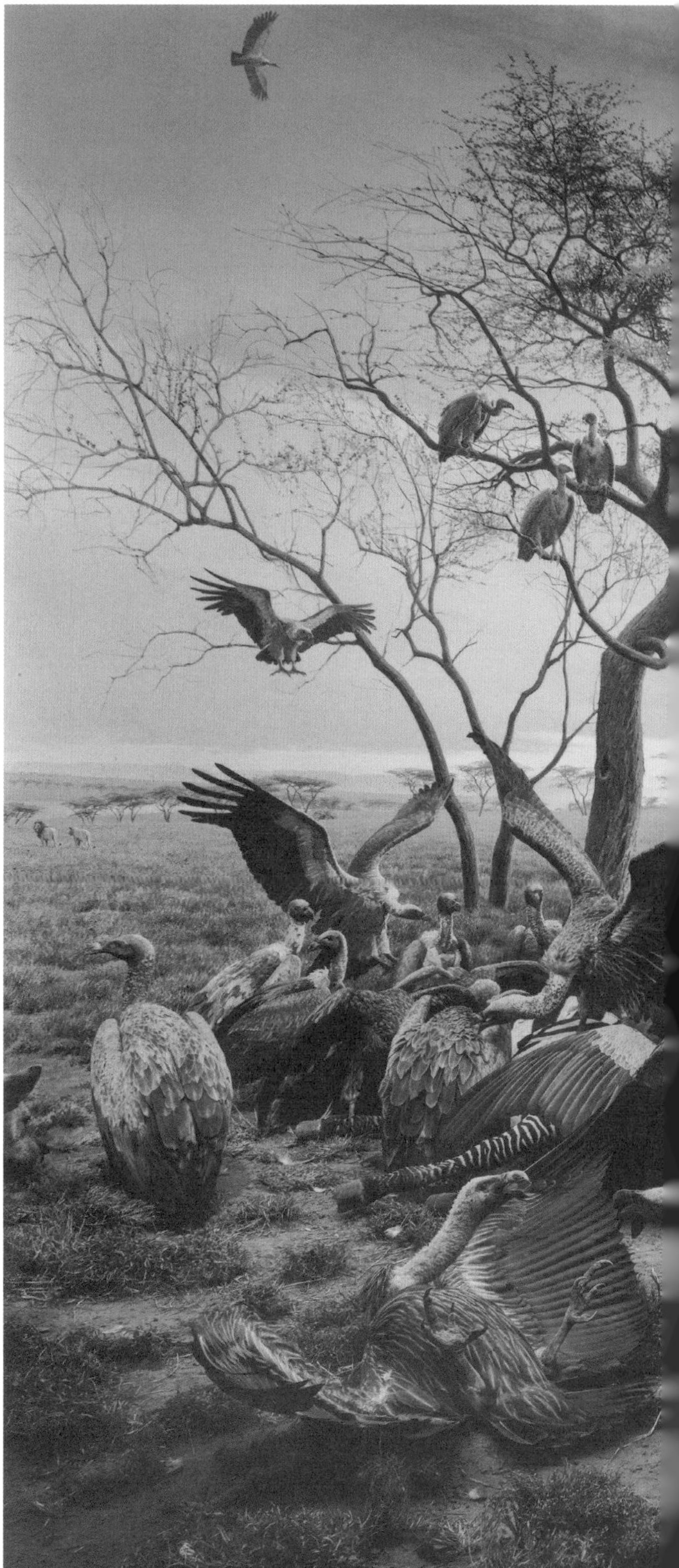

→

For his *Diorama* series, Hiroshi Sugimoto turned his lens to the dioramas at the American Museum of Natural History in New York, USA, where taxidermy set against painted backdrops begins to look real once perspective is flattened through the medium of photography. 'Hyena – Jackal – Vulture' (1976) features, perching on the branch at centre right, a Rüppell's vulture (*Gyps rueppelli*). These carrion-feeding birds of prey have been recorded flying at an altitude of 11,300 metres (37,000 feet), making them the highest-flying bird in existence. Their populations are declining as a result of habitat loss, declining availability of food and poisoning from unregulated use of pesticides.

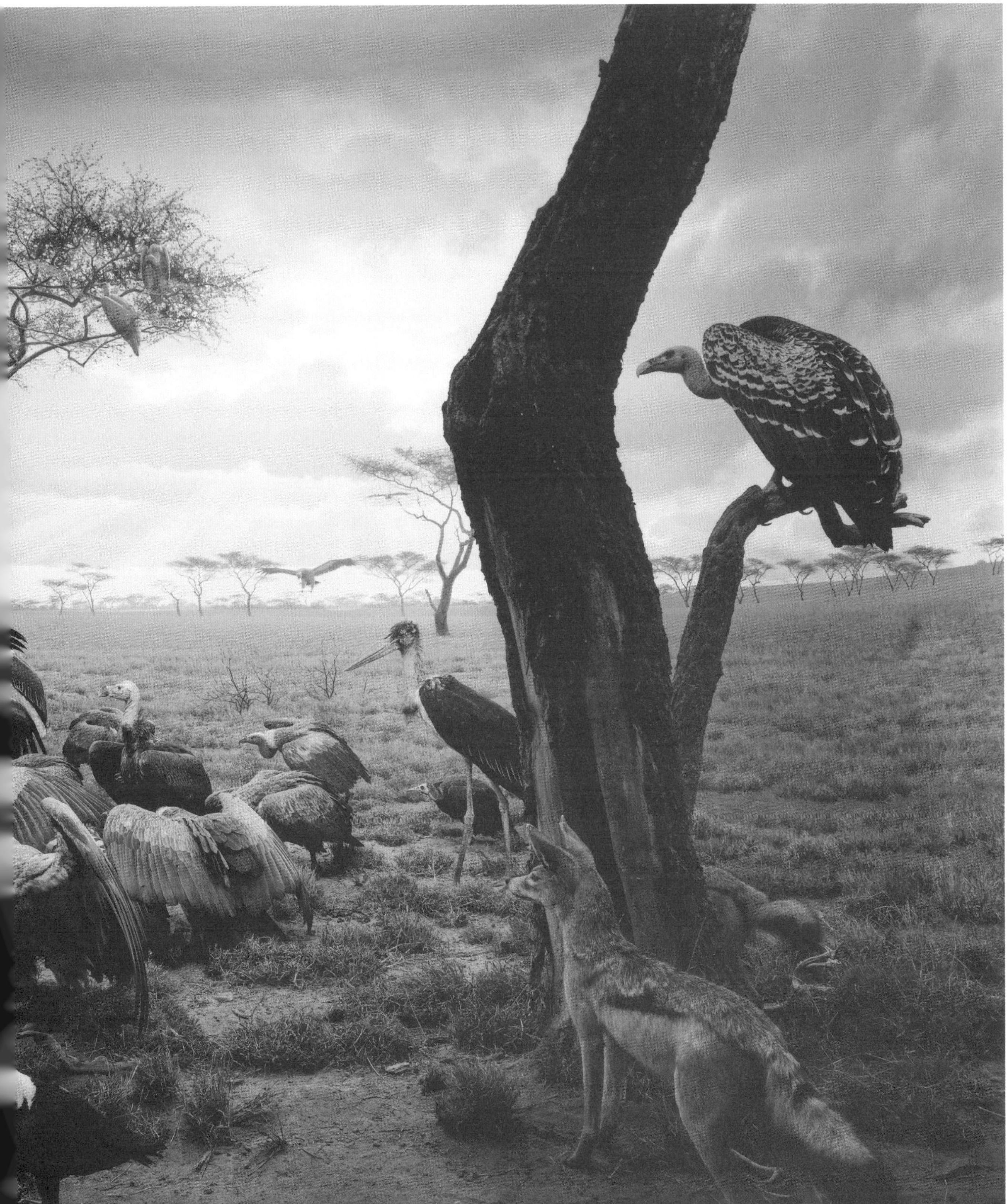

'undaunted on the gold-set cornice;
you eagle,
god's majestic legate'

EURIPIDES, *ION*, WRITTEN 414–412 BCE, TRANSLATED BY H. D.

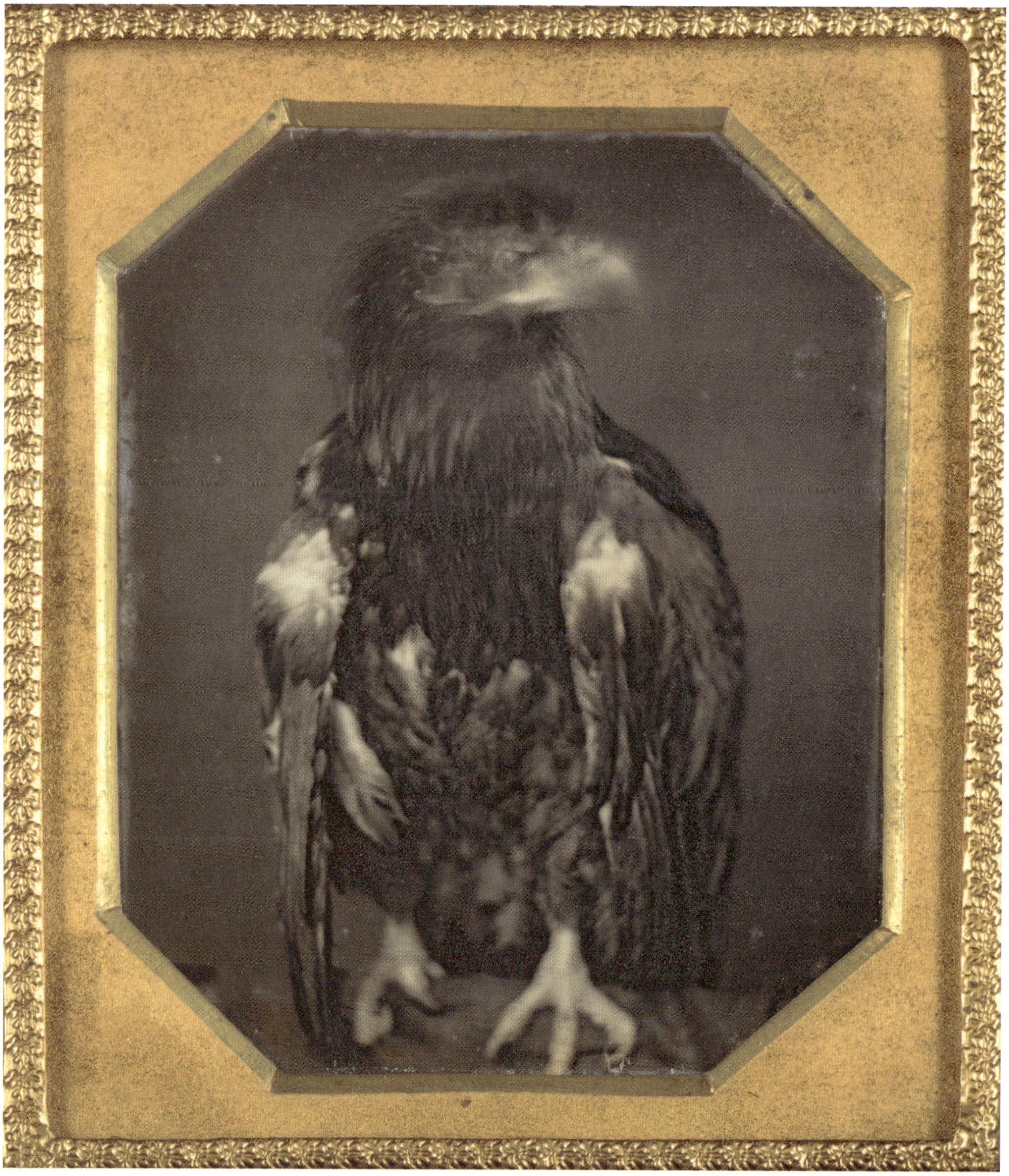

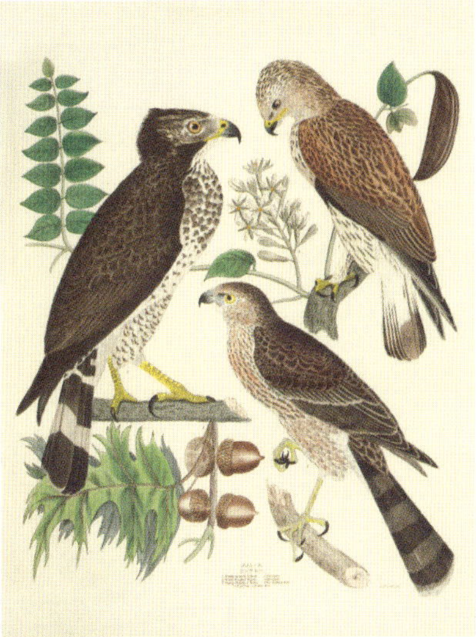

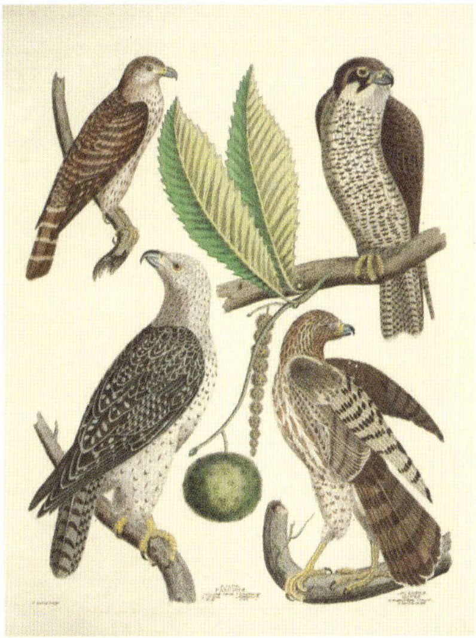

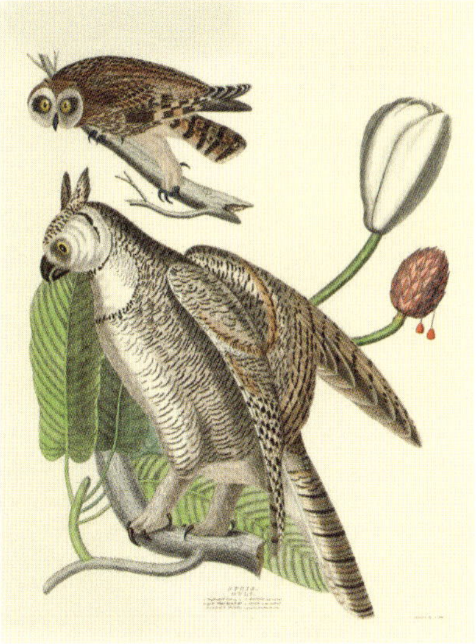

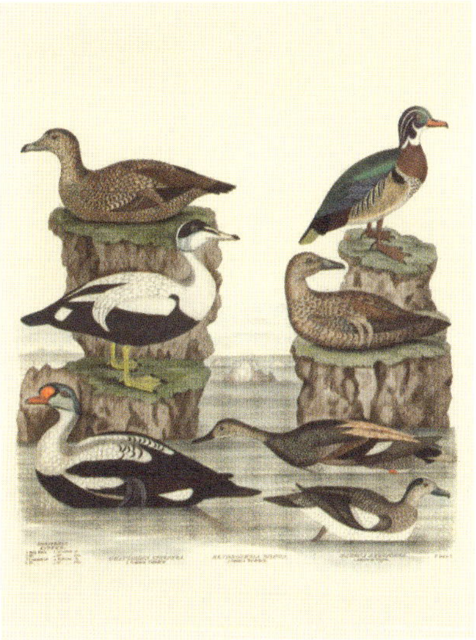

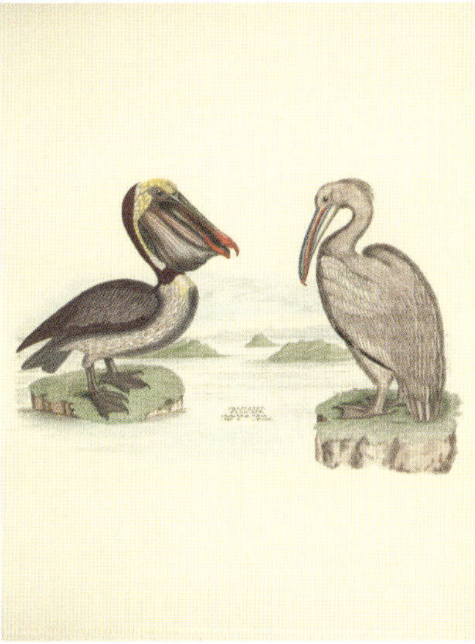

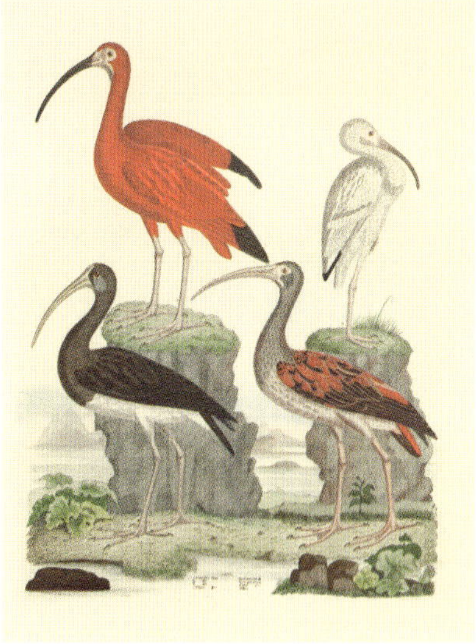

←
This daguerreotype of a juvenile bald eagle (*Haliaeetus leucocephalus*) dates from around 1850. In the past, hunting and pesticide use led to a decline in the bald eagle population, but conservation measures are bringing about a growth in numbers again.

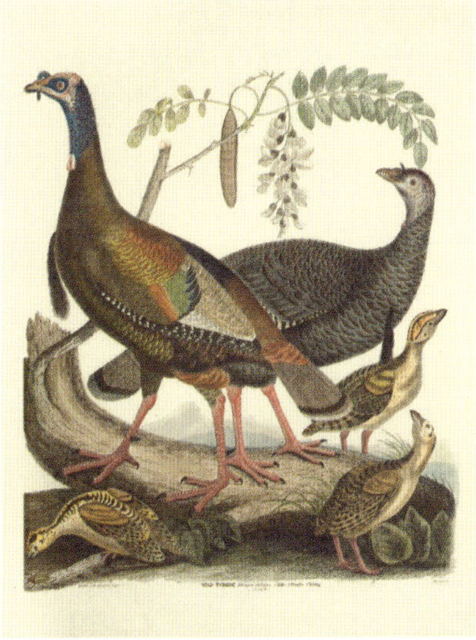

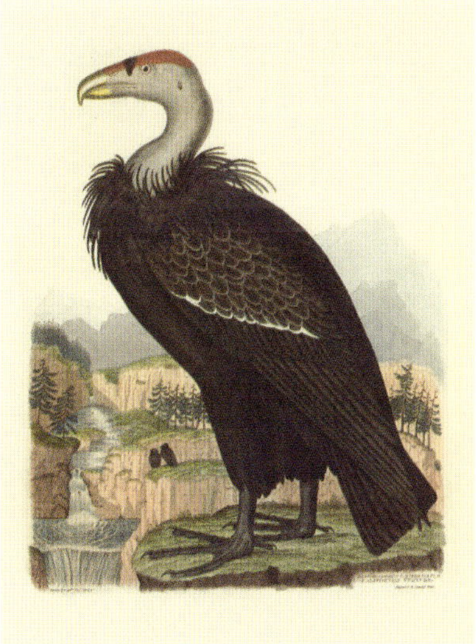

→
⦶ This selection of plates from *American Ornithology* (1835) by Alexander Wilson and Charles Lucian Bonaparte includes hawks, falcons, owls, ducks and ibises. The California condor (*Gymnogyps californianus*), bottom right, became extinct in the 1980s, but since the 1990s captive-bred birds have been released into the wild in parts of California.

(VI)

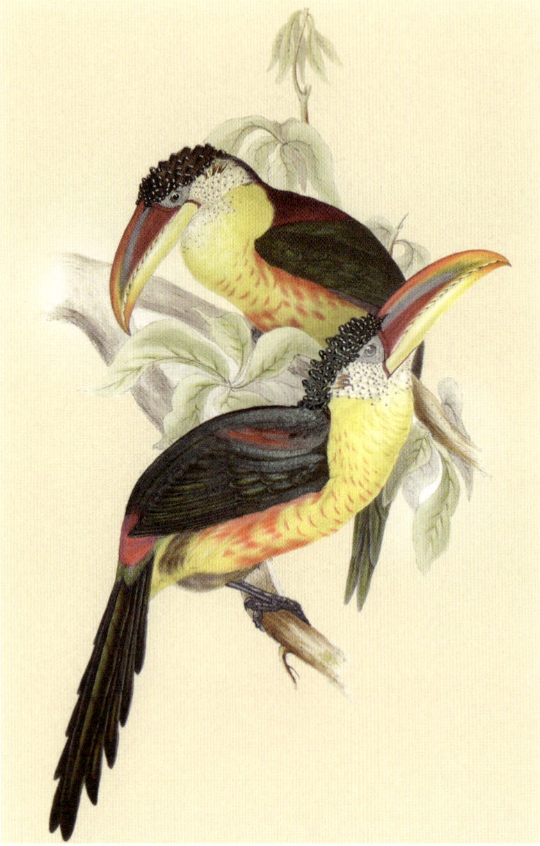

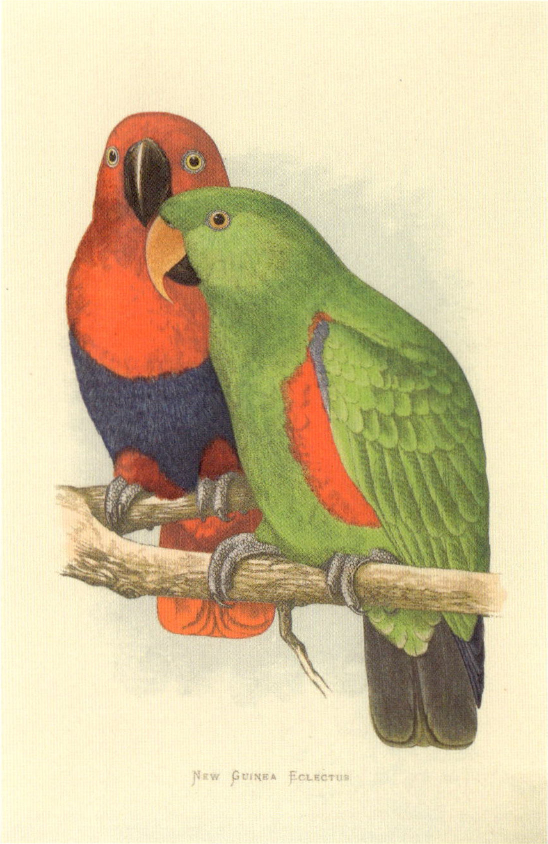

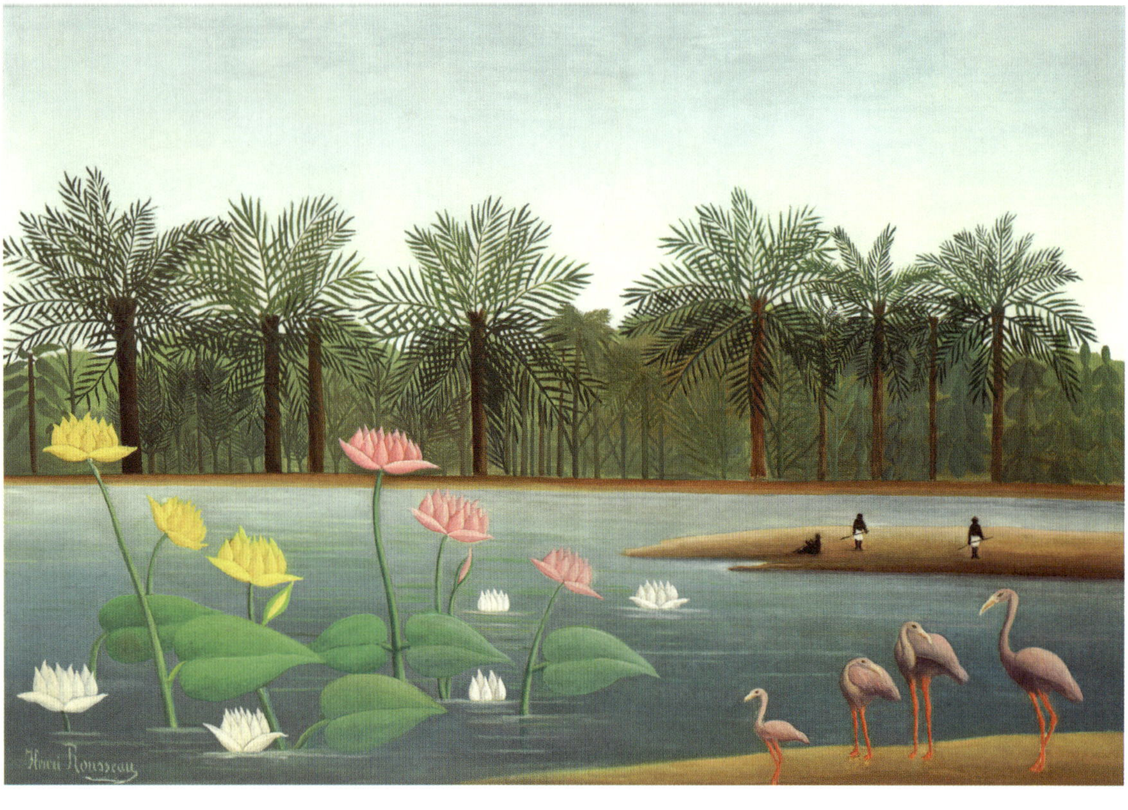

'The most fanciful imagination would fail in picturing to itself anything more odd than the conformation of this bird.'

LOUIS FIGUIER, ON THE FLAMINGO, *REPTILES AND BIRDS*, 1870

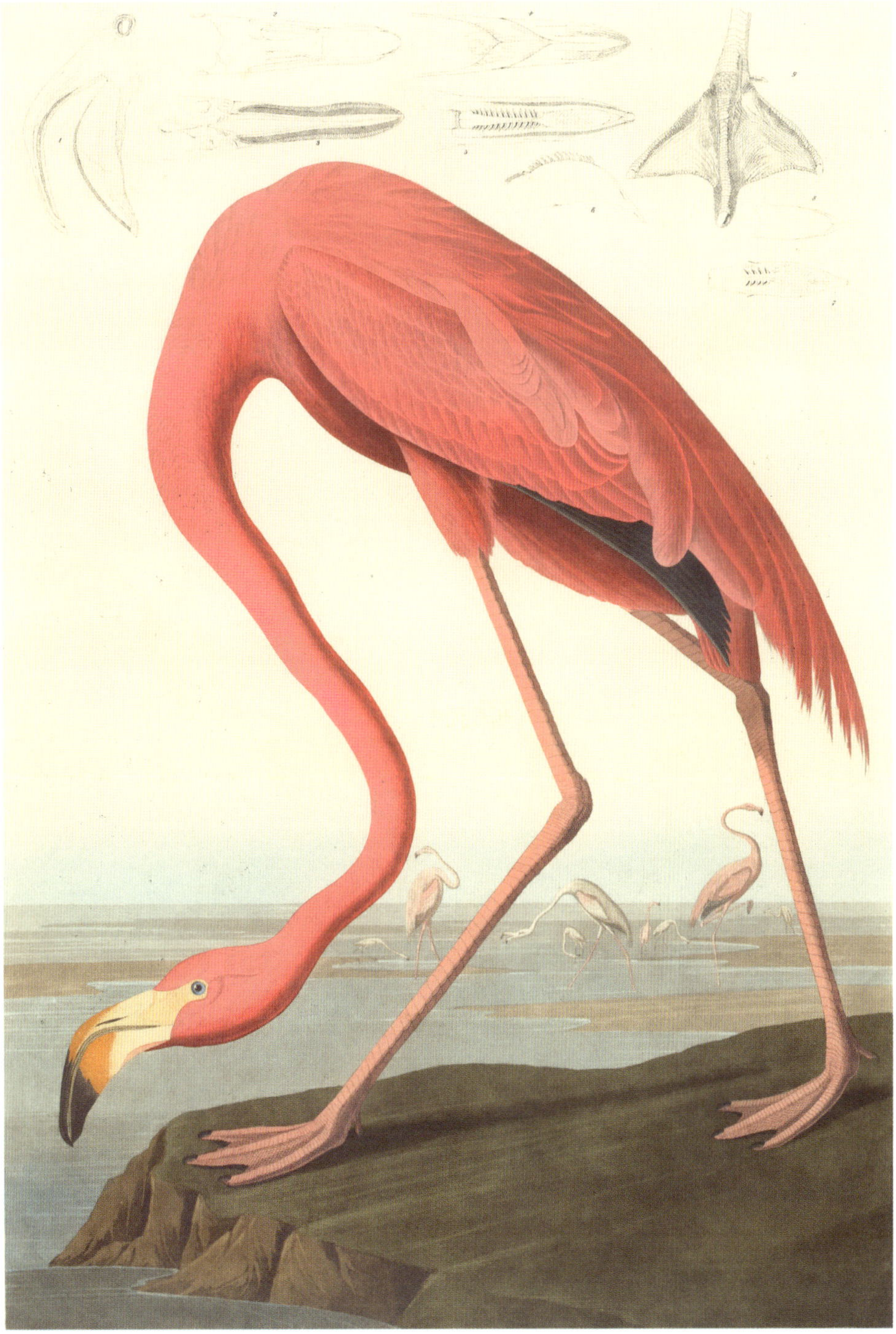

↖
The plate on the left is a hand-coloured lithograph from John Gould's *Monograph of the Ramphastidae, or family of Toucans*, featuring the chestnut-mandibled toucan (*Ramphastos ambiguus swainsonii*). A pair of New Guinea eclectus parrots (*Eclectus polychloros*) are pictured on the image on the right, with a female on the left and a male on the right. The sexes look so different that they were thought to be two species until the late 19th century.

←
Les Flamants ('The Flamingoes') by Henri Douanier Rousseau, painted in 1910, shows a world of his own creation based on illustrations in magazines, encyclopaedias, botanical studies and visits to zoos and botanical gardens.

→
The American flamingo (*Phoenicopterus ruber*), illustrated in Audubon's *Birds of America*, was once native to Florida, but disappeared around the beginning of the 20th century. Of those seen in Florida now, some have escaped from captivity and some may have arrived from other areas in the species' range, which includes the Caribbean, Central America and the Galápagos Islands.

'But now they drift on the still water,
Mysterious, beautiful;
Among what rushes will they build,
By what lake's edge or pool
Delight men's eyes when I awake some day
To find they have flown away?'

W. B. YEATS, 'THE WILD SWANS AT COOLE', 1919

↓
Belgian artist William Degouve de Nuncques was known for nocturnal landscapes, such as *Le Cygne Noir* ('Black Swan'), a pastel drawing from 1895.

→
This illustration, from Thomas Skottowe's 1813 book of animal specimens from New South Wales, shows the superb lyrebird (*Menura novaehollandiae*) fanning its tail (top). This species is a good mimic of sounds in its environment. Also featured in Skottowe's book is the black swan (*Cygnus atratus*) (bottom), a species which has a symbolic role in Australia, and appears on the state flag of Western Australia.

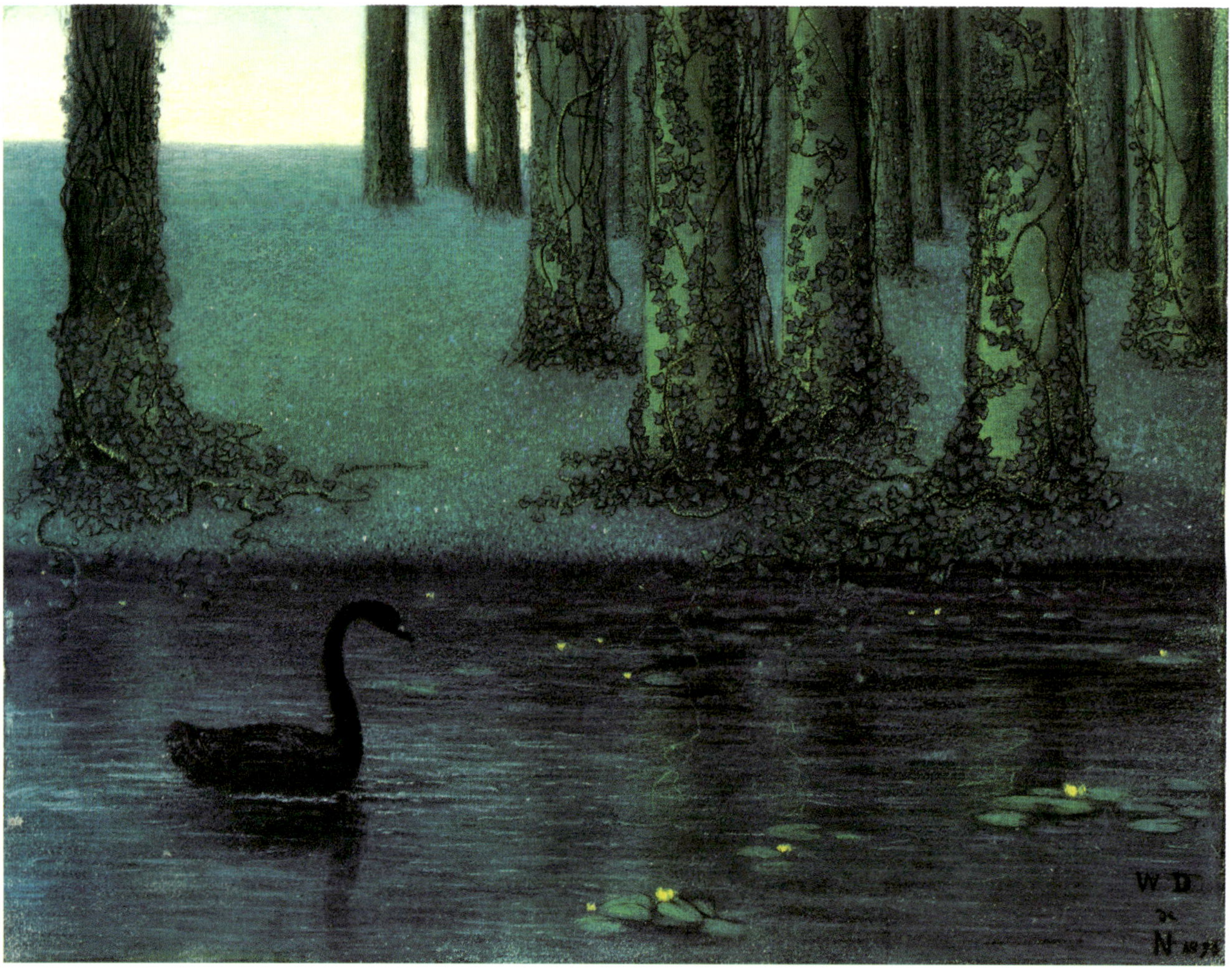

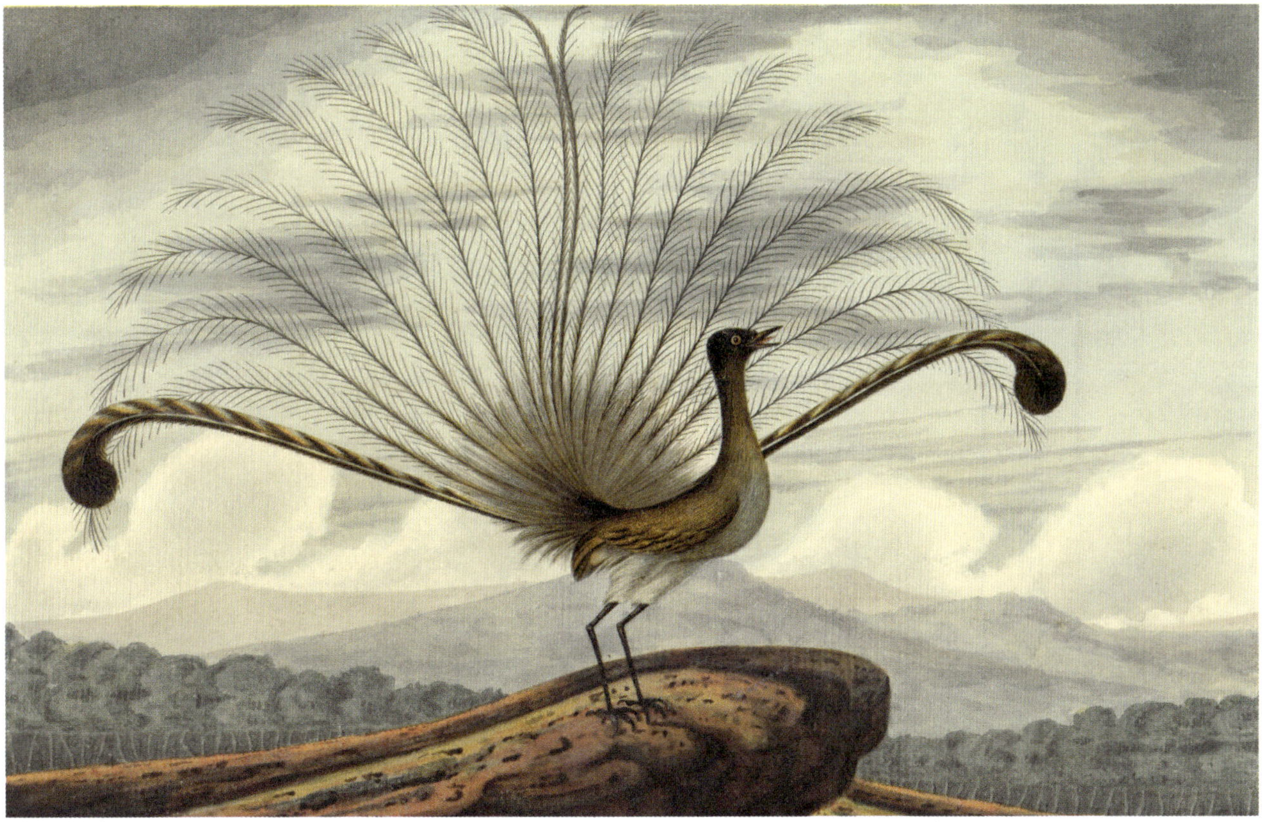

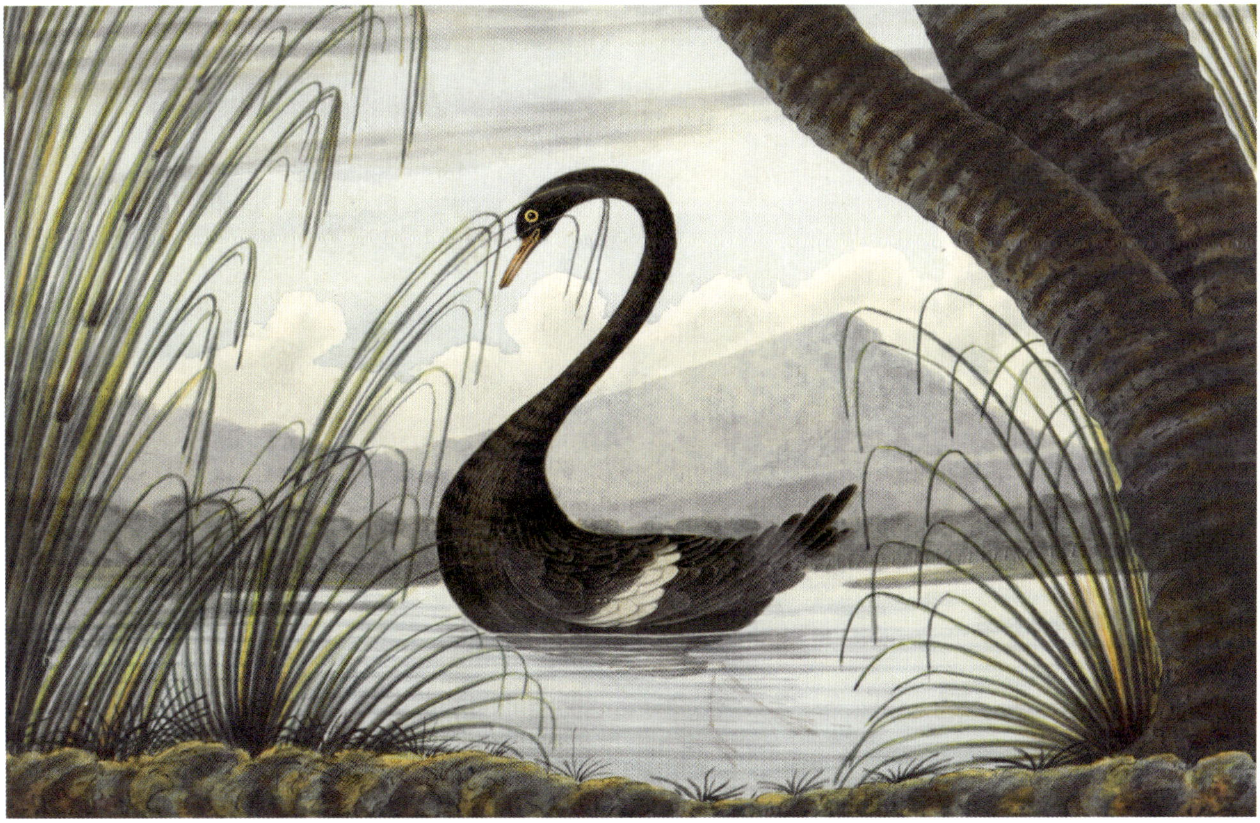

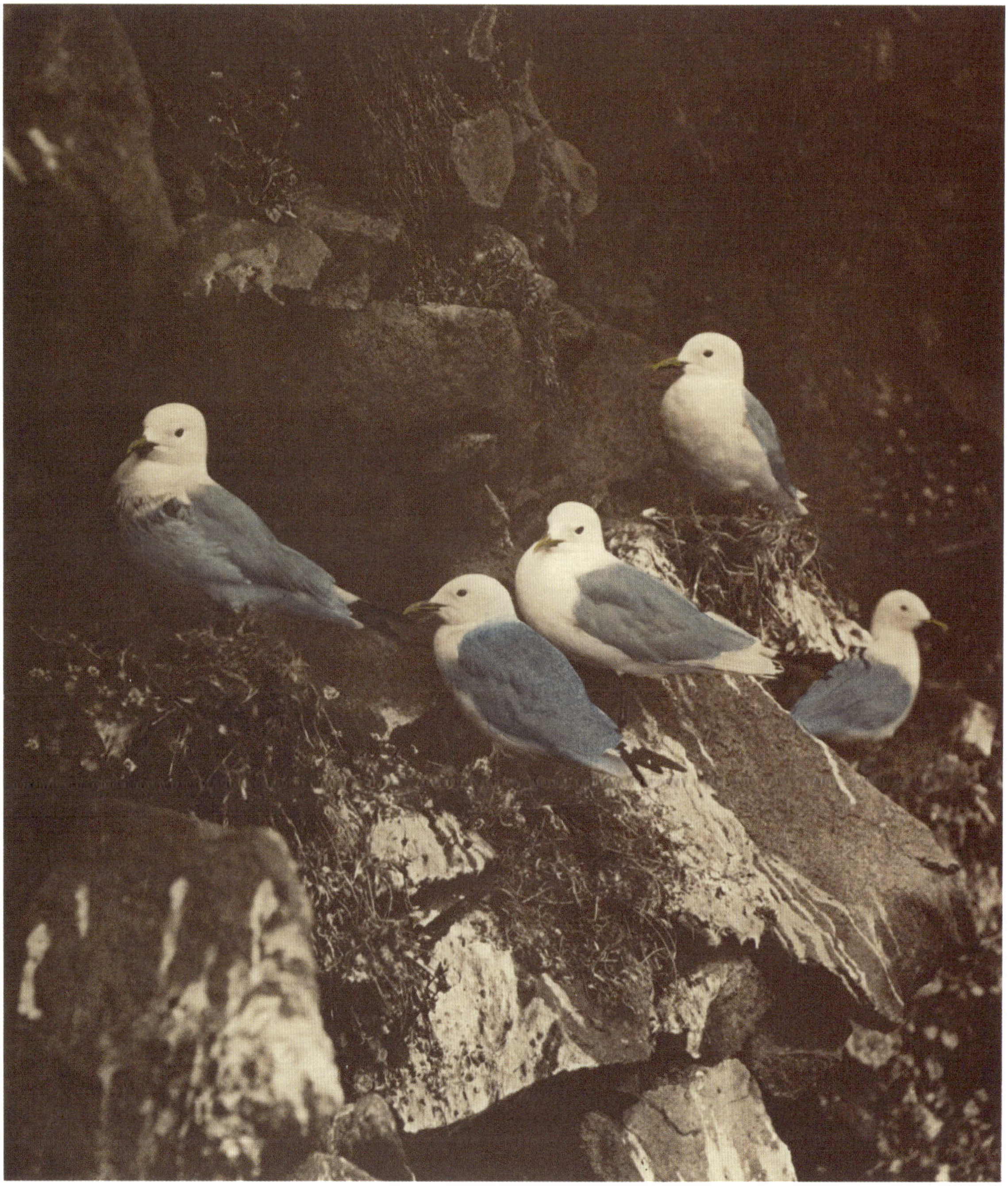

↑
◎ The black-legged kittiwake (*Rissa tridactyla*) is the subject of this hand-coloured print *Kittiwakes at Home* (negative 1896, print 1905) by Cherry Kearton, British naturalist and a pioneer of wildlife photography. Kittiwakes nest around the UK coast in summer, but their numbers are declining due to a shortage of sand eels in UK waters, which are an important part of their diet.

↗
The *Album of Abyssinian Birds and Mammals* (1930), illustrated by Louis Agassiz Fuertes, includes thirty-two paintings of African wildlife, including the silvery-cheeked hornbill (*Bycanistes brevis*).

→
American artist Morris Graves was inspired by nature and Asian art and philosophy. Birds featured regularly among his subjects, as in this watercolour *Shore Birds* painted in 1939.

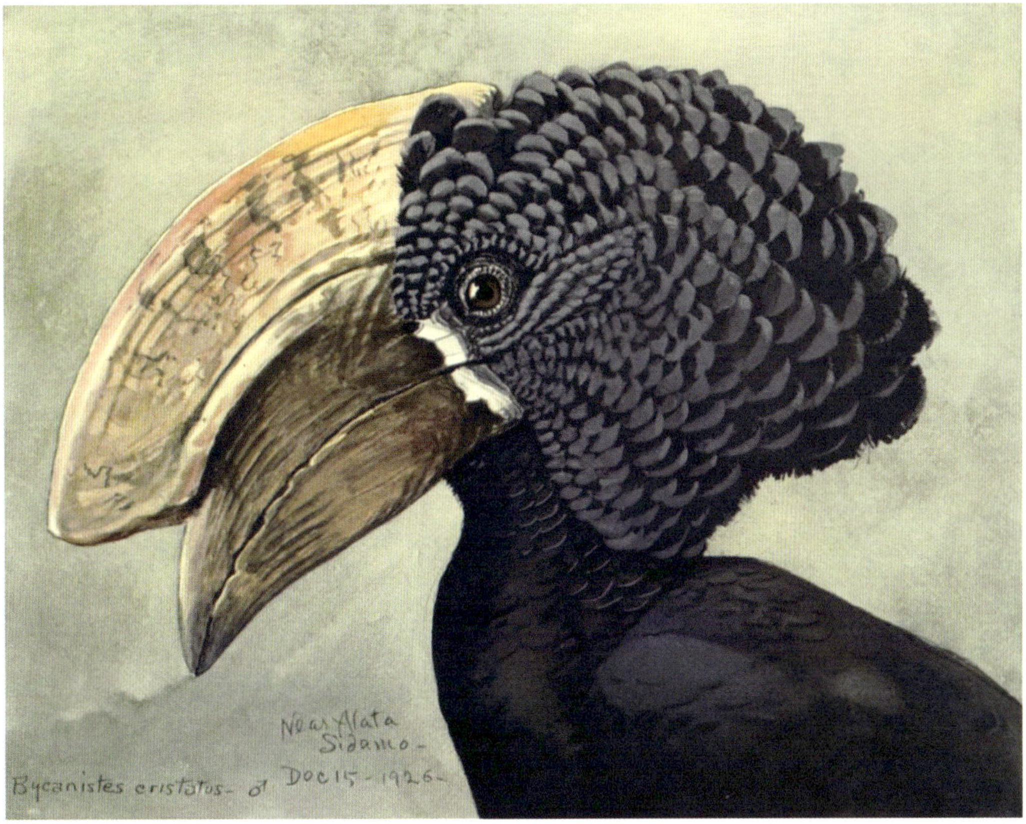
Near Alata
Sidamo
Bycanistes cristatus ♂ Dec 15 - 1926

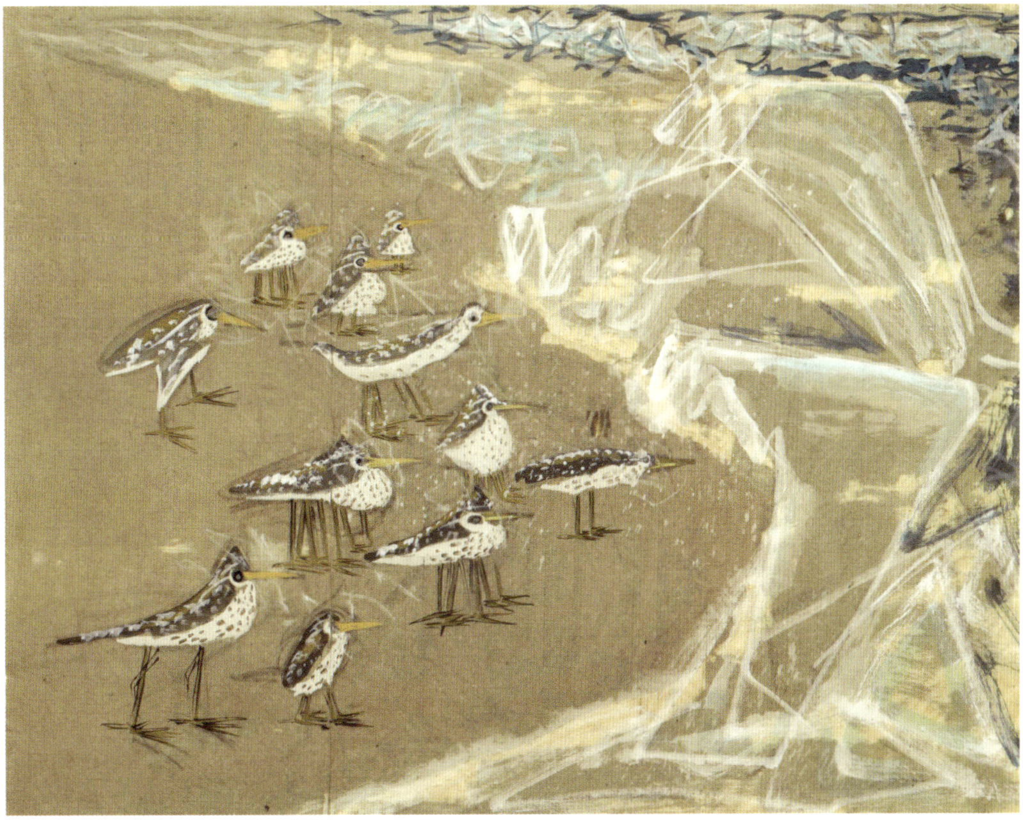

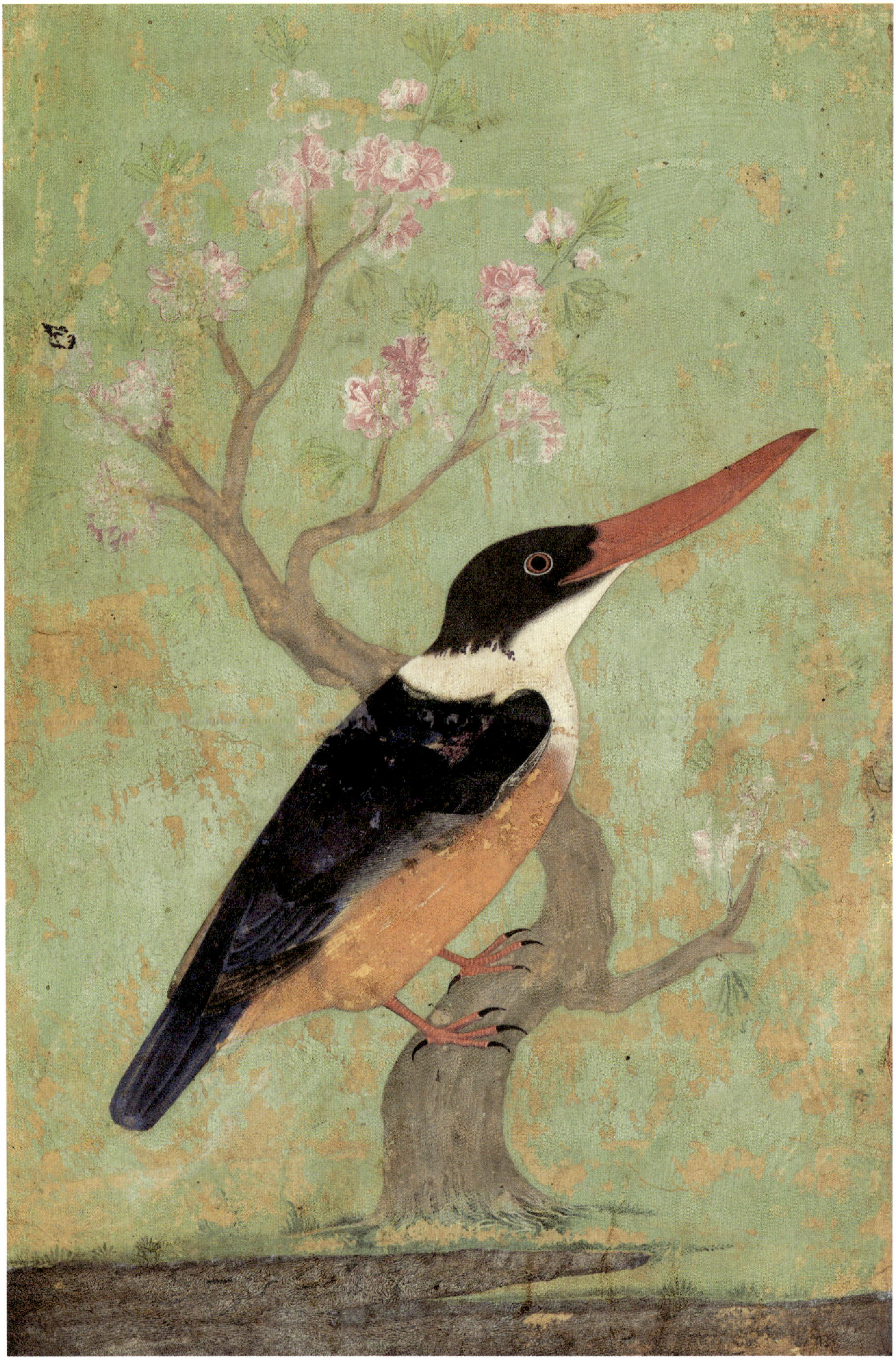

(VI)

←
◎ This watercolour illustration of a black-capped kingfisher (*Halcyon pileata*), which occupies parts of India, China and Southeast Asia, is probably by Mansur, a 17th-century painter at the Mughal court in India.

↘
This silk wall hanging, painted by Tsuruzawa Tansaku Morihiro, is from the second half of the 18th century. Cranes are symbols of longevity and immortality in Japan, and are a protected species.

'Cranes cry from the nine marshbanks and their voices are heard in heaven.'

SHIJING ('BOOK OF ODES'), SECTION II, CHAPTER 3, 11TH–7TH CENTURIES BCE, TRANSLATED BY JOHN THOMPSON

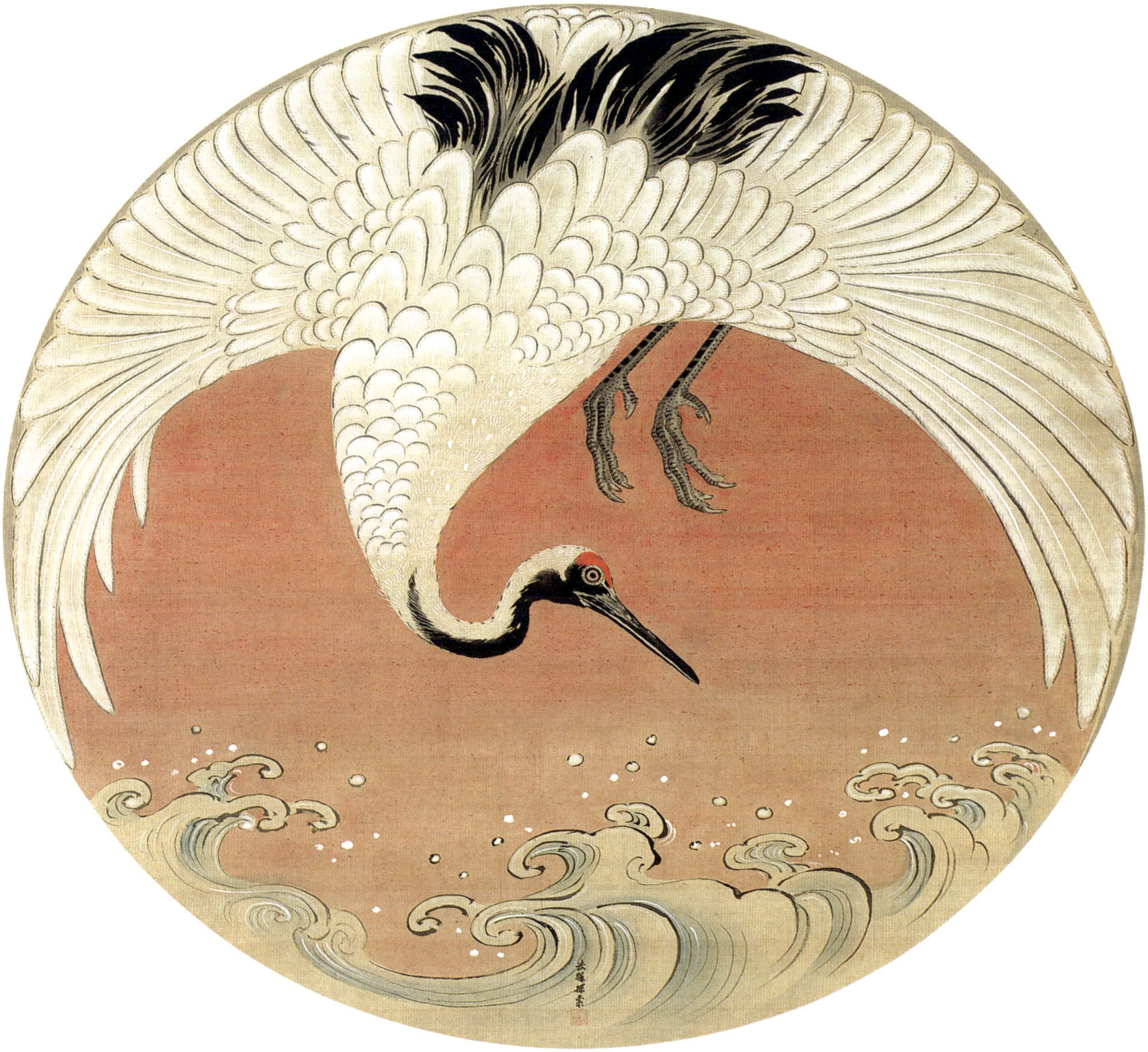

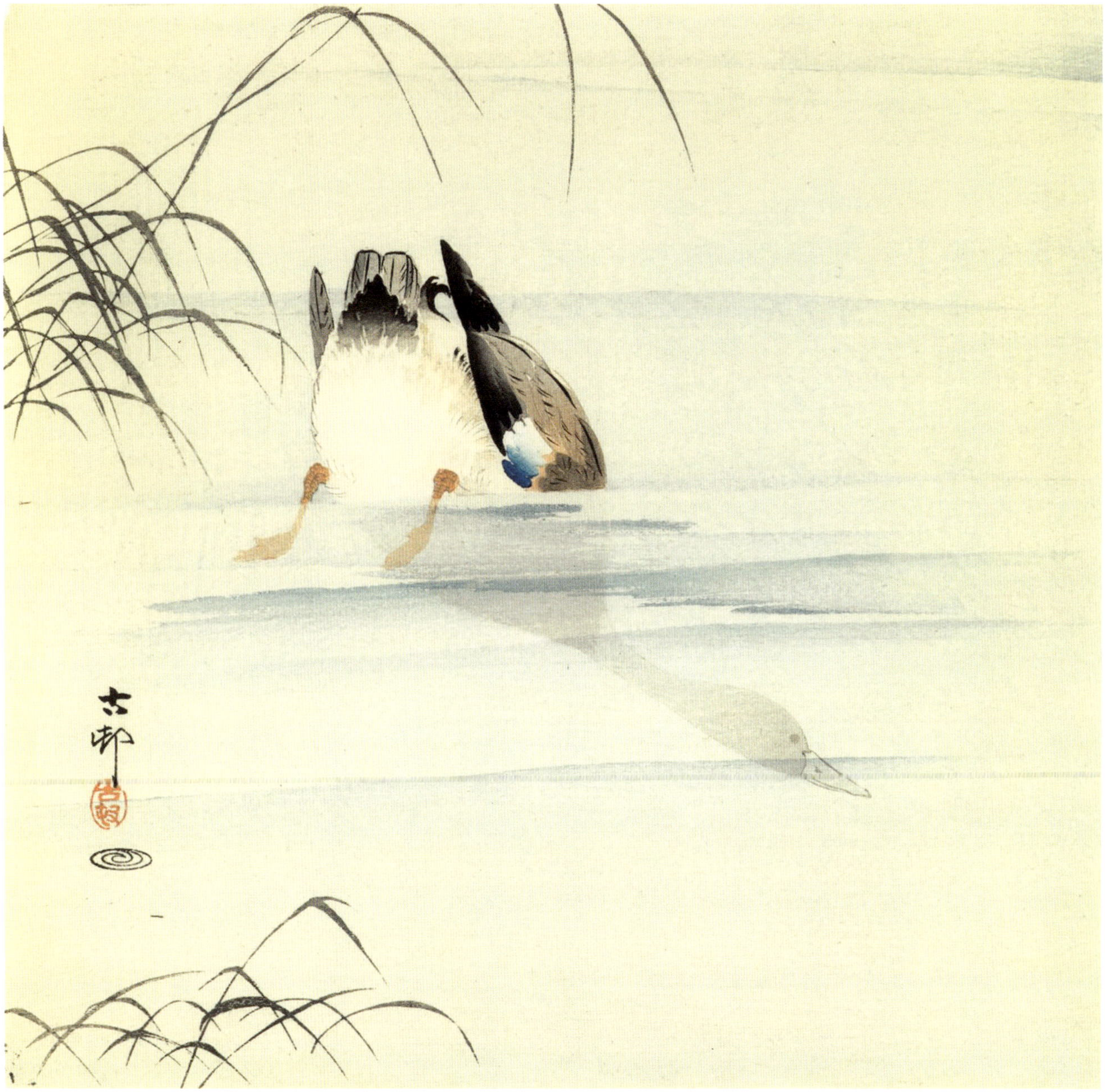

↑
Mallards (*Anas platyrhynchos*) are winter visitors to Japan, captured in this woodblock print, *Diving Mallard* (*c.* 1910), by Ohara Koson. Koson produced prints that have the delicacy of watercolour, while his attention to detail captures the personality of his subjects.

↗
Five Plovers and Waves by Nakamura Hōchū appears in his *Kōrin Gafu* (1826). Hōchū merged the Rimpa style of painting – subjects from nature painted in a soft-edged technique – with the woodblock printing techniques of the Japanese picture-book tradition.

→
Yamamoto Masao's meditative style of photography, exemplified in this image from his *Kawa=Flow* series of 2016, invites the viewer to contemplate nature. He finishes prints by hand.

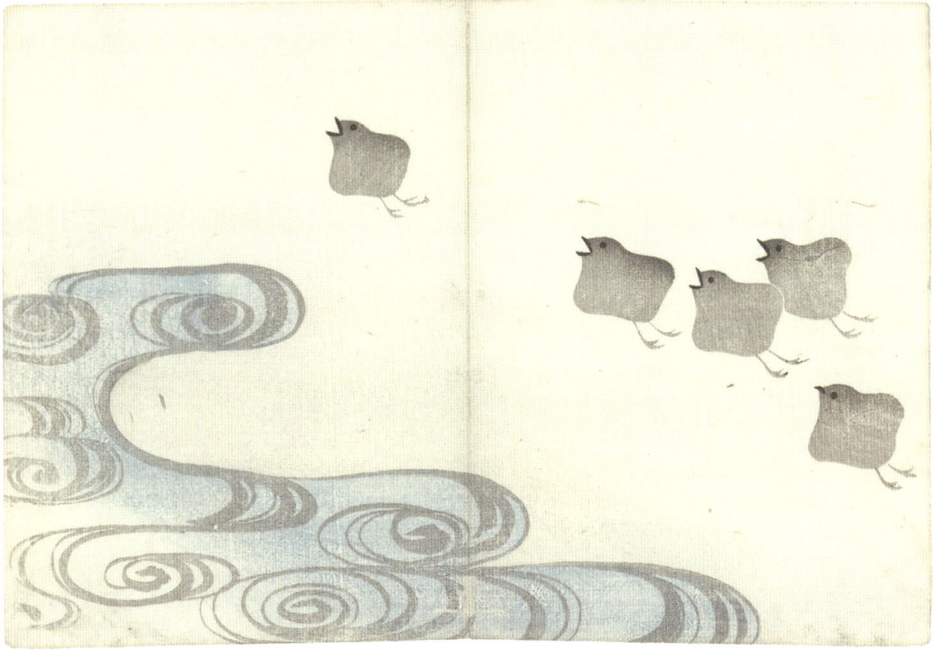

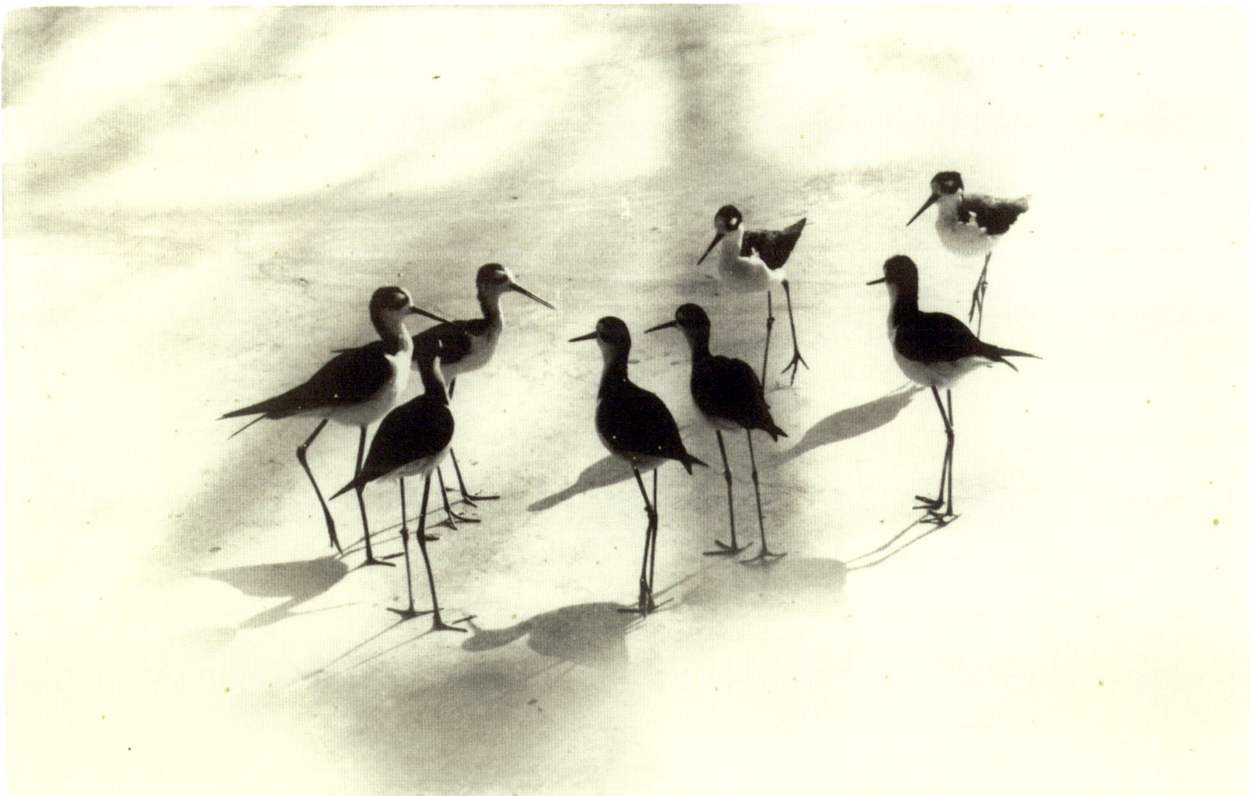

↓
In Renaissance Europe, parrots imported from Asia and South America were popular pets among the wealthy, and they also featured regularly in the art of the time. This pietra dura panel of a parrot and fruit, inlaid with semi-precious stones, is from 17th-century Florence, Italy.

→
Illustrated here in a 17th-century manuscript of Al-Qazwini's encyclopaedia, the Eurasian hoopoe (*Upupa epops*) feeds by probing the ground with their long bills, searching for insects. They can even dislodge large stones with their bills in their hunt for food.

↘
An eagle bites into a human heart in this relief carving from the Toltec culture (10th–12th century) of central Mexico. In Toltec mythology, the eagle symbolized the sun, which needed the nourishment provided by sacrifices for its journey through the night.

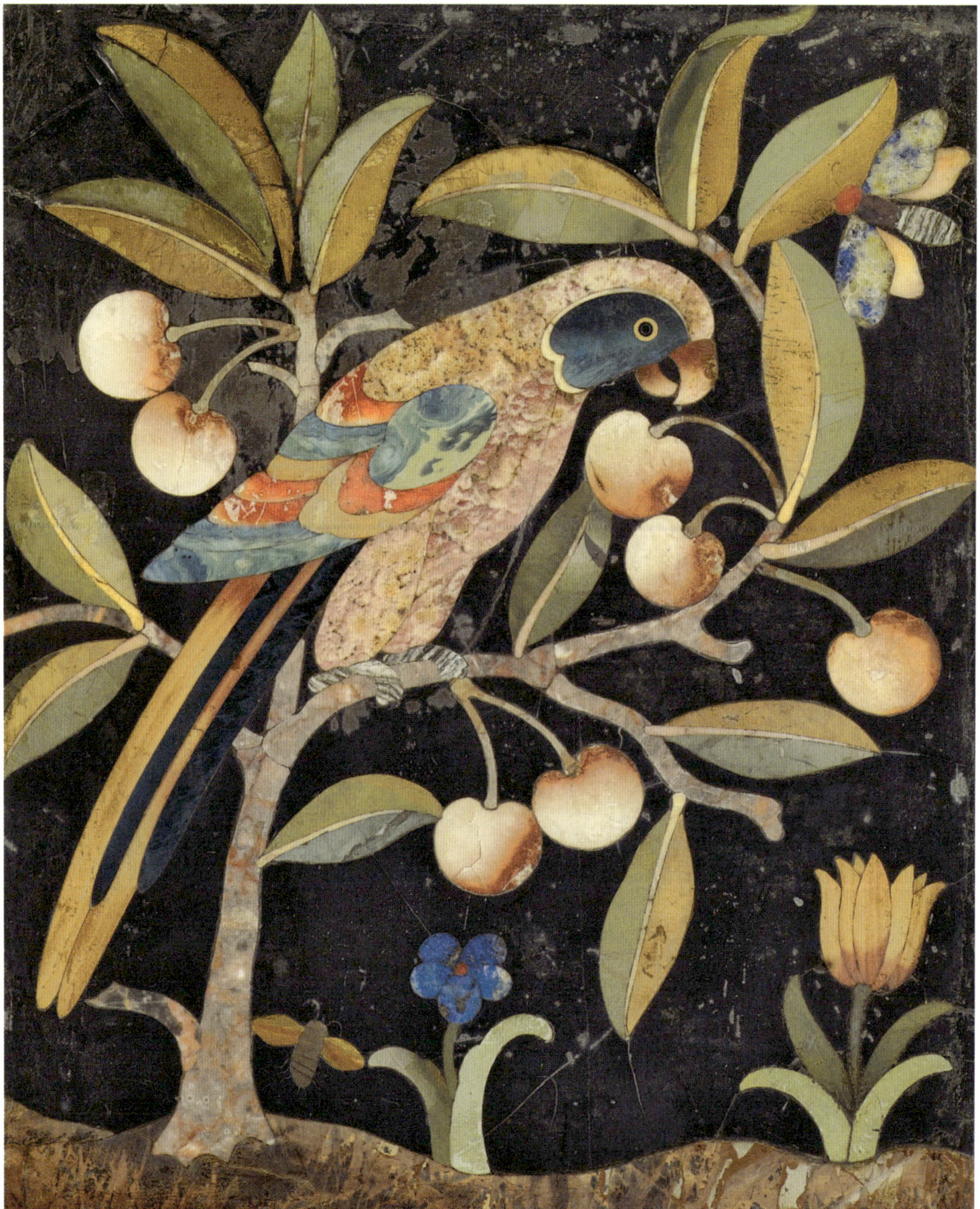

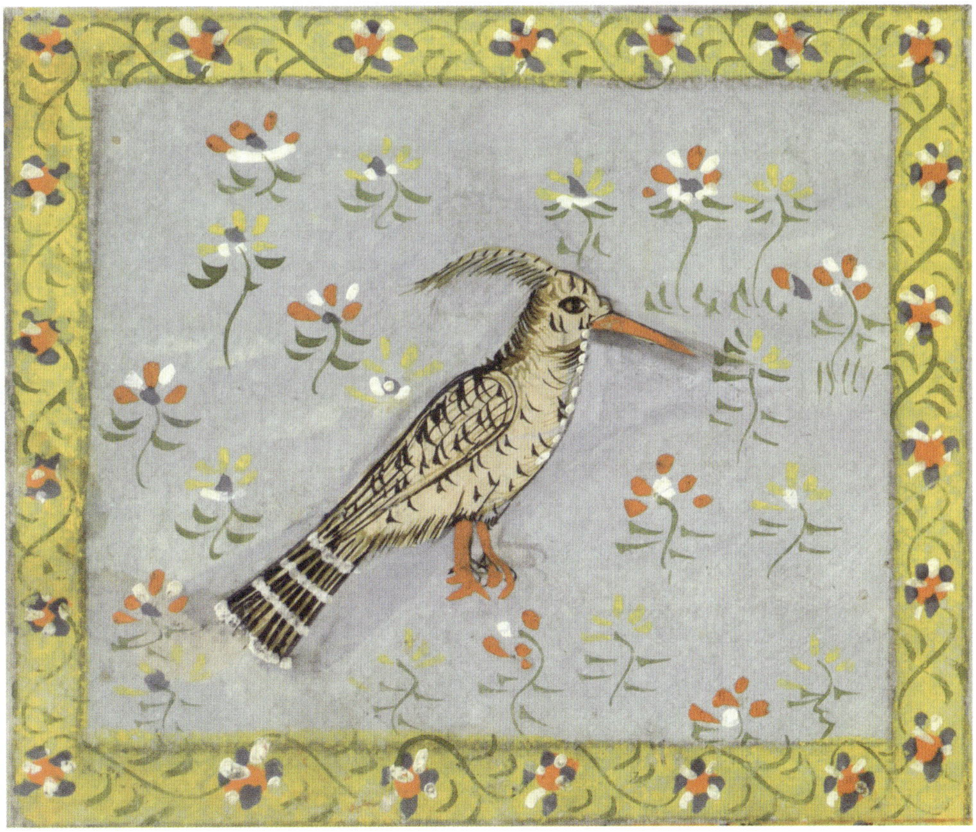

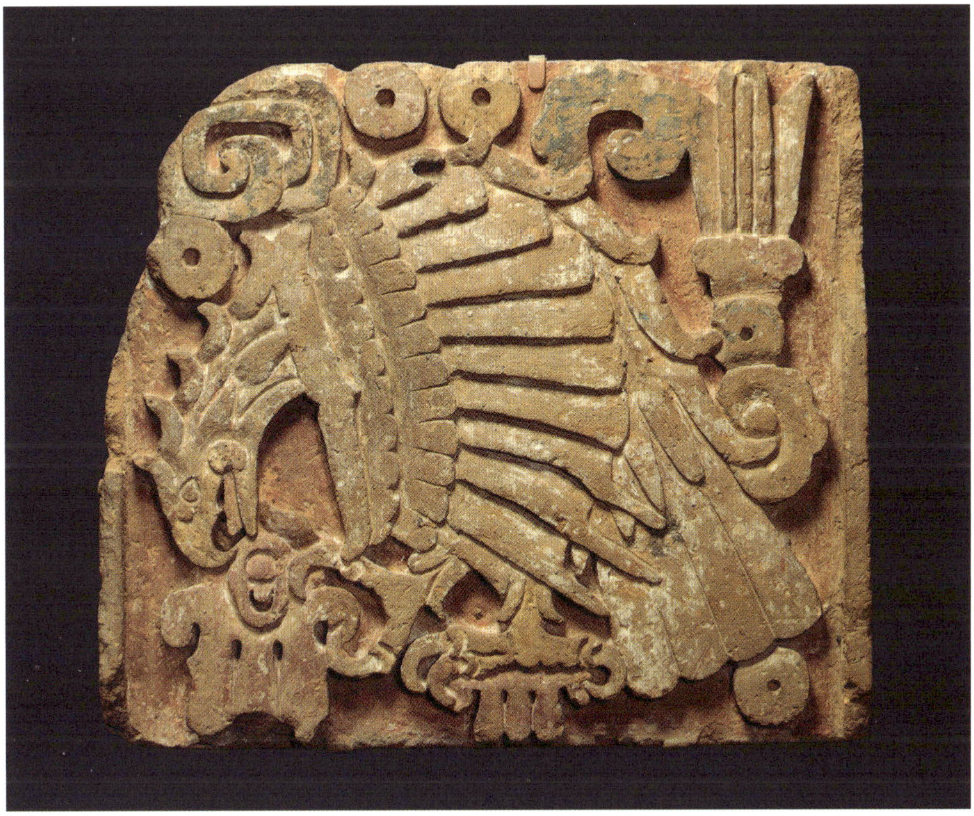

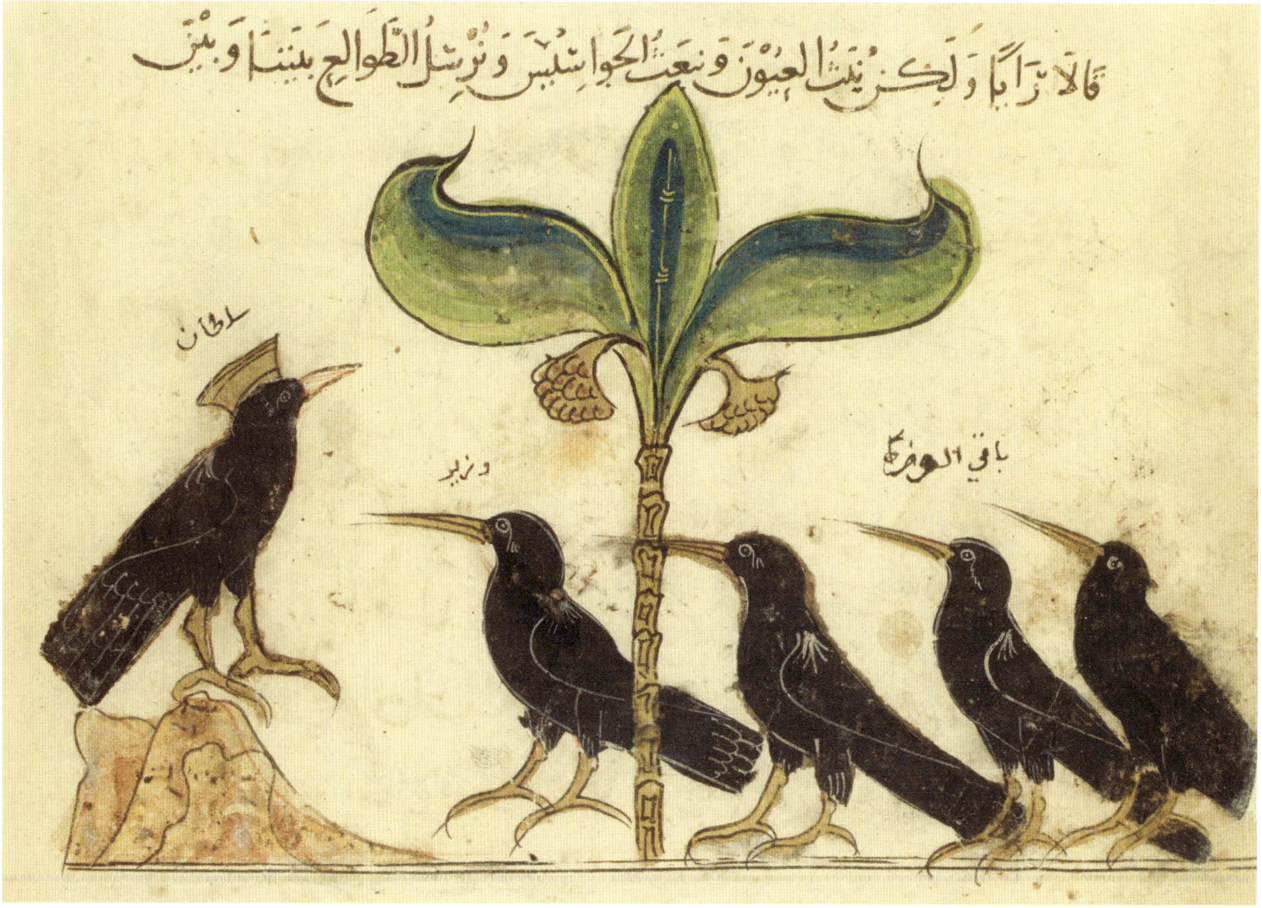

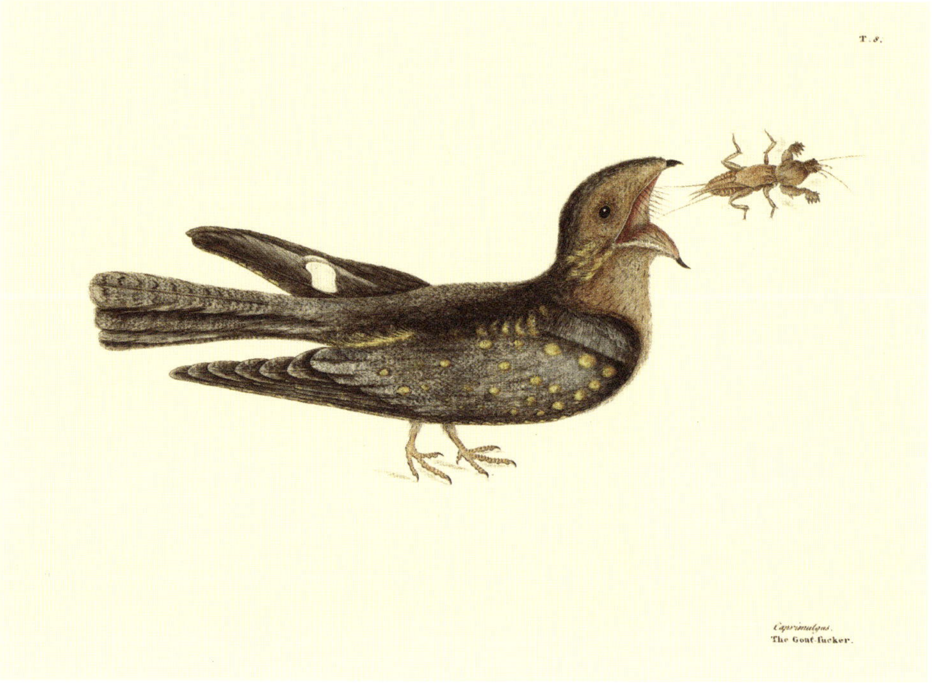

↑
The king of the crows consults his ministers in this 14th-century illustration to a story in the *Kalīla wa-Dimna*, an Arabic version of an ancient Indian collection of animal fables originally written for the instruction of the political elite.

→
⊖ Mark Catesby's 'Goat-Sucker' (1754) is a composite of three nightjar species: the common nighthawk (*Chordeiles minor*) and two near-threatened species, the chuck-will's-widow (*Caprimulgus carolinensis*) and the Eastern whip-poor-will (*Caprimulgus vociferus*). The myth that nightjars drink milk from goats probably arose because they feed on insects that live around livestock.

↳
The White Owl (*c.* 1856) by English artist William J. Webb features a barn owl that has just killed a mouse. Barn owls feed on rats and mice and so are welcomed by farmers.

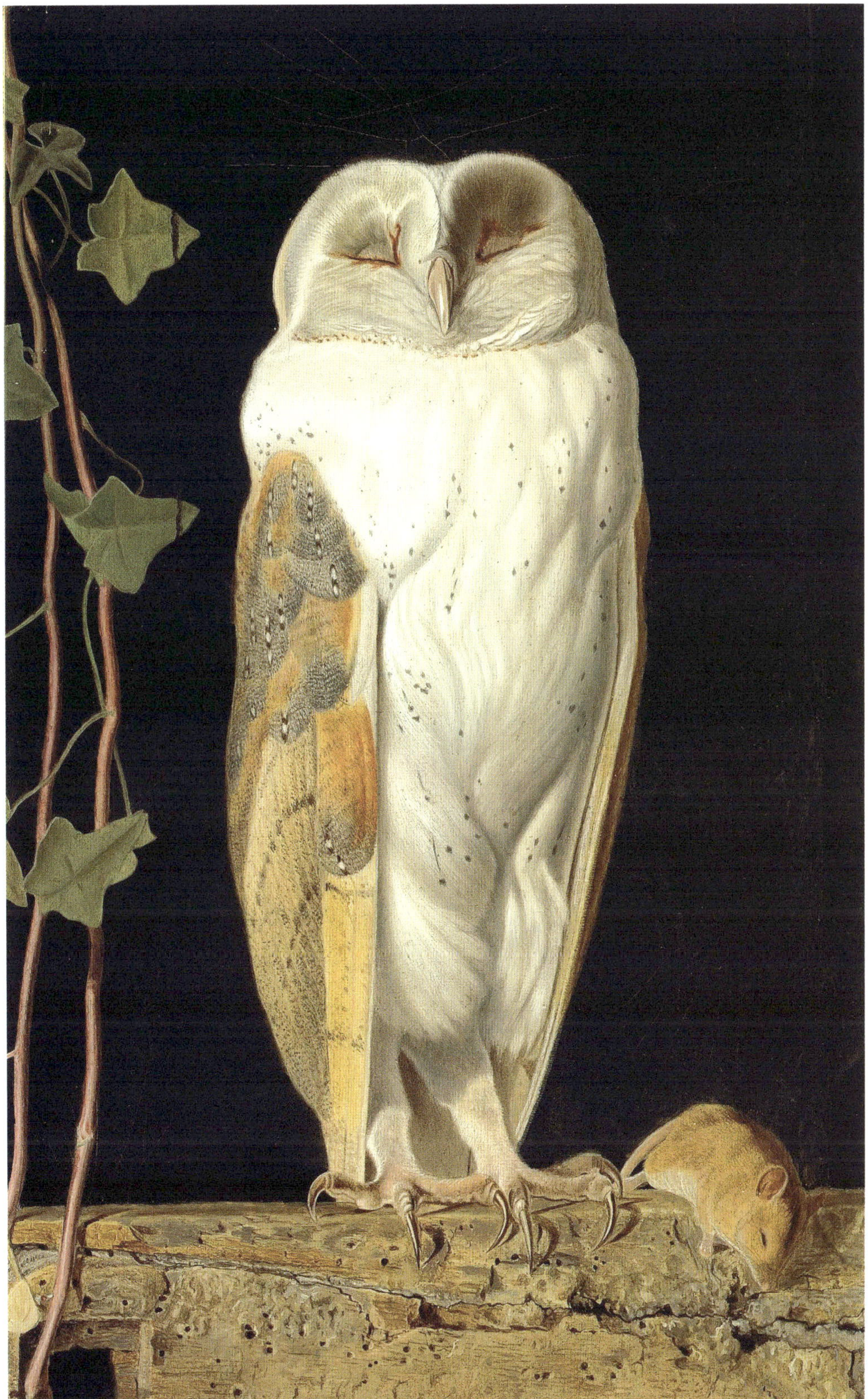

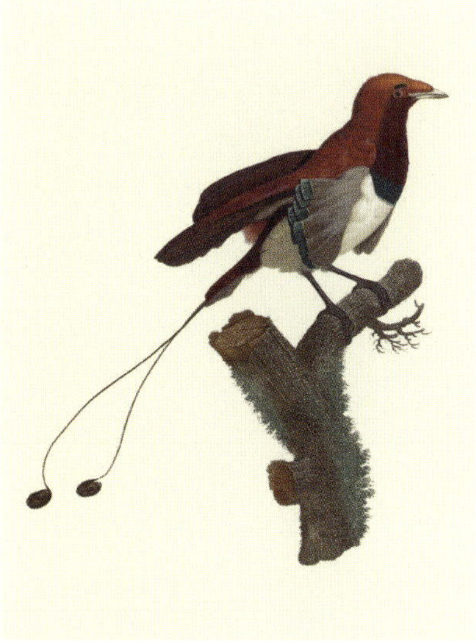
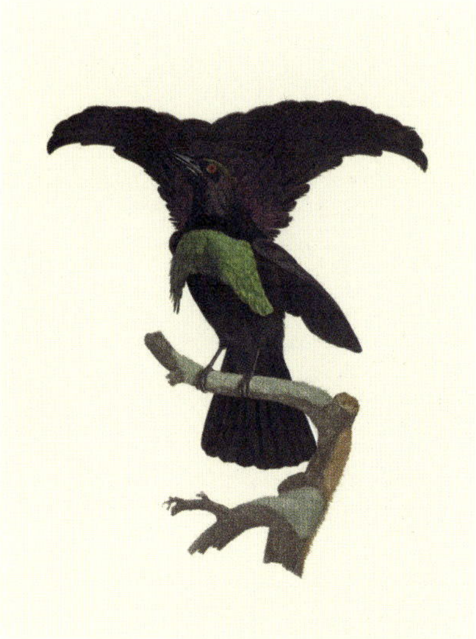
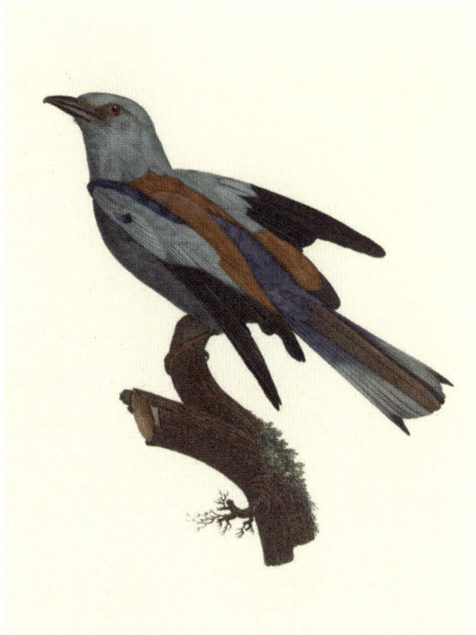

(VI)

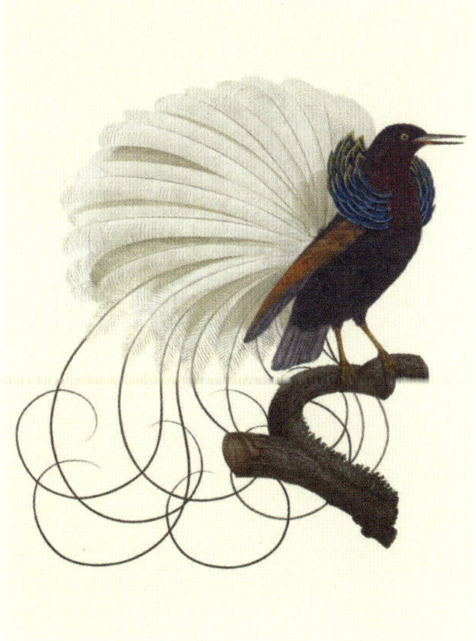
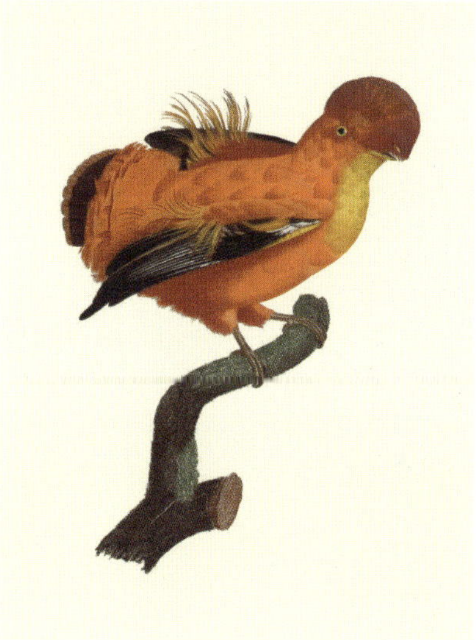
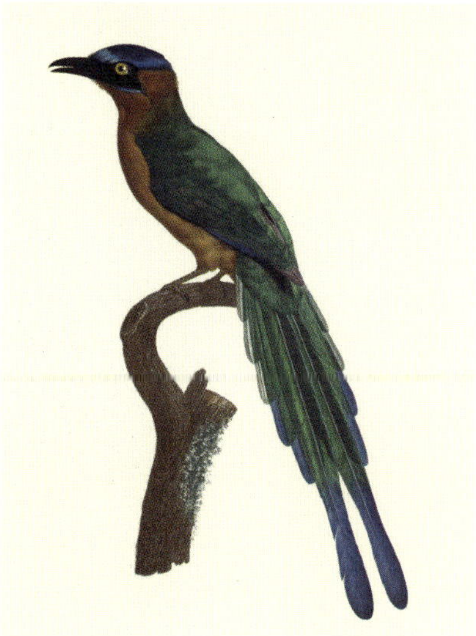
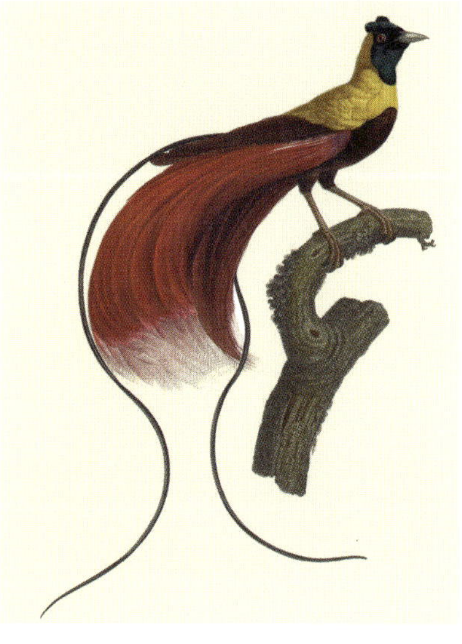
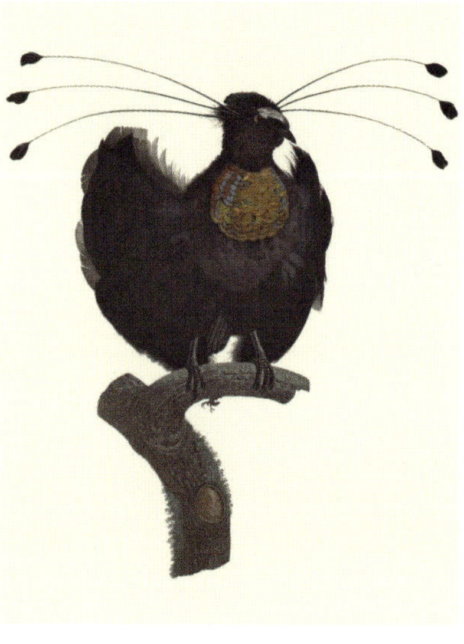
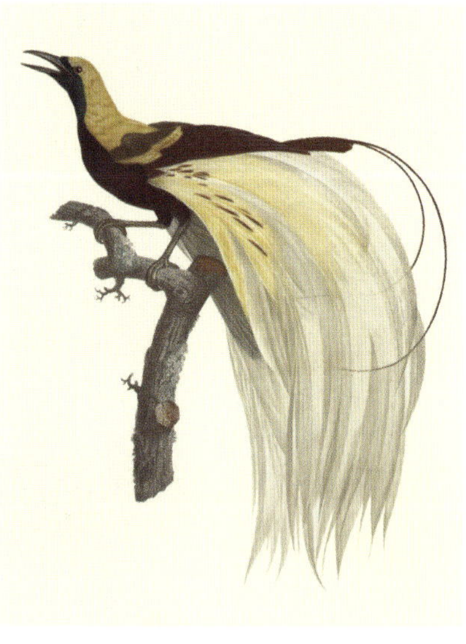

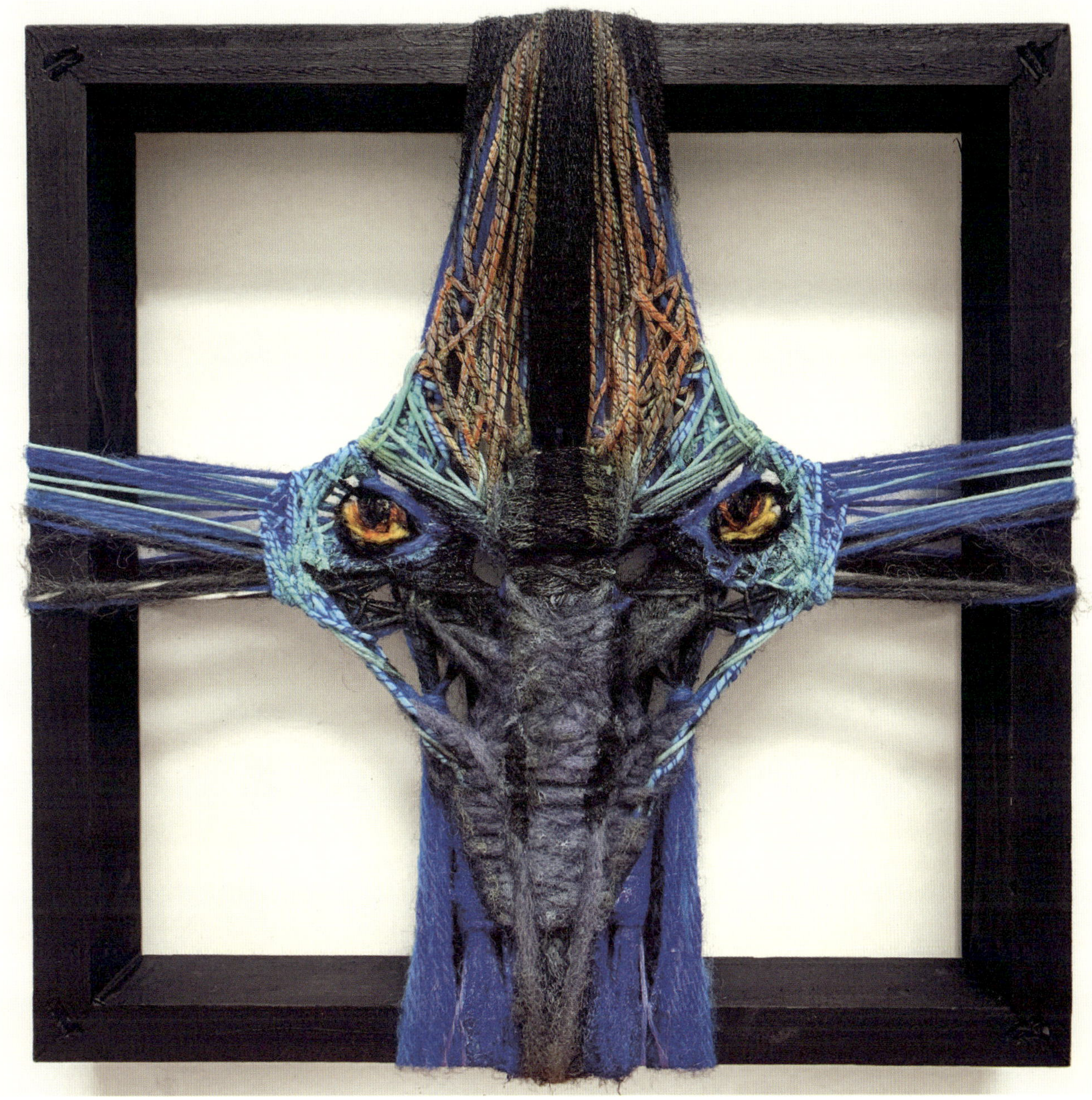

←
⊖ *Histoire naturelle des oiseaux de paradis et des rolliers* by François Le Vaillant (1806) illustrates a collection of bizarre-looking and colourful birds-of-paradise and rollers, including, bottom left, the near-threatened red bird-of-paradise (*Paradisaea rubra*); centre, a Guianian cock-of-the-rock (*Rupicola rupicola*); and, bottom centre, a Western parotia (*Parotia sefilata*).

↑
Latifa Medjdoub's mask-like *Laughing Cassowary* (2024) refers to the tradition of using feathers from cassowaries – large flightless birds of Papua New Guinea – in ceremonial headdresses. It depicts either the Northern (*Casuarius unappendiculatus*) or Southern (*Casuarius casuarius*) cassowary.

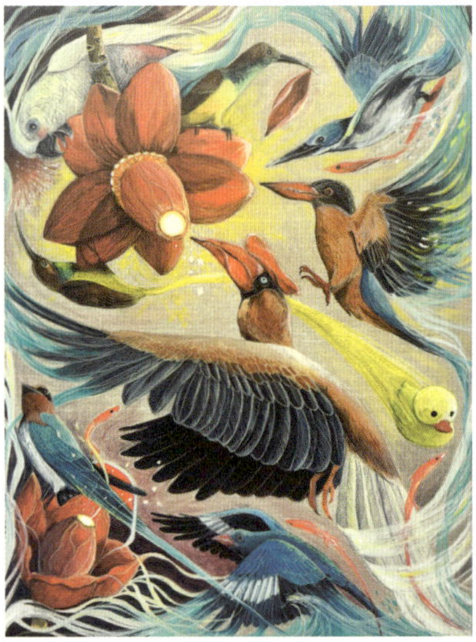
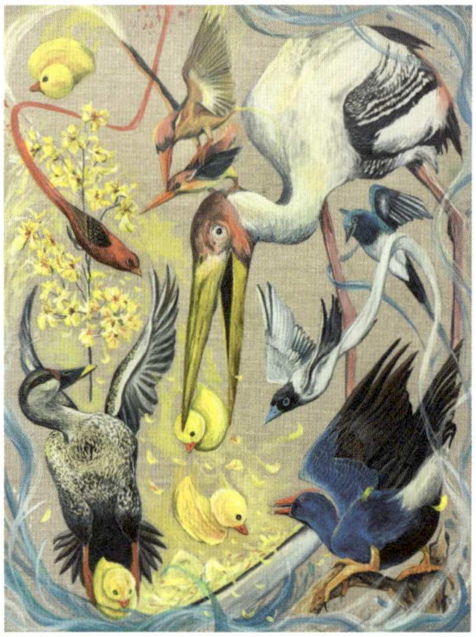
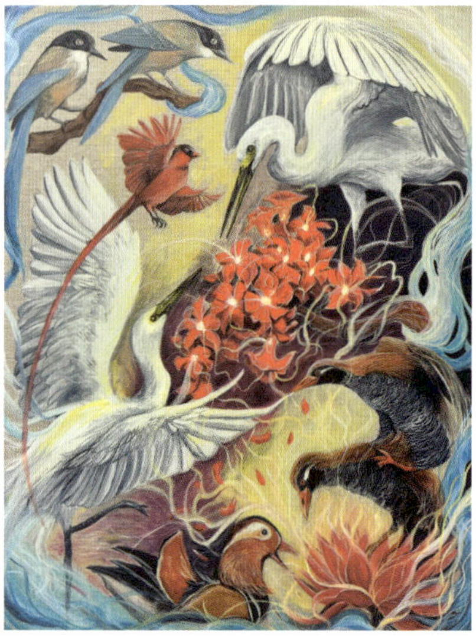

(VI)

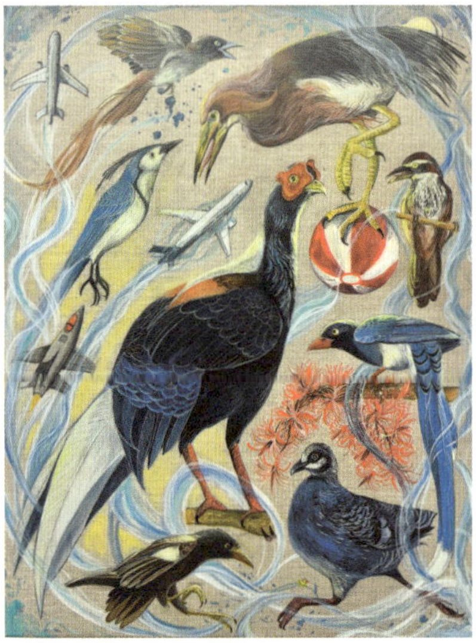
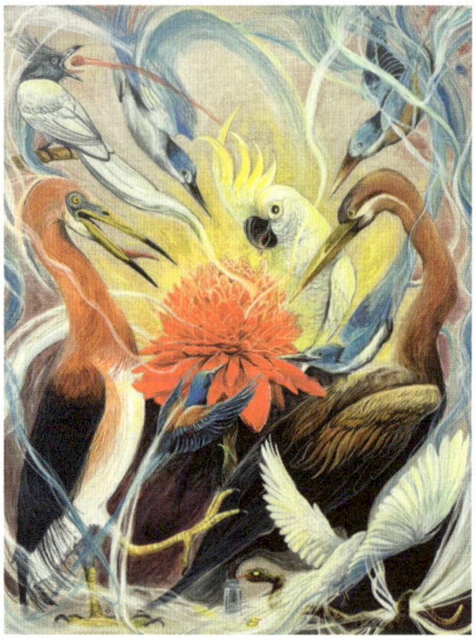
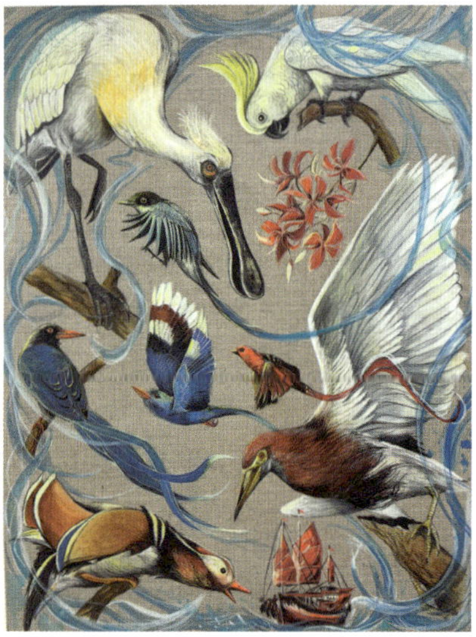
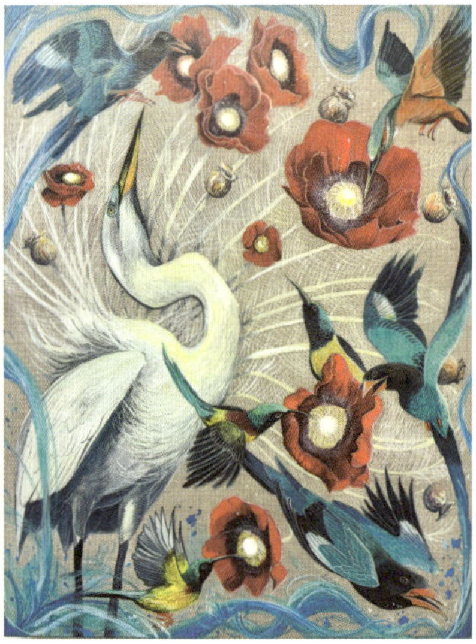
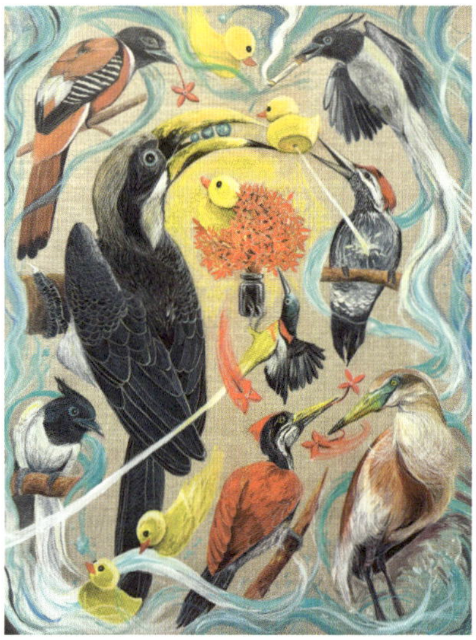
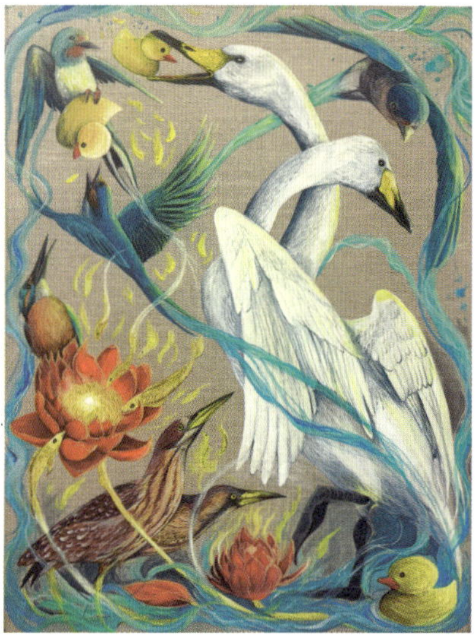

'The strutting peacock with its acid cry, opening its tail. Shot silver and velvet, changing colour before our eyes.'

DEREK JARMAN, 'IRIDESCENCE', *CHROMA: A BOOK OF COLOR*, 1995

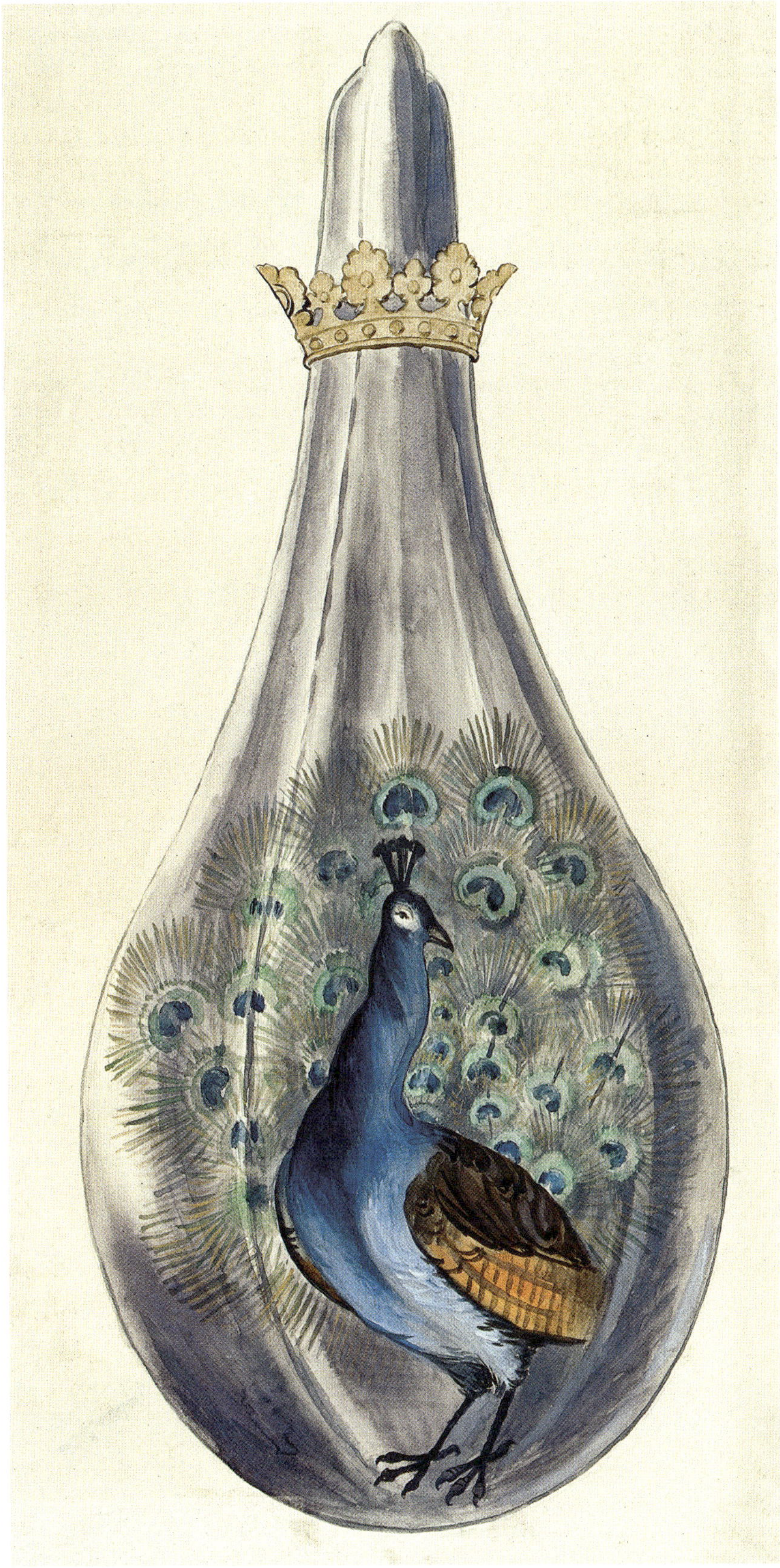

←
Panamanian artist Andrés Silva Vignoli's work blends natural history with urban landscapes, myth and memory. This series is called *Pájaros en la Cabeza* ('The Birds in my Head', 2025).

→
This early 20th-century watercolour by Edith A. Ibbs refers to the 'peacock's tail' phase of the alchemical process, when a heated substance dissolves and produces an array of iridescent colours.

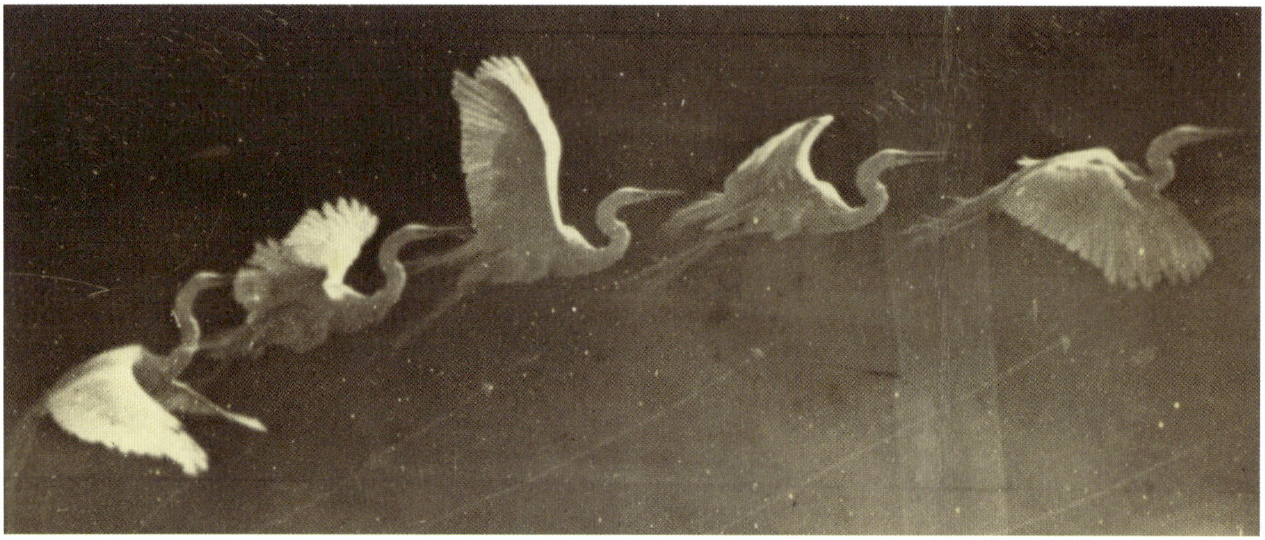

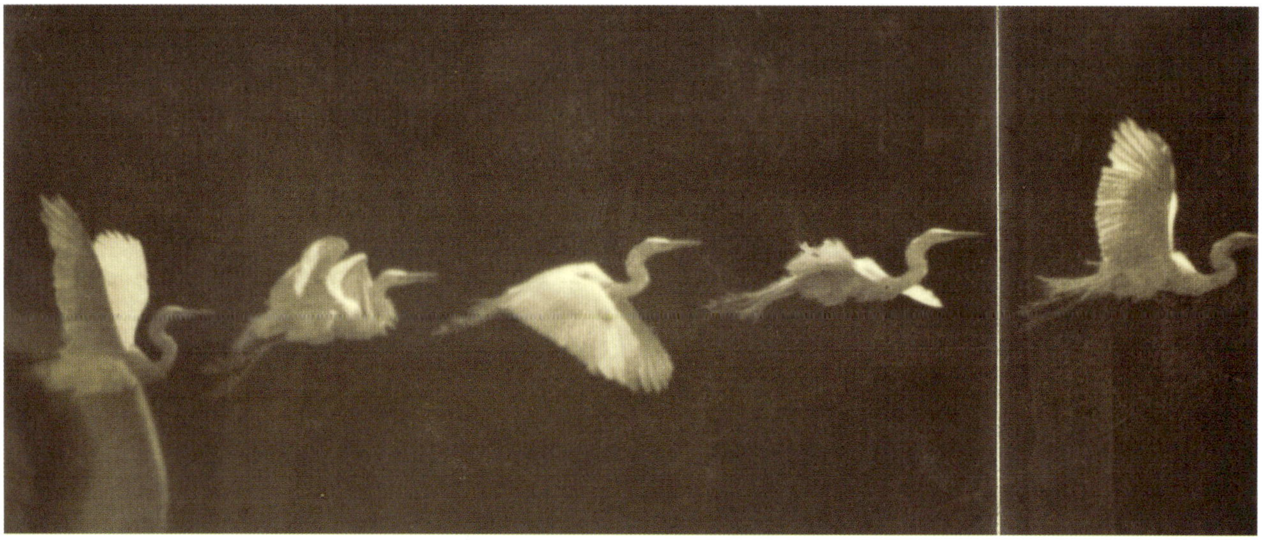

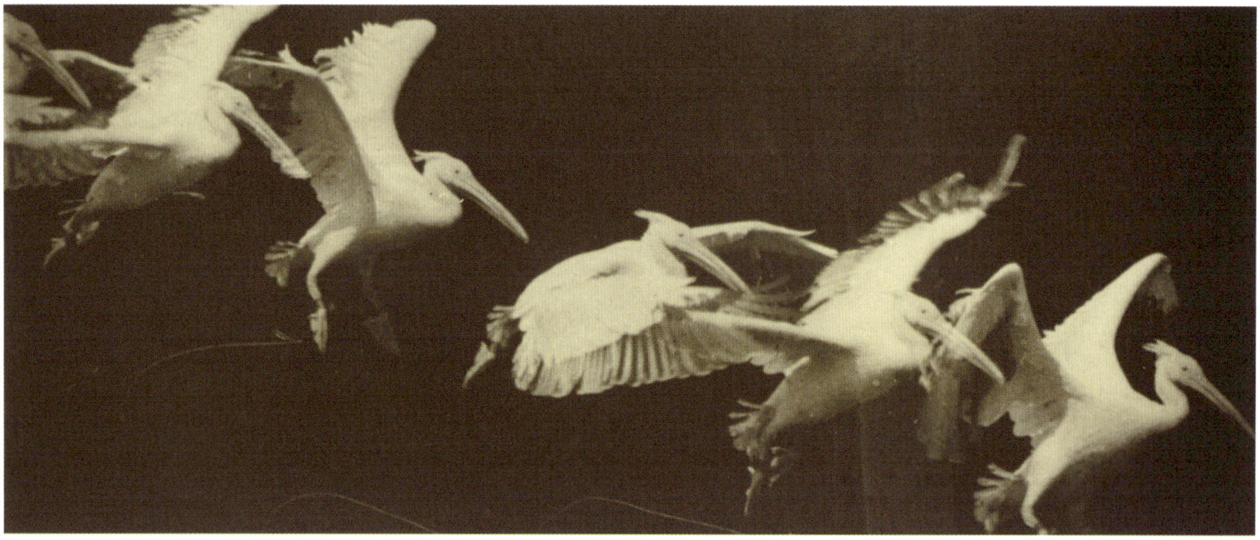

(VI)

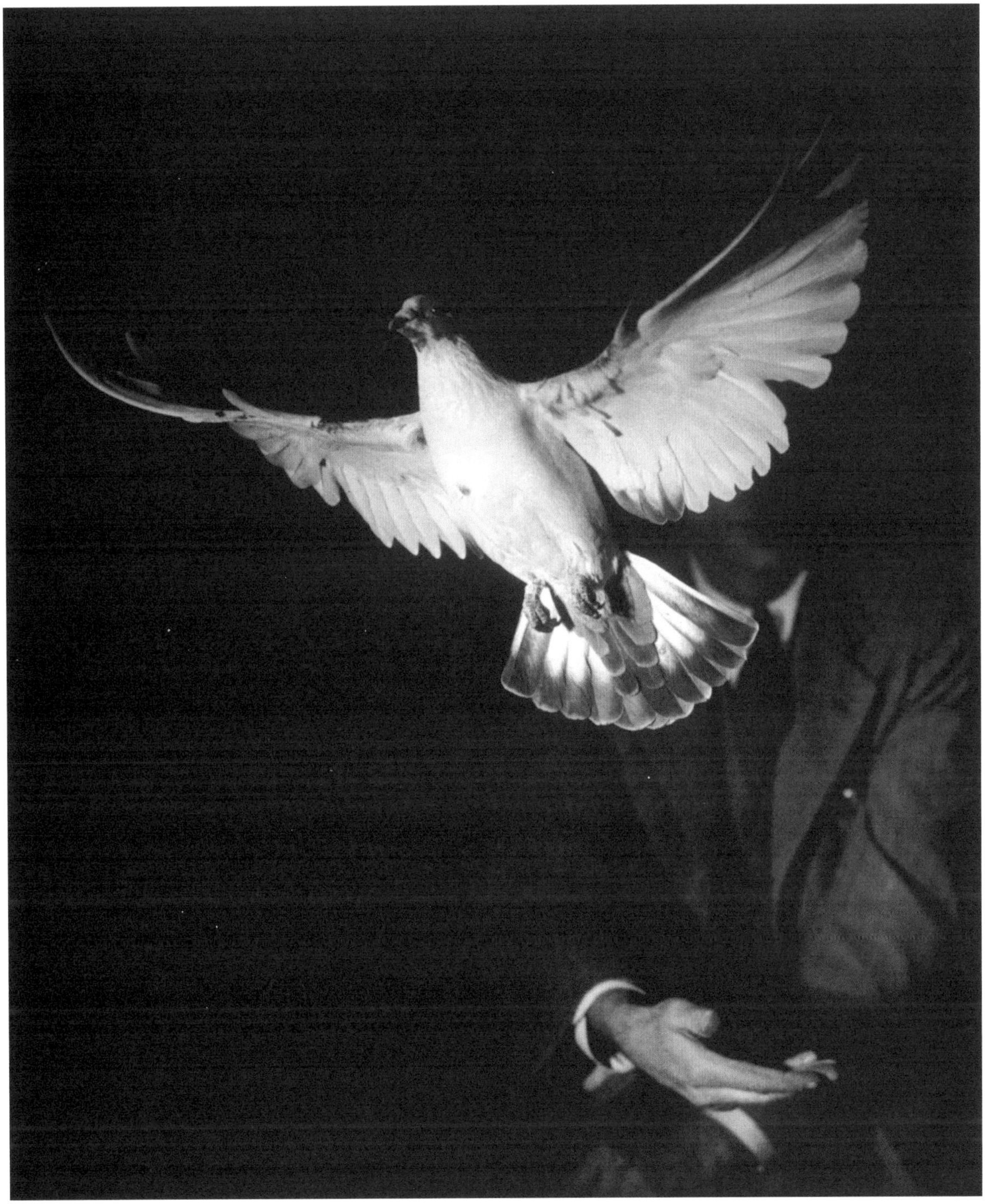

←
Étienne-Jules Marey's photographic sequences of herons in flight (top, *c.* 1883; and, centre, 1887) reveal the extent to which the bird's wings bend and the force they exert when flying. The bottom sequence shows a pelican landing (*c.* 1882).

↑
In the early 1930s, American scientist and researcher Harold Edgerton pioneered the use of stroboscopic lights in photography. He developed a high-speed flash system to record ultra-high-speed motion on film, such as in *Rising Dove* (*c.* 1934), which captures the plumage of the bird's outstretched wings and the aerodynamics of its tail feathers.

CONSERVING THE INFINITE VARIETY OF BIRDS

Stephen Moss

As I write, I am momentarily distracted by a rapid movement outside my office window. A tiny bird materializes briefly, before vanishing moments later. It is a wren – Britain's commonest bird, with roughly ten million breeding pairs – yet they are so fast and elusive I rarely enjoy prolonged views. As always, I am left entranced by this sighting – as I have been with countless encounters with birds during my lifetime. And I am not alone – all around the world, millions of people share my passion.

So why are birds so special, so compelling and so central to our lives? The most obvious reason is their ubiquity: from the poles to the equator, across every habitat on earth, on the land, over the sea and in the air, birds are, quite simply, everywhere. No matter what the season, weather conditions and time of day – or night – you can see or hear birds.

Paradoxically, however, not all birds are easy to find. Some, including the Siberian white crane, are very rare; others, such as the Rio Branco antbird I saw briefly on a recent visit to Guyana, are restricted to a tiny home range; and whole families – such as the South American tapaculos – are so elusive I have yet to see a single one. And even though some birds, such as the barn owl, are relatively common and found throughout much of the world, their nocturnal habits still make them hard to see.

Birds are also beautiful: brightly coloured, with impressive plumage features. On that same trip to Guyana I encountered a pair of sun parakeets, one of the most striking and colourful birds in the world; a sunbittern, with its complex and intricate feather patterns; and an array of hummingbirds, whose grace and beauty – and ability to fly up, down, forwards, backwards – are simply breathtaking. Nor should we forget common and familiar species, such as blue jays and cardinals in North America and robins and goldfinches in Britain and Europe, which we often take for granted.

Birds have fascinating and complex behaviours: the courtship display of the male Indian peafowl (better known as the peacock); the way male bowerbirds of New Guinea and Australia build an arena, decorated with an array of natural and man-made items to attract potential mates; and the intricate, complex and musical song of the nightingale, which can include dozens of different phrases.

Finally, birds are, to borrow a phrase from Shakespeare, of 'infinite variety'.[1] From the tiniest – Cuba's bee hummingbird – to the largest – the ostrich – they are unimaginably different from one another. Even those that to us look very similar reveal subtle differences of shape, plumage or sound; some families, such as the cisticolas of Africa and antbirds of South America, present major identification challenges even to the most experienced birder.

I have been fortunate to travel to all seven of the world's continents, and visit more than fifty countries, to watch, film and write about birds. Over the years, I have grown to appreciate the importance of the links between the local – the birds I encounter around my home – and the wider world.

During spring, near my home in Somerset, south-west England, I can hear – and occasionally see – no fewer than ten different species of warblers. Despite weighing only a few grams, they travel vast distances here each year from all four corners of Africa: chiffchaffs and blackcaps from the north, whitethroats, reed and sedge warblers from the west, lesser whitethroats from the east and willow warblers from the south.

They – along with the swifts, swallows and house martins hawking for insects in the skies above, and the cuckoo with its unmistakeable call – are a traditional sign that spring has finally arrived; a perfect demonstration of the way the natural world intersects with our human one, while at the same time remaining independent of us.

Humanity and birds – and the rest of the natural world – are yoked together. We are all interconnected, and birds play a crucial role in both natural and human ecosystems.

Birds are crucial to how natural environments function: as pest controllers, pollinators and by dispersing seeds. Eurasian jays, for example, bury acorns each autumn to provide a winter food supply, but because they sometimes forget where they have put them, some of these grow into mighty oaks. Vultures act as nature's clean-up squad, scavenging dead animals and so preventing the spread of disease. And changes in the population trends of many birds provide an early-warning system of environmental change, often signalling a problem even before we become aware of it.

Birds have a profound impact on human society, the global economy and each of us as individuals. Both wild and domestic species are, for many, a major source of

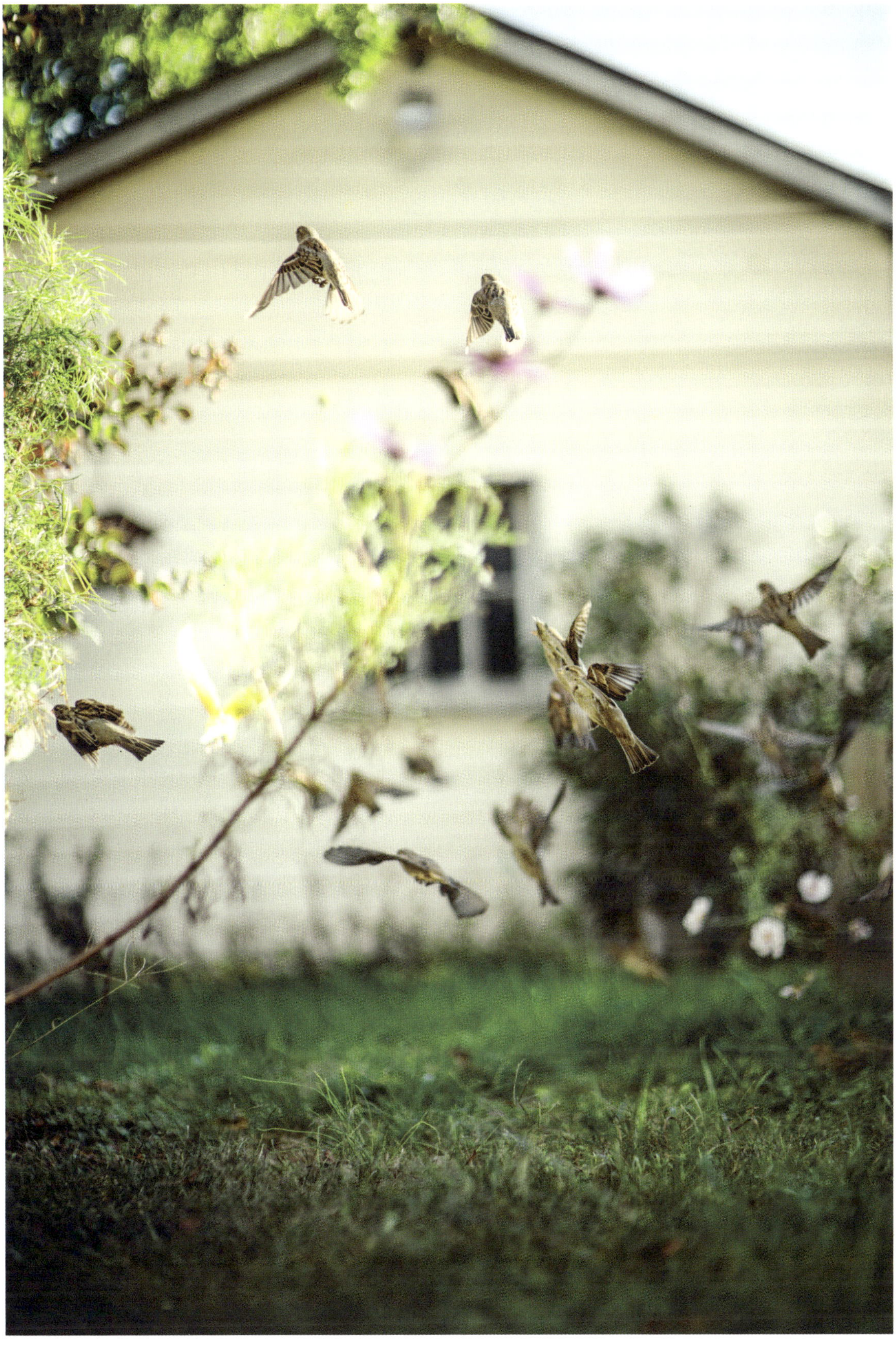

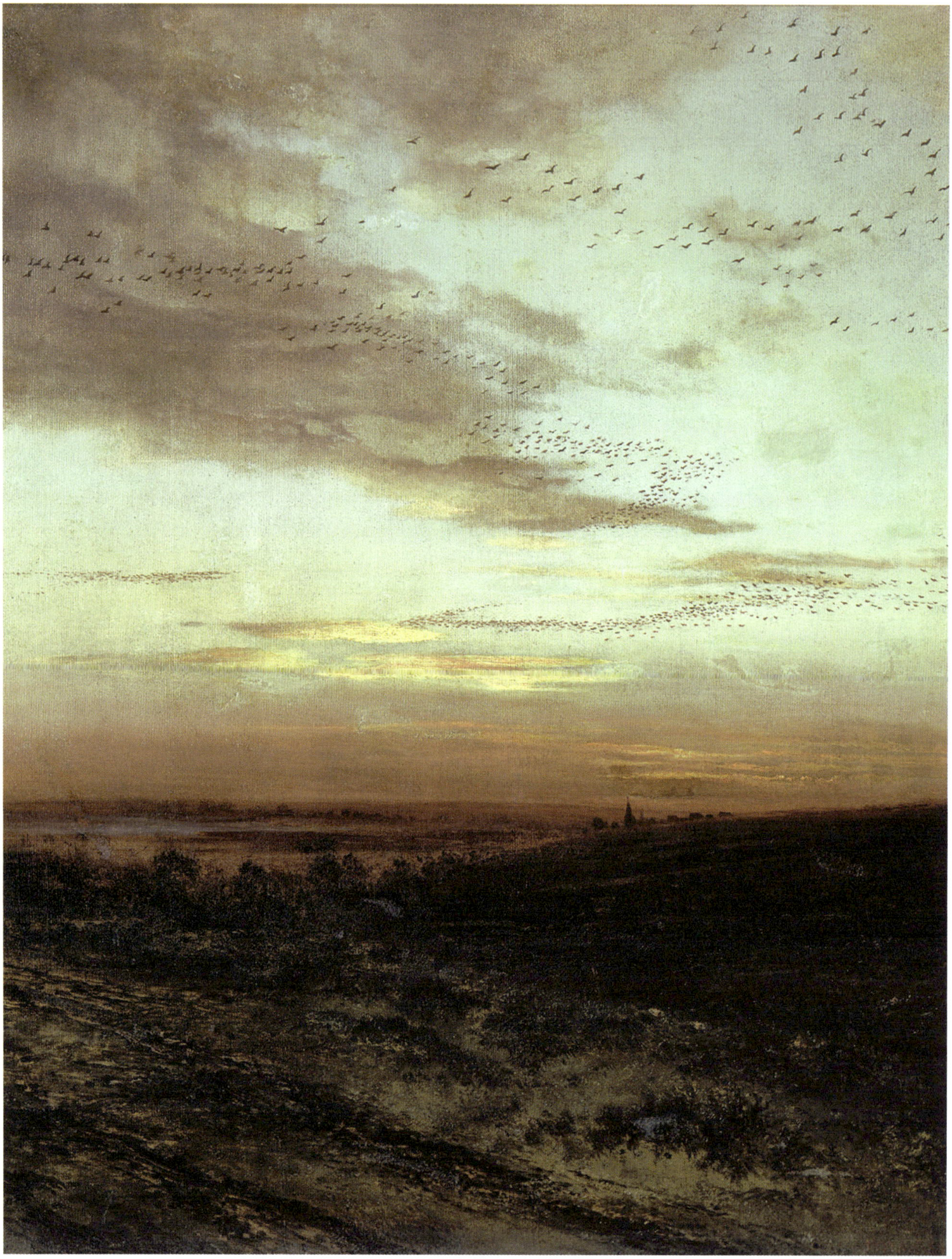

(VI)

food – both meat and eggs – while feathers and down are still collected from slaughtered birds and live wild eider populations and used to fill bedding products and coats. Birding is the leading form of wildlife tourism, providing much-needed income and jobs, especially in developing countries, while cutting-edge design and technology – such as drones – have been inspired by the morphology and behaviour of birds.

Birds are also at the heart of our culture, from art and music to literature and folklore. Well-known examples include Beethoven's Pastoral Symphony and the Beatles' song 'Blackbird', the poetry of John Keats and John Clare, the paintings of Hieronymus Bosch and Pablo Picasso, and a host of children's stories, from Hans Christian Andersen's *The Ugly Duckling* to Otmar Gutmann and Erika Brueggemann's animated series *Pingu*.

In recent years, we have also grown to appreciate that interacting with nature – and especially, because they are so ubiquitous, with birds – provides major benefits for our physical, emotional and mental health, and our overall well-being. Listening to birdsong, in particular, has been shown to reduce stress and anxiety, lift mood and improve focus and concentration.[2]

It is no exaggeration to say that today birds play a greater, more varied and more important part in our lives than at any time in history. Yet, at the same time, birds are under threat as never before. In my own lifetime, the number of individual birds on the planet has fallen by more than half,[3] and although the same period has seen a huge upswell in our passion for the natural world, they continue this seemingly inexorable decline.

According to the global conservation organization BirdLife International, nearly half of all the world's 11,200 or so bird species are currently in decline, while one in eight – about 1,400 in total – are seriously threatened with extinction.

This crisis has come about through what BirdLife describes as 'the human population's increasingly unsustainable use of natural resources'.[4] The key drivers of decline are the expansion and intensification of agriculture; logging and wood harvesting; and the spread of invasive species. Other major threats include habitat loss and degradation from urban development; hunting and trapping for food; pollution of the land and seas; and, most urgently, the negative and unpredictable effects of the current climate crisis. Most birds are impacted by more than one threat: for example, extreme weather events resulting from climate change increase the risk of droughts and wildfires, which lead to the loss of habitats. In spring 2025, extensive fires on moorland in the Welsh uplands destroyed important breeding areas for skylarks, golden plovers and hen harriers.

Our challenge now is to reverse the trend of damage and decline. Some initiatives are already well underway. Global treaties such as the Ramsar Convention on Wetlands (1975), the Convention on International Trade in Endangered Species of Wild Fauna and Flora (1975) and the Convention on the Conservation of Migratory Species of Wild Animals (1983), all help prevent harm to some of the planet's most vulnerable birds.

The designation of key locations as Important Bird and Biodiversity Areas (IBAs) has protected more than 13,000 sites in over 200 countries and territories worldwide. The Preventing Extinctions Programme has had real success in reversing the declines of critically endangered species such as the pink pigeon, Mauritius kestrel and echo parakeet, all of which were predicted to go extinct and are no longer categorized as critically endangered.

Back in the late 1980s, what was then a highly controversial decision to take all remaining wild California condors into captivity and later release captive-bred birds back into the wild, was a huge success, preventing North America's largest raptor from going extinct. More recently, the Albatross Task Force, led by BirdLife International and its UK partner the RSPB, has reduced deaths of these mighty seabirds as a result of being accidentally caught in fishing lines by as much as 99% in the seas off South Africa.[5]

As individuals, we can play our part in saving the world's birds. The annual Global Birdfair, which has taken place in central England since the late 1980s, has raised over £7 million through ticket sales and trade stands for global conservation projects. With matching funding and grants, this total has grown to £42 million.[6] Birders also regularly take part in surveys, which provide conservation organizations with the data they need to track the rises and falls in bird populations, and so better conserve species and habitats.

Saving our planet's birdlife will undoubtedly require considerable imagination, increased resources and collective hard work around the globe. But with passion and determination we can persuade institutions and lawmakers to take action and halt and reverse the decline of bird populations and the destruction of the precious places where they live.

↲
Created in her garden in Washington, DC, Terri Weifenbach's *Des Oiseaux* series reveals the secret world of birds that nest in urban gardens.

←
In *Evening, Migration of Birds* (1874), Russian landscape painter Alexei Savrasov portrayed flocks of birds over the landscape as dusk falls.

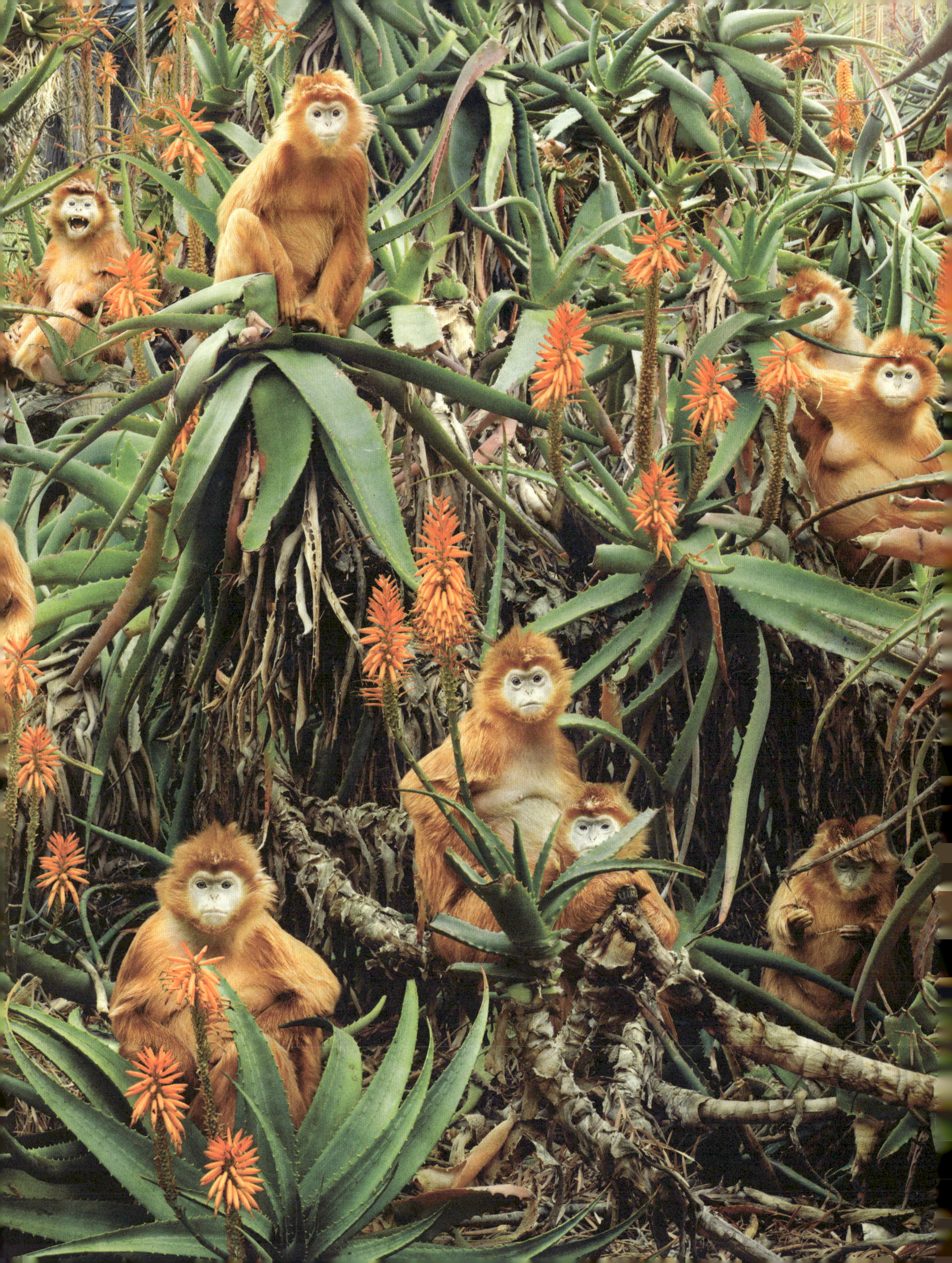

(VII) MAMMALS

GOLDEN SNUB-NOSED MONKEY	HEDGEHOG
MOLE	SQUIRREL
GREATER INDIAN FRUIT BAT	DOG
FOX	WOLF
DEER	GOAT
ASIATIC BLACK BEAR	POLAR BEAR
PANDA BEAR	WHALE
DUGONG	COMMON DOLPHIN
HARBOUR SEAL	WALRUS
PORPOISE	KANGAROO
PLATYPUS	KOALA
ARMADILLO	MARMOSET
HARE	JERBOA
PIG	WILD BOAR
LLAMA	SHEEP
AMERICAN BISON	COW
SPOTTED HYENA	BAIRD'S TAPIR
HORSE	BULL
CAMEL	DONKEY
ORYX ANTELOPE	ADDAX
ANTEATER	MOUSE
LION	ASIATIC LION
MASAI GIRAFFE	INDIAN RHINOCEROS
ZEBRA	HIPPOPOTAMUS
ASIAN ELEPHANT	TIGER
SNOW LEOPARD	CAT
GIBBON	ORANGUTAN
AZARA'S NIGHT MONKEY	GORILLA

MAMMALS are a diverse group of warm-blooded vertebrates, defined by their ability for their females to produce milk to feed their young. Typically, they give birth to live young, but a few, including the echidnas (or spiny anteaters), lay eggs. Because they evolved from land-dwelling ancestors, all mammals, even those such as whales and dolphins who exist solely in the water, breathe air through lungs. They inhabit every environment that animals are able to exist in, with the exception of the deepest reaches of the ocean.

↓

⊖ In medieval times, it was believed that hedgehogs would sometimes climb grape vines, shake the grapes onto the ground and then roll on the floor to stab the grapes onto their spines so that they could carry their bounty back to their burrows for their hoglets. Bestiary writers often assigned spiritual meanings to animals; in this example, from the *Rochester Bestiary* (*c.* 1230), the hedgehog rolling on grapes was an allegory for man bearing the fruit of his worldly sins, as well as for Christ bearing the sins of man, or the devil. Today, we know that the common hedgehog (*Erinaceus europaeus*) rarely consumes fruit and if they ever appear to, it is likely because they are trying to eat the insects that are feasting on the fruit.

↓↓

◎ Anna Pugh's *Mole in a Hole* (2016) depicts moles in their burrows, surrounded by their favourite food: earthworms. It is thought that their saliva contains a toxin that can paralyse earthworms, allowing moles to store live worms for later consumption.

→

Though *Squirrels in a Plane Tree* (1605–8) is Mughal, the squirrels depicted are likely red squirrels, a species that is not found in India. The presence of the hunter is also unusual as squirrels weren't commonly hunted in India at the time and he doesn't appear to be Indian. The artist, likely Abu'l Hasan, may therefore have borrowed his subject matter from European images that were traded in the Mughal empire.

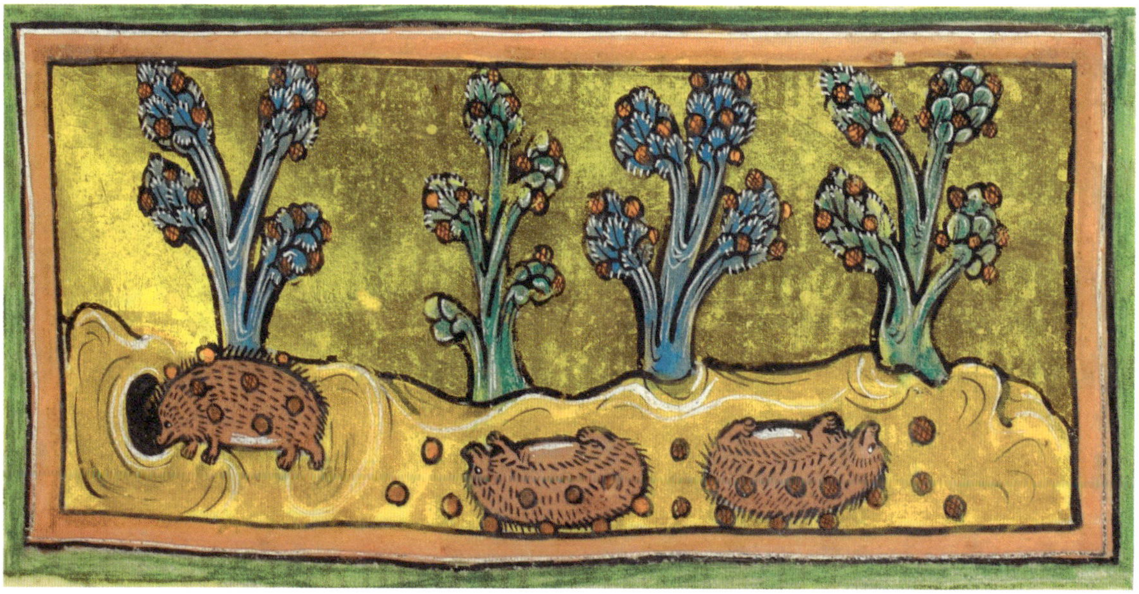

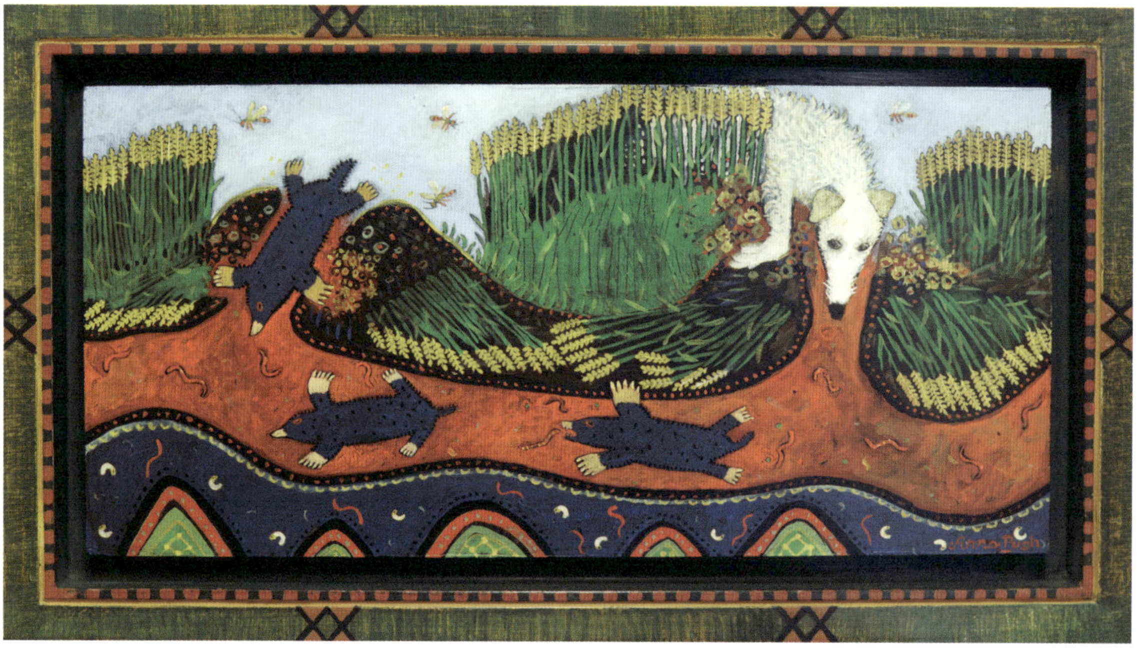

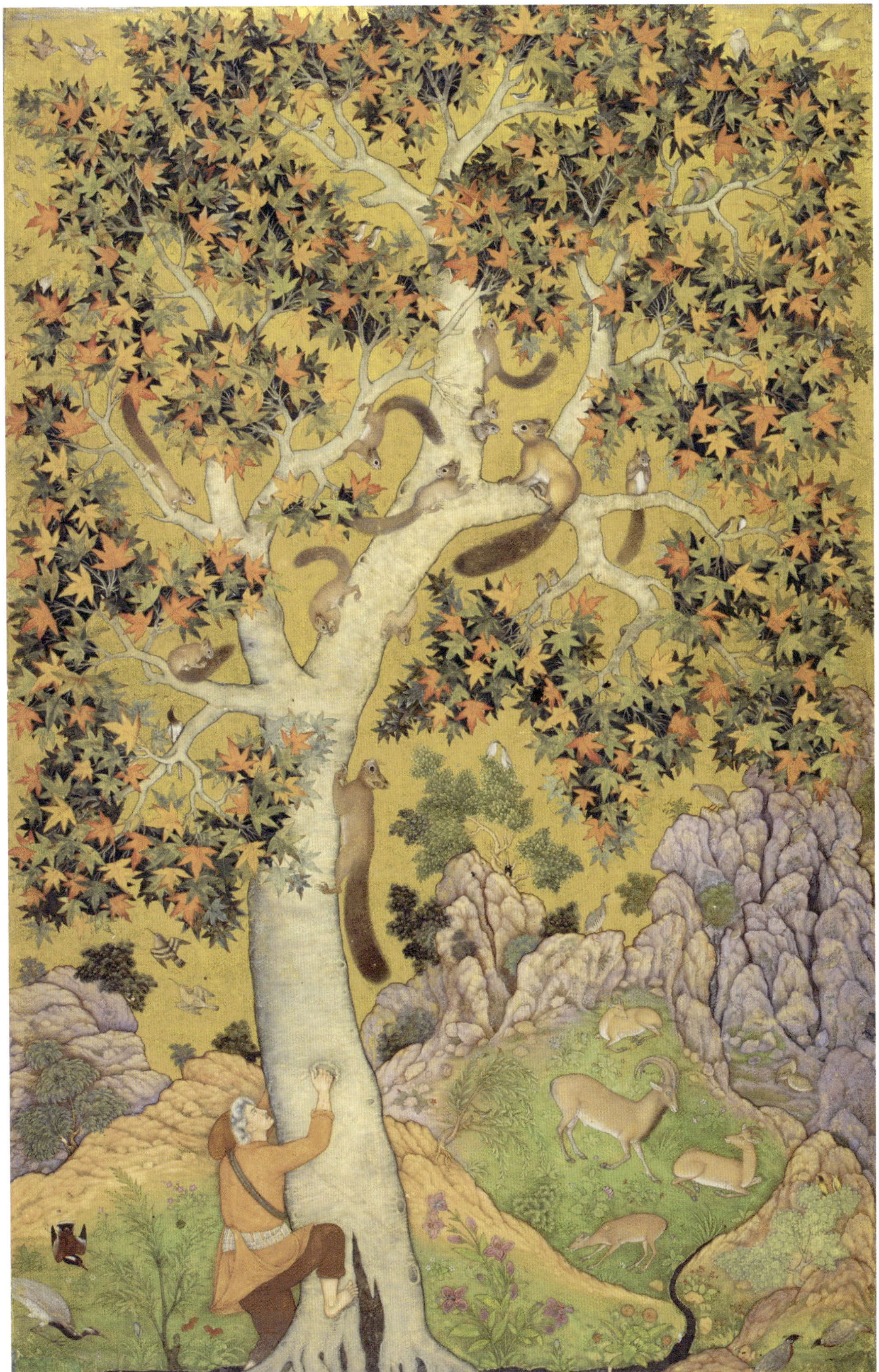

'They class themselves in our mind with those few animals which seem to belong to a bygone age .... There is somewhat reptilian to them – dragonlike .... If a bat, a mammal, can fly with no feathers, with wings fastened to his arms, why, indeed, can't I?'

GEORGE J. BREED AND WILLIAM G. BARTON, 'NOTES FROM THE FIELD – JULY', *SONGS AND SAUNTERINGS*, 1892

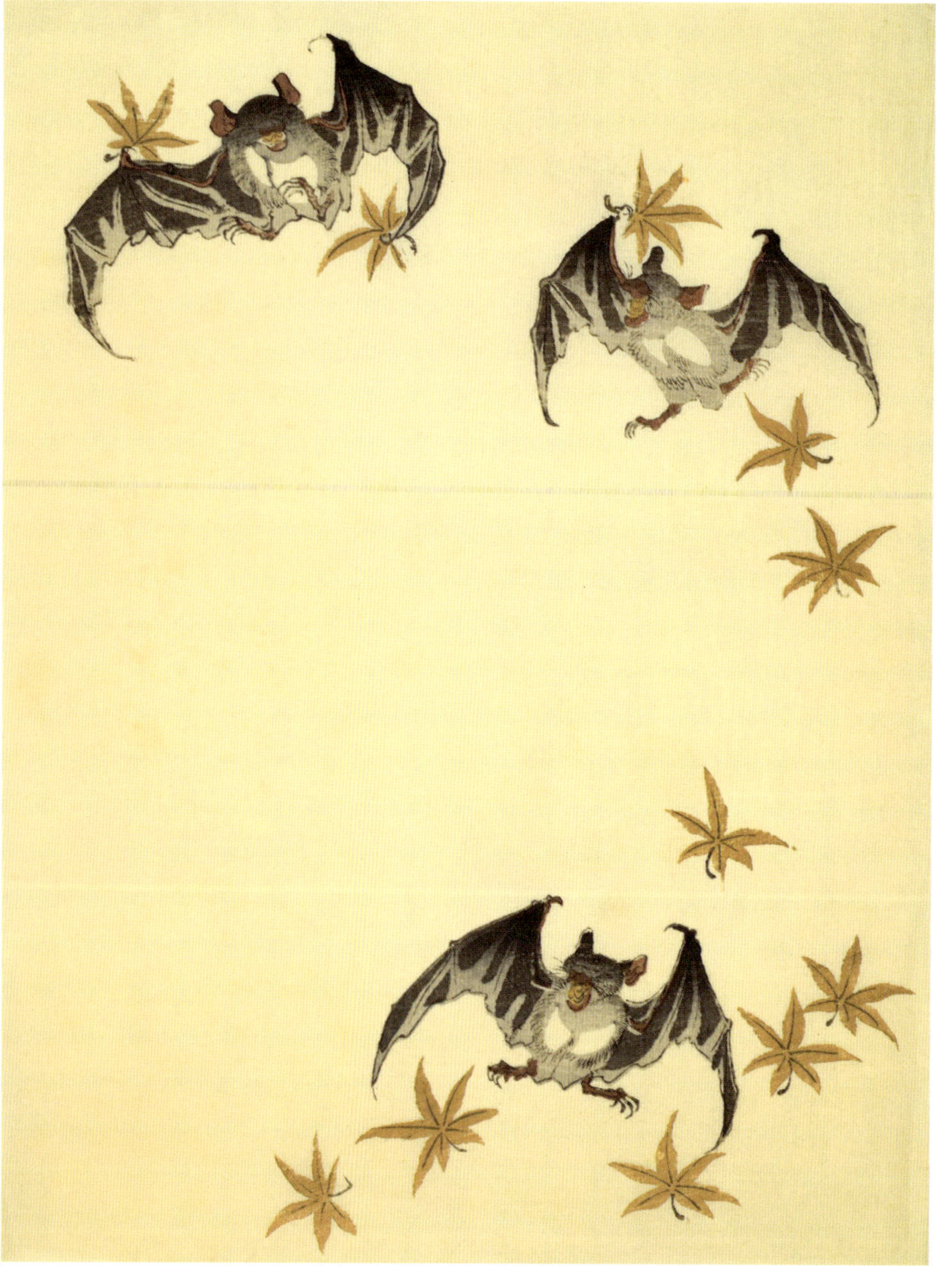

←
A *c.* 1900 Japanese woodblock print shows bats, which are a symbol of happiness and prosperity, surrounded by leaves of the Japanese maple (*Acer palmatum*).

↗
In Japanese folklore the bat is a symbol of longevity and good luck, as is the moon. Both are celebrated in Takahashi Biho's *Bat in Moon* (*c.* 1905).

→
⊖ This illustration (*c.* 1777–82) of a greater Indian fruit bat (*Pteropus medius*), attributed to Bhawani Das, emphasizes its wide wingspan, which can reach 1.5 metres (4 feet 11 inches). The bat feeds on fruit and is important in seed dispersal as well as pollination.

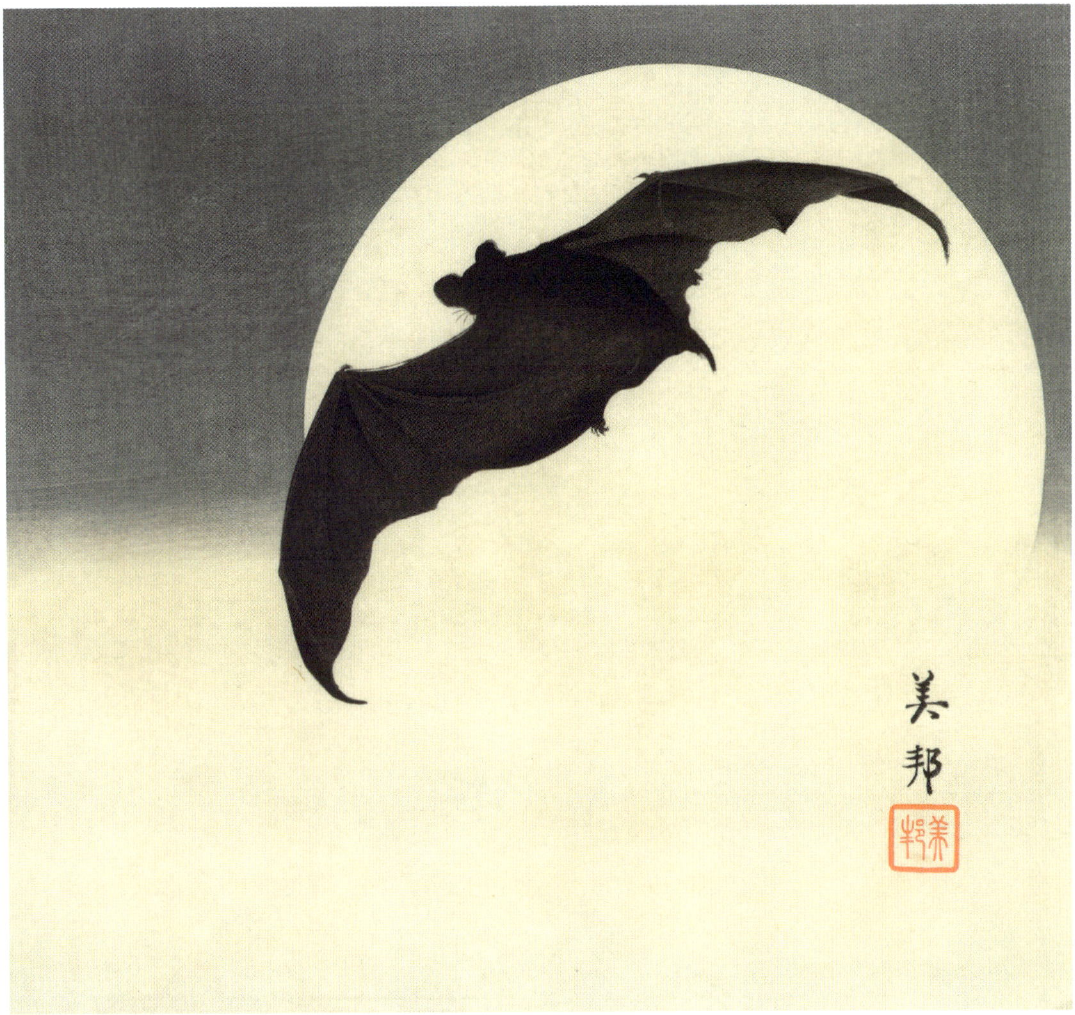
美邦

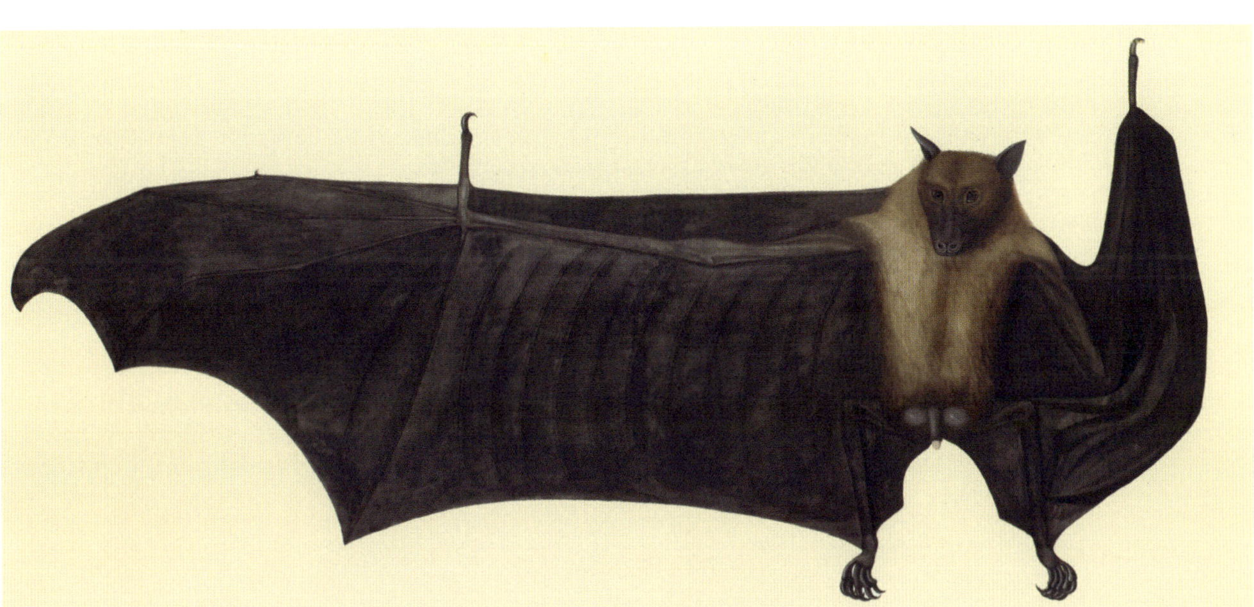

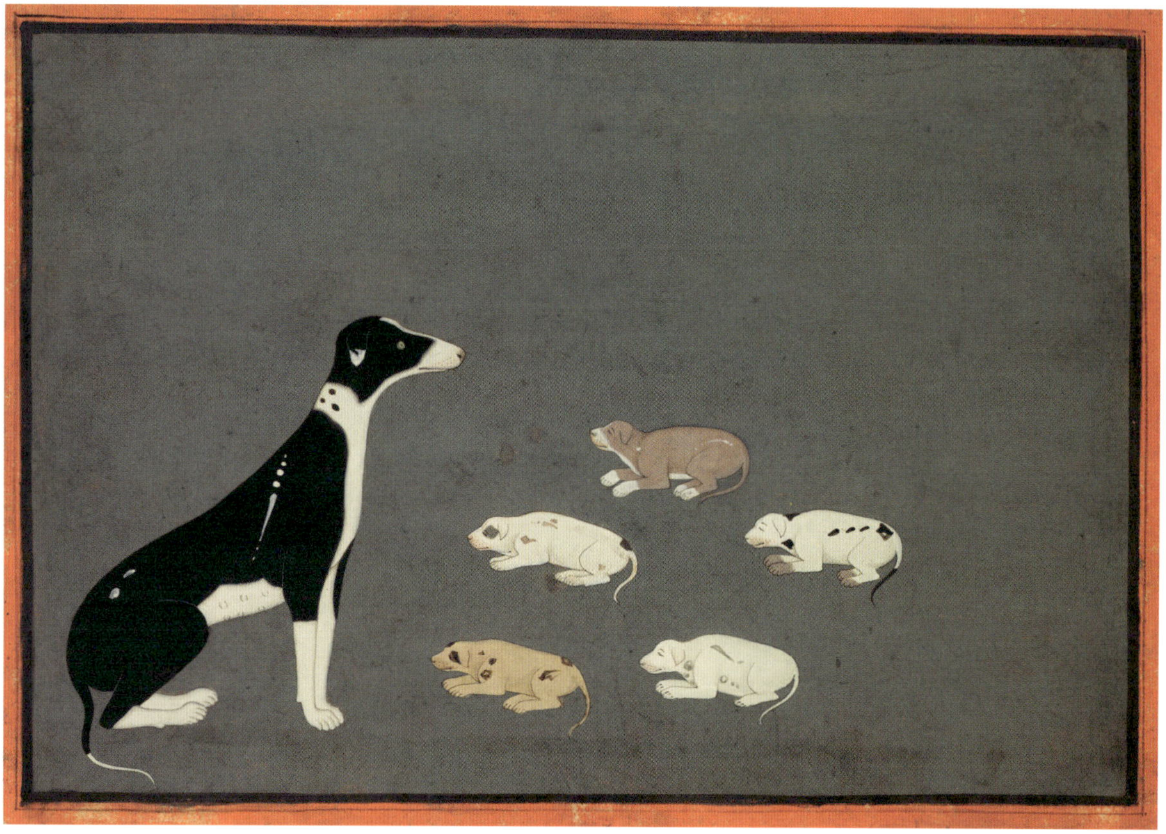

(VII)

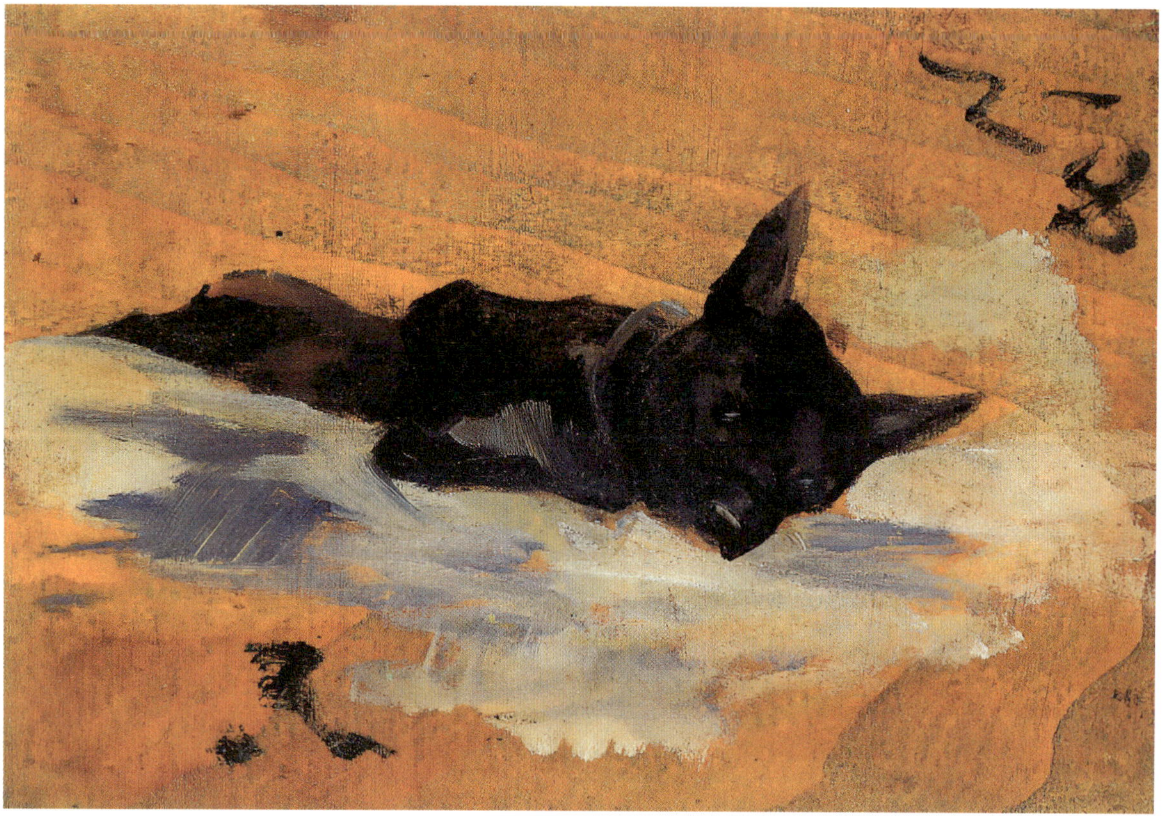

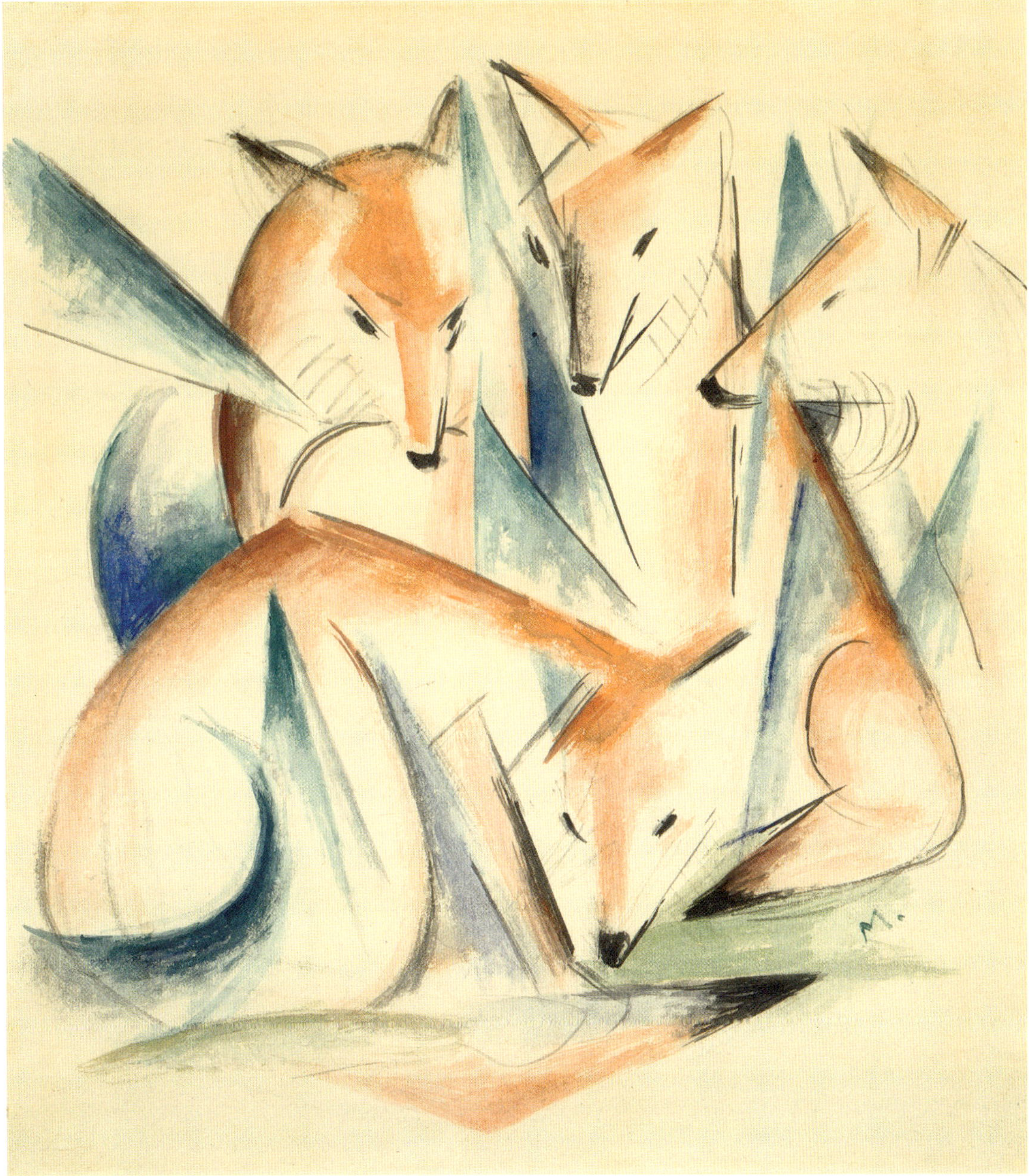

↖
As evidenced in this *c.* 1780 illustration, dogs were popular pets among the elite of Mughal India. They were particularly popular with the emperors, including Akbar, who imported breeds from abroad, and Jahangir, who requested the British ambassador to supply him with English hunting dogs.

←
In *Little Dog* (1888), Henri de Toulouse-Lautrec has used a few bold strokes to capture the dog's relaxed attitude. Dogs are the earliest domesticated species of animal; evidence suggests that they have acted as companions to humans for more than 14,000 years.

↑
For German artist Franz Marc, animals symbolized a pre-industrial age when people lived in harmony with nature. In *Four Foxes* (1913), executed in watercolour and blue chalk, he has used a simplified, abstract approach to express the essential nature of his subject.

(VII)

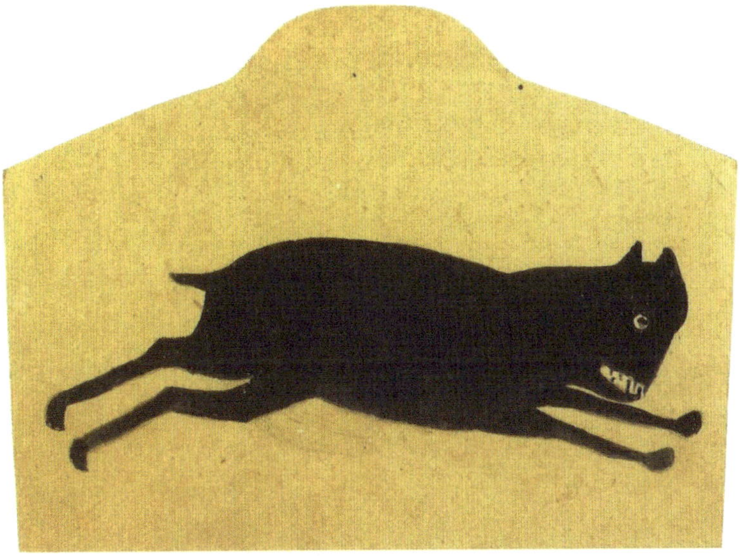

↲
In this slice of painted plaster, from a series of murals by Francisco de Goya known as the Black Paintings (1819–23), a dog gazes up at the sky. Is it sinking into quicksand or is it concealed behind a wall?

←
American self-taught artist Bill Traylor drew subjects from daily life based on observation and memory, including *Black Dog Running* (*c.* 1939–42), which was painted on cardboard.

↓
American painter Frederic Remington has captured a lone wolf in *Moonlight, Wolf* (*c.* 1904). Wolves form a vital part of the ecosystem as they keep populations of prey species under control. They had come close to extinction in the US by the mid-20th century, but conservation measures have led to rising numbers again in some areas.

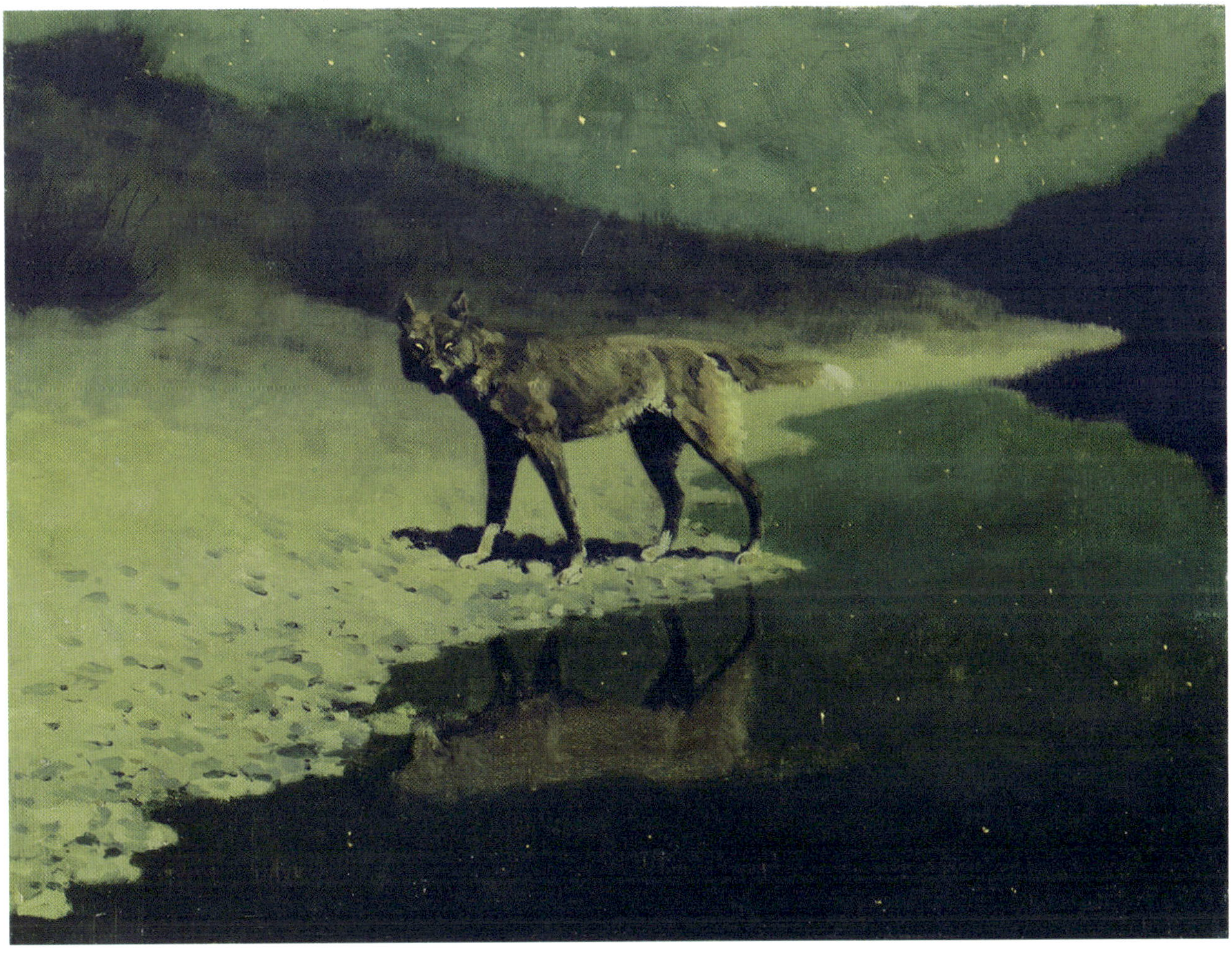

(VII)

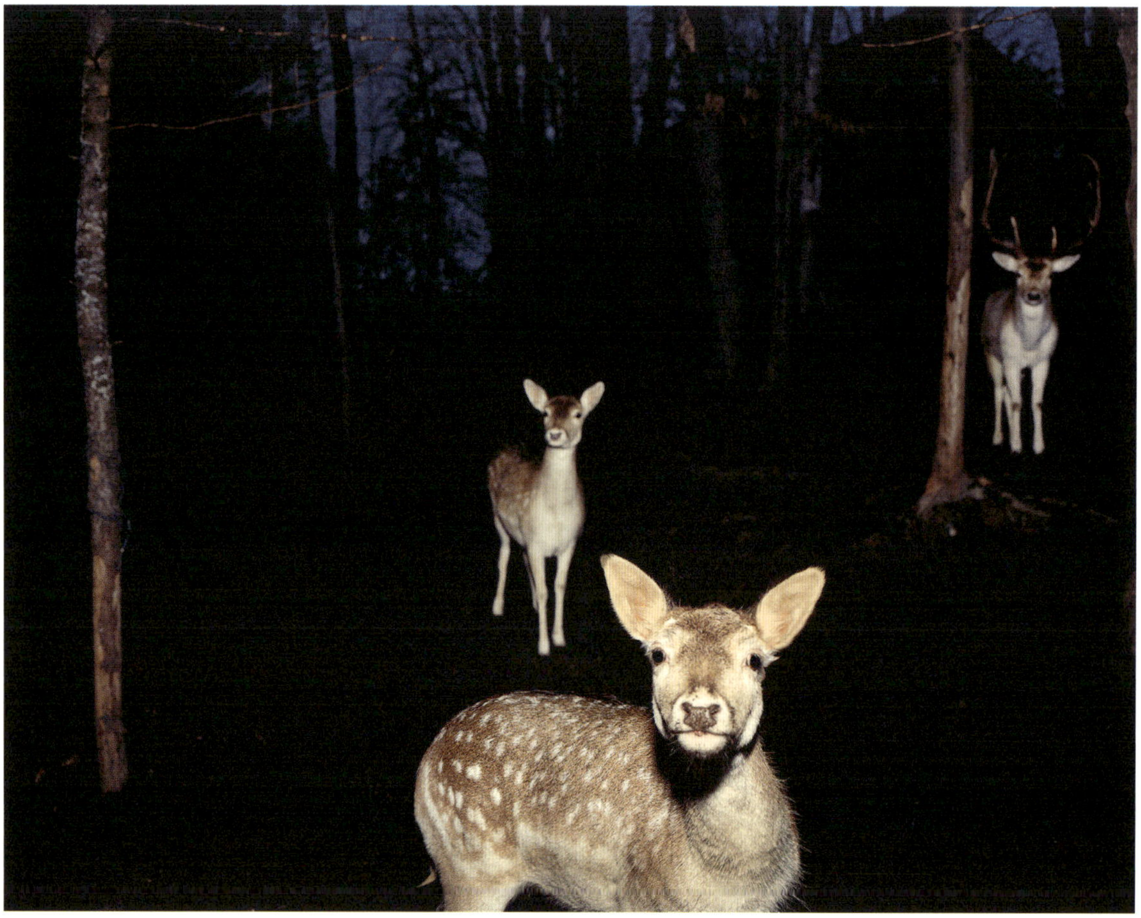

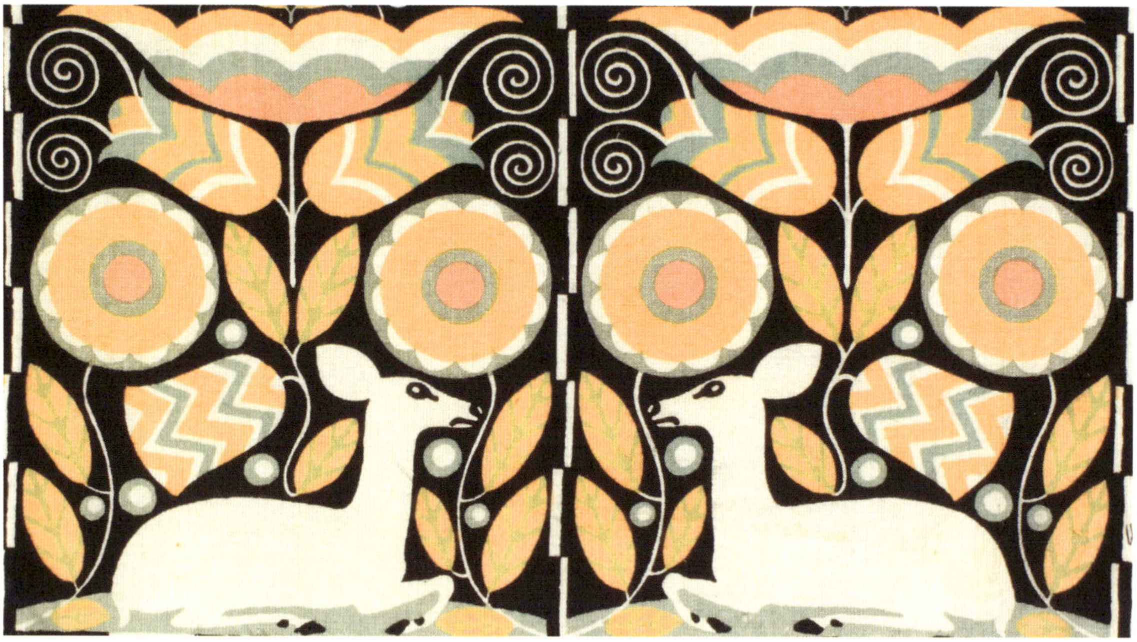

'There never was a deer which was not an unalloyed thing of beauty alive – there never was a deer dead which was not a reproach to the brutality of some man.'

WILLIAM ELLIS, 'IN THE SMOKING ROOM', 1899

←
Jocelyn Lee's 'Deer at Night, Maine' (2002) is from her photographic series, *The Physical World*. Lee's work captures the fragility of the living world.

↙
Entitled *Waldidyll* ('Forest Idyll'), this *c.* 1910 textile was designed by Austrian Carl Otto Czeschka, who intended it to be used for dress or furnishing fabric.

↓
This medieval illumination of billy goats is from a 1430–40 manuscript of *Livre de la Chasse* ('Book of the Hunt'), which was originally composed between 1387 and 1389 by Gaston Fébus, a self-proclaimed master of the blood sport.

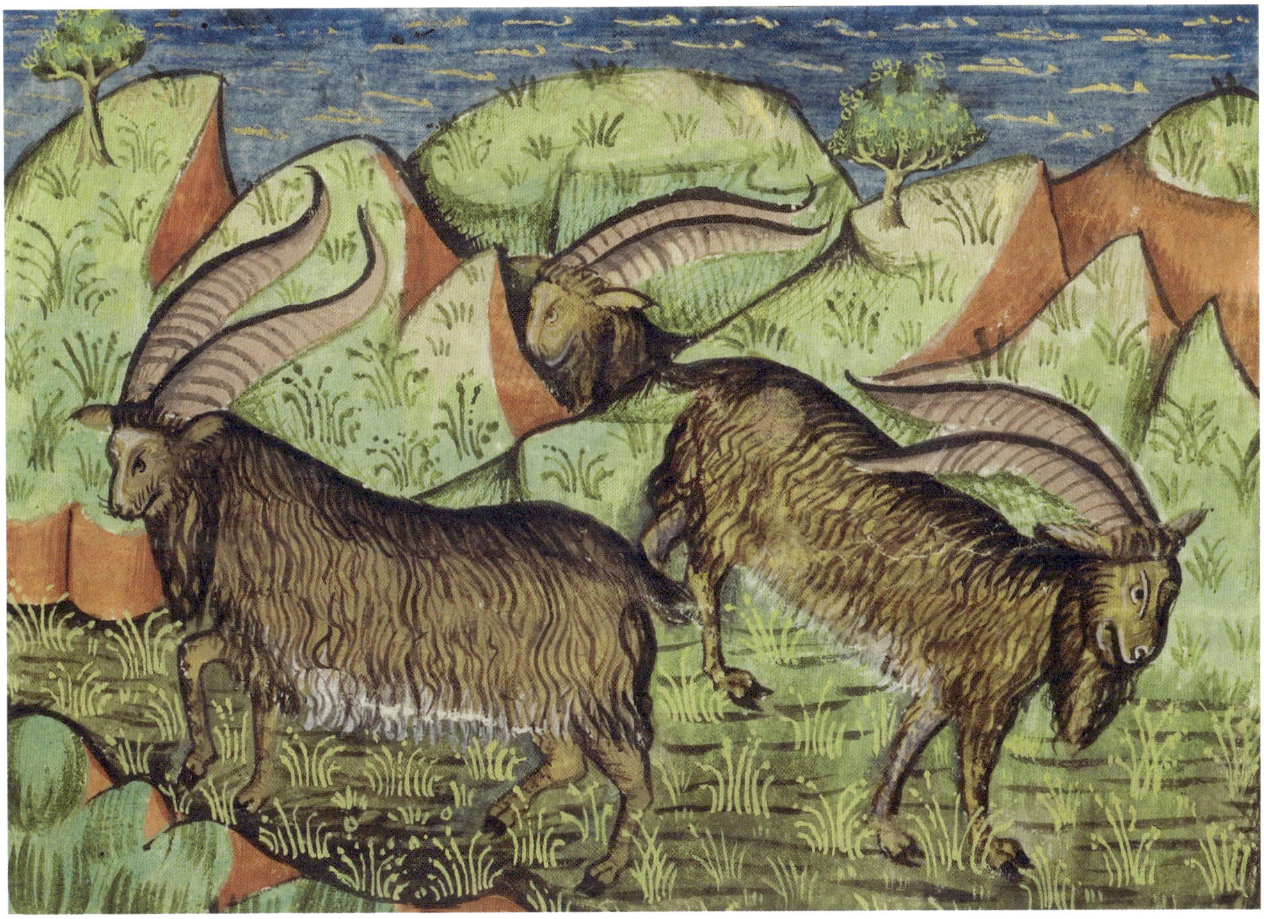

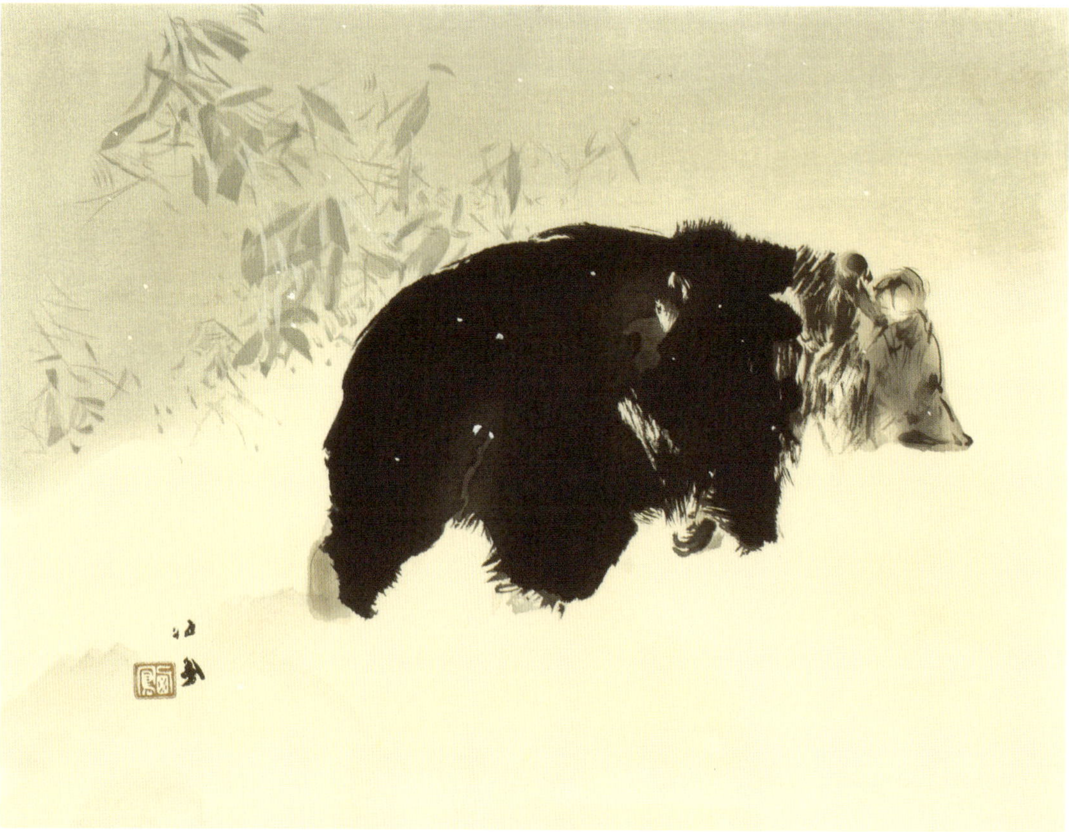

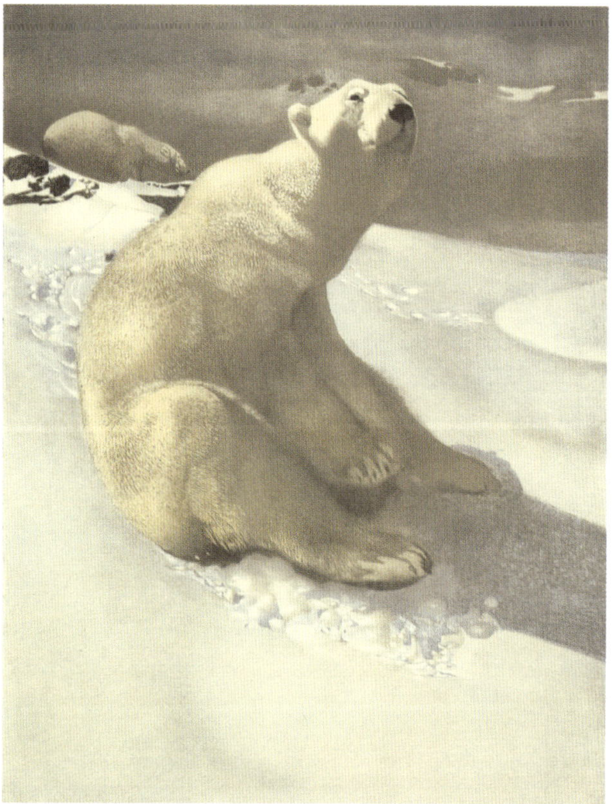

↑
◎ Takeuchi Seiho's *Bear in Snow* (1940) conveys the power and vitality of the bear. Japan's Asiatic black bear (*Ursus thibetanus*) population is currently growing.

←
◎ Austrian painter Carl Ederer has captured a polar bear (*Ursus maritimus*) seated on snow while another bear walks in the background. Polar bears are the largest carnivore on land and an apex predator, but theirArctic sea-ice habitat is disappearing due to climate change. Estimates of their numbers in the wild range from 22,000 to 31,000.

↗
This wooden carving of a howling wolf from southern Siberia dates from *c.* 500–200 BCE. Wolves howl to communicate their location, to bring the pack together, to warn off rival packs, and sometimes to express affection towards another pack member.

→
◎ The giant panda (*Ailuropoda melanoleuca*), illustrated here in a publication dating from between 1868 and 1874, is one of the most threatened species in the world as the bamboo forests where they live become increasingly fragmented. Pandas are seed dispersers and their presence helps vegetation to flourish, which in turn helps to maintain a high level of biodiversity.

'Did the polar bear leave his field of ice and journey toward the tropics? How did he know where the Ark was?'

ROBERT GREEN INGERSOLL, *SOME MISTAKES OF MOSES*, 1879

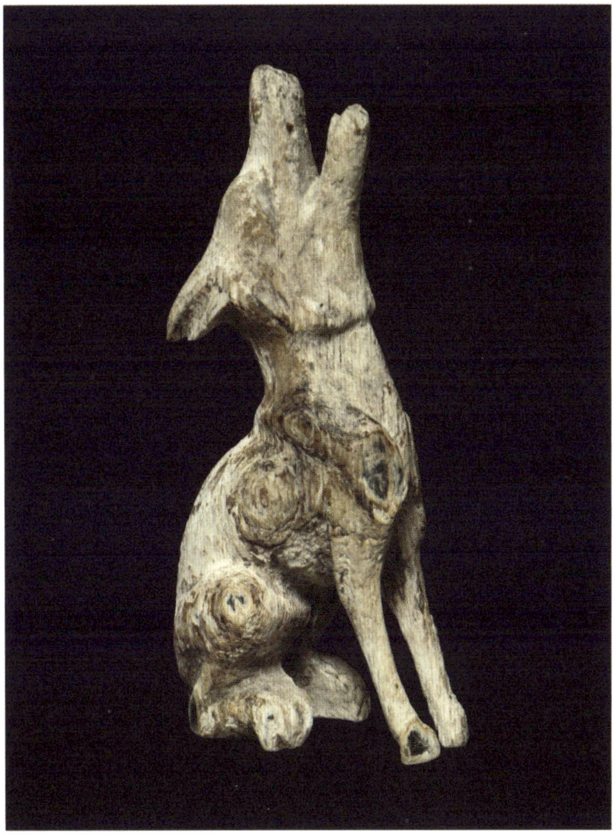

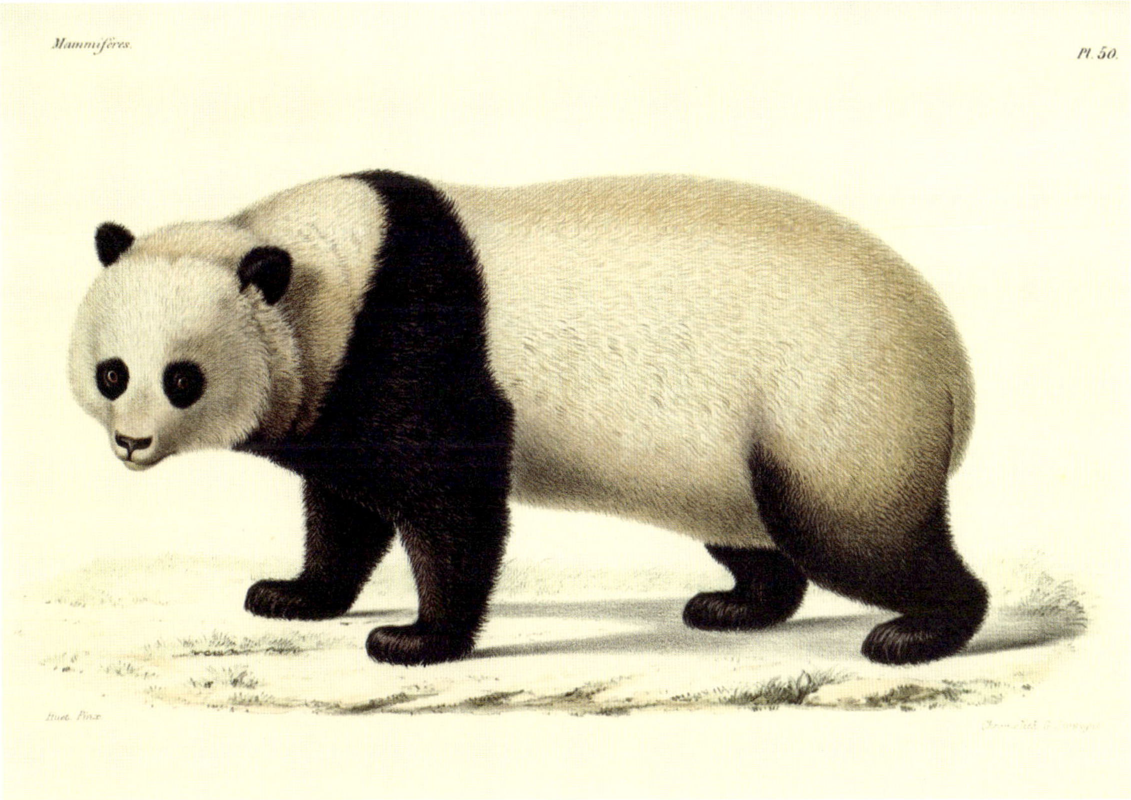

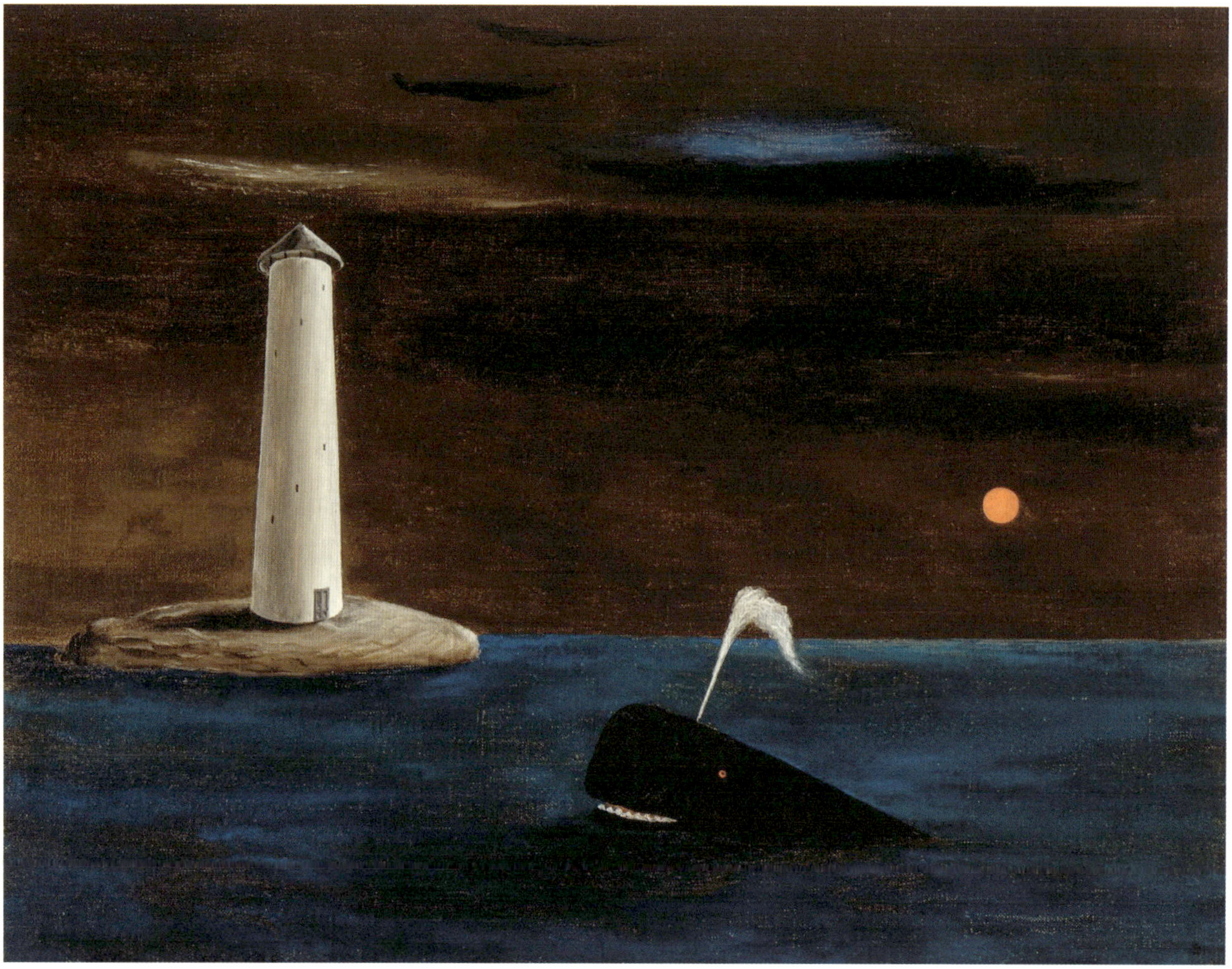

'The whale is the biggest, the meekest, and the most interesting of living animals. As we go north, we readjust our ideas of distance and immensity. Rivers are longer, lakes are more majestic, and whales bigger than we have ever dreamed.'

AGNES DEANS CAMERON, *THE NEW NORTH: BEING SOME ACCOUNT OF A WOMAN'S JOURNEY THROUGH CANADA TO THE ARCTIC*, 1910

↑
American painter Gertrude Abercrombie made autobiographical works that were inspired by Surrealism, putting herself into her paintings through the use of a symbolic language of animals, plants and female figures, as in *Untitled (Lighthouse, Whale and Black Sea)*, dating from 1948.

↗
◎ The dugong (*Dugong dugon*), illustrated here by James Stewart around 1837, reproduces slowly and has a long lifespan of up to seventy years, making the species particularly vulnerable to extinction.

→
◎ This frieze – which was reconstructed in 1995 using conserved 13th-century BCE fragments from the Mycenaean acropolis of Gla in Boeotia, Greece – depicts the common dolphin (*Delphinus delphis*), once an abundant species in the Mediterranean. Since the 1960s, numbers have declined rapidly due to human activity and climate change, and the species is now endangered in the Mediterranean.

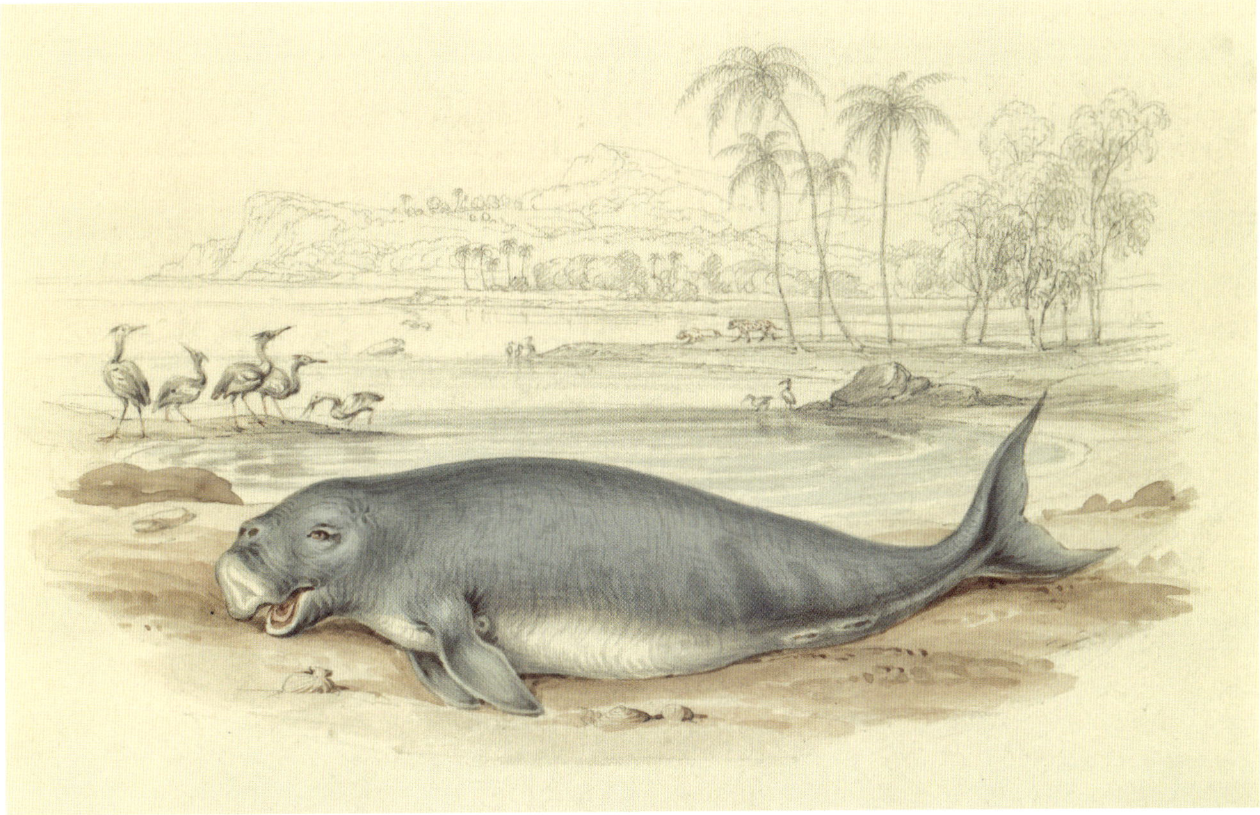

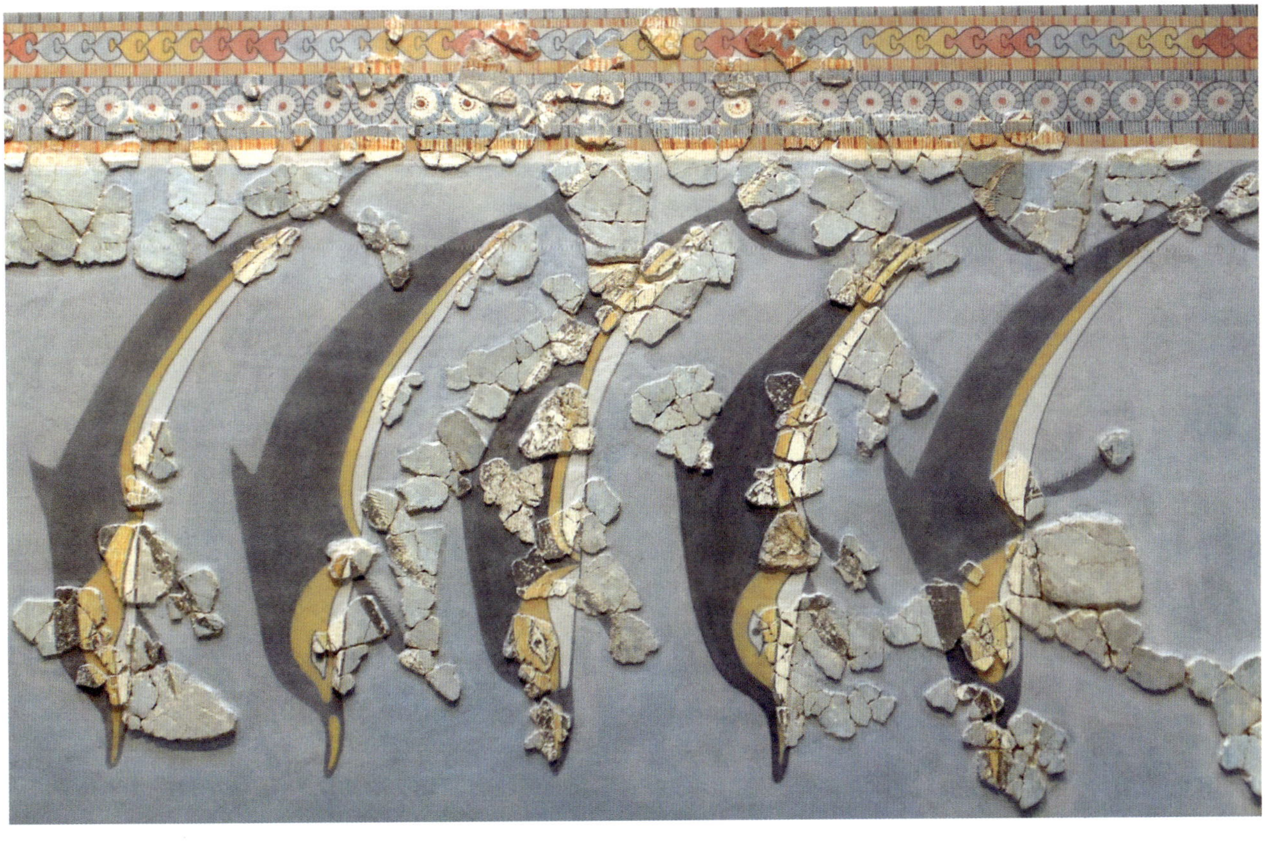

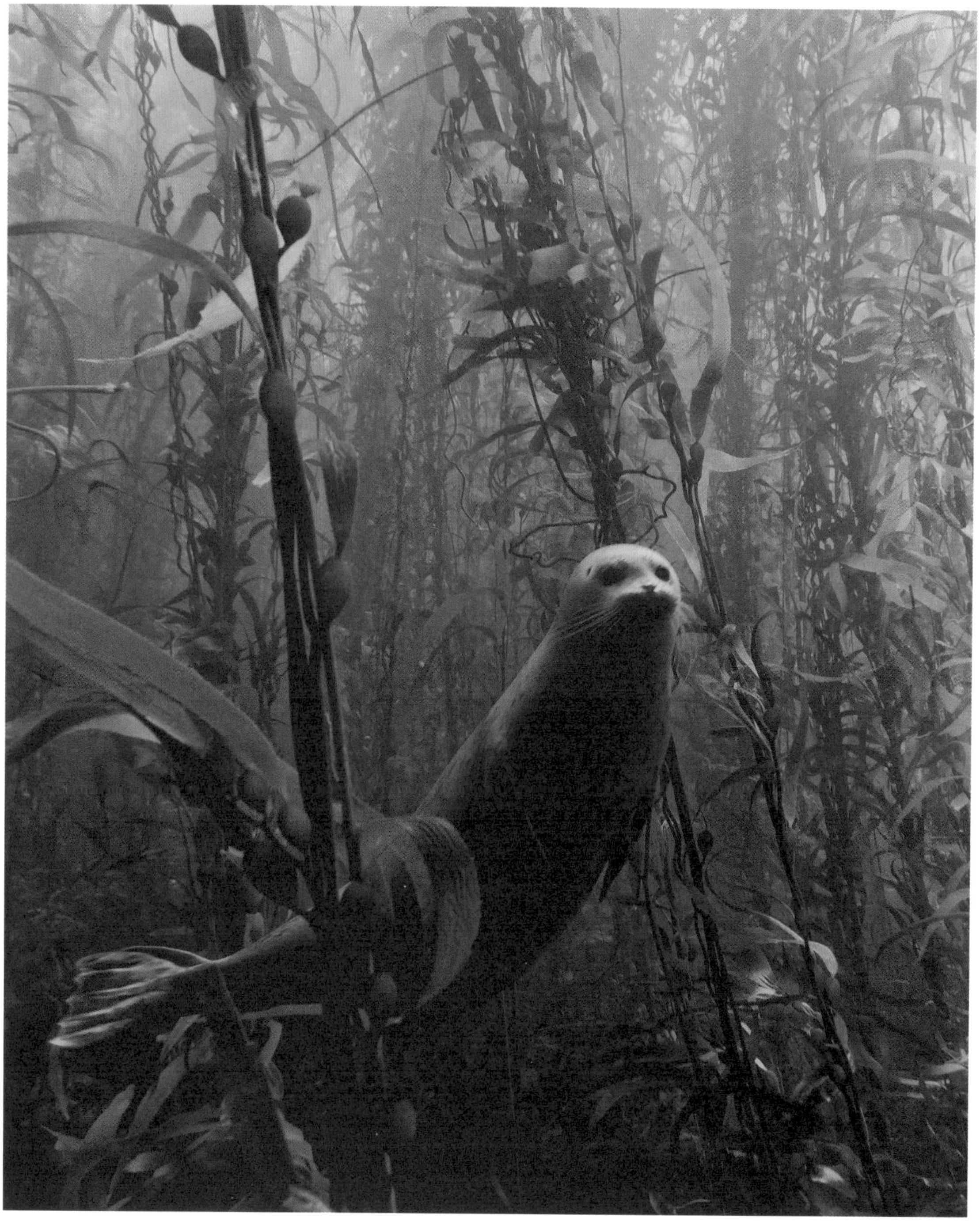

'The lives of walruses, like those of polar bears and seals, are changing. All are living at the frontier of climate change, and all are suffering as a consequence.'

SIR DAVID ATTENBOROUGH, IN THE BBC TV SERIES *OUR PLANET*, 2019

←
Kelp forests provide a home for the harbour seal (*Phoca vitulina*). This example was photographed by cinematographer Kyle McBurnie off the coast of California in 2012.

→
◎ Based on a photograph by Dr James Murie, this engraving shows a walrus (*Odobenus rosmarus*) that was purchased by the Zoological Society of London (now London Zoo) and survived in its gardens for just a few weeks. Typically, walruses have tusks that can grow forever, sometimes reaching a length of 1 metre (3 feet 3 inches). As Murie observed in 1870, however, this walrus's tusks 'had little more than protruded beyond the mandibular lips.'

↓
British naturalist Edward Roper spotted these marine mammals in the second half of the 19th century in what he describes as the doldrums or equatorial calms, the near-windless belt across the earth's equator. He described them as a school of porpoises, though the beaked shape of their heads is more akin to that of a dolphin.

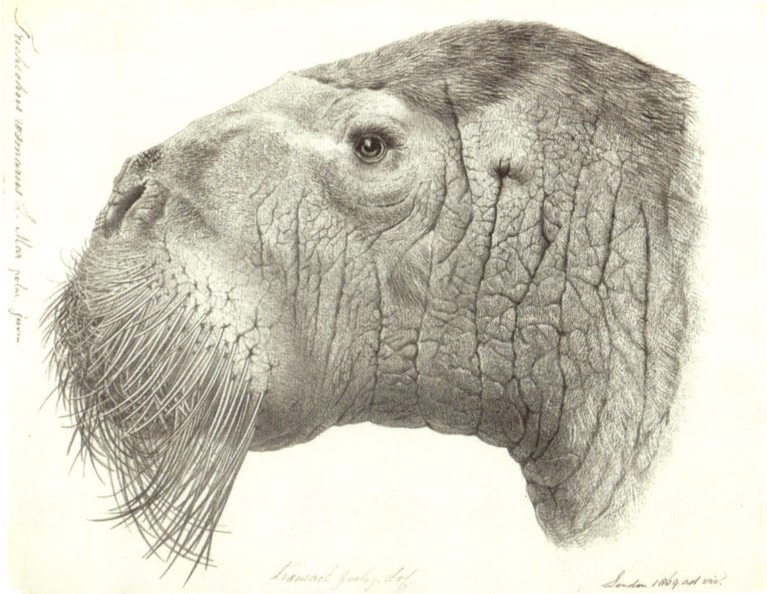

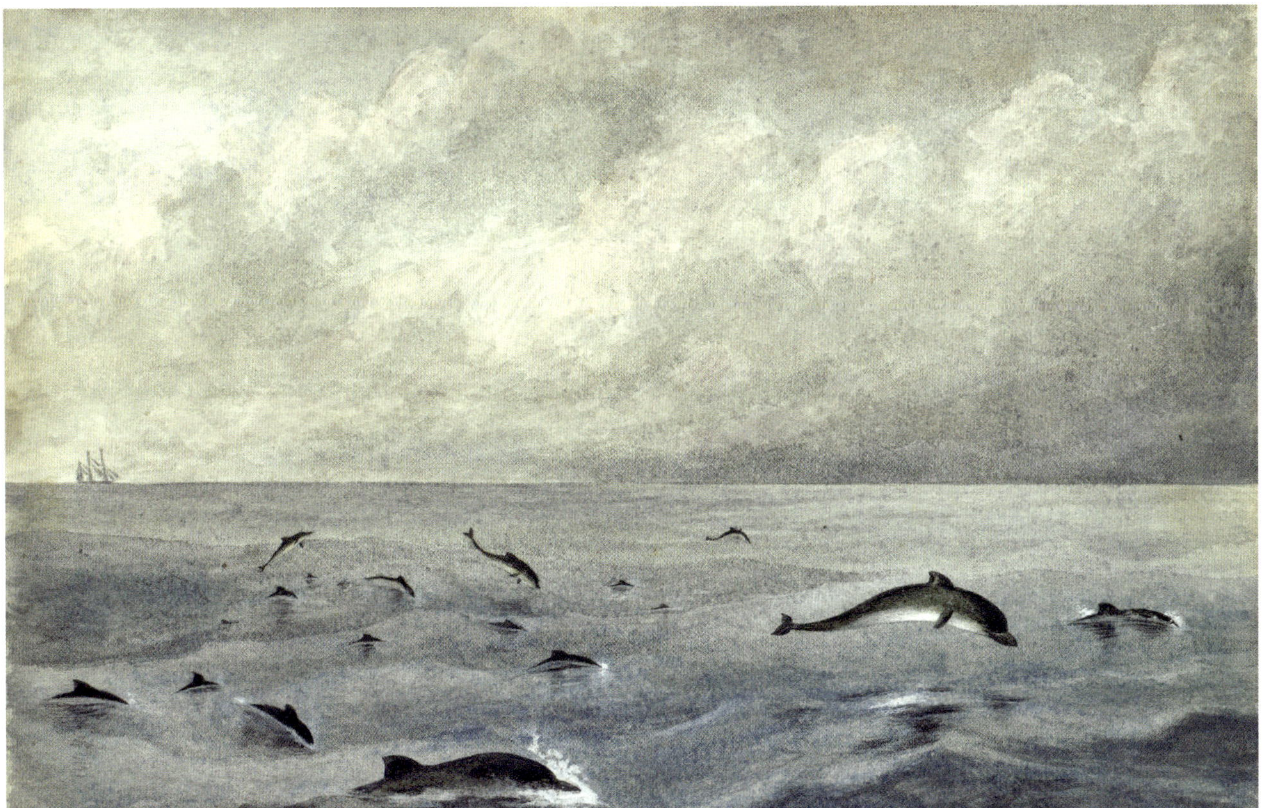

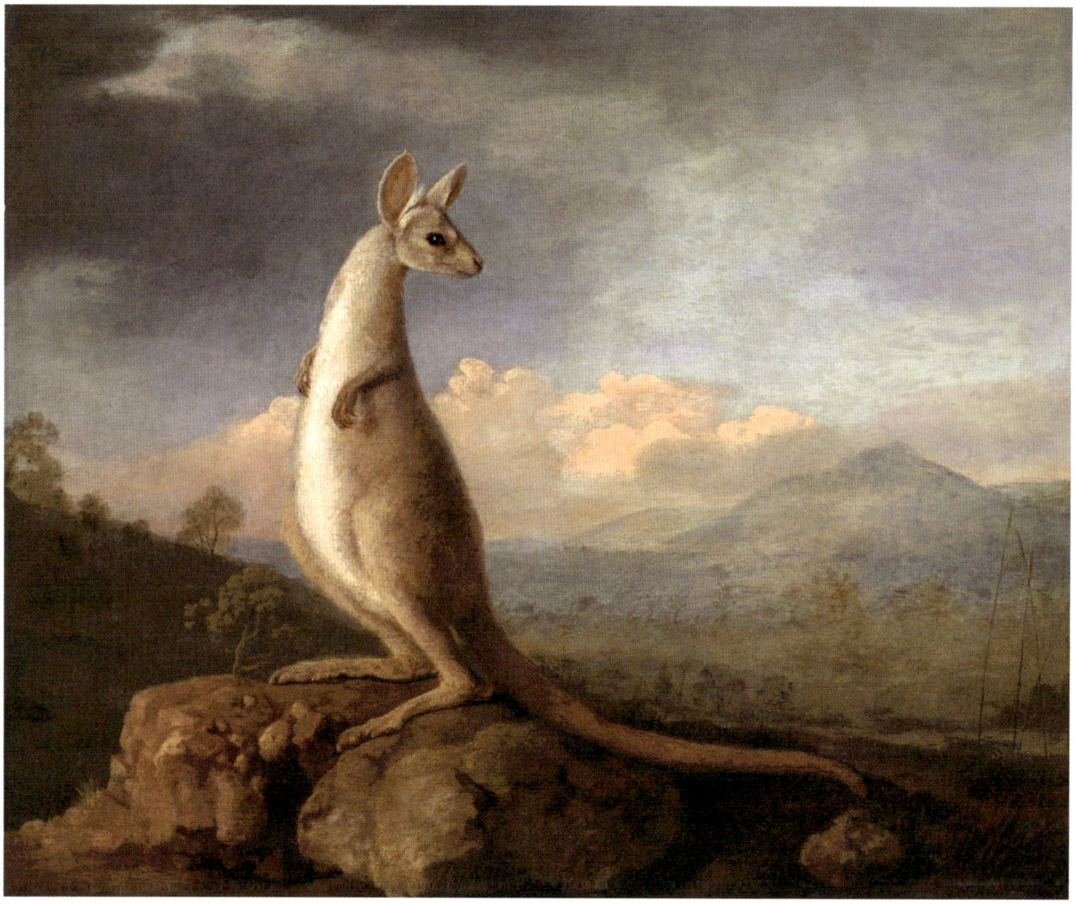

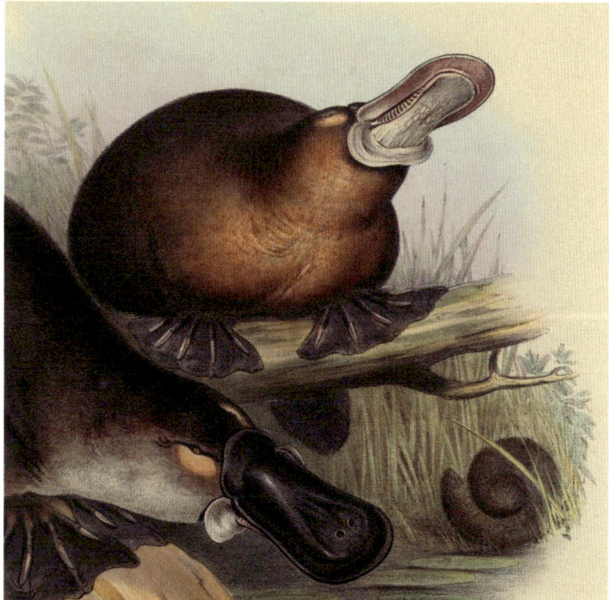

↑
George Stubbs had never seen a kangaroo when he painted *The Kongouro from New Holland* in 1772. He worked from verbal and written descriptions, a few pencil sketches by Sydney Parkinson and a stuffed kangaroo pelt.

←
⊖ When scientists in Europe first saw illustrations of the platypus (*Ornithorhynchus anatinus*), they thought the strange-looking, semi-aquatic creature with a bill like a duck and an animal's body, was a hoax. The platypus can detect electric signals given off by the muscles in its prey.

→
◎ Originally titled 'a monkey or bear of New South Wales', this drawing of a koala is from William Govett's notes and sketches made during a surveying expedition from 1830 to 1835. Land clearance, logging, urbanization and climate change all threaten koala numbers today.

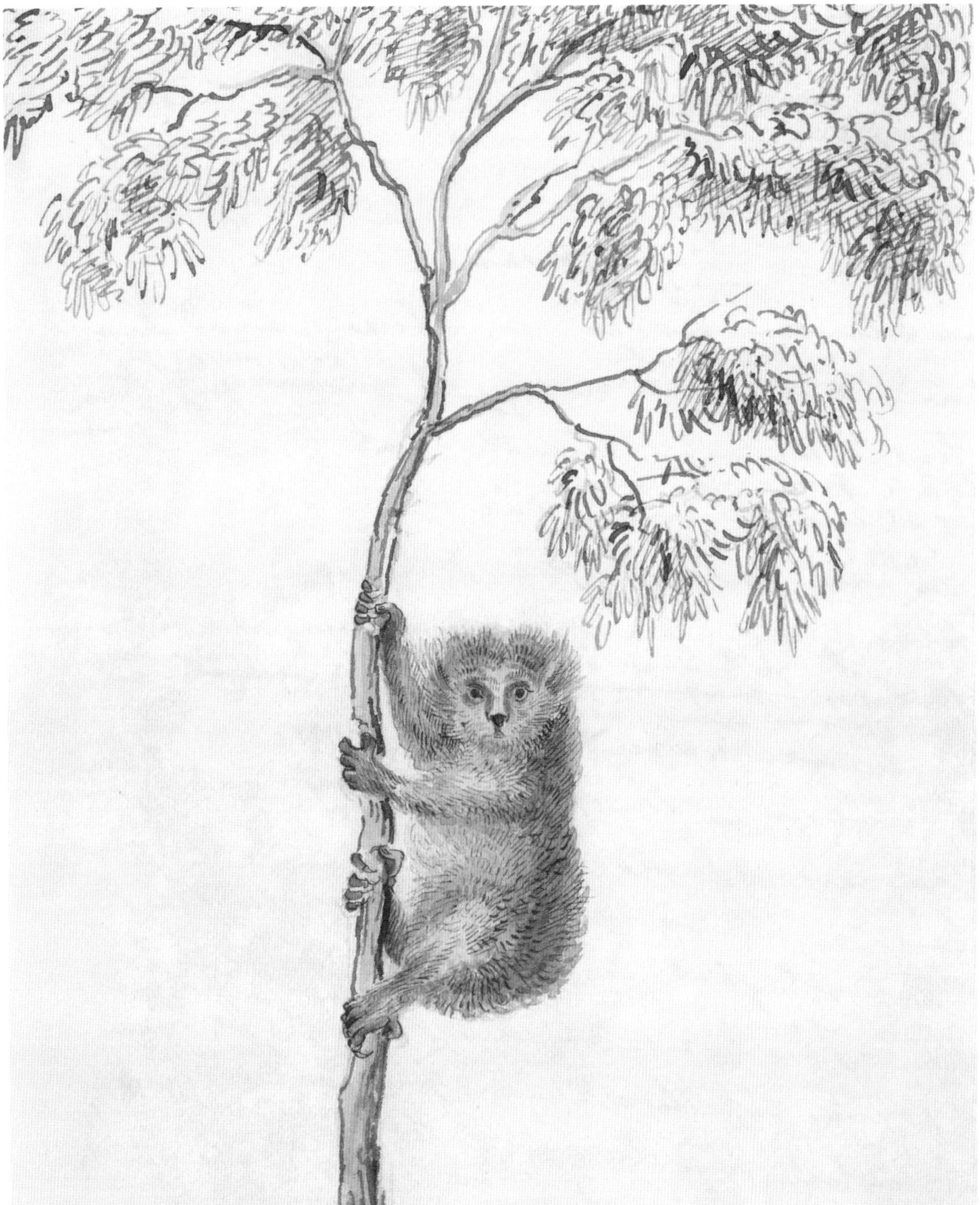
The Monkey or Bear of NSWales.

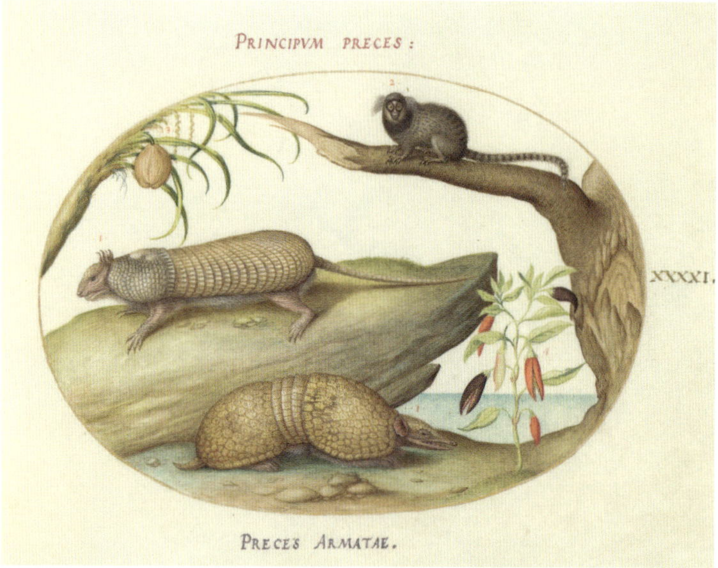

←
Armadilloes, illustrated here in Flemish artist Joris Hoefnagel's *Two Types of Armadillos with a Marmoset, a Coconut Palm, and a Pepper Plant* (late 16th century), cross rivers by walking underwater on the riverbed. They can inflate their stomachs to become buoyant if necessary.

↓
Hans Hoffmann's *A Hare in the Forest* (*c.* 1585) shows a hare surrounded by insects and plants, including plantain, thistle and harebell flowers. None of these plants co-exist in the wild.

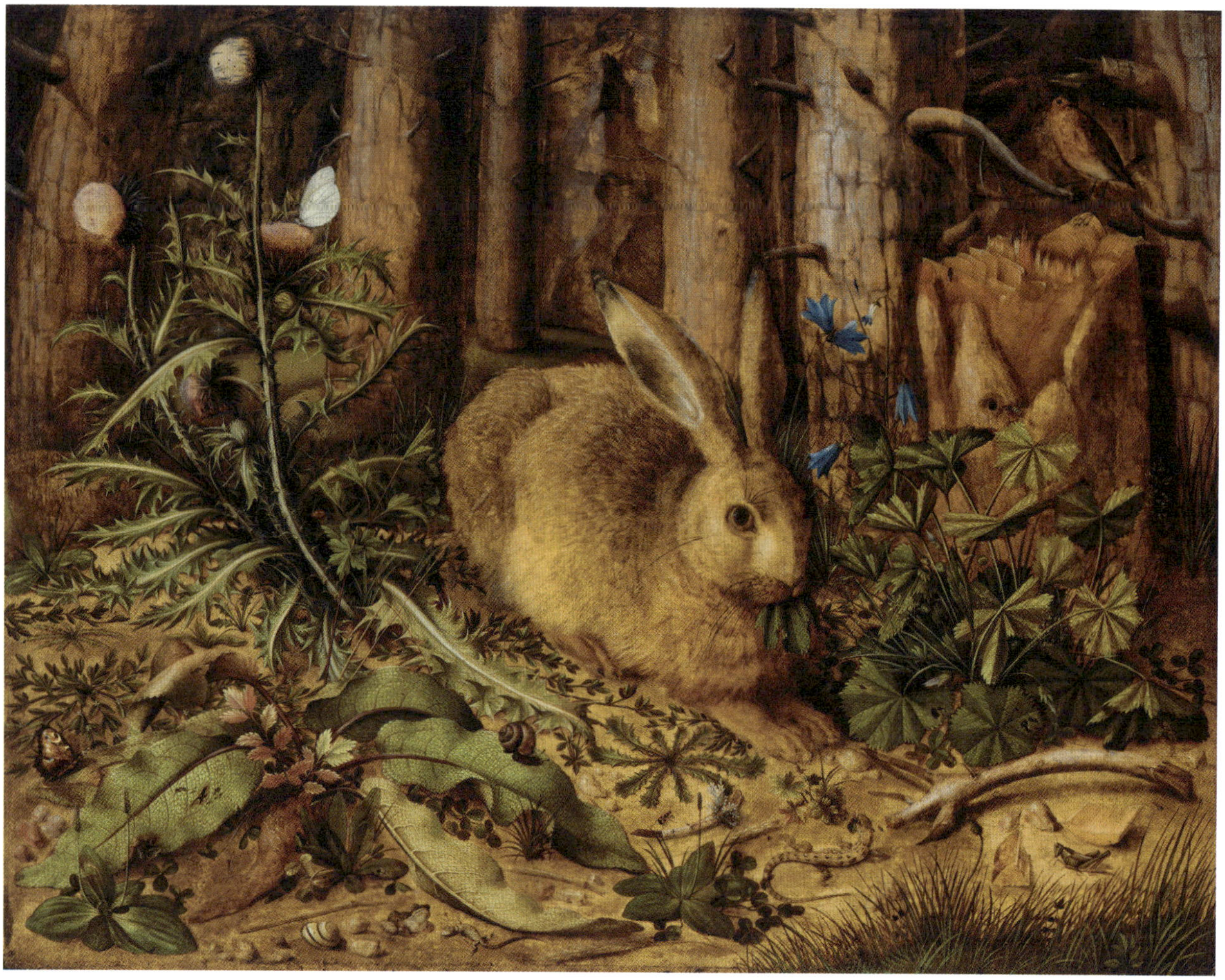

'His nose upon his two fore-feet did lye, With his grey eyes he glared obliquely; His head he always set against the wind'

MARGARET CAVENDISH, 'THE HUNTING OF THE HARE', 1664

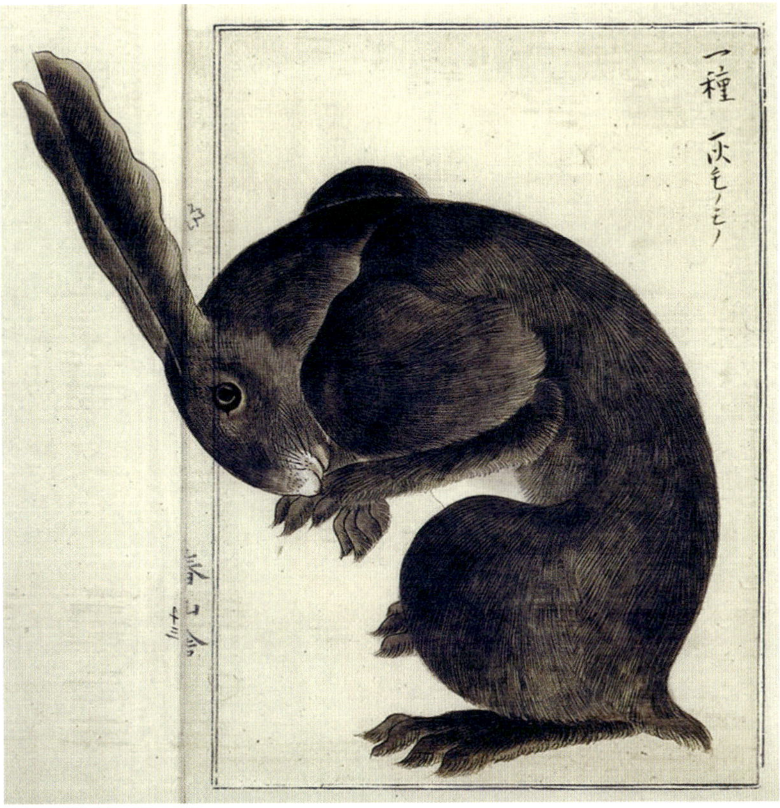

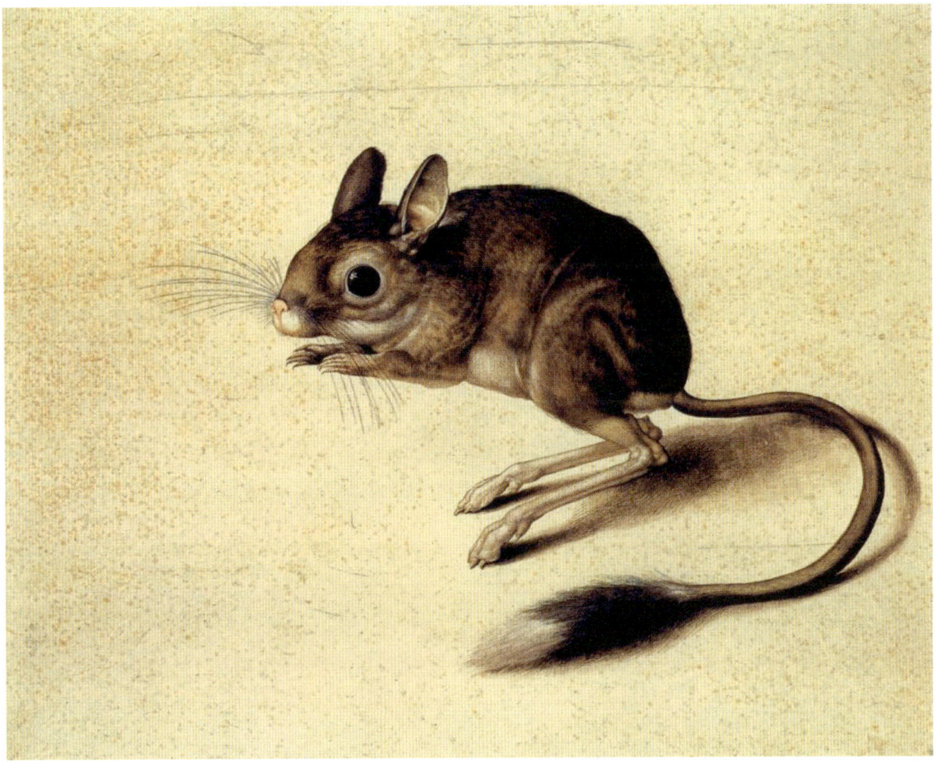

↗
In Japan, hares are symbols of good fortune, fertility and cleverness, and they feature in several myths. Takagi Shunzan's 1850s woodblock print is from the *Honzō Zusetsu*, a collection of natural history illustrations.

→
This late 16th-century drawing of a jerboa is by Italian artist Jacopo Ligozzi, known for his drawings from nature. Jerboas create complex tunnel systems, which protect them from weather extremes.

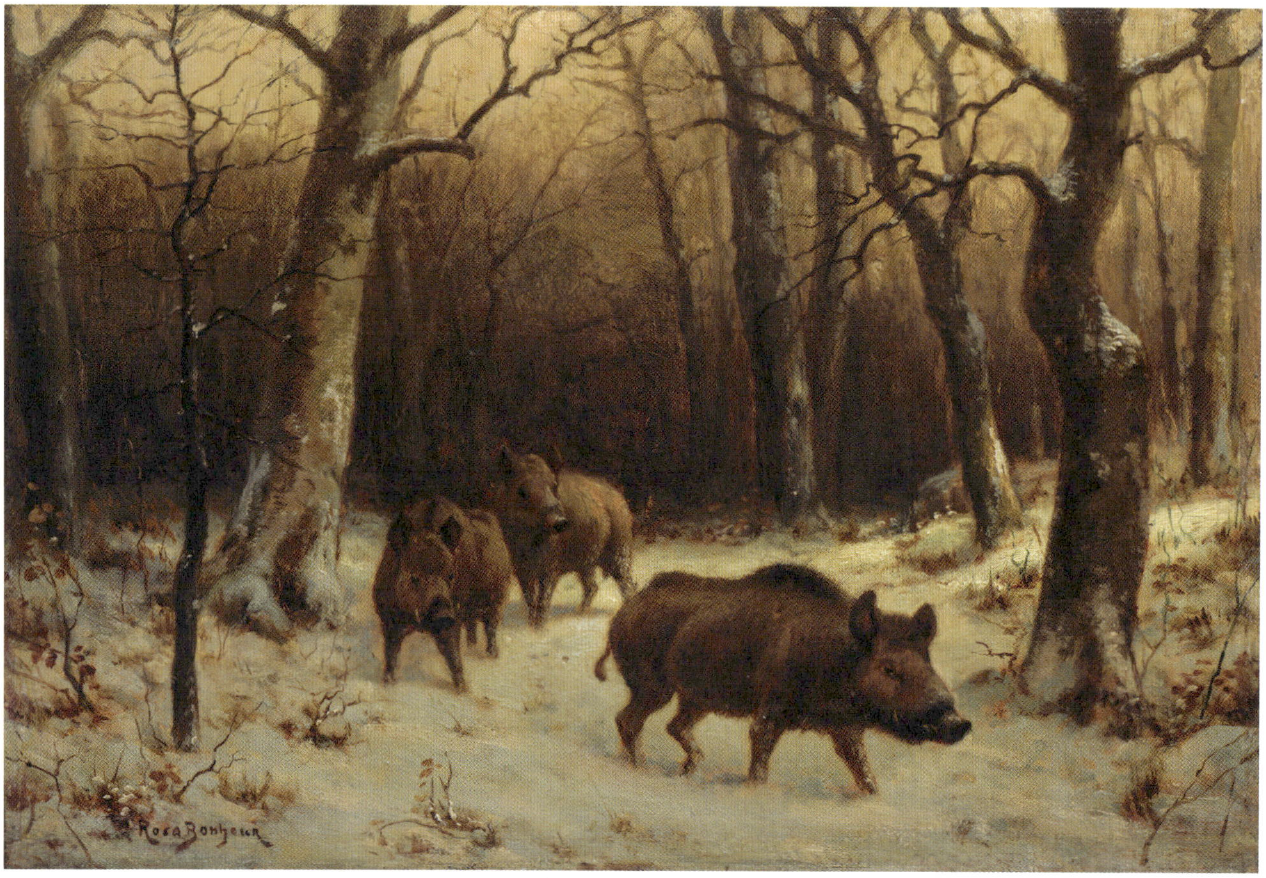

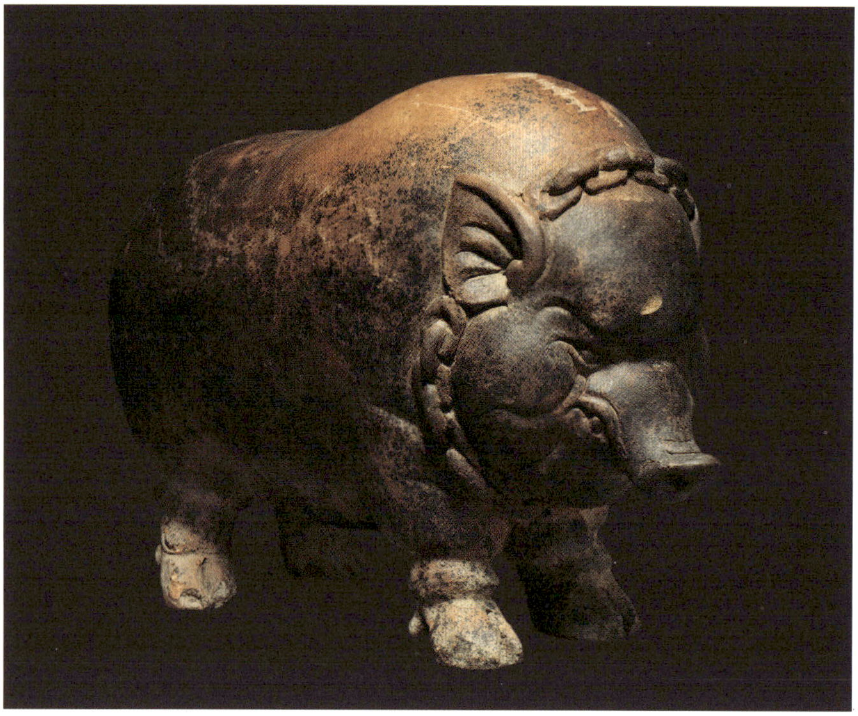

↑
Wild boar (*Sus scrofa*), featured in this *c.* 1872–77 oil painting by Rosa Bonheur, are timid forest dwellers by nature, but they can become aggressive if they feel threatened. Wild boar were first domesticated to become farmed pigs around 9,000 years ago, but some of those animals escaped and created wild populations that sometimes bred with the undomesticated boar.

←
This early example of a piggy bank (14th–15th century) is from the Majapahit kingdom of eastern Java. The idea for piggy banks probably came from the fact that pigs scavenge for food wherever they can find it, and in so doing accumulate wealth for their owners as they put on weight. Their owners can then cash them in by selling them for slaughter.

→
The wild boar piglet's distinctive, humbug-like stripy markings, illustrated in this 1578 drawing by German artist Hans Hoffmann, help camouflage it in woodland.

↓
Pigs have sophisticated spatial understanding and good memories, making them excellent foragers. Throughout the 17th century, when Dutch artist Adriaen Verdoel painted *Sleeping Pig*, they often roamed free, in forests and sometimes streets, to find their own supplementary food sources.

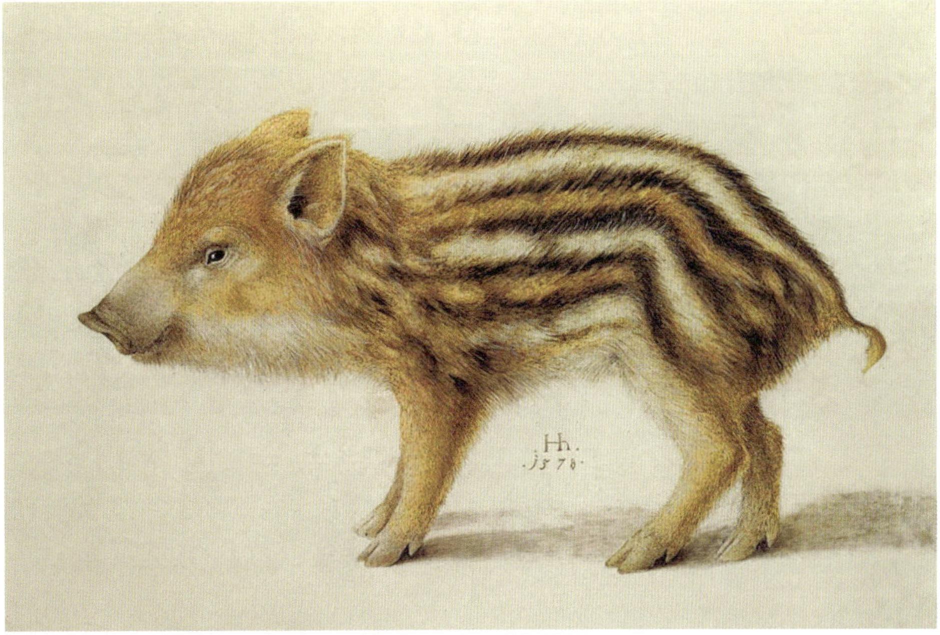

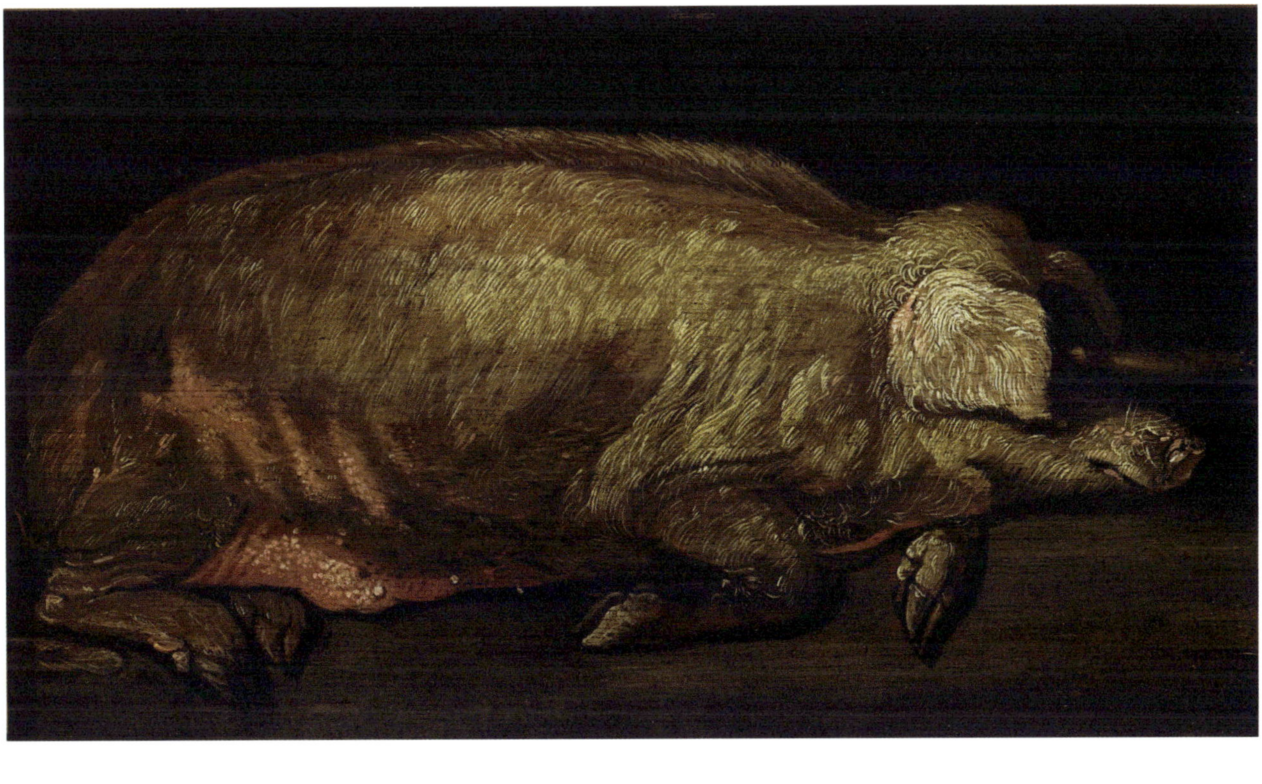

'Love the animals: God has given them the rudiments of thought and untroubled joy. So do not trouble it, do not harass them.'

FYODOR DOSTOYEVSKY, *THE BROTHERS KARAMAZOV*, 1879–80, TRANSLATED BY CONSTANCE GARNETT

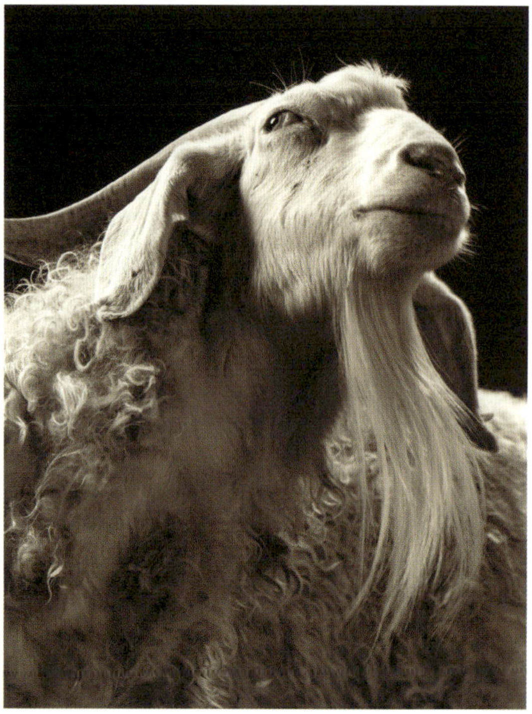

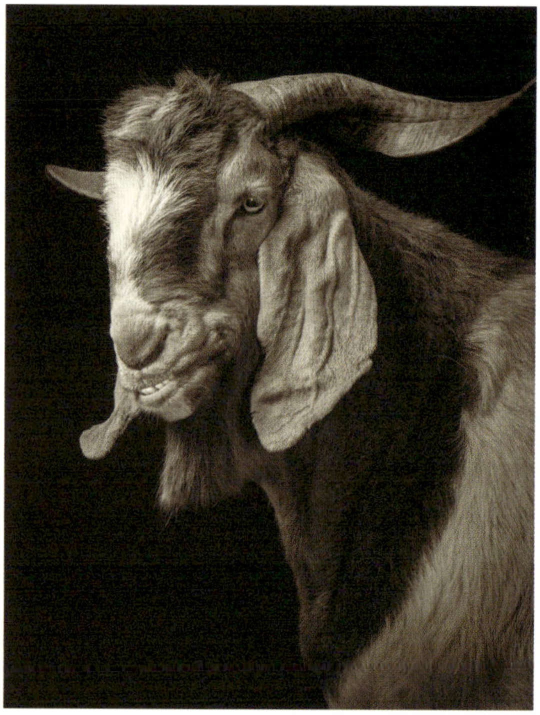

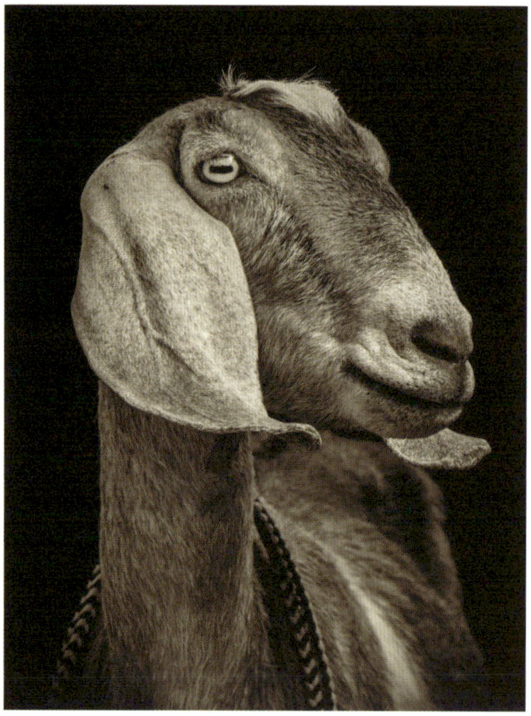

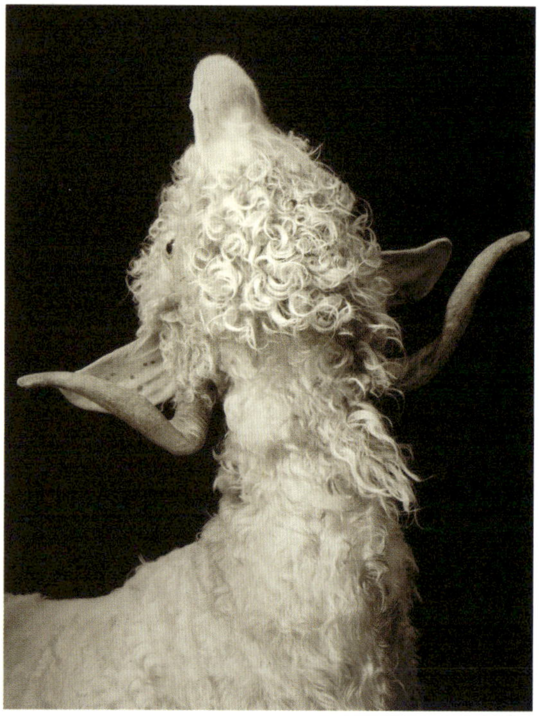

↙
American photographer Kevin Horan's four portraits of goats are from his *Chattel: A Portrait Study* series (2012–14). The goats appear to be posing for the camera. Each animal is named: clockwise from top left, Sydney, Jake, Sherlock and Briede.

→
Llamas are of vital importance in Andean cultures, providing meat, wool and transport. Miniature effigies of camelids, such as this llama figurine dating from between 1400 and 1535 CE, were buried by the Inca at sacred sites, together with other items, to mark state occasions.

↓
Sculptor Henry Moore made drawings of sheep from his studio window in the English countryside, including *Sheep in Snow Scene* (1974).

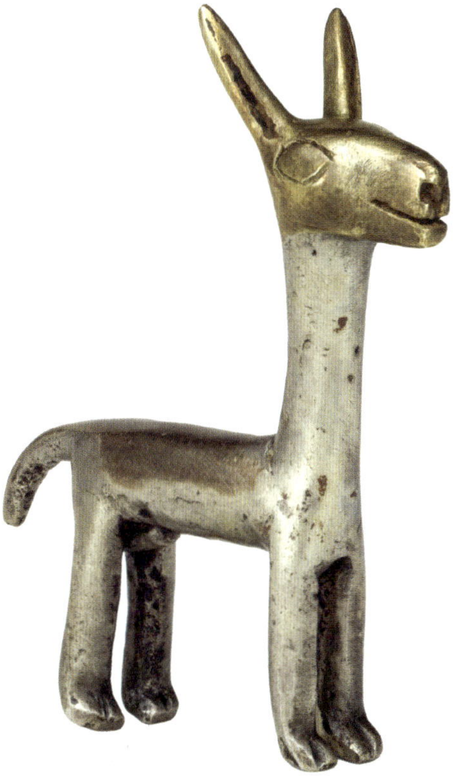

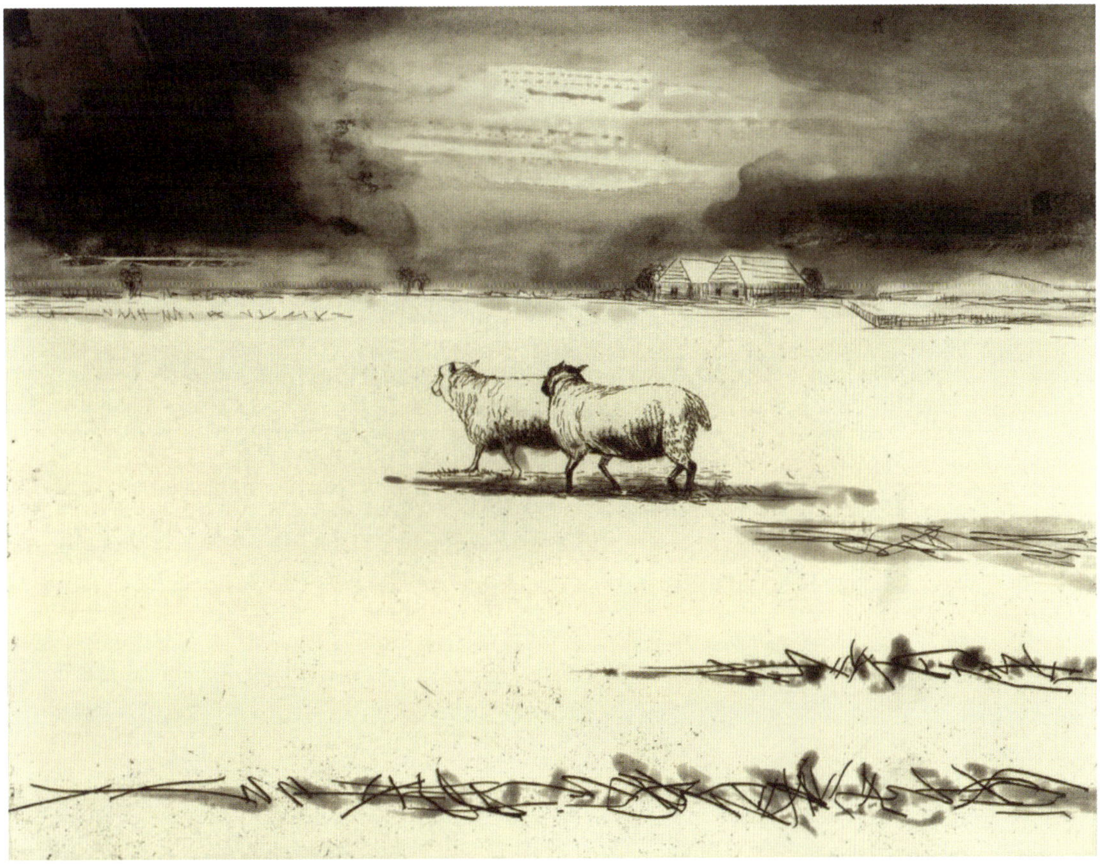

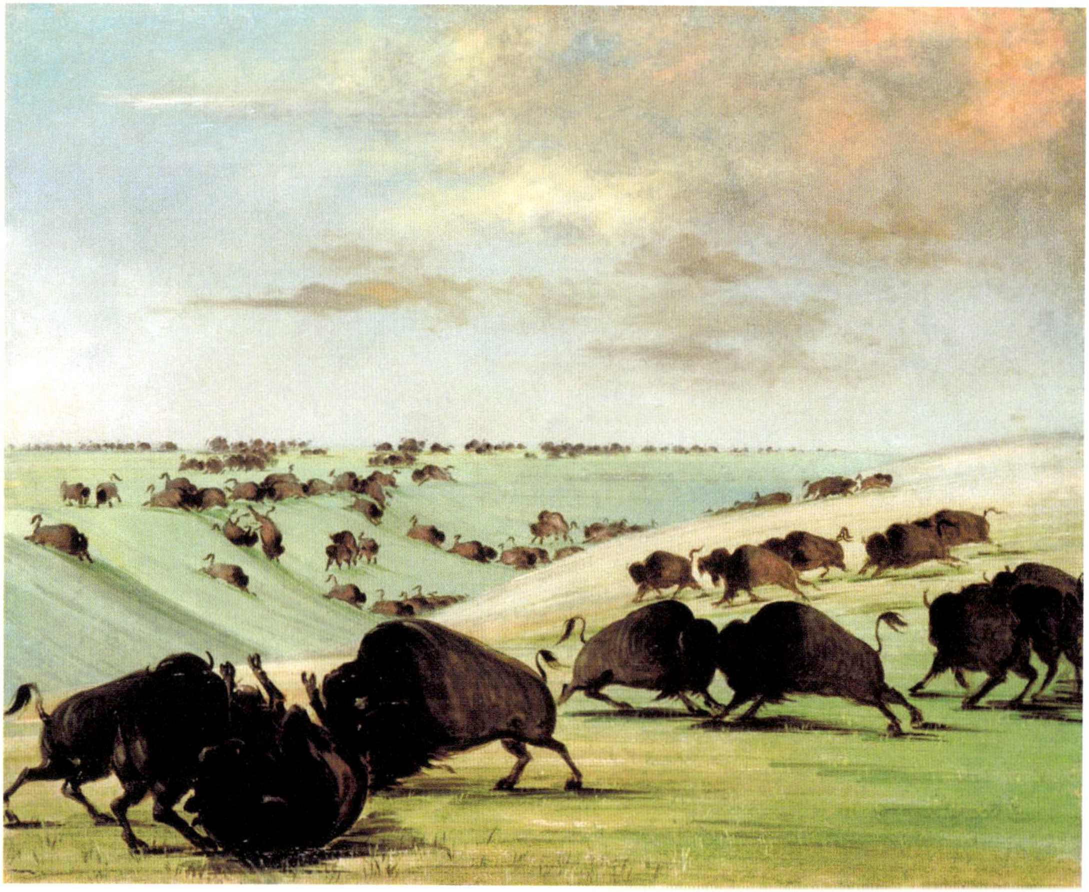

(VII)

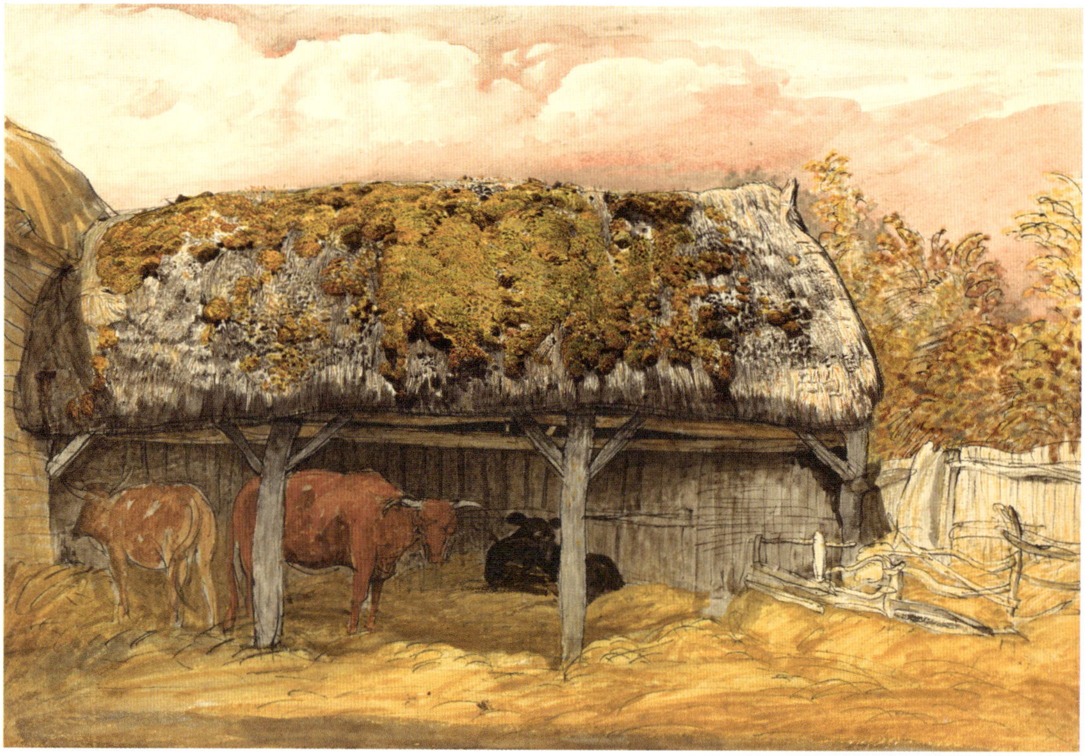

←
⊖ American bison (*Bison bison*), commonly known as buffalo, have lived in North America since prehistoric times, but the sight of large herds roaming the plains, with males battling each other during the mating season, as portrayed in George Catlin's *Buffalo Bulls Fighting in Running Season, Upper Missouri* (1837–39), is long gone.

↙
Samuel Palmer's *A Cow Lodge with a Mossy Roof* (*c*. 1829) presents a tranquil pastoral scene. The typical docility of cows is one of the products of 10,000 years of selective breeding. More than 1,000 breeds of cattle (*Bos taurus*) can be traced to a common bovine ancestor, the auroch (*Bos primigenius*), a powerful, horned species that weighed up to a tonne.

→
Dutch artist Kees Timmer's 1960 painting, *Hyena*, features the spotted hyena (*Crocuta crocuta*). These intelligent animals live in large clans. They have several different vocalizations, including the 'laugh' for which they are known, which is a sign of nervousness or submission.

↓
◎ This 1877 print likely depicts the Baird's tapir (*Tapirus bairdii*), a species that inhabits rainforests and woodlands in Central America. Tapirs are seed dispersers and help forests to regenerate. They are threatened by land clearance for soya production and factory farming.

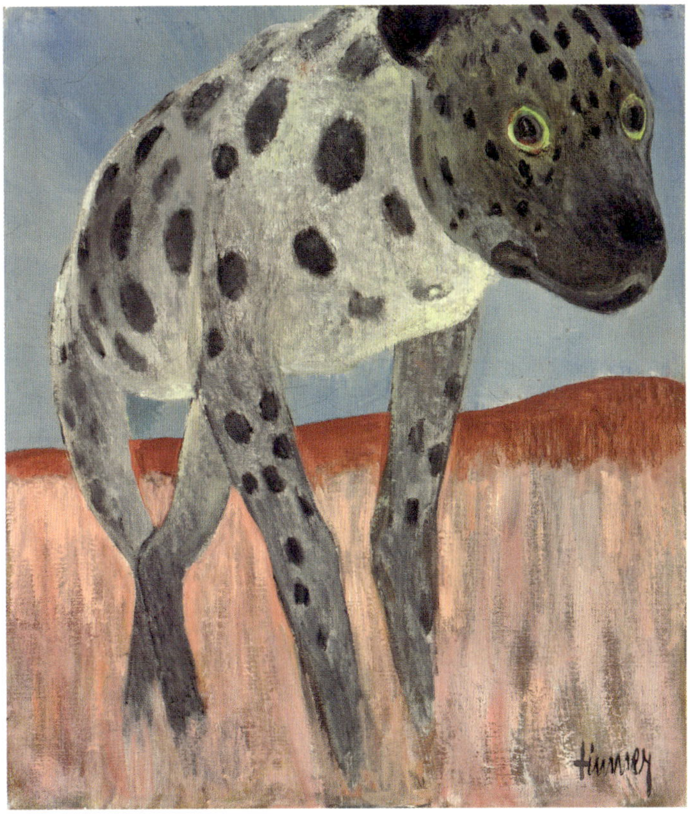

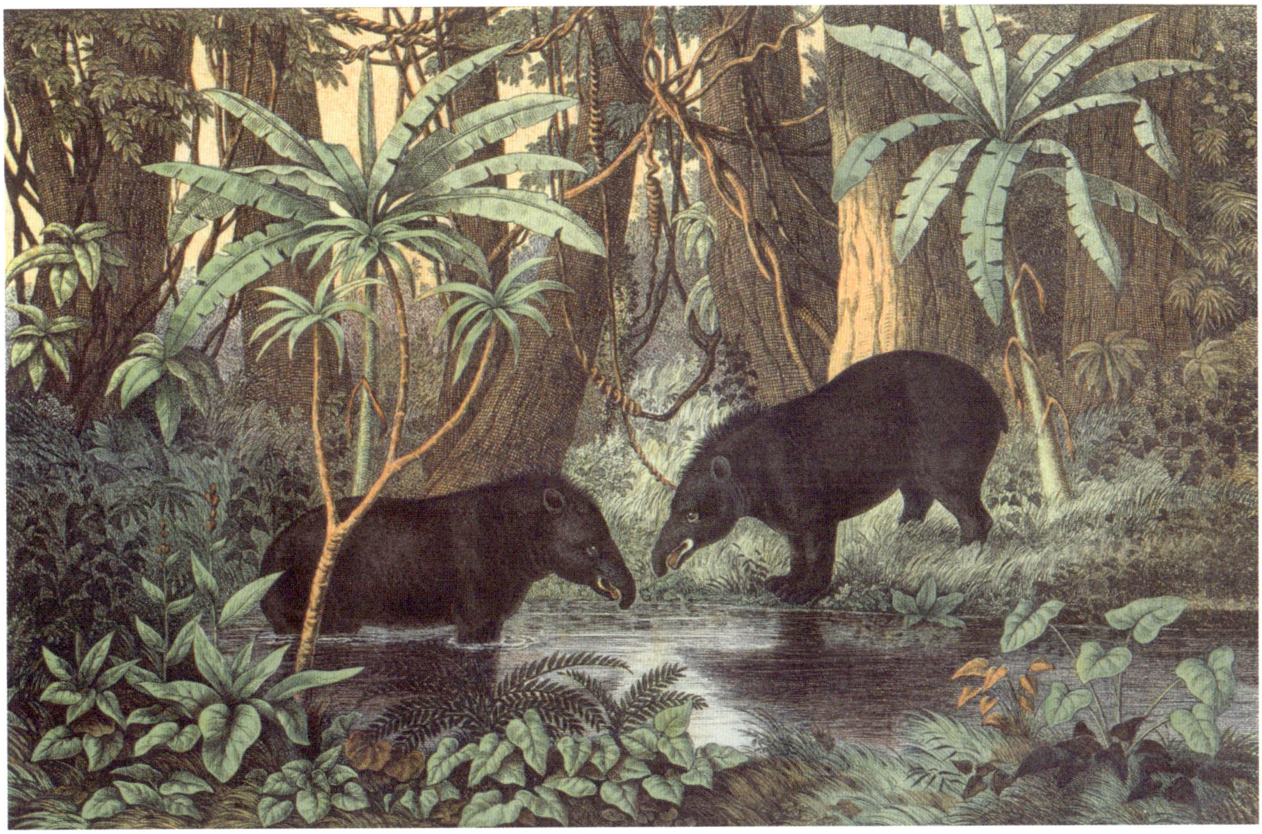

'High-arch'd his neck and tapering is his head;
His back is plump, his barrel curtly drawn;
His fiery breast luxuriates with brawn.'

VIRGIL, *THE GEORGICS*, BOOK III, 29 BCE, TRANSLATED BY R. D. BLACKMORE

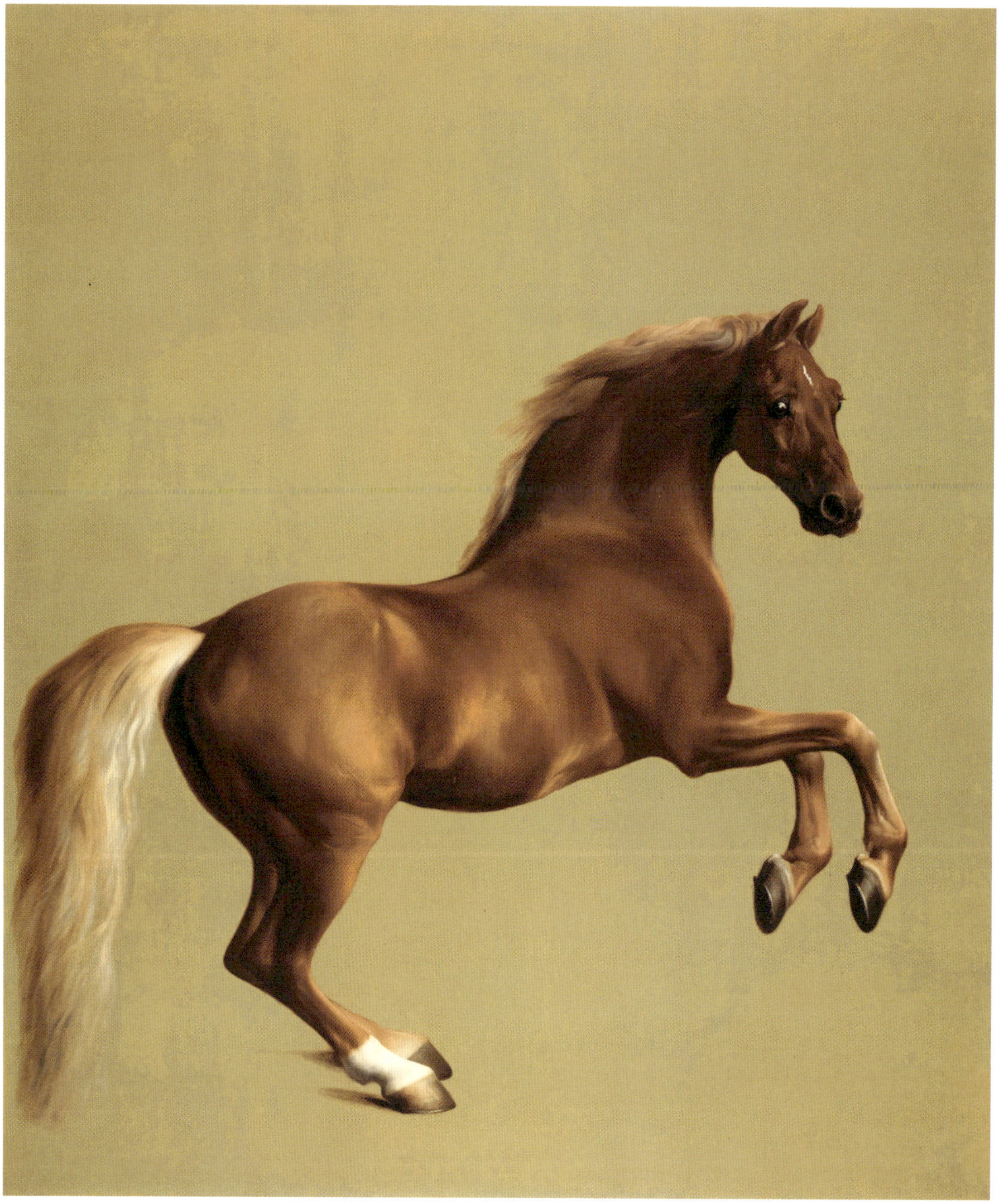

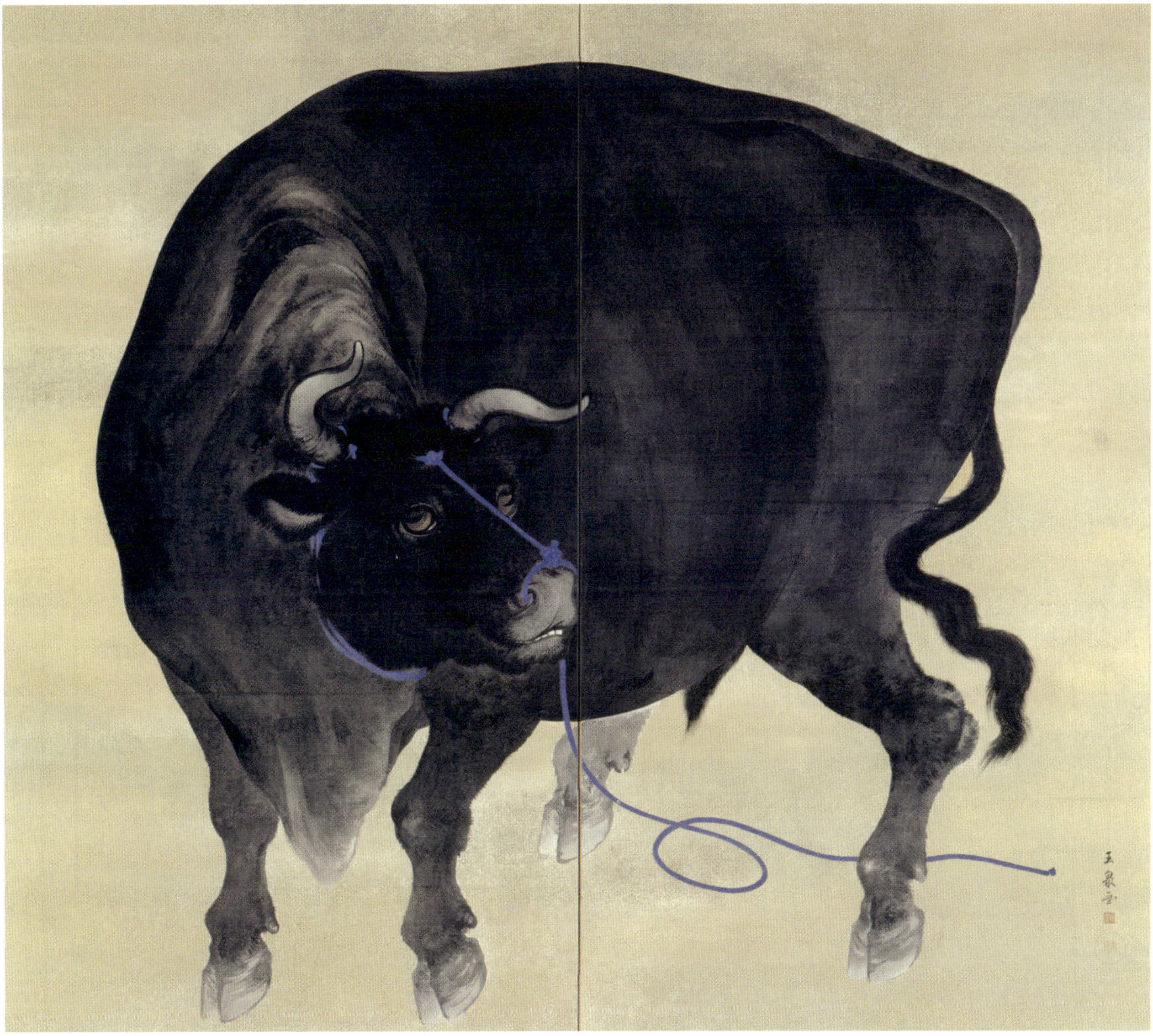

←
George Stubbs's life-size portrait of *Whistlejacket* (*c.* 1762) celebrates a famous racehorse of the time. Stubbs has captured the energy and power of the stallion, who was reputed to be hard to handle.

↑
Mochizuki Gyokusen's *Black Bull* (1850–1900) adorns a two-panel folding screen. The bull is one of the twelve animals in the Chinese zodiac, symbolizing strength, determination and honesty.

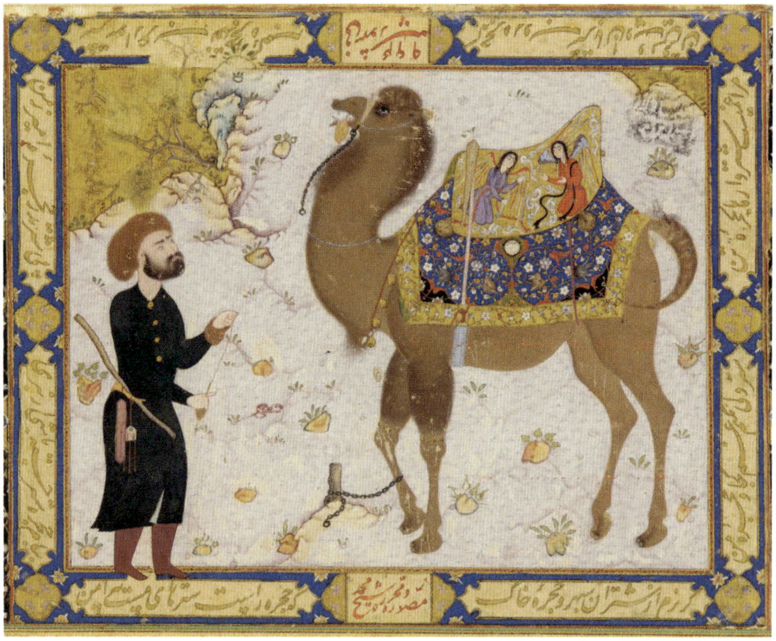

←
Persian poet and miniaturist Shaykh Muhammad's *Camel and Keeper* (1556–57) is framed by a poem about the taming of a haughty camel.

↓
In '*Über das Tier in der Kunst*' ('On Animals in Art'), German Expressionist Franz Marc wrote, 'I am trying to heighten my sense of the organic rhythm of all things, trying to empathize pantheistically with the shivering and coagulating of blood in nature, in trees, in animals, in the air.' This 'animalization' of art was best achieved, he said, through painting animals, as seen in *The Blue Horses (The Horses in the Pasture)* (1910).

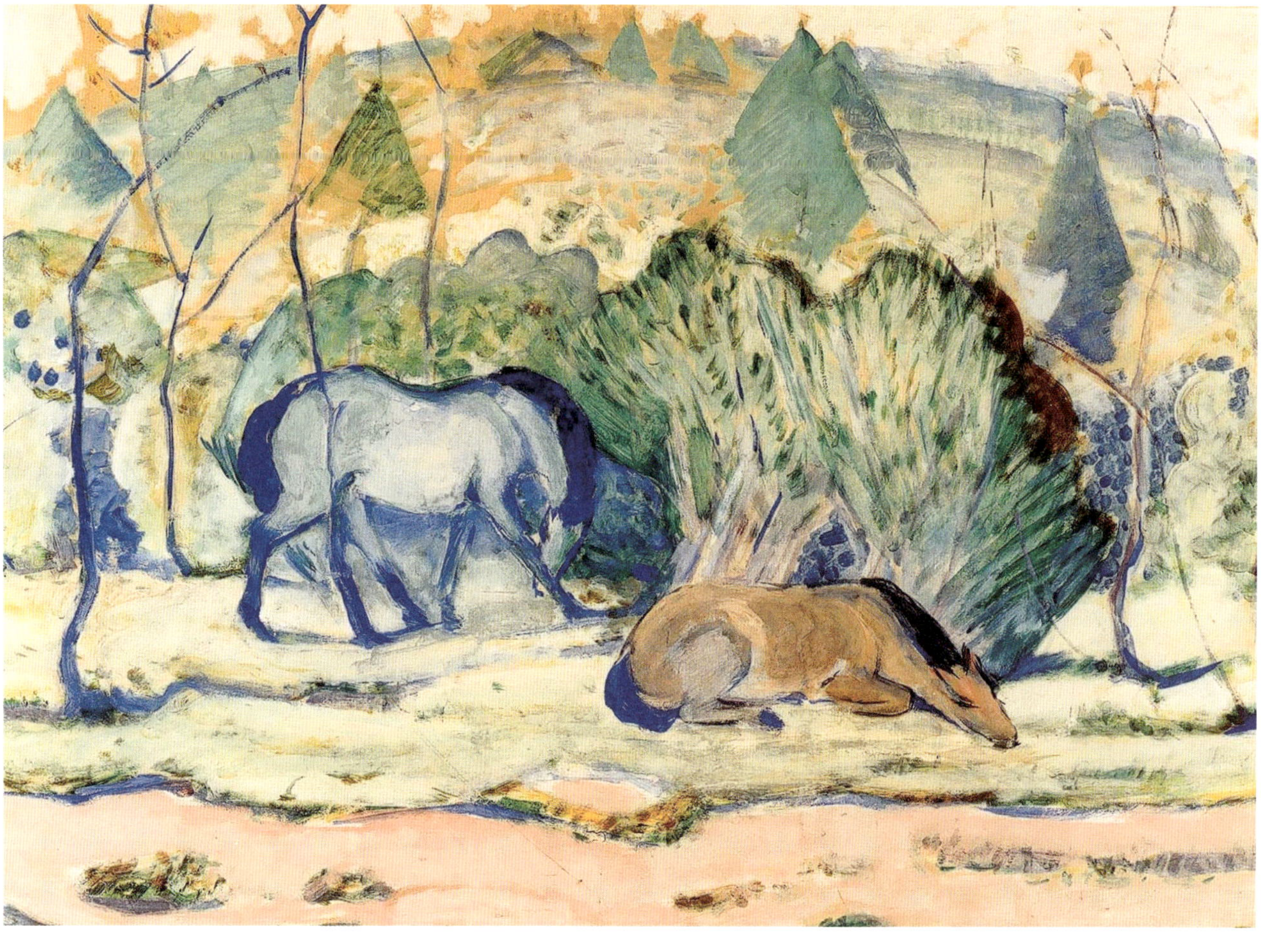

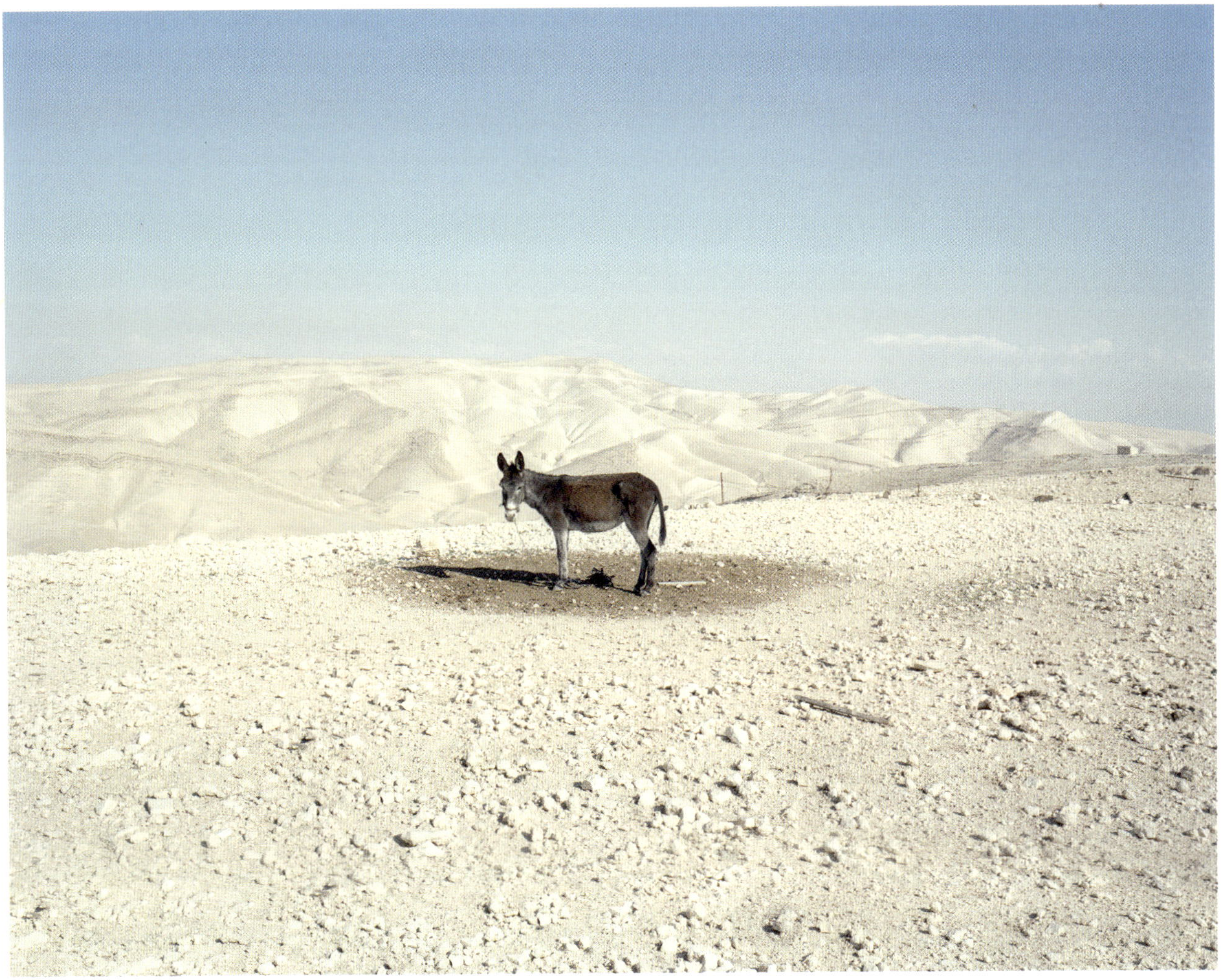

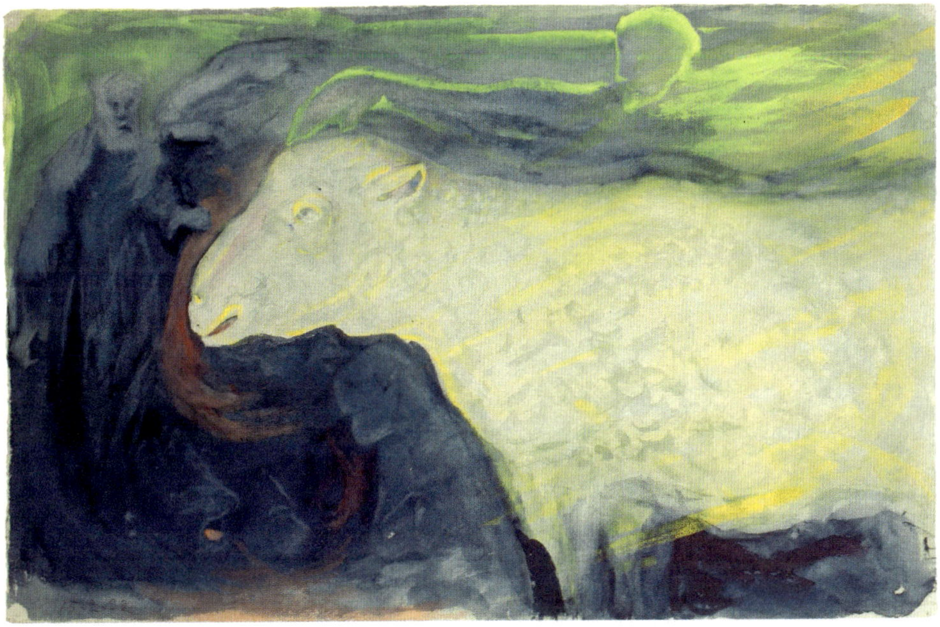

↑
Greek photographer Paris Petridis presents animals as beings that share the environment with us; they are not just part of the backdrop. *Bethlehem* (2012) features a donkey tethered in a desolate landscape.

→
Better known for her abstract paintings, Swedish artist and mystic Hilma af Klint also painted plants, animals and landscapes, such as this watercolour of a sheep, painted on 15 February 1932.

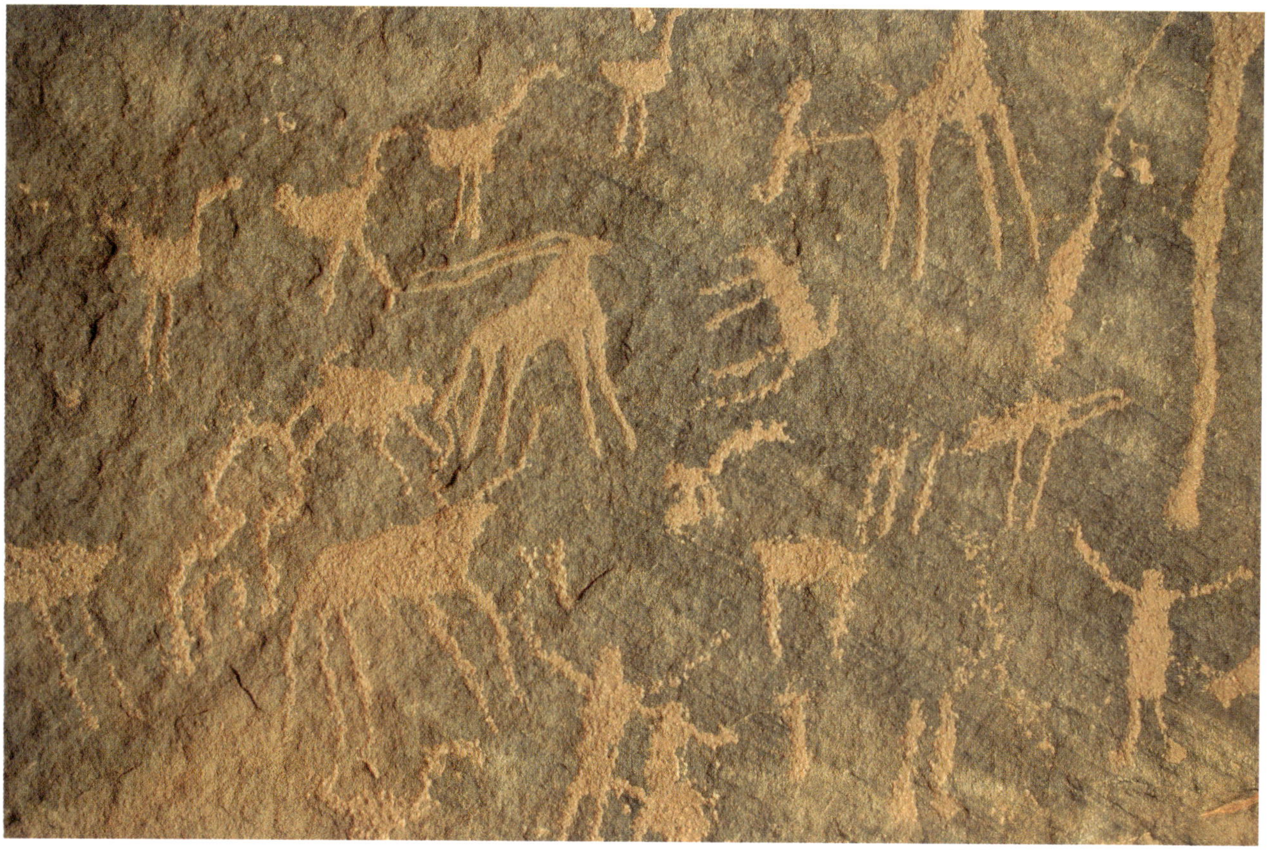

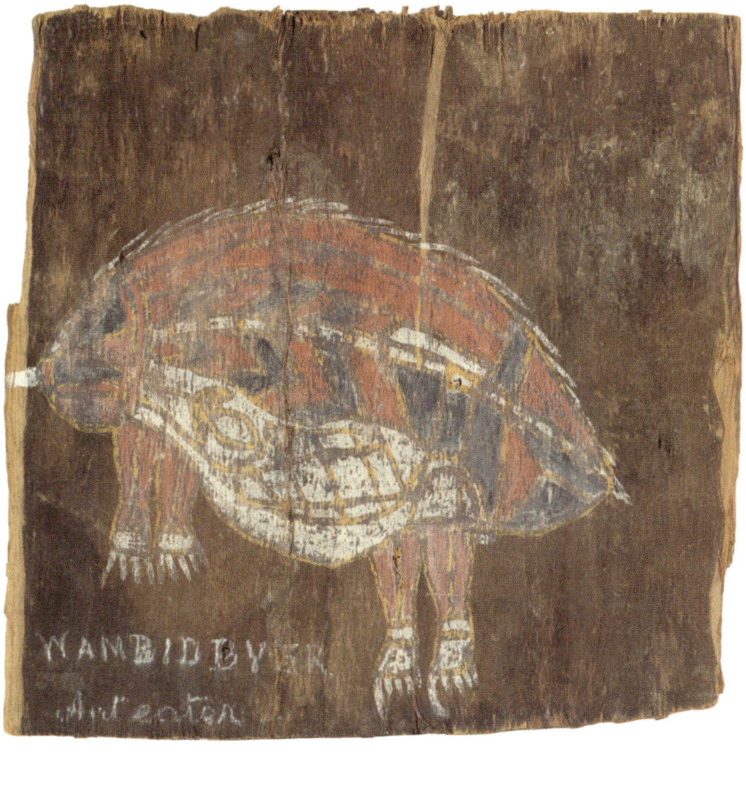

'A great number of the Saharan rock images seem to have been only motivated by a desire to accurately depict nature. But the absence of landscapes or vegetation ... and the choice of animal species ... make this whole ensemble a veritable mineral bestiary.'

JEAN-LOÏC LE QUELLEC, *ROCK ART IN AFRICA*, 2004

←
Ⓓ Ostriches, giraffes, dogs and antelopes resembling the oryx genus and the critically endangered addax (*Addax nasomaculatus*) are represented alongside people in this 8000–3000 BCE engraving on rock in Karkur Talh, Sudan.

↙
This *c.* 1912 representation of a *wambiddyer*, or anteater, by a Kunwinjku artist from the Northern Territory of Australia, is painted in natural earth pigments on eucalyptus bark.

→
The South American Nazca culture (100 BCE–800 CE) is known for its pottery, such as this small bowl decorated with mice (100–500 CE). Mice were feared by the Nazca as they destroyed crops.

↓
A sleeping antelope is incised into a rock surface in the Tassili n'Ajjer National Park, Algeria. This prehistoric animal petroglyph is one of many in the region, and reflects a time when the Sahara was grassland and large animals thrived there.

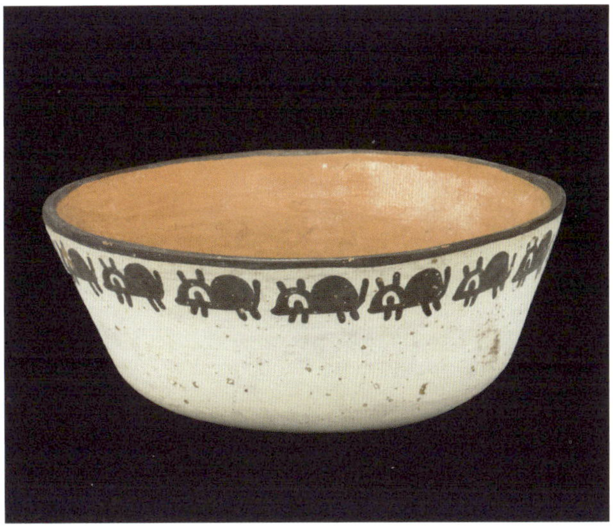

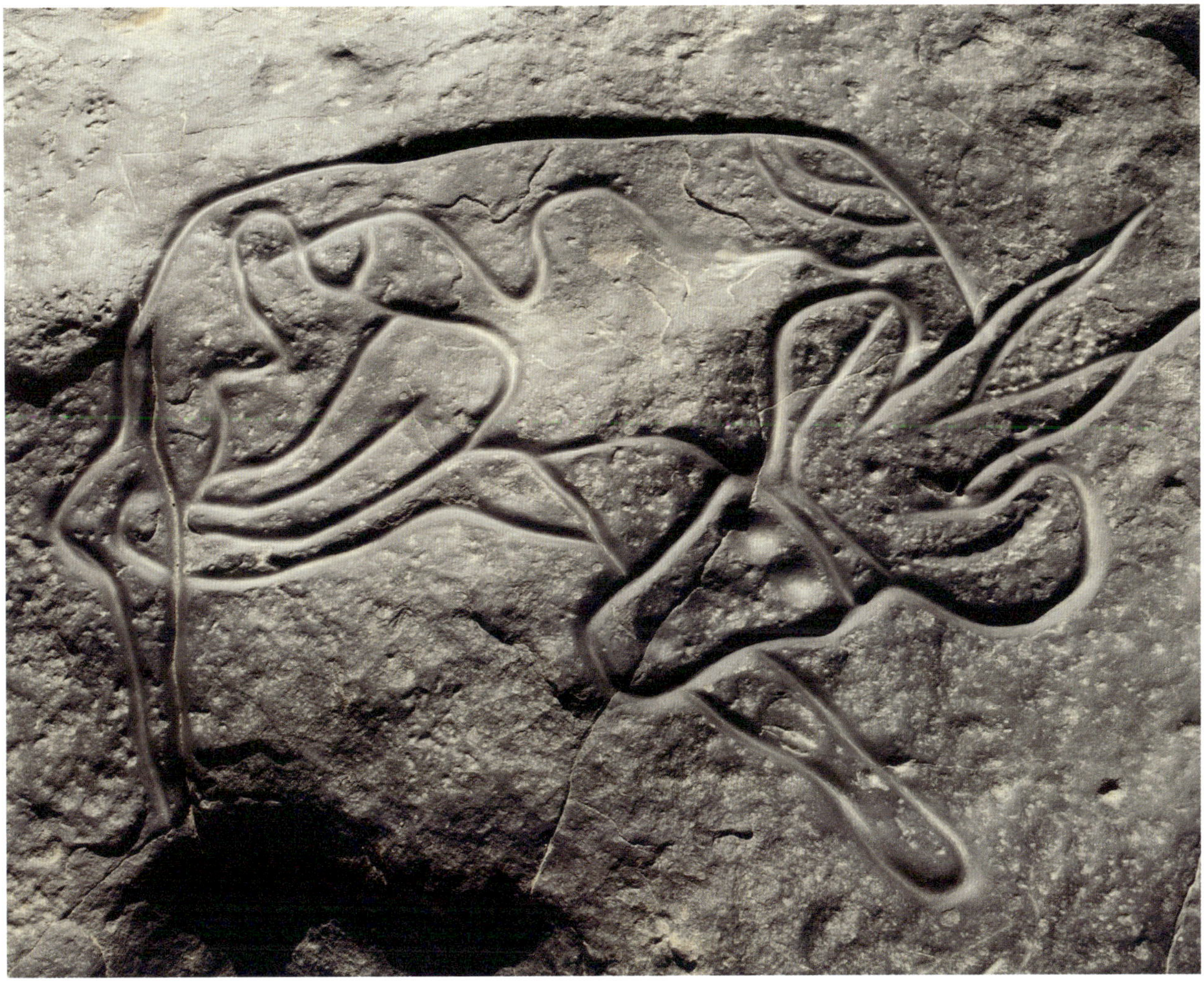

'When you see the fangs of lions exposed, do not suppose that the lion is smiling'

AL-MUTANABBĪ (915–965), 'THE POET REPROACHES SAYF AL-DAWLA FOR PREFERRING ANOTHER ABOVE HIM', TRANSLATED BY A. J. ARBERRY

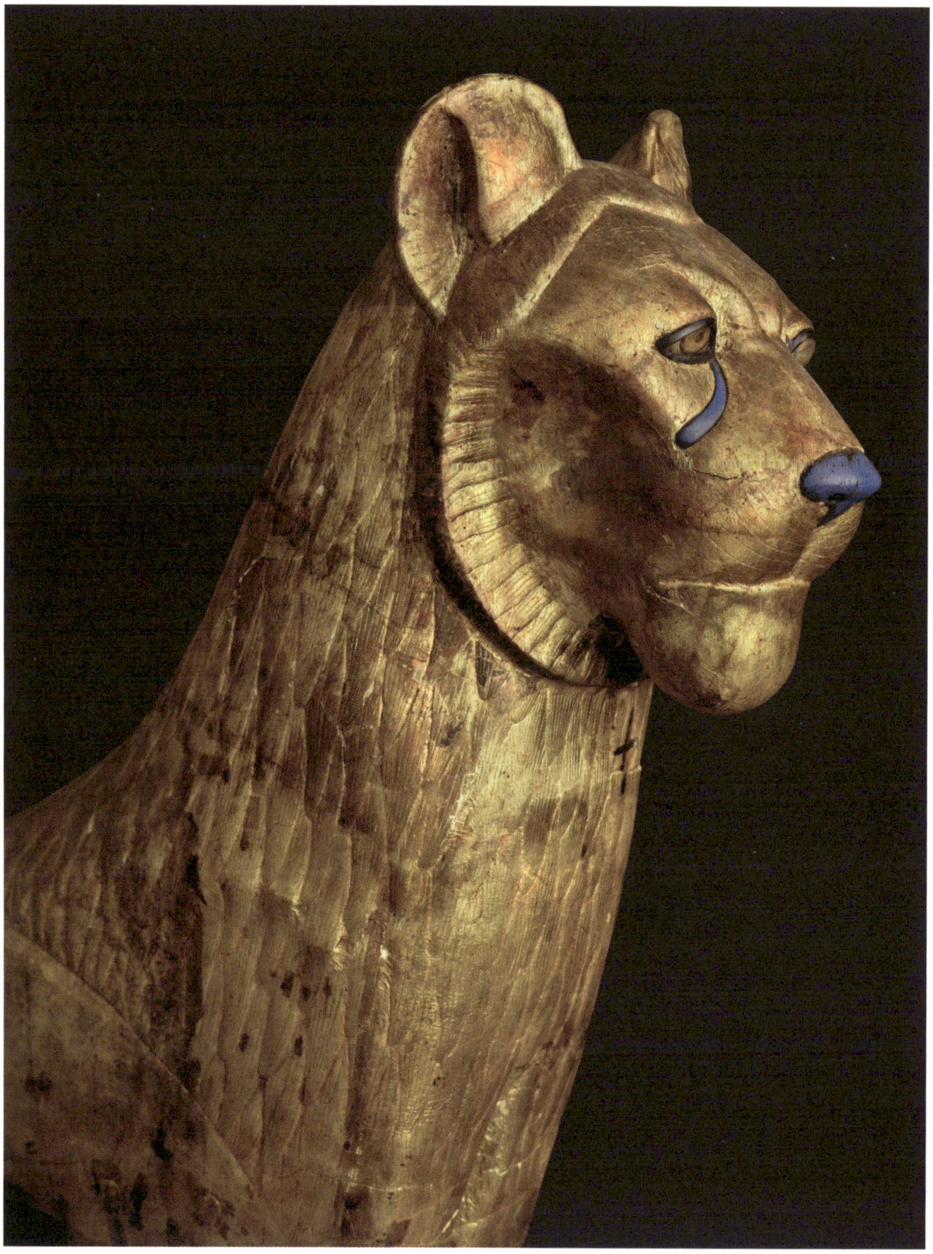

←
◎ This gilded, lion-headed bedpost is from a ritual bed that was discovered by Howard Carter in Tutankhamun's (*c.* 1341–*c.* 1323 BCE) tomb. In ancient Egypt, the lion (*Panthera leo*) symbolized strength and virility, and often decorated furniture legs, denoting the support and protection that this powerful creature offered to the pharaoh, although in this case the figures may represent lionesses.

→
◎ This 17th-century drawing likely depicts four Asiatic lions (*Pantera leo leo*). Lions were a national symbol of the Iranian Safavid Dynasty and symbolized royal courage and divine authority.

↓
◎ In 17th-century Amsterdam, lions transported from Africa were kept in private menageries, giving Rembrandt van Rijn the opportunity to draw them from observation, creating studies such as this (*c.* 1645–50).

↳
◎ This Masai giraffe (*Giraffa tippelskirchi*) was photographed in the Smithsonian's National Zoo, Washington, DC, by Candida Höfer in 1992. It is titled *Zoologischer Garten Washington DC IV 1992*. The view of the giraffe is criss-crossed with cable fencing and the mural it stands in front of – an imitation of its native East African habitat – is framed by painted architectural columns and arches. The space is illusory and inadequate to the viewer and the giraffe. The abundance of life and expanse of savannah promised by the painted backdrop is, likewise, false: the species is endangered, its habitat under threat.

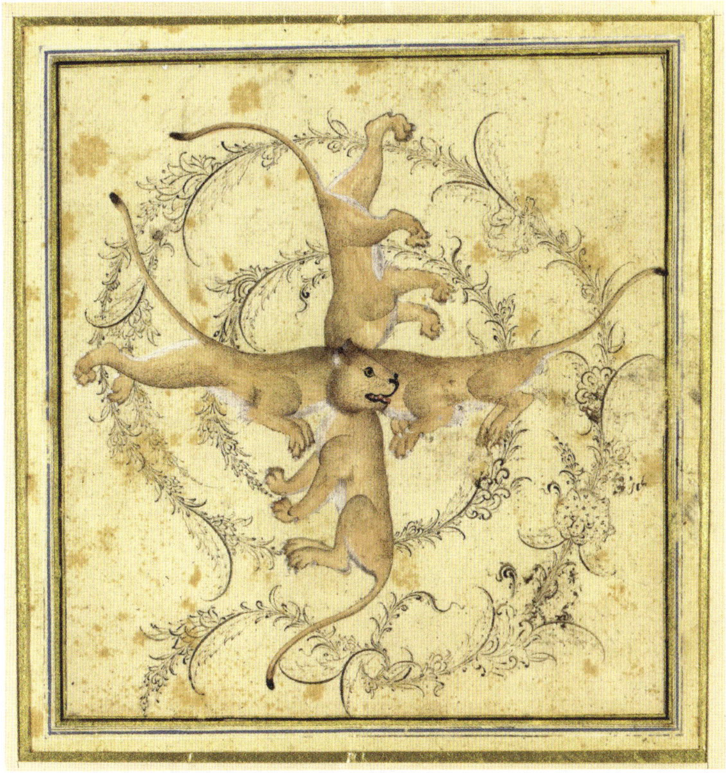

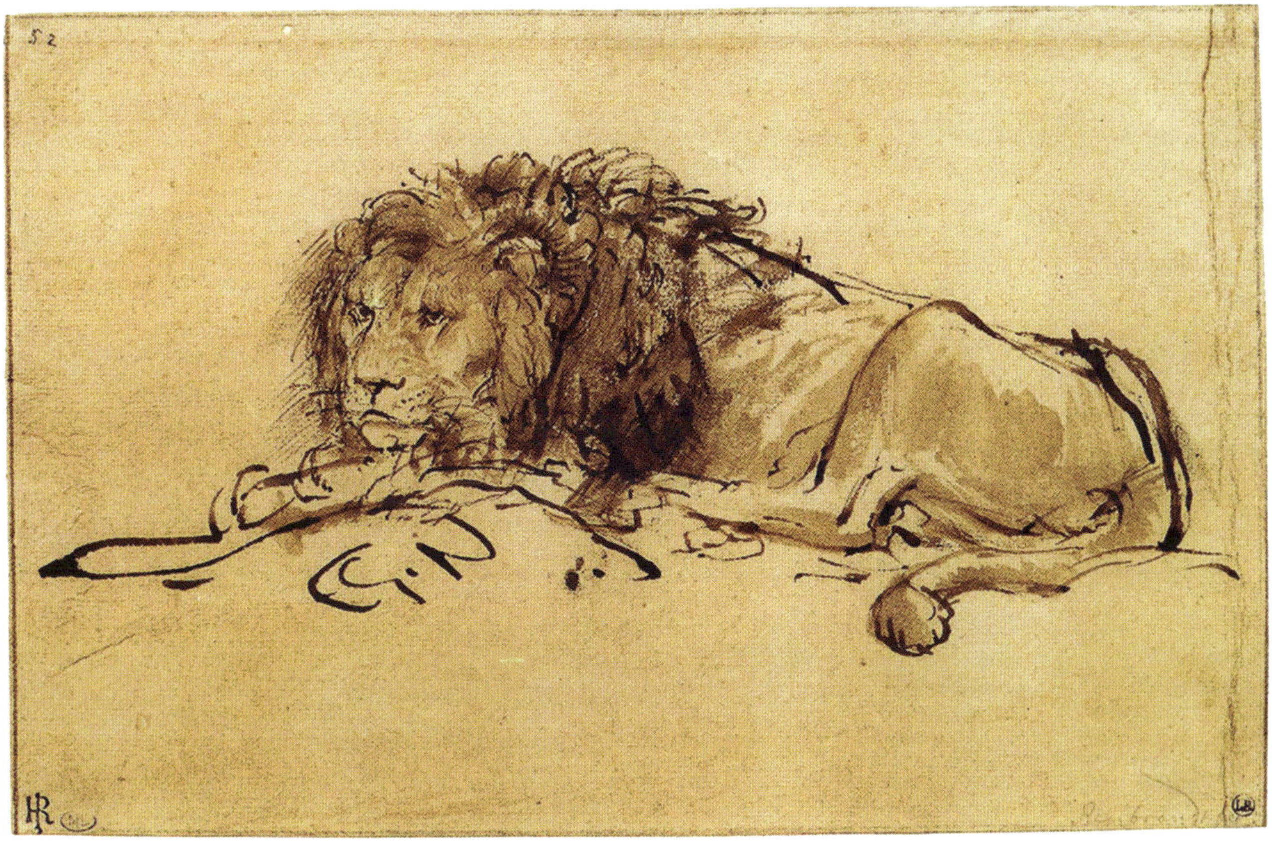

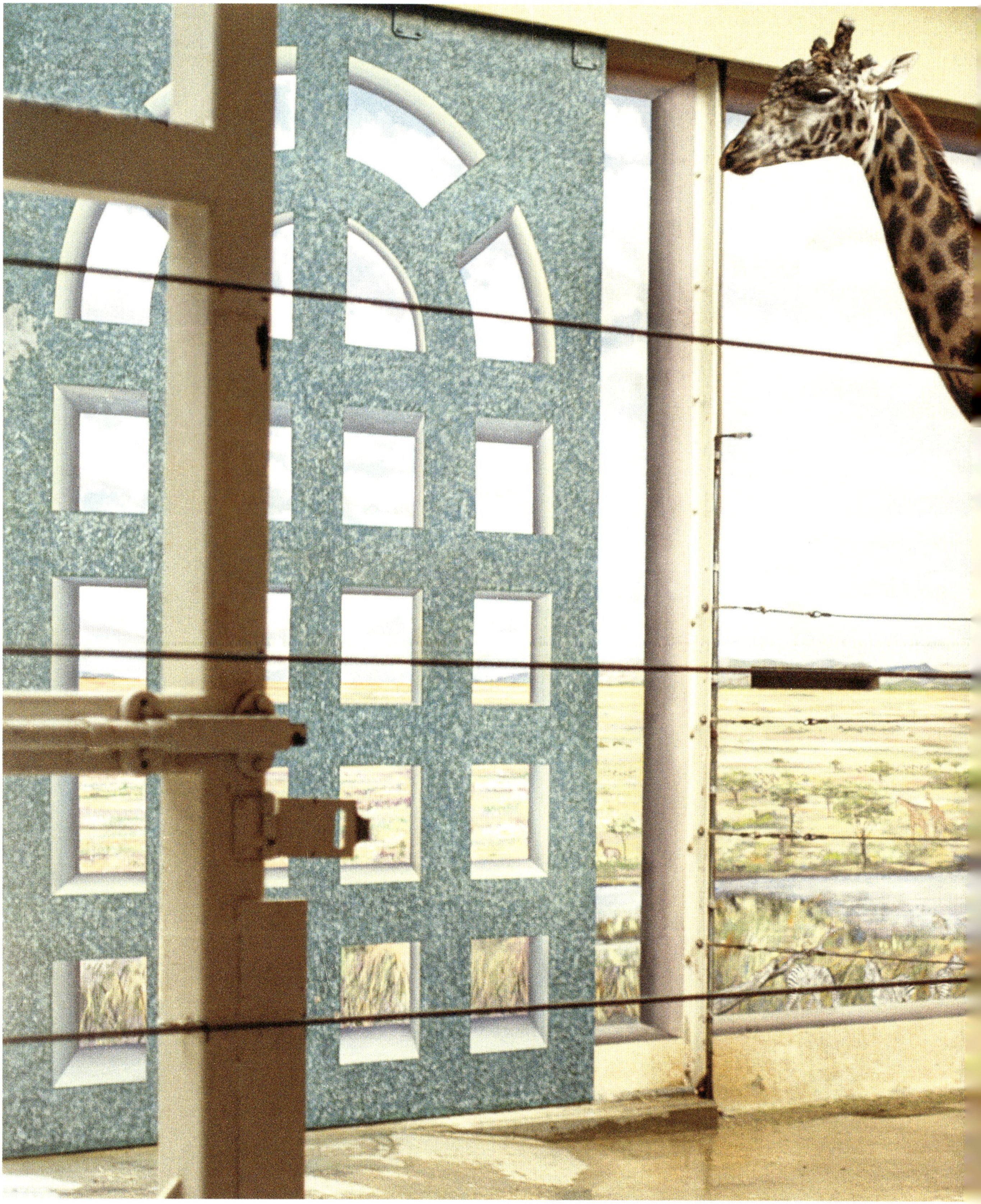

(VII)

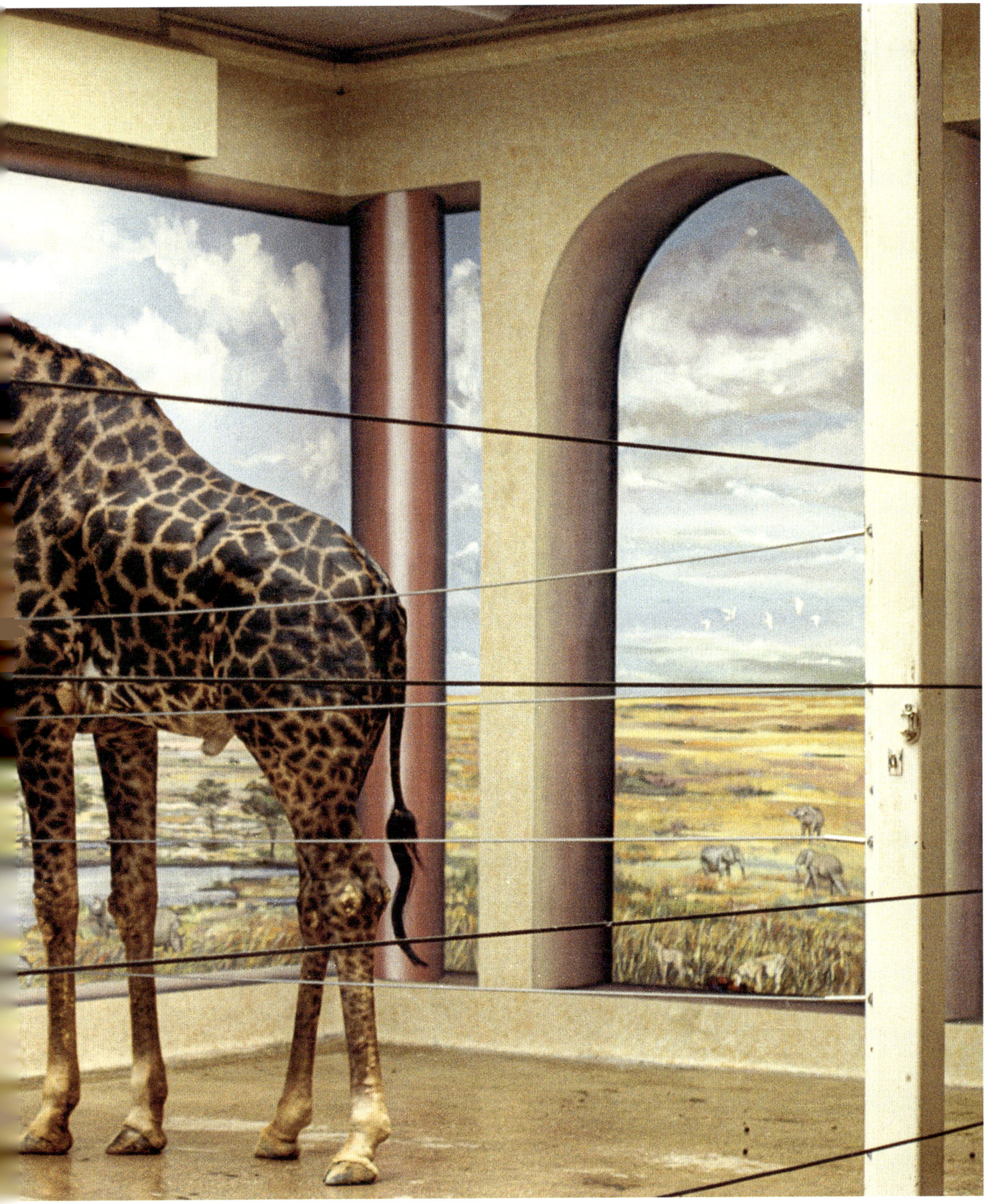

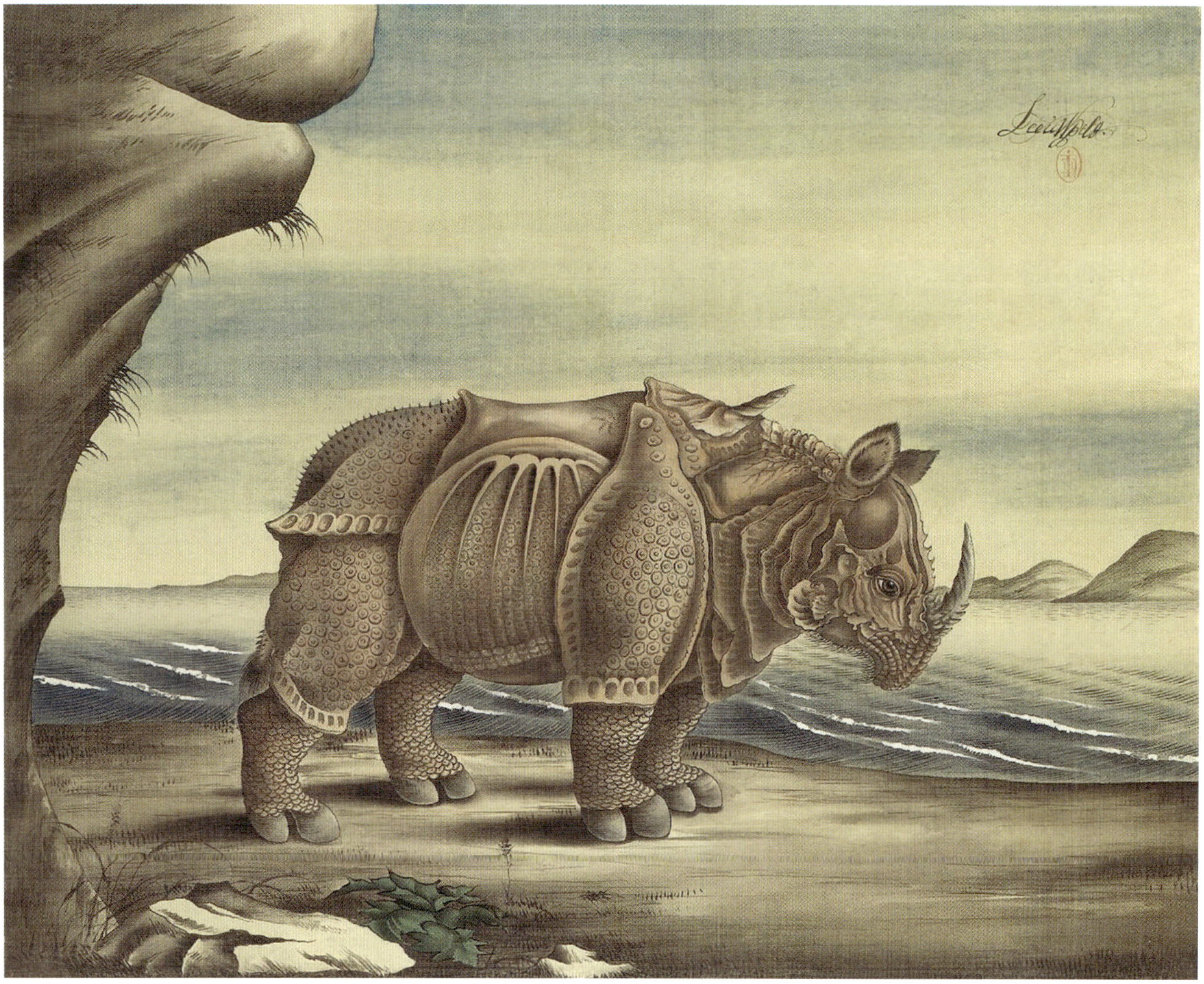

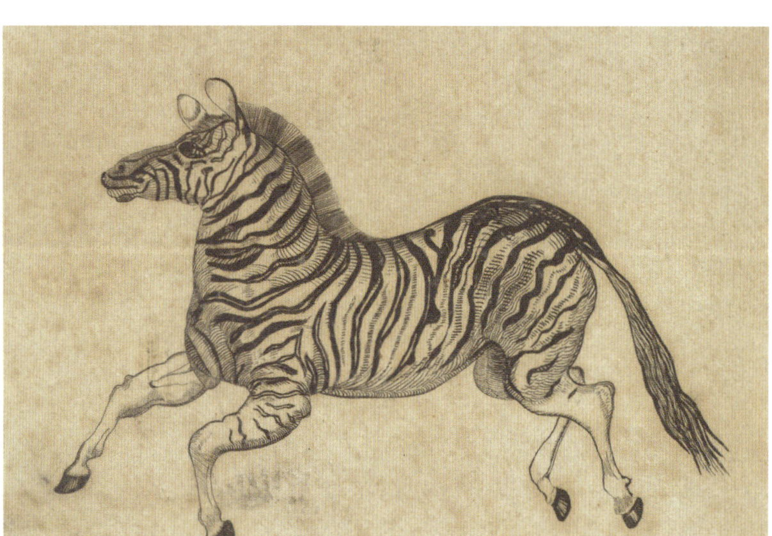

'Of all wild animals the gentlest is the elephant. ... It is very sensitive and possessed of an intelligence superior to that of other animals.'

ARISTOTLE, *HISTORIA ANIMALIUM*, BOOK IX, MID 4TH CENTURY BCE, TRANSLATED BY D'ARCY WENTWORTH THOMPSON

←
◎ An Indian rhinoceros (*Rhinoceros unicornis*) adorns this 18th-century Japanese scroll by Ishikawa Mōkō. The Indian rhinoceros was once widespread across northern India and Nepal, but came close to extinction at the beginning of the 20th century. It is now protected, and the population has increased to over 3,500.

↙
This proof of a banknote vignette (*c.* 1824–37), shows a zebra. On the African plains, zebras are 'pioneer grazers', meaning that they feed on long grass, preparing the way for species that need to feed on short, nutritious grasses.

→
A hippopotamus is represented on this ancient Egyptian ostrakon dated *c.* 1479–1425 BCE. Ostraca were pieces of stone or pottery that were used as writing or drawing material, sometimes as preparatory studies for tomb paintings.

↓
◎ A captive Asian elephant (*Elephas maximus*) is shown in this Rajasthan-school illustration (*c.* 1800). Elephants have long been used in battle and in temple ceremonies, and some run amok due to aggression or suddenly taking fright at something in their surroundings.

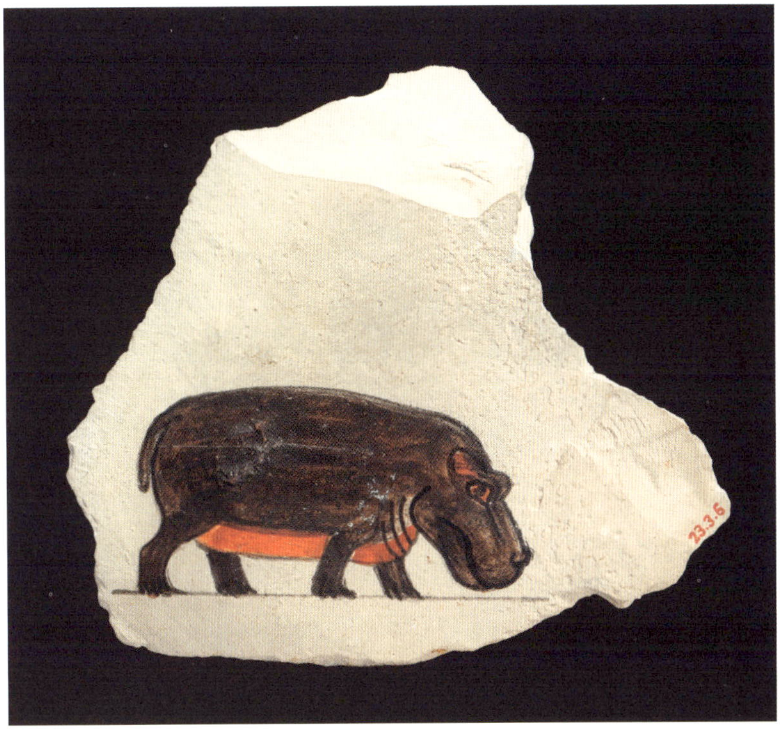

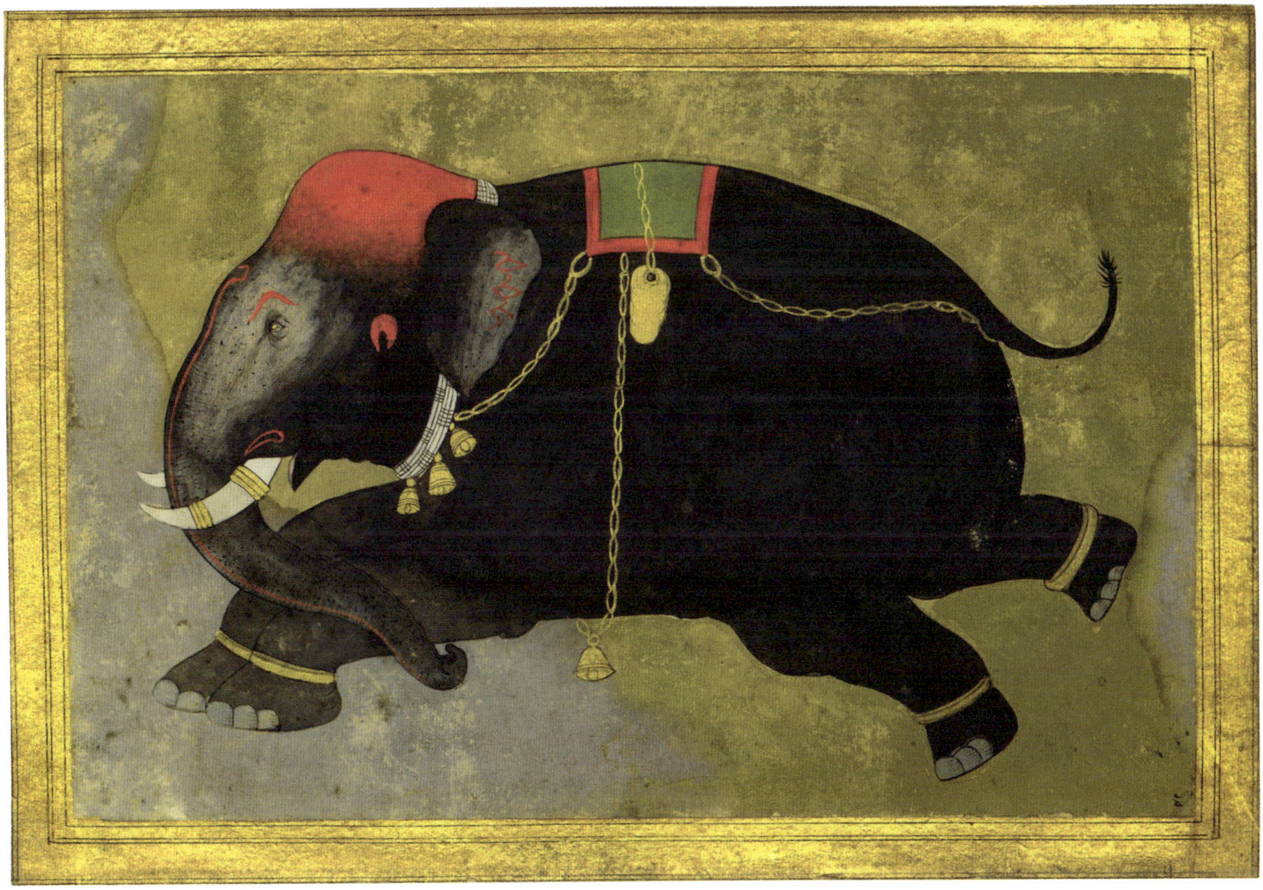

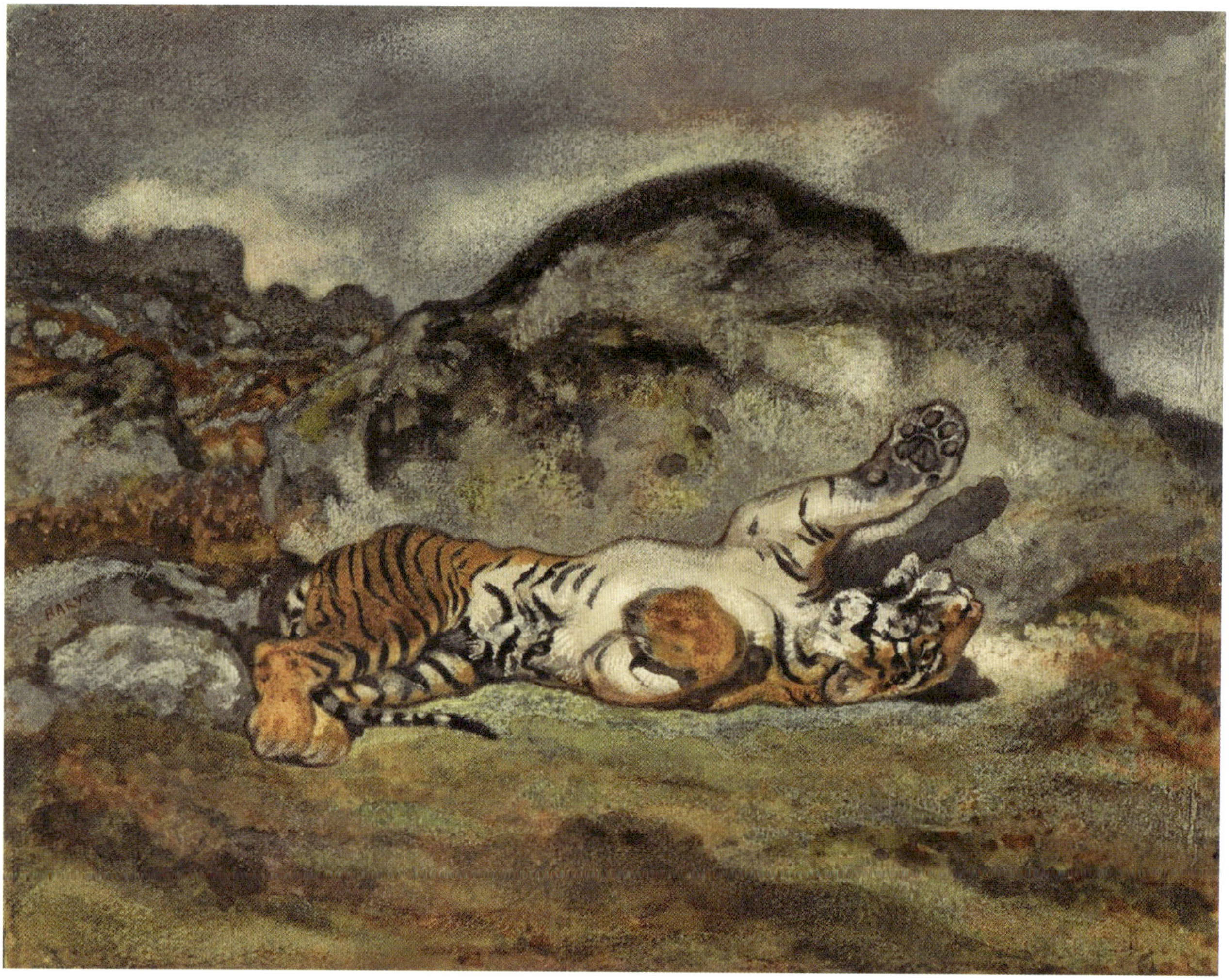

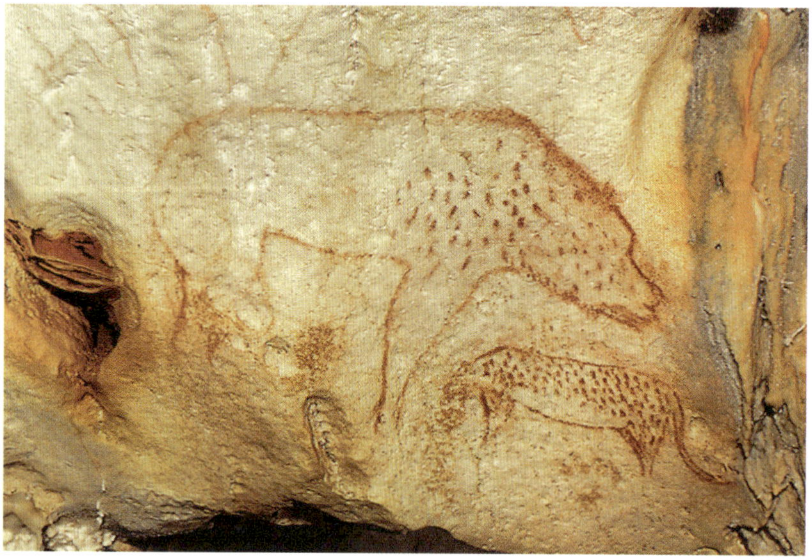

↑
◎ *Tigre se roulant* (*Rolling Tiger*, *c.* 1840) was likely based on a tiger (*Panthera tigris*) in the collection of the Jardin des Plantes in Paris, France, which artist Antoine-Louis Barye frequently visited with his friend Eugène Delacroix.

←
This painting of a bear facing a panther is from the Chauvet cave, France. Created with natural iron oxide pigment 43,000 to 28,000 years ago, the panther is likely a now-extinct European subspecies of leopard (*Panthera pardus spelaea*) that, unlike its modern leopard relatives, has no spots on its belly.

→
◎ Created in Kangra, India, around 1800, this painting depicting a tiger (*Panthera tigris*) represents the zodiac sign of Leo, or Simha.

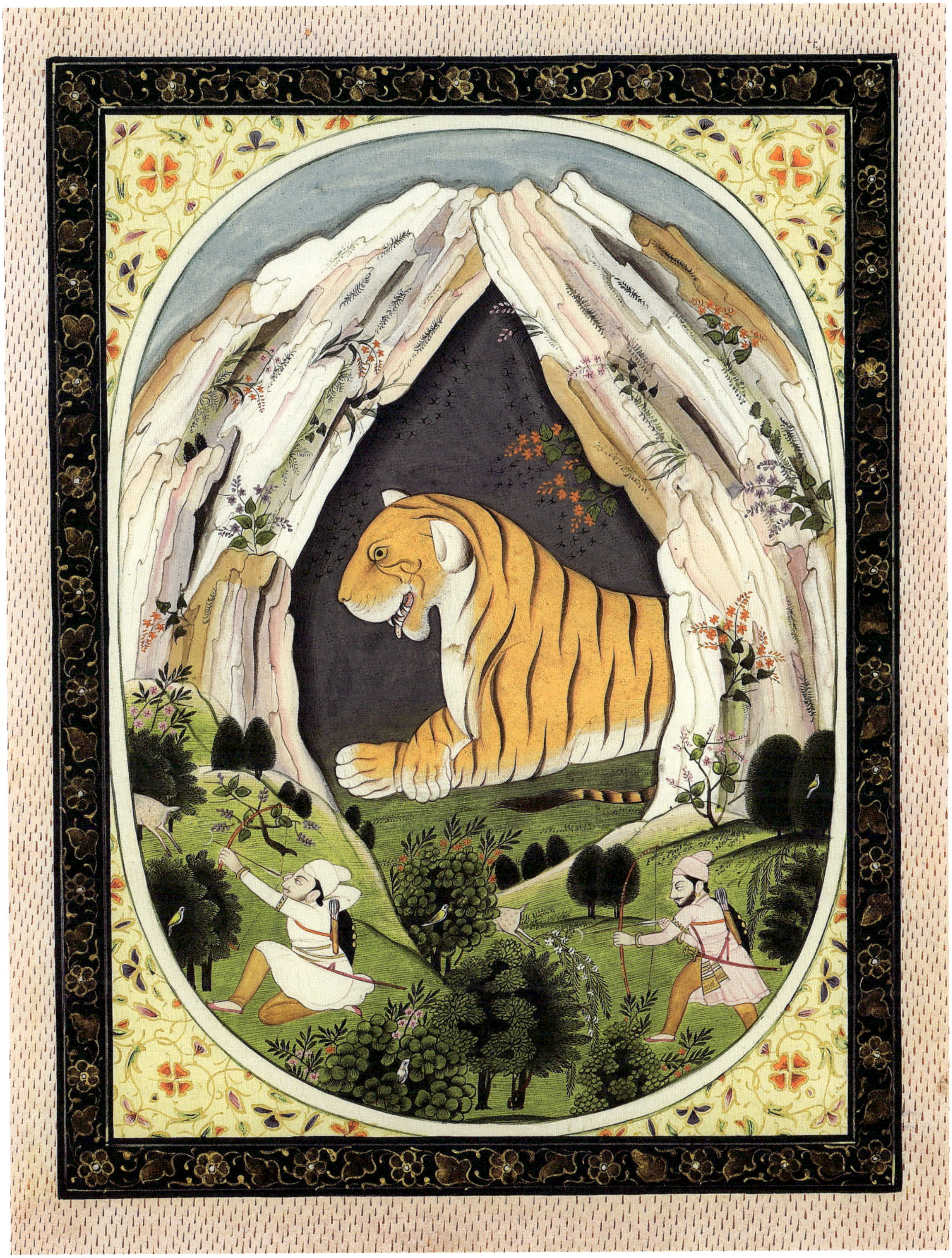

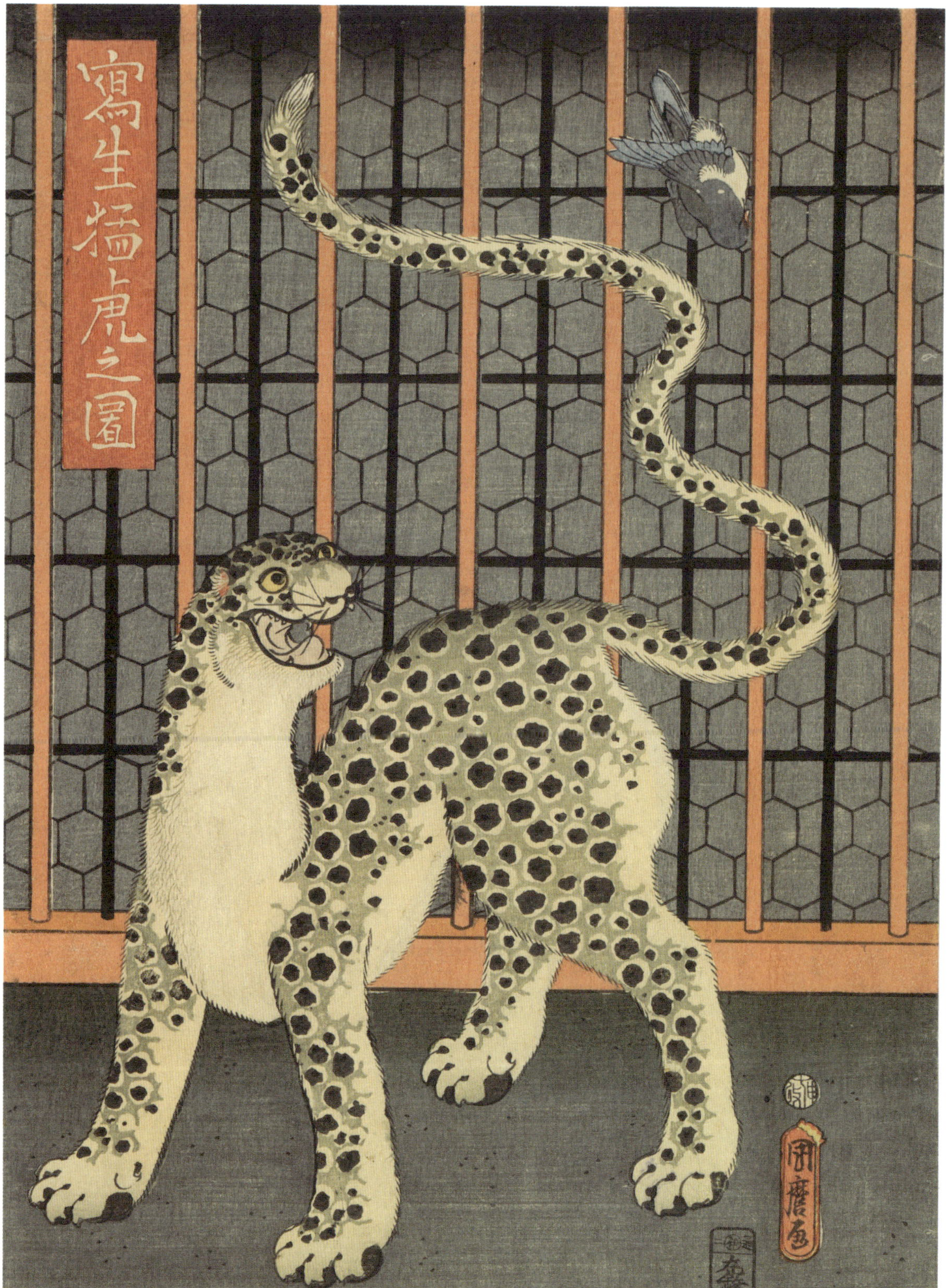

(VII)

←
◎ The elusive snow leopard (*Panthera uncia*), likely the species illustrated in this 19th-century print by Utagawa Kunimaro, inhabits the mountain ranges of Central Asia. Although much of its territory is difficult to access, its habitat is reducing in size due to hunting, human encroachment and the effects of climate change.

→
British artist Tracey Emin's drawing of her two cats, *Teacup and Pancake* (2025), produced for a charity Christmas card, celebrates the bond between humans and their pets.

↓
Duncan Grant's *The Cat, Opussyquinusque* (*c.* 1932) portrays the resident cat at Charleston, the Bloomsbury Group's rural retreat in East Sussex, England.

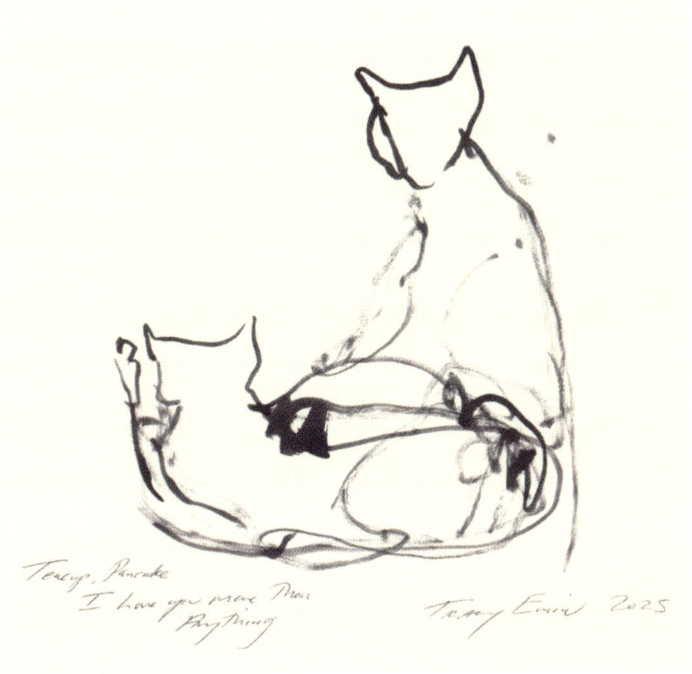

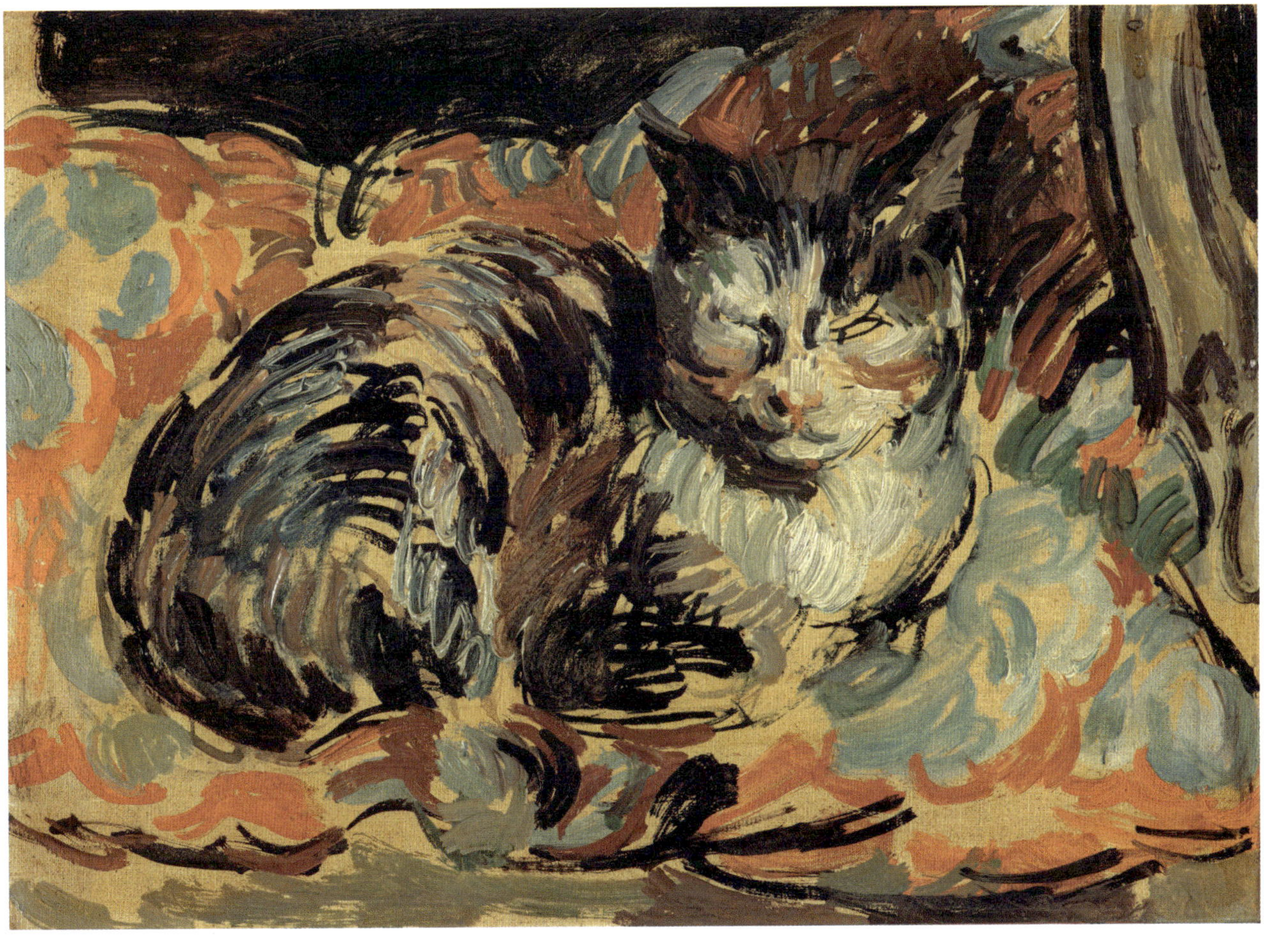

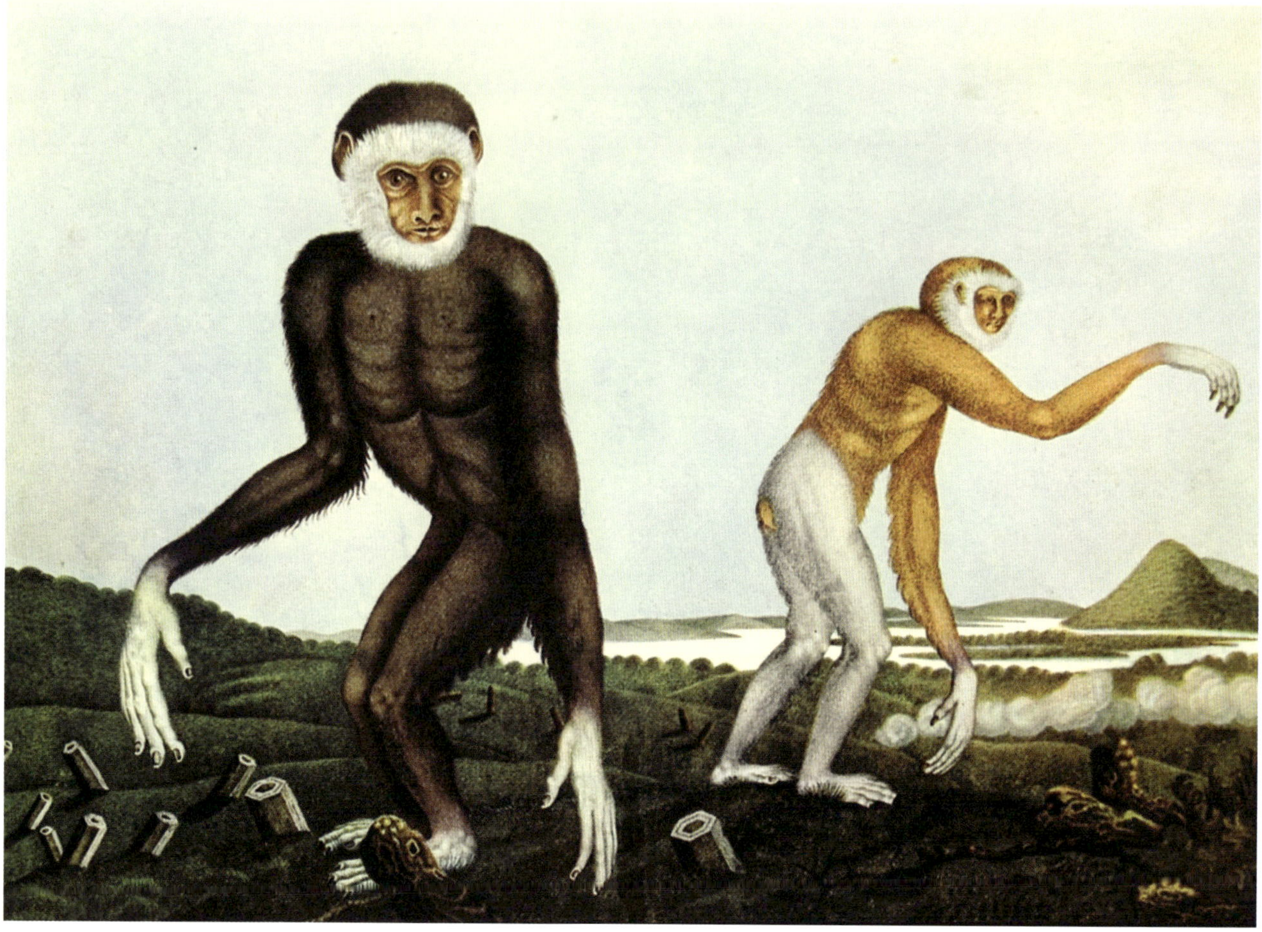

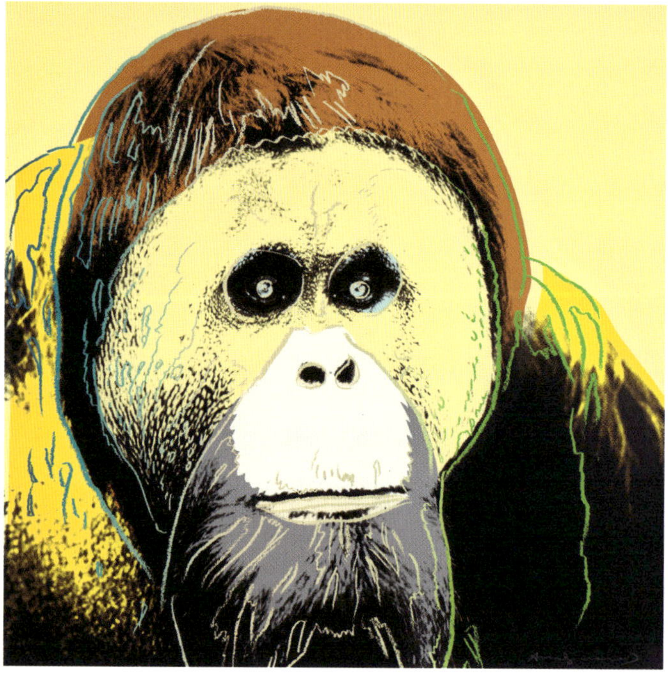

↑
◎ These gibbons, illustrated by Aloys Zötl in 1833, are probably hoolock gibbons, of which there are three species. Bangladesh's Western hoolock has seen a drastic fall in numbers due to deforestation and hunting for food and the pet trade.

←
◍ Shocked to discover how many well-known species were classified as endangered, Andy Warhol produced an *Endangered Species* series of screenprints, which included *Orangutan* (1983). There are three species of orangutan, all of which are critically endangered.

↗
The monkey in Edward Lear's illustration from *Gleanings from the Menagerie and Aviary at Knowsley Hall* (1846) is an Azara's night monkey (*Aotus azarae*), a South American forest-dweller. Lear worked directly from his observations of animals in Edward Stanley's menagerie.

→
Henri Rousseau painted imaginary jungle scenes, such as *Tropical Forest with Monkeys* (1910), inspired by his visits to botanical gardens and zoos, and observations from postcards and books.

'There is, nevertheless, a certain respect, a general duty of humanity, not only to beasts that have life and sense, but even to trees and plants. We owe justice to men and graciousness and benignity to other creatures that are capable of it; there is a certain ... mutual obligation betwixt them and us.'

MICHEL DE MONTAIGNE, 'OF CRUELTY', 1580, TRANSLATED BY CHARLES COTTON

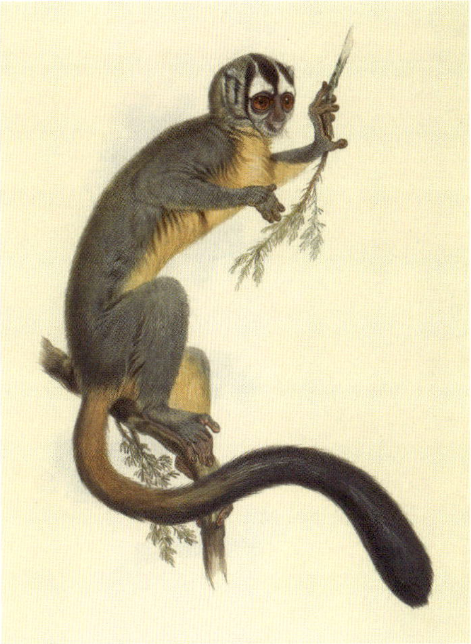

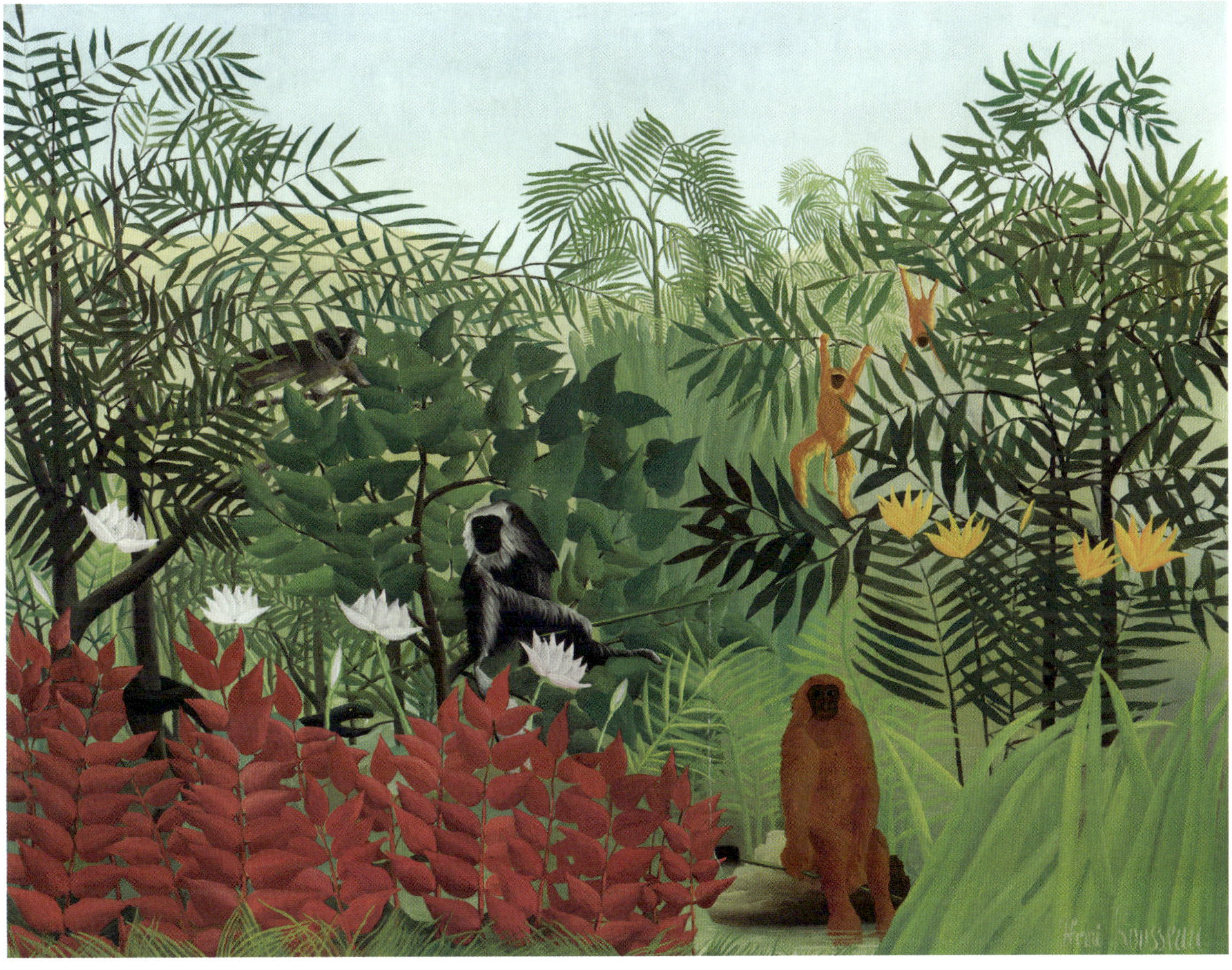

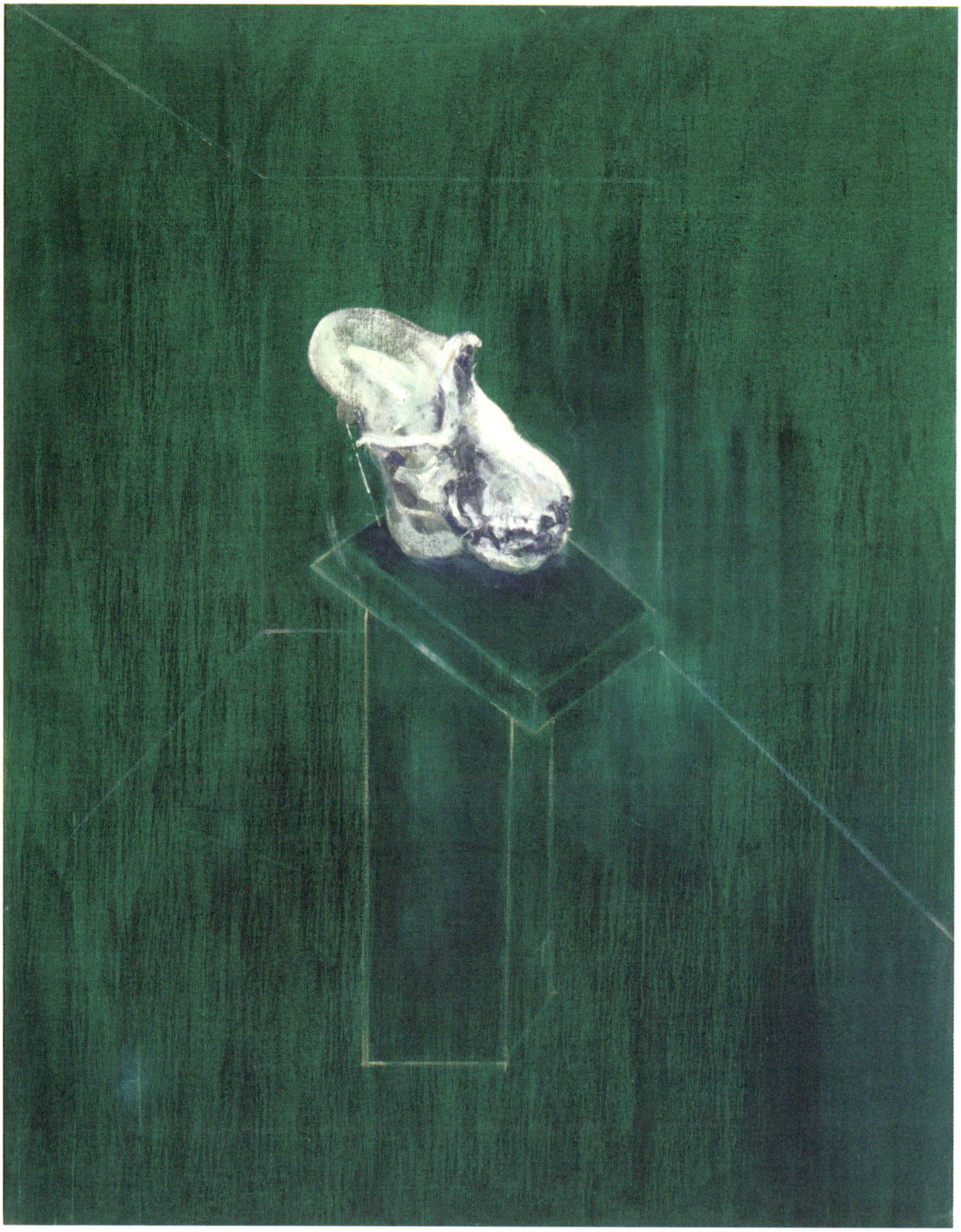

↑
Ⓓ Francis Bacon's *Skull of a Gorilla*, painted in 1957, can be seen as a reflection on mortality. Both the Western gorilla (*Gorilla gorilla*) and Eastern gorilla (*Gorilla beringei*) are critically endangered, and their slow reproduction rate makes their potential recovery from population decline particularly difficult.

→
Ⓓ In *Baby Gorilla* (1900), Scottish artist William Walls has captured a sense of watchfulness in the subject. Primate infants learn through observation, and researchers have found that adult gorillas moderate their style of nonverbal communication to be more repetitive and tactile when playing with infants.

'How I hate the man who talks about the "brute creation", with an ugly emphasis on brute. … I am proud of my close kinship with other animals. I take a jealous pride in my Simian ancestry.'

W. N. P. BARBELLION (BRUCE F. CUMMINGS), 'OUR SIMIAN ANCESTRY' (1910), *THE JOURNAL OF A DISAPPOINTED MAN*, 1919

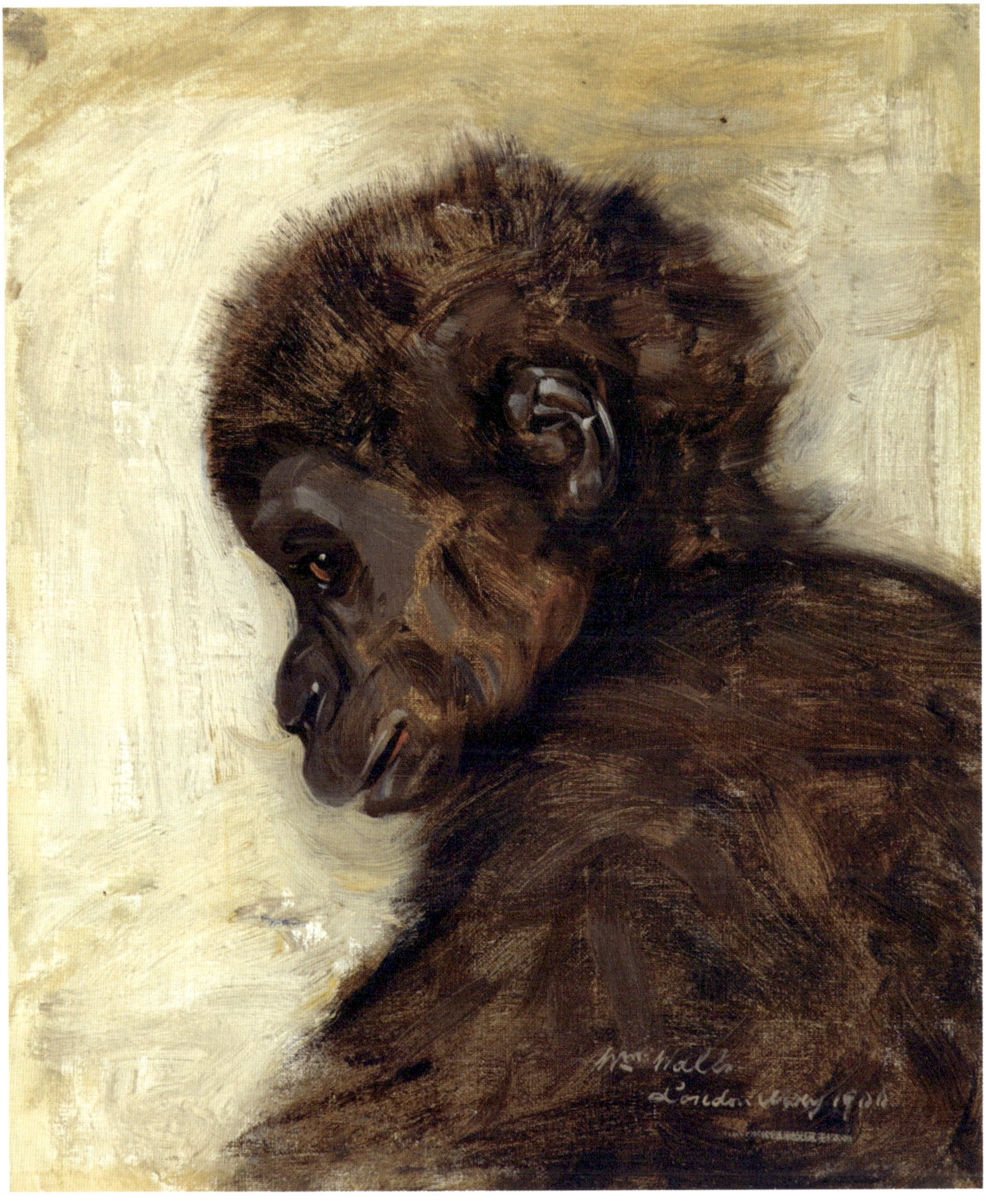

CODA: ENDLESS FORMS MOST BEAUTIFUL AND MOST WONDERFUL[1]

Stephen Ellcock

The construction of the Ark was a gigantic public spectacle, carried out over the course of many decades in full view of a society that regarded Noah and his family as dangerous fanatics, unhinged outcasts, even terrorists.

The entire monumental undertaking, from the felling of the timber to the final gathering of every animal, bird and 'creeping thing' upon the earth,[2] unfolded before an astonished and increasingly hostile populace, who viewed the insanely ambitious enterprise as little more than a colossal folly, a blasphemous act of inconceivable hubris; the eccentric pipedream of a lunatic.

The Ark was intended as a warning of impending disaster, a rebuke to a world sinking under the weight of its own depravity, pride and wickedness. It was both a portent of doom and a symbol that offered the promise of redemption, a last-minute reprieve from a watery grave. Noah's contemporaries faced a stark and unmistakable choice: repent or face certain destruction. Complacency won the day and the ensuing deluge swept the old world away.

The choices we now face, as inheritors of the world vouchsafed to Noah, are equally stark. We have been indoctrinated to believe that the earth and all its creatures, treasures and precious resources are ours to plunder, subjugate and dominate. We have anointed ourselves kings and queens of creation, our hallowed status at the apex of the great chain of being secured by violence, coercion and brute, pitiless, force. It is the fate of every living thing to suffer and wither under the yoke of our domination and greed. Every single day, across every land and sea, millions of creatures are subjected to unimaginable cruelty – driven from their natural habitats, hunted for profit to the edge of extinction, forced to endure lives of confinement, terror and pain. Cruelty toward our fellow beings – and towards the earth herself – corrodes the human spirit, leaving us bereft and impoverished, fatally estranged from the very source of life on which we depend.

And yet, even now, as we hurtle headlong towards catastrophe, with environmental Armageddon looming ever closer, our understanding of the astonishing abundance and variety of life on earth is still in its infancy. Scientists have catalogued only a small number of the millions of species that dwell here. The vast majority remain a mystery. And in this book, this *Ark* of images, we have been able to display only a fraction of those, many of which are now under severe threat.

Life is the universe's rarest accident, a miracle born of cosmic chance; its fate is in all our hands. It falls to us to seize the levers of power from those intent on wreaking havoc, to bring the relentless juggernaut of unchecked growth grinding to a halt, to build a life around conservation, compassion, mutual care and renewed reverence for all creation. We are facing a tipping point, a moment of truth, of maximum peril. If we fail to act, if we fail to adapt or mend our ways, if we hesitate or waver, then oblivion beckons.

According to the book of Genesis, after the Flood receded, God placed a rainbow in the sky as a sign of his covenant with his chosen people that he would never abandon them. It is a symbol of hope and mercy. Today, rainbows often appear in our crowded skies, reminding us that there is always hope of a new beginning, of peace after the storm. However, to achieve a complete reset, a comprehensive new beginning, we need to make a binding covenant with ourselves to henceforth respect the earth and every living creature that lives alongside us.

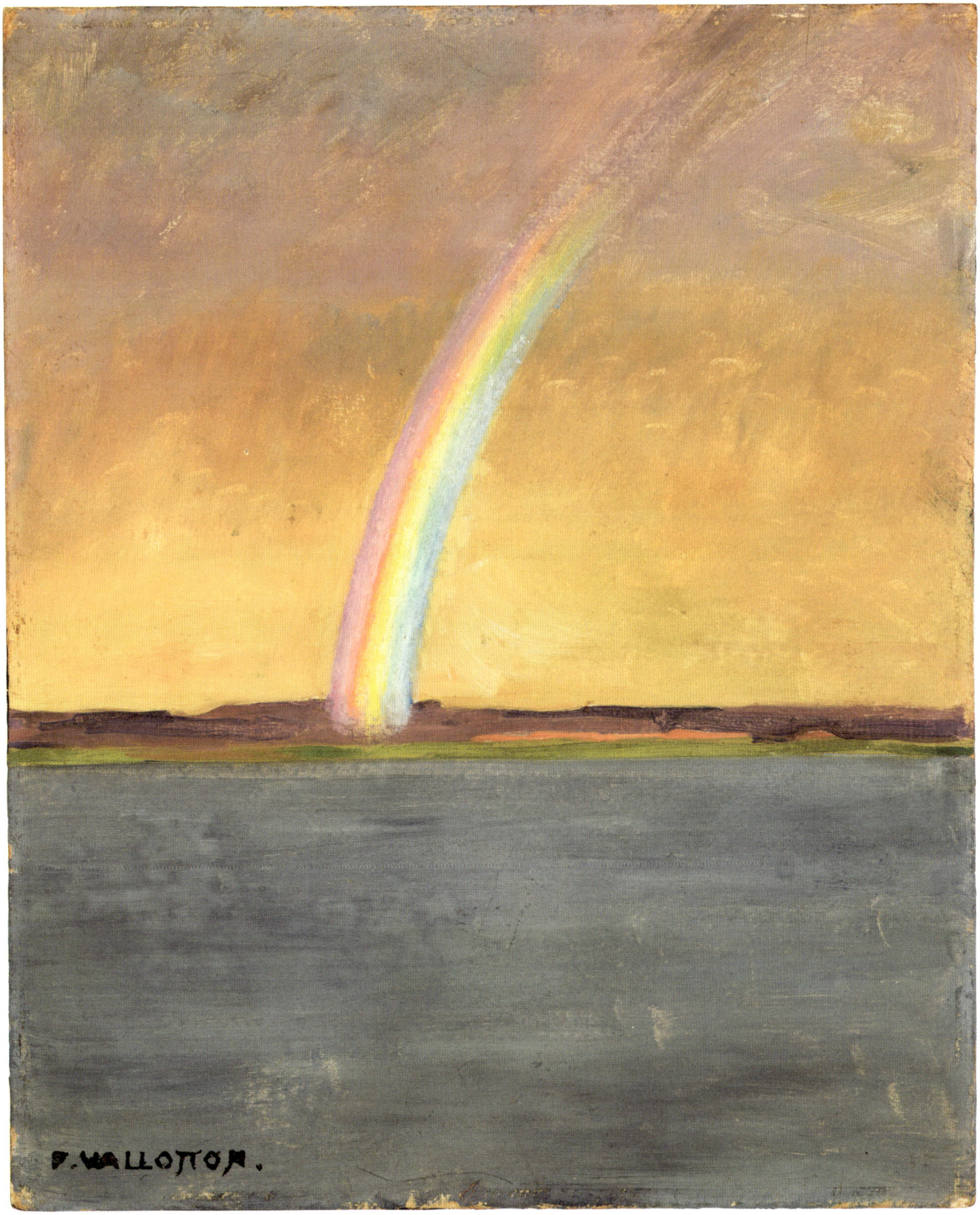

↑
Félix Vallotton painted dreamlike landscapes. In *L'Arc-en-ciel* ('Rainbow', *c.* 1909), the rainbow can be seen as a bridge between heaven and earth.

↳
George Frederic Watts's *After the Deluge* (1885–86) shows the sun returning after the rain that led to the Flood described in the book of Genesis.

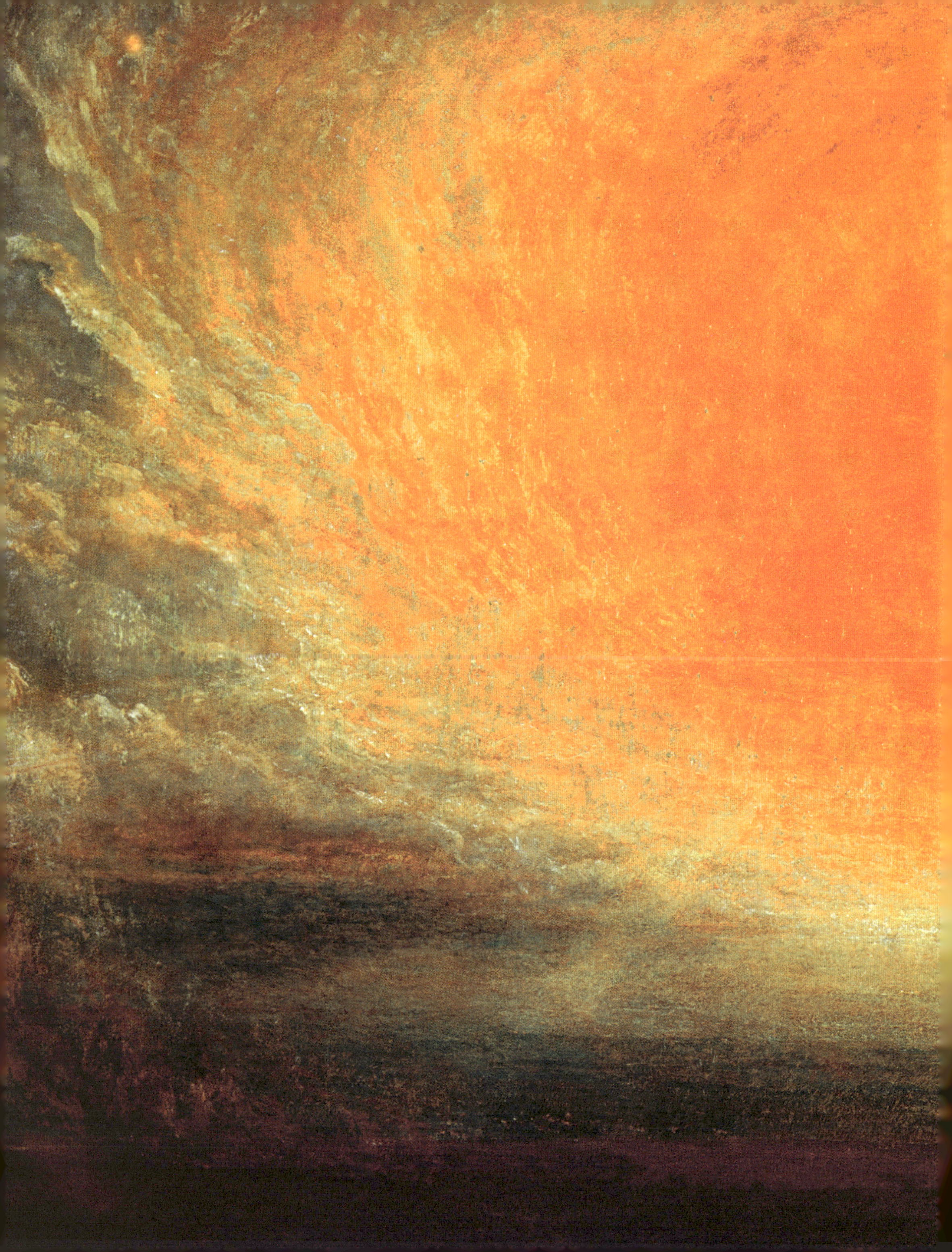

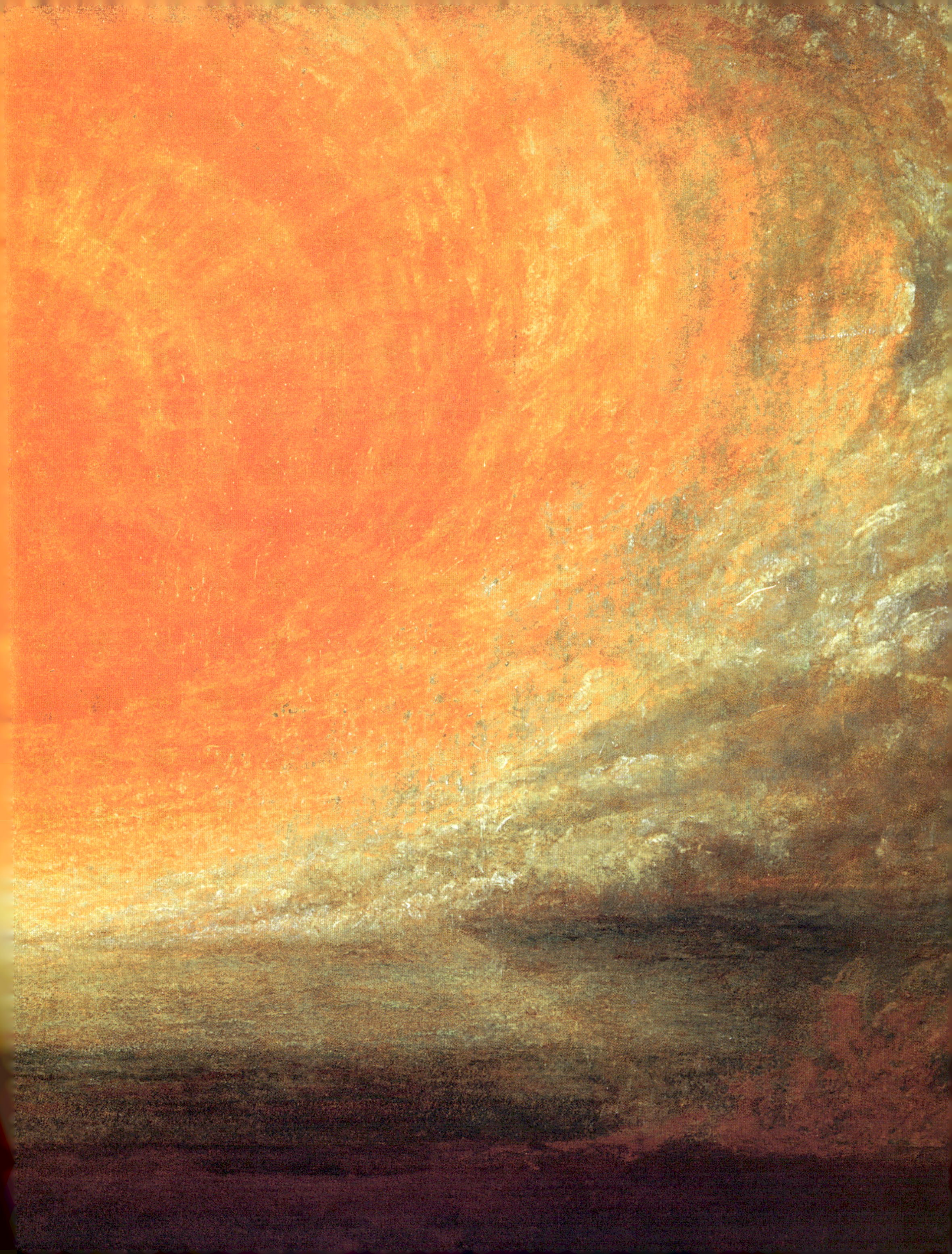

Endnotes

FOREWORD BY ROBERT MACFARLANE (PP. 6–9)

1 Tammana Begum, 'What is a Mass Extinction and are we Facing a Sixth One?', Natural History Museum (19 May 2021; updated 21 February 2023), www.nhm.ac.uk/discover/what-is-mass-extinction-and-are-we-facing-a-sixth-one.html [accessed 5 February 2026]; Elizabeth Kolbert, *The Sixth Great Extinction* (London: Bloomsbury, 2014).
2 Ursula K. Le Guin, *The Word for World is Forest* (London: Gollancz, 2015 [1972]), p. 83.
3 Stephen Jay Gould, *Wonderful Life: The Burgess Shale and the Nature of History* (London: Vintage Publishing, 2000 [1989]).
4 John Muir, Sierra Journal, Summer of 1869 [*c.* 1887], Vol. II (8 July 1869–27 July 1869), p. 130; John Muir Papers, https://scholarlycommons.pacific.edu/jmn-sj2/69/ [accessed 29 January].
5 Aristotle, *The Works of Aristotle*, Volume IV: Historia Animalium, trans. D'Arcy Wentworth Thompson (Oxford: Clarendon Press, 1910).
6 Isaac Newton, Letter to Richard Bentley (25 February 1692/93), 189.R.4.47, f. 7r; The Newton Project, www.newtonproject.ox.ac.uk/view/texts/normalized/THEM00258 [accessed 29 January 2026].
7 The Bible, Genesis 1:28, King James Bible; www.biblehub.com [accessed 29 January 2026].
8 Carl Linnaeus, *Carolus Linnaeus Systema Naturae, 1735*, trans. Maria Sara Johanna Engel-Ledeboer and Hendrik Engel (Nieuwkoop: Hes & De Graff Publishers, 1964).
9 E. O. Wilson, 'Beware the Age of Loneliness', *The Economist* (18 November 2013).

INTRODUCTION: UTOPIA OR OBLIVION? BY STEPHEN ELLCOCK (P. 10)

1 Title courtesy of: R. Buckminster Fuller, *Utopia or Oblivion: The Prospects for Humanity* (New York: Bantam Books, Inc., 1969).

ON THE ARK BY SARAH PERRY (PP. 14–17)

1 The Bible, Genesis 8:10, King James Bible; www.biblehub.com [accessed 21 January 2026].
2 Talmud, Tractate Sanhedrin 96a; Sefaria Library, www.sefaria.org/Sanhedrin.96a.1 [accessed 9 March 2026].
3 *The Babyloniaca of Berossus*, Book 2, section 2, trans. Stanley Mayer Burstein (Malibu, CA: Undena Publications, 1978).
4 'Calamo Currents', *New Zealand Herald*, Vol. XX, Issue 6667, supplement (31 March 1883), p. 1.
5 Gerard Manley Hopkins, 'Inversnaid', *Poems of Gerard Manley Hopkins* (Oxford: Oxford University Press, 1918), p. 95.
6 Alfred, Lord Tennyson, *In Memoriam A. H. H.* (London: The Bankside Press, 1900 [1850]), canto LIV, p. 53.

TREES: ARCHITECTS OF THE NATURAL WORLD BY HARRIET RIX (PP. 82–5)

1 Yorick P. Veenma, Neil S. Davies, Kenneth T. Higgs and William J. McMahon, 'Biogeomorphology of Ireland's Oldest Fossil Forest: Plant-sediment and Plant-animal Interactions Recorded in the Late Devonian Harry Lock Formation, Co. Wexford', *Palaeogeography, Palaeoclimatology, Palaeoecology*, Vol. DCXXI (July 2023).
2 Nancy L. Harris, David A. Gibbs, Alessandro Baccini et al., 'Global Maps of Twenty-First Century Forest Carbon Fluxes', *Nature Climate Change*, Vol. XI (2021), pp. 234–40.
3 Ruth J. Mitchell, Paul E. Bellamy, Christopher J. Ellis et al., 'Collapsing Foundations: The Ecology of the British Oak, Implications of its Decline and Mitigation Options', *Biological Conservation*, Vol. CCXXXIII (May 2019), pp. 316–27.
4 Sou Fujimoto, quoted on an exhibition display panel: *The Architecture of Sou Fujimoto: Primordial Future Forest*, Mori Art Museum, Tokyo (2 July–9 November 2025).

KEEPING THE BEAUTIFUL SEA ALIVE BY HELEN SCALES (PP. 154–7)

1 Yves Cousteau, *Life and Death in the Coral Sea* (Garden City, NY: Doubleday & Company, Inc., 1971), p. 13.
2 Rishi Sharma, Manuel Barange, Vera Agostini et al. (eds.), *Review of the State of World Marine Fishery Resources – 2025*. FAO Fisheries and Aquaculture Technical Paper, No. 721 (Rome: Food and Agriculture Organization of the United Nations, 2025).
3 Xia Zhu, Chelsea M. Rochman, Britta Denise Hardesty, Chris Wilcox, 'Plastics in the Deep Sea – A Global Estimate of the Ocean Floor Reservoir', *Deep Sea Research Part I: Oceanographic Research Papers*, Vol. CCVI (April 2024); La Daana K. Kanhai, Katarina Gardfeldt, Thomas Krumpen et al., 'Microplastics in Sea Ice and Seawater Beneath Ice Floes from the Arctic Ocean', *Scientific Reports*, Vol. X, art. 5004 (2020).
4 Karl Mauritsson, Jean-Pierre Desforges, Karin C. Harding, 'Maternal Transfer and Long-Term Population Effects of PCBs in Baltic Grey Seals Using a New Toxicokinetic-Toxicodynamic Population Model', *Archives of Environmental Contamination and Toxicology*, Vol. LXXXIII, no. 4 (October 2022), pp. 376–94; Jean-Pierre Desforges, Ailsa Hall, Bernie McConnell et al., 'Predicting Global Killer Whale Population Collapse from PCB Pollution', *Science*, Vol. CCCLXI, no. 6409 (2018), pp. 1373–6.
5 Sheree Curry, 'New Research: Earth Is Hotter Than Ever Recorded', *Newsroom: University of St. Thomas* (11 January 2022), https://news.stthomas.edu/earth-is-hotter-than-ever-and-humans-are-to-blame/ [accessed 30 January 2026].
6 Philip J. Landrigan, Hervé Raps, Maureen Cropper et al., 'The Minderoo-Monaco Commission on Plastics and Human Health', *Annals of Global Health*, Vol. LXXXIX, no. 1 (2023), p. 23.
7 Deep-Sea Mining Moratorium, https://deep-sea-conservation.org/solutions/no-deep-sea-mining/ [accessed 30 January 2026].

CONSERVING THE INFINITE VARIETY OF BIRDS BY STEPHEN MOSS (PP. 222–5)

1 William Shakespeare, *Antony and Cleopatra*, ed. W. J. Rolfe (New York: American Book Company, 1909 [*c.* 1607]), Act II, scene 2, l. 238, p. 64.
2 Eleanor Ratcliffe, 'Restorative Perceptions and Outcomes Associated with Listening to Birds', Doctoral Thesis, University of Surrey (8 September 2015).
3 'State of the World's Birds 2022 Paints Most Concerning Picture for Nature Yet', *BirdLife International* (28 September 2022), www.birdlife.org/news/2022/09/28/state-of-the-worlds-birds-2022-paints-most-concerning-picture-for-nature-yet/ [accessed 30 January 2026].
4 'A Range of Threats Drives Declines in Bird Populations', *DataZone by BirdLife* (1 January 2004; updated 9 December 2022), www.datazone.birdlife.org/articles/a-range-of-threats-drives-declines-in-bird-populations [accessed 30 January 2026].
5 'The Albatross Task Force: Reducing Albatross Deaths by 99%', *BirdLife International* (2021), www.birdlife.org/projects/albatross-task-force/ [accessed 30 January 2026].
6 Stephen Moss, 'Brilliant Birdfair', *BirdGuides* (1 December 2024), www.birdguides.com/articles/comment/stephen-moss-brilliant-birdfair/ [accessed 30 January 2026].

CODA: ENDLESS FORMS MOST BEAUTIFUL AND MOST WONDERFUL BY STEPHEN ELLCOCK (P. 274)

1 Title courtesy of the concluding paragraph of Charles Darwin, *The Origin of Species*, ed. Archibald G. B. Russell (London: John Murray, 1859), p. 490.
2 The Bible, Genesis 8:19, King James Bible; www.biblehub.com [accessed 23 January 2026].

Further Reading

Abbey, Edward, *Desert Solitaire* (New York: McGraw-Hill, 1968).
Adams, Douglas, and Carwardine, Mark, *Last Chance to See* (London: Heinemann, 1990).
Attenborough, David, *Life on Earth* (London: William Collins and BBC, 1979).
Attenborough, David, *A Life on Our Planet: My Witness Statement and a Vision for the Future* (London: Ebury Press, 2020).
Baker, J .A., *The Peregrine* (London: William Collins, 1967).
Baker, J. A., *The Hill of Summer* (London: William Collins, 1969).
Ballard, J. G., *The Drowned World* (London: Victor Gollancz, 1962).
Barnett, Lincoln, and the editors of *Life*, *The Wonders of Life on Earth* (New York: Golden Press, 1960).
Berger, John, *Why Look at Animals?* (Great Ideas) (London: Penguin Books, 2009).
Bookchin, Murray, *Toward an Ecological Society* (Montreal, QC: Black Rose Books, 1980).
Bookchin, Murray, *The Ecology of Freedom: The Emergence and Dissolution of Hierarchy* (Palo Alto, CA: Cheshire Books, 1982).
Brand, Stewart, (ed.), *The Last Whole Earth Catalog* (Menlo Park, CA: Portola Institute, 1969).
Brusatte, Steve, *The Rise and Reign of the Mammals: A New History, from the Shadow of the Dinosaurs to Us* (London: Picador, 2022).
Butler, Octavia, *Parable of the Sower* (New York: Four Walls Eight Windows, 1993).
Carson, Rachel, *The Sea Around Us* (Oxford: Oxford University Press, 1951).
Carson, Rachel, *Silent Spring* (New York: Houghton Mifflin, 1962).
Carton, Wim, and Malm, Andreas, *The Long Heat: Climate Politics When It's Too Late* (London: Verso Books, 2025).
Casey, Susan, *The Underworld: Journeys to the Depths of the Ocean* (London: Penguin Books, 2024).
Clare, John, *The Shepherd's Calendar*, eds Eric Robinson and Geoffrey Summerfield (Oxford: Oxford University Press, 1964 [1827]).
Clare, John, *Major Works* (Oxford: Oxford University Press, 2004).
Coe, Sue, and Coe, Mandy, *Meat: Animals and Industry* (North Vancouver, BC: Gallerie Publications, 1991).
Coe, Sue, *Dead Meat* (New York: Four Walls Eight Windows, 1996)
Coe, Sue, *Cruel: Bearing Witness to Animal Exploitation* (New York: OR Books, 2012).
Coe, Sue, and Eisenman, Stephen, *The Ghosts of Our Meat* (Carlisle, PA: Trout Gallery, Dickinson College, 2013).
Coe, Sue, *The Animals' Vegan Manifesto* (New York: OR Books, 2017).
Coe, Sue, *Zooicide: Seeing Cruelty, Demanding Abolition* (New York: OR Books, 2018).
Cousteau, Jacques-Yves, and Dumas, Frédéric, *The Silent World: A Story of Undersea Discovery and Adventure* (New York: Harper & Brothers, 1953)
Cousteau, Jacques-Yves, and Schiefelbein, Susan, *The Human, the Orchid, and the Octopus: Exploring and Conserving Our

Natural World (London: Bloomsbury Publishing, 2007).
Darwin, Charles, *The Origin of Species and The Voyage of the Beagle* (London: Penguin Books, 2003).
Davis, Mike, *The Ecology of Fear: Los Angeles and the Imagination of Disaster* (New York: Metropolitan Books, 1998).
Davis, Mike, *Dead Cities* (New York: The New Press, 2002).
Davis, Mike, *Planet of Slums* (New York: Verso, 2006).
Dawson, Ashley, *Extinction: A Radical History* (New York: OR Books, 2016).
Dickinson, Emily, *Complete Poems* (London: Faber & Faber, 2016).
Dillard, Annie, *Pilgrim at Tinker Creek* (New York: Harper's Magazine Press, 1974).
Dillard, Annie, *Teaching a Stone to Talk* (New York: Harper & Row, 1982).
Dochartaigh, Kerri ní, *Thin Places* (London: Canongate, 2021).
Earle, Sylvia, *Sea Change: A Message of the Oceans* (New York: Ballantine Books, 1995).
Earle, Sylvia, *The World Is Blue: How Our Fate and the Ocean's Are One* (Washington, DC: National Geographic Books, 2009).
Eiseley, Loren, *The Immense Journey* (New York: Random House, 1957).
Eiseley, Loren, *The Unexpected Universe* (New York: Harcourt, Brace and World, 1969).
Eiseley, Loren, *The Night Country* (New York: Charles Scribner's Sons, 1971).
Emerson, Ralph Waldo, *Nature and Selected Essays*, ed. Larzer Ziff (London: Penguin Books, 2003).
Evans, George Ewart, *The Pattern Under the Plough* (London: Faber & Faber, 1966).
Flyn, Cal, *Islands of Abandonment Life in the Post-Human Landscape* (London: HarperCollins Publishers, 2021).
Foer, Jonathan Safran, *Eating Animals* (London: Penguin Books, 2011).
Fortey, Richard, *Life: A Natural History of the First Four Billion Years of Life on Earth* (New York: Knopf, 1998).
Fortey, Richard, *Close Encounters of the Fungal Kind: In Pursuit of Remarkable Mushrooms* (London: William Collins, 2024).
Fossey, Dian, *Gorillas in the Mist* (Boston, MA: Houghton Mifflin, 1983)
Foster, John Bellamy, Clark, Brett, and York, Richard, *The Ecological Rift: Capitalism's War on the Earth* (New York: Monthly Review Press, 2010).
Fressoz, Jean-Baptiste, and Locher, Fabien, *Chaos in the Heavens: The Forgotten History of Climate Change* (London: Verso Books, 2024).
Ghosh, Amitav, *The Great Derangement: Climate Change and the Unthinkable* (Chicago: The University of Chicago Press, 2016).
Giono, Jean, *The Man Who Planted Trees*, trans. Barbara Bray, intro. Richard Mabey (London: Harvill Press, 1995 [1953]).
Godfrey-Smith, Peter, *Other Minds: The Octopus and the Evolution of Intelligent Life* (London: Harper Collins, 2017).
Goodall, Jane, *In the Shadow of Man* (Boston, MA: Houghton Mifflin, 1971)
Goodall, Jane, and Abrams, Douglas, *The Book of Hope: A Survival Guide for an Endangered Planet* (New York: Viking Books, 2022).
Gould, Stephen Jay, *Wonderful Life: The Burgess Shale and the Nature of History* (New York: W. W. Norton, 1989).
Gould, Stephen Jay, *Life's Grandeur: The Spread of Excellence from Plato to Darwin* (London: Penguin Books, 1997).
Guillibert, Paul, *Anthropocene Communism: Land and Capital in the Age of Disaster*, trans. Matt Reeck (London: Verso Books, 2025).
Halliday, Thomas, *Otherlands: A World in the Making* (London: Allen Lane, 2022).
Hardy, Thomas, *The Collected Poems of Thomas Hardy* (London: Macmillan, 1952).
Hay, John, *The Immortal Wilderness* (New York: W. W. Norton & Company, 1987).
Hoare, Philip, *Leviathan, or The Whale* (London: Fourth Estate, 2008).
Hopkins, Gerard Manley, *The Poems of Gerard Manley Hopkins* (London: Oxford University Press, 1930 [1918]).
Hudson, W. H., *Far Away and Long Ago* (London: J. M. Dent & Sons, 1918).
Hudson, W. H., *Adventures Among Birds* (London: J. M. Dent & Sons, 1933).
Hughes, Ted, *Crow* (London: Faber & Faber, 1970).
Humboldt, Alexander von, *Selected Writings*, ed. Andrea Wulf (London: Everyman Library, 2018).
Hurston, Zora Neale, *Mules and Men* (Philadelphia, PA: J. B. Lippincott & Co., 1935).
Jamie, Kathleen, *Sightlines* (London: Sort of Books, 2012).
Jefferies, Richard, *Nature Near London* (London: Chatto & Windus, 1883).
Jefferies, Richard, *The Story of My Heart* (London: Longmans, Green and Co., 1883).
Jefferies, Richard, *After London* (London: Cassell & Company, 1885).
Johnson, Ayana Elizabeth, and Wilkinson, Katharine K. (eds), *All We Can Save: Truth, Courage and Solutions for the Climate Crisis* (New York: Penguin Random House, 2020).
Kimmerer, Robin Wall, *Braiding Sweetgrass* (Minneapolis, MN: Milkweed Editions, 2013).
Kincaid, Jamaica, *Among Flowers: A Walk in the Himalaya* (Washington, DC: National Geographic, 2005).
Kolbert, Elizabeth, *The Sixth Extinction* (New York: Henry Holt & Co., 2014).
Kurlansky, Mark, *Cod: A Biography of the Fish that Changed the World* (London: Vintage, 1997).
Laing, Olivia, *The Garden Against Time: In Search of a Common Paradise* (London: Picador, 2024).
Le Guin, Ursula K., *The Word for World is Forest* (New York: Berkeley Publishing, 1976).
Leopold, Aldo, *A Sand County Almanac* (Oxford: Oxford University Press, 1949).
Lewis-Stempel, John, *Meadowland: The Private Life of an English Field* (London: Doubleday, 2014).
London, Jack, *White Fang* (London: Penguin Books, 2008).
Lopez, Barry, *Arctic Dreams* (New York: Charles Scribner's Sons, 1986).
Lopez, Barry, *Apologia* (San Antonio, TX: Trinity University Press, 1998).
Lovelock, James E., *Gaia: A New Look at Life on Earth* (Oxford: Oxford University Press, 1979).
Lovelock, James E., *The Revenge of Gaia: Earth's Climate Crisis & The Fate of Humanity* (London: Penguin Books, 2007).
Löwy. Michael, *Ecosocialism: A Radical Alternative to Capitalist Catastrophe* (Chicago, IL: Haymarket Books, 2015).
McCarthy, Cormac, *The Road* (New York: Alfred A. Knopf, 2006).
McGregor, James H. S., *Back to the Garden: Nature and the Mediterranean World from Prehistory to the Present* (New Haven, CT: Yale University Press, 2015).
Mabey, Richard, *The Cabaret of Plants: Botany and the Imagination* (London: Profile Books, 2015).
Macfarlane, Robert, *The Wild Places* (London: Granta Books, 2007).
Macfarlane, Robert, *Landmarks* (London: Penguin Books, 2015).
Macfarlane, Robert, *Is a River Alive?* (London: Penguin Books, 2025).
Magnason, Andri Snær, *On Time and Water* (London: Profile Books, 2021).
Malm, Andreas, *The Progress of This Storm: Nature and Society in a Warming World* (London: Verso Books, 2020).
Malm, Andreas, *How to Blow Up a Pipeline: Learning to Fight in a World on Fire* (London: Verso Books, 2021).
Malm, Andreas, *Fighting in a World on Fire: The Next Generation's Guide to Protecting the Climate and Saving Our Future* (London: Verso Books, 2023).
Marsh, George Perkins, *Man and Nature, or, Physical Geography as Modified by Human Action* (London: Sampson Low, Son & Marston, 1864).
Matthiessen, Peter, *Blue Meridian: The Search for the Great White Shark* (New York: Random House, 1971).
Matthiessen, Peter, *The Snow Leopard* (New York: Viking Press, 1978).
Matthiessen, Peter, *African Silences* (New York: Knopf Doubleday, 1991).
Meadows, Donella, et al., *The Limits to Growth* (New York: Potomac Associates /Universe Books, 1972).
Melville, Herman, *Moby Dick* (London: Everyman Library, 1991)
Merchant, Carolyn, *Radical Ecology: The Search for a Livable World* (London: Routledge, 1992).
Merwin, W. S., *The Rain in the Trees* (New York: Alfred A. Knopf, 1988).
Mies, Maria, and Shiva, Vandana, *Ecofeminism* (London: Zed Books, 1993).
Moore, Marianne, *Collected Poems* (London: Faber & Faber, 1951).
Morton, Timothy, *Dark Ecology* (New York: Columbia University Press, 2016).
Moss, Stephen, *Ten Birds That Changed the World* (London: Faber, 2023).
Moss, Stephen, and Westwood, Brett, *Wonderland: A Year of Britain's Wildlife, Day by Day* (London: John Murray, 2017).
Muir, John, *The Mountains of California* (New York: The Century Company, 1894).
Muir, John, *My First Summer in the Sierra* (Boston, MA: Houghton Mifflin, 1911).
Næss, Arne, *Ecology, Community and Lifestyle* (Cambridge: Cambridge University Press, 1989).
Nash, Roderick Frazier, *Wilderness and the American Mind* (New Haven, CT: Yale University Press, 1967).
Nixon, Rob, *Slow Violence and the Environmentalism of the Poor* (Cambridge, MA: Harvard University Press, 2011).
Olson, Roberta, New York Historical Society, et al., *Audubon's Aviary* (New York: Rizzoli, 2012).
Perry, Sarah, *Enlightenment* (London: Jonathan Cape, 2024).
Pliny the Elder, *Natural History*, trans. John Healey (London: Penguin Books, 1991).
Powers, Richard, *The Overstory* (New York: W. W. Norton, 2018).
Proulx, Annie, *Bird Cloud* (New York: Scribner, 2011).
Proulx, Annie, *Fen, Bog & Swamp: A Short History of Peatland Destruction and Its Role in the Climate Crisis* (New York: Scribner, 2022).
Quammen, David, *The Song of the Dodo: Island Biogeography in an Age of Extinctions* (London: Pimlico Books, 1997).
Regan, Tom, *The Case for Animal Rights* (Berkeley, CA: University of California Press, 2004).
Rix, Harriet, *The Genius of Trees: How Trees Mastered the Elements and Shaped the World* (London: Bodley Head, 2025).
Roberts, Callum, *The Unnatural History of the Sea* (London: Allen Lane, 2007).
Roberts, Callum, *The Ocean of Life: The Fate of Man and the Sea* (London: Allen Lane, 2012).
Rundell, Katherine, *The Golden Mole: and Other Living Treasure* (London: Faber & Faber, 2022).
Ryōkan, *Sky Above, Great Wind: The Life and Poetry of Zen Master Ryōkan*, trans. and ed. Kazuaki Tanahashi (London: Penguin Books, 2011).
Salgado, Sebastião, *Genesis* (Cologne: Taschen, 2013).
Sapolosky, Robert, *A Primate's Memoir* (New York: Simon & Schuster, 2002).
Scales, Helen, *Spirals in Time: The Secret Life and Curious Afterlife of Seashells* (London: Bloomsbury, 2015).
Scales, Helen, *The Brilliant Abyss: True Tales of Exploring the Deep Sea, Discovering Hidden Life and Selling the Seabed* (London: Bloomsbury, 2021).
Scales, Helen, *What the Wild Sea Can Be: The Future of the World's Ocean* (London: Grove Press, 2024).
Schell, Jonathan, *The Fate of the Earth* (New York: Alfred A. Knopf, 1982).
Schlanger, Zoë, *The Light Eaters: The New Science of Plant Intelligence* (London: Fourth Estate, 2024).
Scranton, Roy, *Learning to Die in the Anthropocene: Reflections on the End of a Civilization* (San Francisco, CA: City Lights Books, 2015).
Sebald, W. G., *The Rings of Saturn* (London: Harvill Press, 1995).
Sheldrake, Merlin, *Entangled Life: How Fungi Make Our Worlds, Change Our Minds and Shape Our Futures* (London: Bodley Head, 2020).
Singer, Peter, *Animal Liberation Now* (London: Bodley Head, 2023).
Snyder, Gary, *Turtle Island* (New York: New Directions, 1974).

Solnit, Rebecca, *A Field Guide to Getting Lost* (New York: Viking, 2005).

Sōseki, Natsume, *Kusamakura*, trans. Meredith McKinney (London: Penguin Books, 2008).

Tagore, Rabindranath, *The Crescent Moon* (London: Macmillan, 1913).

Taylor, Bron, *Dark Green Religion: Nature Spirituality and the Planetary Future* (Berkeley, CA: University of California Press, 2010).

Thomas, Edward, *Collected Poems* (London: Faber & Faber, 2004).

Thoreau, Henry David, *Walden; or, Life in the Woods* (Boston, MA: Ticknor and Fields, 1854).

Thoreau, Henry David, *The Maine Woods* (London: Penguin Books, 1988).

Thunberg, Greta, *Our House is on Fire: Scenes of a Family and a Planet in Crisis* (London: Penguin Books, 2020).

Thunberg, Greta (ed.), *The Climate Book* (London: Allen Lane, 2022).

Thwaite, Anne, *Glimpses of the Wonderful: The Life of Philip Henry Gosse* (London: Faber & Faber, 2002).

Tsing, Anna Lowenhaupt, *The Mushroom at the End of the World: On the Possibility of Life in Capitalist Ruins* (Princeton, NJ: Princeton University Press, 2015).

Tudge, Colin, *The Tree: A Natural History of What Trees Are, How They Live and Why They Matter* (New York: Crown Publishing, 2006).

White, Gilbert, *The Natural History of Selborne* (London: Penguin Books, 1994 [1789]).

White, T. H., *The Goshawk* (London: Jonathan Cape, 1951).

Whitman, Walt, *Collected Poems* (London: Penguin Books, 1986).

Williams, Heathcote, *Whale Nation* (London: Jonathan Cape, 1988).

Williams, Heathcote, *Sacred Elephant* (London: Jonathan Cape, 1989).

Williamson, Henry, *Tarka the Otter: His Joyful Water Life and Death in the Country of the Two Rivers* (London: G. P. Putnam's Sons, 1927).

Williamson, Henry, *Salar the Salmon* (London: Faber & Faber, 1935).

Wilson, Edward O., *Biophilia* (Boston, MA: Harvard University Press, 1990).

Wilson, Edward O., *The Diversity of Life* (Boston, MA/London: Harvard University Press and Allen Lane, 1992).

Winters, Ed, *This is Vegan Propaganda (And Other Lies the Meat Industry Tells You)* (London: Penguin Books, 2022).

Wohlleben, Peter, *The Hidden Life of Trees: What They Feel, How They Communicate* (London: HarperCollins, 2016).

Yong, Ed, *An Immense World: Discover How Animals Perceive the World and See Nature as you Never Have Before* (London: Bodley Head, 2022).

Young, Andrew, *Selected Poems* (Manchester: Carcanet, 1998).

Further Viewing

73 Cows, dir. Alex Lockwood (Lockwood Film, 2018).

Aguirre, Wrath of God, dir. Werner Herzog (Werner Herzog Filmproduktion, 1972).

All That Breathes, dir. Shaunak Sen (HBO Documentary films, 2022).

The Animals Film, dir. Victor Schonfeld and Myriam Alaux (Beyond the Frame and Slick Pix, 1981)

Anima Mundi, dir. Godfrey Reggio (Studio Equatore and Bulgari, 1991).

Anthropocene: The Human Epoch, dir. Jennifer Baichwal, Nicholas de Pencier and Edward Burtynsky (Mercury Films and Seville International, 2018).

The Assassin, dir. Hou Hsiao-hsien (Spot Films et al., 2015).

Âtman, dir. Toshio Matsumoto (Matsumoto Production Company, 1975).

Au hasard Balthazar, dir. Robert Bresson (Argos Films et al., 1966)

Baraka, dir. Ron Fricke (Magidson Films, 1992).

Becoming Cousteau, dir. Liz Garbus (Story Syndicate et al., 2021).

Birdwatchers, dir. Marco Bechis (Classic, 2008).

Blue Planet I and *II*, presented by David Attenborough (BBC, 2001; 2017).

Call of Life: Facing the Mass Extinction, dir. Monte Thompson (Species Alliance, 2010).

Chasing Coral, dir. Jeff Orlowski (Doc Society and Exposure Labs, 2017).

Chasing Ice, dir. Jeff Orlowski (Doc Society and Exposure Labs, 2012).

Children of Men, dir. Alfonso Cuarón (Universal Pictures and Strike Entertainment, 2006).

The Cove, dir. Louie Psihoyos (Diamond Docs et al., 2009).

Cow, dir. Andrea Arnold (Doc Society, BBC Film and Halcyon Pictures, 2021).

Cowspiracy: The Sustainability Secret, dir. Kip Andersen and Keegan Kuhn (A.U.M. Films and First Spark Media, 2014).

Daughters of the Dust, dir. Julie Dash (Geechee Girls et al., 1991).

David Attenborough: A Life on Our Planet, dir. Alastair Fothergill, Jonathan Hughes and Keith Scholey (Altitude Film Entertainment et al., 2020).

Dersu Uzala, dir. Akira Kurosawa (Atelier 41 et al., 1975).

Dominion, dir. Chris Delforce (Farm Transparency Project, 2018).

Don't Look Up, dir. Adam MacKay (Hyperobject Industries, 2021).

Dream of Light, dir. Victor Erice (Euskal Media et al., 1992).

Dreams, dir. Akira Kurosawa and Ishirô Honda (Akira Kurosawa USA, 1990).

Earth (*Zemlya*) , dir. Alexander Dovzhenko (VUFKU, 1930).

Earthlings, dir. Shaun Monson (Nation Earth, 2005).

Embrace of the Serpent, dir. Ciro Guerra (Buffalo Films et al., 2015).

The Emerald Forest, dir. John Boorman (Christel Films and Embassy Pictures, 1985).

Encounters at the End of the World, dir. Werner Herzog (Discovery Films and Creative Differences Productions, 2007).

The End of the Line, dir. Rupert Murray (Doc Society et al., 2009).

Evil Does Not Exist, dir. Ryūsuke Hamaguchi (Fictive and NEOPA, 2023).

Fantastic Fungi, dir. Louie Schwartzberg (Moving Art Studio and Reconsider, 2019).

First Reformed, dir. Paul Schrader (Killer Films et al., 2017).

The Flowers of St. Francis, dir. Roberto Rossellini (Cineriz and Rizzoli Film, 1950).

Forest Murmurs, dir. Slavko Vorkapich (1947).

The Fountain, dir. Darren Aronofsky (Warner Bros. et al., 2006).

Ghost Elephants, dir. Werner Herzog (Sobey Road Entertainment, 2025).

The Gleaners and I, dir. Agnès Varda (Ciné-tamaris, 2000).

Gone to Earth, dir. Michael Powell and Emeric Pressburger (Selznick International Pictures, 1950).

Gorillas in the Mist, dir. Michael Apted (Universal Pictures et al., 1988).

Grass: A Nation's Battle For Life, dir. (and produced by) Ernest B. Schoedsack and Merian C. Cooper (1925).

The Green Planet, presented by David Attenborough (BBC, 2022).

Grizzly Man, dir. Werner Herzog (Lionsgate et al., 2005).

Gunda, dir. Viktor Kossakovsky (Artemis Rising Foundation et al., 2020).

The Hellstrom Chronicle, dir. Walon Green and Ed Spiegel (Wolper Pictures, 1971).

The Host, dir. Bong Joon Ho (Nintendo Entertainment et al., 2006).

How to Blow Up a Pipeline, dir. Daniel Goldhaber (Chrono, Lyrical Media, Spacemaker Productions, 2022).

I Am Greta, dir. Nathan Grossman (B-Reel Films, 2020).

Inhabitants, dir. Artavazd Peleshian (Belarusfilm, 1970).

Ivan's Childhood, dir. Andrei Tarkovsky and Eduard Abalov (Mosfilm and Trete Tvorcheskoe Obedinenie, 1962).

Jane, dir. Brett Morgen (National Geographic Studios and Public Road Productions, 2017).

Journey to Avebury, dir. Derek Jarman (1971).

Kifaru, dir. David Hambridge (Vs. Goliath Visual and Ragtag Tribe Films, 2019).

Koyaanisqatsi, dir. Godfrey Reggio (American Zoetrope et al., 1982).

Leviathan, dir. Véréna Paravel and Lucien Castaing-Taylor (Sensory Ethnography Lab, 2013).

The Life Collection box set, presented by David Attenborough (BBC, 2018).

Microcosmos, dir. Claude Nuridsany and Marie Pérennou (Galatée Films et al., 1996).

Mission Blue, dir. Robert Nixon and Fisher Stevensis (Insurgent Media and True Blue Films, 2014).

My Octopus Teacher, dir. Pippa Ehrlich and James Reed South Africa (Off The Fence and The Sea Change Project, 2020).

Neon Genesis Evangelion, dir. Hideaki Anno (Gainax et al., 1995–96).

Neon Genesis Evangelion: The End of Evangelion, dir. Hideaki Anno (Gainax et al., 1997).

Noah, dir. Darren Aronofsky (Paramount Pictures et al., 2014).

Our Daily Bread, dir. King Vidor (Viking Productions, 1934).

Our Daily Bread, dir. Nikolaus Geyrhalter (Nikolaus Geyrhalter Filmproduktion et al., 2005).

Our Planet I and *II*, presented by David Attenborough (Netflix, 2021; 2023).

The Outlaw and His Wife, dir. Victor Sjöström (Svenska Biografteatern AB, 1918).

'Penda's Fen', episode of *Play for Today*, dir. Alan Clarke (BBC, 1974).

Planet Earth I, II and *III*, presented by David Attenborough (BBC, 2006; 2016; 2023).

Princess Mononoke, dir. Hayao Miyazaki (Studio Ghibli et al., 1997).

Project Nim, dir. James Marsh (Red Box Films et al., 2011).

Le Quattro Volte, dir. Michelangelo Frammartino (Vivo Film et al., 2010).

Racing Extinction, dir. Louie Psihoyos (Oceanic Preservation Society and Okeanos, 2015).

Requiem for a Village, dir. David Gladwell (BFI, 1975).

The Road, dir. John Hillcoat (Dimension Films et al., 2009).

The Sacrifice, dir. Andrei Tarkovsky (Faragó Film and Stiftelsen Svenska Filminstitutet, 1986).

The Salt of the Earth, dir. Wim Wenders and Juliano Ribeiro Salgado (Decia Films et al., 2014).

Samsara, dir. Ron Fricke (Magidson Films et al., 2011).

Le Sang des Bêtes, dir. Georges Franju (Forces et voix de la France, 1949).

Science Is Fiction: 23 Films by Jean Painlevé (Criterion Collection, 2017).

Seasons of the Year, dir. Artavazd Peleshyan (Armenfilm, 1975).

Seaspiracy, dir. Ali Tabrizi (A. U. M. Films and Disruptive Studios, 2021)

Shadows of Forgotten Ancestors, dir. Sergei Parajanov (Dovzhenko Film Studios, 1965).

Sharkwater, dir. Rob Stewart (Diatribe Pictures and SW Productions, 2006).

Silent Running, dir. Douglas Trumbull (Universal Pictures and Trumbull/Gruskoff Productions, 1972).

The Silent World, dir. Jacques Cousteau and Louis Malle (FSJYC Production et al., 1956).

Snowpiercer, dir. Bong Joon Ho (CJ Entertainment et al., 2013).

Soylent Green, dir. Richard Fleischer (MGM, 1973).

Spring, Summer, Fall, Winter… and Spring, dir. Kim Ki-duk (Korea Pictures et al., 2003).

The Territory, dir. Alex Pritz (Doc Societty et al., 2022).

The Thin Red Line, dir. Terrence Malick (Fox 2000 Pictures et al., 1998).

The Tree of Life, dir. Terrence Malick (Cottonwood Pictures et al., 2011).

Tropical Malady, dir. Apichatpong Weerase-thakul (Downtown Pictures et al., 2004).

Uncle Boonmee Who Can Recall His Past Lives, dir. Apichatpong Weerasethakul (Kick the Machine et al., 2010).

Virunga, dir. Orlando von Einsiedel (Doc Society et al., 2014).

Wall-E, dir. Andrew Stanton (Walt Disney Pictures and Pixar, 2008).

Week End, dir. Jean-Luc Godard (Comacico et al., 1967)

The Wind, dir. Victor Sjöström (MGM, 1928).

Winged Migration, dir. Jacques Cluzaud, Michel Debats and Jacques Perrin (Galatée Films et al., 2001).

Winstanley, dir. Kevin Brownlow and Andrew Mollo (BFI, 1975).

Yeelen, dir. Souleymane Cissé (Les Films Cissé et al., 1987).

Sources of Quotations

INTRODUCTION

13 The Bible, Genesis 6:19–21, King James Version, www.biblehub.com [accessed 23 January 2026].

MICROORGANISMS & FUNGI

24 Dan Rokhsar, quoted in Sandra Hines, 'Tiny diatom has global environmental impact', *UW News* (7 October 2004), www.washington.edu/news/2004/10/07/tiny-diatom-has-global-environmental-impact/ [accessed 23 January 2026].

28 Jacques-Yves Cousteau, 'Ocean Policy and Reasonable Utopias', *The Forum*, Vol. XVI, no. 5 (Summer 1981), pp. 897–910, pp. 899–900.

31 Steven L. Stephenson, *The Kingdom Fungi: The Biology of Mushrooms, Molds, and Lichens* (Portland, OR: Timber Press Inc., 2010), p. 145.

33 John Muir, *My First Summer in the Sierra* (Boston, MA: Houghton Mifflin Company, 1917), p. 157.

34 Henry David Thoreau, Manuscript Volume 27 (July 9–November 9, 1858), *The Writings of Henry D. Thoreau*, thoreau.library.ucsb.edu/writings_journals27.html [accessed 23 January 2026].

39 P. H. Dudley, 'Woods and their Destructive Fungi II', *Popular Science Monthly*, Vol. XXIX (September 1886), pp. 604–17, p. 604.

43 John Clare, *Poems by John Clare* (London: Henry Frowde, 1908), p. 178.

44 Richard Jefferies, *The Life of the Fields* (London: Chatto & Windus, 1884), p. 72.

45 Quotation within caption: 'Painter's Purpose: Interview with Eliot Hodgkin', *The Studio*, 1957; cited on: Eliot Hodgkin, www.eliothodgkin.com/article/painters-purpose-interview-with-eliot-hodgkin/ [accessed 26 February 2026].

46 Julia McNair Wright, *Nature Readers: Sea and the Way-Side*, No. 3 (Boston, MA: D. C. Heath & Co., 1888), p. 24.

48 William Henry Hudson, *The Book of a Naturalist* (New York: George H. Doran Co., 1919), p. 337.

50 Pierre de Ronsard, *Songs and Sonnets of Pierre de Ronsard*, trans. Curtis Hidden Page (Boston, MA: Houghton, Mifflin & Co., 1903), p. 39.

55 Mrs Beeton, 'The Potato as an Article of Human Food', *The Book of Household Management* (London: S. O. Beeton, 1861).

57 Cited in: Mu-Chou Poo, *Wine and Wine Offering in the Religion of Ancient Egypt* (London: Routledge, 2009), p. 102.

63 Ralph Waldo Emerson, 'Hamatreya' (1846), *Works of Ralph Waldo Emerson* (London: George Routledge and Sons, Ltd, 1897), p. 518.

70 John Muir, 'The American Forests', *The Atlantic* (August 1897), p. 157.

74 William Blake, *The Letters of William Blake* (London: Methuen & Co., 1906), p. 62.

76 Anton Chekhov, *The Life and Letters of Anton Chekhov*, trans. S. S. Koteliansky and Philip Tomlinson (New York: Benjamin Blon, Inc., 1965), p. 123.

INVERTEBRATES

88 Emily Dickinson, *Poems*, Book III: Nature, poem V, ed. Mabel Loomis Todd and T. W. Higginson (Boston, MA: Roberts Brothers, 1890), p. 73.

90 Matsuo Bashō, Haiku 913, *Bashō: The Complete Haiku*, trans. Jane Reichhold (Tokyo: Kodansha International Ltd, 2008), p. 215.

92 Louisa May Alcott, *Little Men* (Garden City, NY: Nelson Doubleday Inc., 1871), p. 303.

97 Sir David Attenborough, Episode 5: 'Supersocieties', *Life in the Undergrowth*, BBC (21 December 2005); cited on: IMDB, www.imdb.com/title/tt0760222/quotes/ [accessed 26 Jan 2025].

98 Katherine Mansfield, *Poems* (London: Constable & Co. Ltd, 1923), p. 45.

102 Charles Darwin, entry 36, Notebook N [Metaphysics and Expression] (1838–1839), eds Paul Barrett and John van Wyhe; *The Complete Work of Charles Darwin Online*, http://darwin-online.org.uk/ [accessed 26 January 2026].

105 Leslie Brunetta and Catherine L. Craig, *Spider Silk* (New Haven, CT: Yale University Press, 2010), p. 190.

108 Walt Whitman, 'Whispers of Heavenly Death', *The Broadway: A London Magazine* (October 1868), pp. 21–2, p. 21.

115 William Wordsworth, 'Despondency corrected', *The Excursion: Being a Portion of The Recluse* (London: Longman, Hurst, Rees, Orme and Brown, 1814).

116 Dante Gabriel Rossetti, *Poems* (London: Ellis & White, 1881), p. 255.

119 Arthur Arnold, 'On our Fisheries', *Once a Week*, No. 198 (11 April 1863), p. 443.

121 Edward Forbes, *A Monograph of the British Naked-Eyed Medusae* (London: Printed for the Ray Society, 1848), p. 32.

125 Rachel L. Carson, 'Undersea', *The Atlantic* (September 1937), p. 322.

FISH

129 Harriet Osgood Taylor, *Japanese Gardens* (New York: Dodd, Mead & Company, 1912), p. 102.

130 John Ruskin, *The Two Paths* (London: Smith, Elder and Co., 1859), p. 243.

136 *The Little Flowers of Saint Francis*, trans. and ed. Rev. H. E. Manning (London: Burns and Lambert, 1884), pp. 97–8.

142 Albert Carl Ludwig Gotthilf Günther, *An Introduction to the Study of Fishes* (Edinburgh: Adam and Charles Black, 1880), p. 622.

144 John Steinbeck, *The Log from the Sea of Cortez* (London: Penguin Books, 1995), p. 179.

147 Terry Waghorn, 'Can We Eat Our Fish and Protect Them Too?', *Forbes* (21 February 2012), www.forbes.com/sites/terrywaghorn/2012/02/21/can-we-eat-our-fish-and-protect-them-too/ [accessed 26 January 2026].

149 *T'ai-p'ing kuang-chi*, compiled by Li Fan et al., completed 978; cited in: Anne Birrell, *Chinese Mythology: An Introduction* (Baltimore, MD: Johns Hopkins University Press, 1999), p. 242.

152 C. F. Holder 'Fishes and their Young', *St. Nicholas*, Vol. XIII, no. 8 (June 1886), p. 602.

AMPHIBIANS & REPTILES

163 Matsuo Bashō, cited in: *A History of Japanese Literature*, trans. William George Aston (London: William Heinemann, 1899), p. 295.

164 Thomas R. Henry, *The Strangest Things in the World: A Book About Extraordinary Manifestations of Nature* (Washington, DC: Public Affairs Press, 1958), p. 168.

167 Kobayashi Issa, trans. David G. Lanoue. Accessed via: haikuguy.com/issa/search.php [accessed 26 February 2026].

169 'definition of "mim": the breathing holes of long-necked turtles, crabs etc.', Dictionary of the Bininj Kunwok language of West Arnhem Land, https://njamed.com/entry/mim/ [accessed 26 February 2026].

172 D. H. Lawrence, *Pansies* (New York: Alred A. Knopf Inc., 1929), p. 148.

177 Tertullian, 'On the Pallium', Chapter 3: Beasts Similarly Subject to the Law of Mutation, Ante-Nicene Fathers, Vol. IV: *Tertullian* (Edinburgh: T. and T. Clark, 1867–73).

178 John Greenleaf Whittier, 'The Double-Headed Snake of Newbury' (1859), *The Works of John Greenleaf Whittier*, Vol. I (Boston, MA: Houghton, Mifflin and Company, 1893), p. 192.

180 Sebastião Salgado, quoted in: Veronica Espasito, 'Sebastião Salgado: "I was transformed into an environmentalist"', *The Guardian* (26 September 2022), www.theguardian.com/artanddesign/2022/sep/25/sebastiao-salgado-photography-exhibition-sothebys-new-york [accessed 26 February 2026].

BIRDS

187 Charles Baudelaire, 'The Owls' (1857), trans. Clark Ashton Smith, *Weird Tales*, Vol. XXXVI, no. 2 (November–December 1941), p. 120.

195 Cited in: Marcus Porcius Cato and Marcus Terentius Varro, *Roman Farm Management: The Treatises of Cato and Varo*, trans. Fairfax Harrison (New York: The Macmillan Company, 1913), p. 304.

198 Henry David Thoreau, *Walden; or, Life in the Woods* (Boston, MA: Ticknor and Fields, 1854), p. 339.

200 H. D., 'The Bird-Chorus of Ion', *Heliodora and Other Poems* (Boston, MA: Houghton Mifflin Co., 1924), p. 124.

203 Louis Figuier, *Reptiles and Birds* (New York: D. Appleton and Co., 1870), p. 317.

204 W. B. Yeats, *The Wild Swans at Coole* (New York: The Macmillan Company, 1919), pp. 2–3.

209 *Shijing*, cited in: John Thompson, *On the Guqin*, www.silkqin.com [accessed 27 January 2026]

219 Derek Jarman, *Chroma: A Book of Color* (Woodstock, NY: The Overlook Press, 1995), p. 145.

MAMMALS

230 George J. Breed and William G. Barton, 'Notes from the Field – July', *Songs and Saunterings* (Salem, MA: The Salem Press Publishing and Printing Co., 1892), pp. 152–77, p. 172.

237 William Ellis, 'In the Smoking Room', *The Philosopher*, Vol. VI, no. 1 (July 1899), pp. 186–92, p. 190.

239 Robert Green Ingersoll, *Some Mistakes of Moses* (Washington, DC: C. P. Farrell, 1879), p. 149.

240 Agnes Deans Cameron, *The New North: Being Some Account of a Woman's Journey through Canada to the Arctic* (New York: D. Appleton and Co., 1910), p. 281.

242 Sir David Attenborough, Season 1, Episode 2: 'Frozen Worlds', *Our Planet*, BBC (5 April 2019).

243 Quotation within caption: Dr James Murie, 'On the Walrus', *Transactions of the Zoological Society of London*, Vol. VII (London: 1872), p. 419.

247 Margaret Cavendish, 'The Hunting of the Hare', *Poems and Fancies* (London: William Wilson, 1664), p. 134.

250 Fyodor Dostoyevsky, *The Brothers Karamazov*, trans. Constance Garnett (New York: The Lowell Press, 1912), p. 355.

254 Virgil, *The Georgics of Virgil*, Book III, trans. R. D. Blackmore (London: Sampson Low, Son and Marston, 1871), p. 74, ll. 97–9.

256 Quotation within caption: Franz Marc, 'Über das Tier in der Kunst', translated in Vivian Endicott Barnett (ed.), *Franz Marc and August Macke 1909–1914* (Munich, London and New York, 2018), p. 48.

259 Jean-Loïc Le Quellec, *Rock Art in Africa: Mythology and Legend*, trans. Paul Bahn (Paris: Flammarion, 2004), p. 14.

260 Al-Mutanabbī, 'Poem 12: the poet reproaches Sayf al-Dawla for preferring another above him', *Poems of Al-Mutanabbi: A Selection*, trans. A. J. Arberry (Cambridge: Cambridge University Press, 1967), p. 72.

264 Aristotle, *The Works of Aristotle*, Volume IV: Historia Animalium, trans. D'Arcy Wentworth Thompson (Oxford: Clarendon Press, 1910), Book IX, pp. 630a–b.

271 Michel de Montaigne, *Essays of Michel de Montaigne*, Vol. I, Book 2, Chapter 11: On Cruelty, 1580, trans. Charles Cotton, ed. William Carew Hazlitt (New York: A. L. Burt, 1892), pp. 404–19, p. 419.

273 W. N. P. Barbellion (Bruce F. Cummings), *The Journal of a Disappointed Man* (New York: George H. Doran, 1919), p. 27.

Sources of Illustrations

Every effort has been made to locate and credit copyright holders of the material reproduced in this book. The author and publisher apologize for any omissions or errors, which can be corrected in future editions.

a = above, **b** = below, **c** = centre, **l** = left, **r** = right

Dimensions of works are given in centimetres and inches, height before width before depth.

2 Roelandt Savery, *Noah's Ark*, *c.* 1628 (detail). Oil on canvas, 174.5 × 212 (68 ¾ × 83 ½). National Museum in Warsaw (M. Ob.579 MNW) / **4** Jethro Buck, *Genesis*, 2015. Pigment paint on Sanganir paper, 90 × 60 (35 ½ × 23 ⅝). Courtesy Jethro Buck / **7** Gabriel Mejia, *Burning Lands, Talacogon, Agusan Del Sur, Philippines*, March 2024. © Gab Mejia / **8** Joseph Stella, *Tree of My Life*, 1919 (detail). Oil on canvas, 213.4 × 193 (84 × 76). Private Collection / **11** Kahn & Selesnick, *The Augury of the Ark*, 2016–17. Archival ink jet print. © Kahn & Selesnick / **12** *Noah's Ark*, miniature from Arabic manuscript from the Seljuk era, 13th century. Photo Emmanuel Lattes/Alamy / **13** *God Closing the Door of Noah's Ark*, French, *c.* 1190. Stained glass window, 76.2 × 37.8 (30 × 14 ⅞). The Cloisters Collection, 1925. The Metropolitan Museum of Art, New York (25.120.394a) / **15** *The Dove and the Raven* (detail), watercolour, *Holkham Bible Picture Book*, England, *c.* 1320–30. Photo The British Library archive/ Bridgeman Images / **16** Andy Goldsworthy, *Sycamore leaves edging the roots of a sycamore tree*, Berrydown Foundation, Hampshire, 1 November 2013. © Andy Goldsworthy / **18–19** Cai Guo-Qiang, *Heritage*, 2013. 99 life-sized replicas of animals. Polystyrene, gauze, resin and hide. QAGOMA Collection. Commissioned 2013 with funds from the Josephine Ulrick and Win Schubert Diversity Foundation through and with the assistance of the Queensland Art Gallery, Brisbane. Photo Natasha Harth, courtesy Queensland Art Gallery. © Cai Guo-Qiang / **20** Dasha Plesen, *Orange and blue*, 2021 (detail). Nutrient, pigments, spores, 9 (3 ⅝). © Dasha Plesen / **22** Charles A. Kofoid and Olive Swezy's *The Free-Living Unarmored Dinoflagellata*, Berkeley, University of California Press, 1921 / **23** Courtesy Noé & Christian Sardet/ Plankton Chronicles / **24** Archive of Josef Reischig. Courtesy the Estate of Doc. RNDr. Josef Reischig, CSc. / **25a** Photo Frank Fox/Science Photo Library / **25b** Photo Jan Hinsch/Science Photo Library / **26, 27** Courtesy Antonio Guillén/Proyecto Aqua/Water Project / **28** Jan C. Schlegel, *Phytoplankton, Life on Earth* series, 2024. Handmade platinum prints. © Jan C. Schlegel / **29** Anne Arden McDonald, *Lichen Alphabet*, 2014. Collage, silver gelatin photograph, 38.1 × 27.9 (15 × 11). © Anne Arden McDonald / **30** Katie Scott and Ester Gaya's *Fungarium*, *Welcome to the Museum* series. First printed by Big Picture Press, an imprint of Bonnier Books UK, 2019. Illustration © 2019 Katie Scott / **31** Eliot Porter, *Lichen Behind Skogar, Iceland, July 6, 1972*. Dye imbibition print, 26.8 × 20.6 (10 9/16 × 8 ⅛). Bequest of the artist. © 1990 Amon Carter Museum of American Art, Fort Worth (P1990.51. 2777.1) / **32** *Marvels of the Universe*, London, Hutchinson & Co., 1912 / **33** Koloman Moser, *Schwämme*, 1899 (detail). Decorative fabric, wool, silk and cotton, 94.5 × 119 (37 ¼ × 46 ⅞). Johann Backhausen & Söhne, Vienna, 1900. MAK, Wien (WI 56) / **34** Maruyama Okyo, *Mushrooms and Flowers*, from his sketchbook, *c.* 1770 (detail). Colour on paper, 38.5 × 44.5 (15 ¼ × 17 ⅝). Tokyo National Museum (C0045514). Photo TNM Image Archives / **35a** J. V. von Krombholz's *Naturgetreue Abblidungen und Beschreibungen der essbaren, schädlichen und verdächtigen Schwämme*, Praga, 1831–46 / **35b** Philippe Ferdinand de Hamilton, *Studies of Fungi*, *c.* 1736. Oil on paper laid on panel, 22.2 × 30.2 (8 ¾ × 12). Private Collection / **36–7** Peter Fischli and David Weiss, *Untitled (Flowers)*, 1997–98. Inkjet print, 74 × 107.4 (29 ⅛ × 42 ⅓). © Peter Fischli and David Weiss, Courtesy Matthew Marks Gallery / **38** Photo Keith Burdett/Alamy / **39** After Prieur, *Champignons Suspects*, *c.* 1827. Lithograph with watercolour by A. Cornillon. The Wellcome Collection, London (20596i). / **40** Jan Davidsz de Heem, *Vase of Flowers*, *c.* 1660 (detail). Oil on canvas, 69.6 × 56.5 (27 ⅜ × 22 ¼). Andrew W. Mellon Fund. National Gallery of Art, Washington, DC (1961.6.1). / **42** Andrew Zuckerman, *Seeds (Fast Lady 141, Honeypatch Squash 206 45, Beauregarde 383, Kogniut 218, Cherry Belle Radish 211, Golden Boy 260)*, 2024. Archival pigment prints, 152.4 × 114.3 (60 × 45) each. © 2025 Andrew Zuckerman / **43** Albarrán Cabrera, *The Mouth of Krishna #60198*, 2016. Pigments, gampi paper and gold leaf, 48 × 32 (19 × 12 ⅝). © Albarrán Cabrera / **44** Jonas Frei, *Walnut cross-sections (Juglans and Carya genera)*, photograph, 2024. © Jonas Frei / **45a** Pieter Bruegel the Elder, *The Harvesters*, 1565. Oil on wood, 119 × 162 (46 ⅞ × 63 ¾). Rogers Fund, 1919. The Metropolitan Museum of Art, New York (19.164) / **45b** Eliot Hodgkin, *Two Dead Leaves No. 2*, 1969. Tempera on board, 16 × 16 (6 ⅜ × 6 ⅜). Private Collection. © The Estate of Eliot Hodgkin / **46** Louis van Houtte's *Flore des serres et des jardins de l'Europe*, Gand, chez L. van Houtte, éd., 1845 / **47** Carl Julius Fritzsche's *Ueber den Pollen*, St Petersburg, Kais. Academie der Wissenschaften, 1837 / **48** Gustav Klimt, *Sunflower*, 1907–8. Oil and gold leaf on canvas, 110 × 110 (43 ⅜ × 43 ⅜). Bequest of Peter Parzer, Vienna, 2012. Oberes Belvedere, Wien (10500) / **49a** Claude Monet, *The Water Lily Pond*, 1900. Oil on canvas, 90.2 × 92.7 (35 ½ × 36 ½). Given in memory of Governor Alvan T. Fuller by the Fuller Foundation. Museum of Fine Arts, Boston (61.959) / **49b** Jean-François Millet, *Dandelions*, 1867–68. Pastel on tan wove paper, 40.6 × 50.2 (16 × 19 ¾). Gift of Quincy Adams Shaw through Quincy Adams Shaw, Jr., and Mrs. Marian Shaw Haughton. Museum of Fine Arts, Boston (17.1524) / **50a** Henri Fantin-Latour, *Plate of Peaches*, 1862. Oil on canvas, 18.1 × 32 (7 ⅛ × 12 ⅝). M. Theresa B. Hopkins Fund. Museum of Fine Arts, Boston (60.792) / **50b** Édouard Manet, *Plums*, *c.* 1880. Oil on canvas, 19.2 × 25 (7 9/16 × 9 13/16). Gift of Ann Eden Woodward. The Museum of Fine Arts, Houston (79.182) / **51al** Ellen Isham Schutt, *Prunus persica: Elberta Cling*, 1908. USDA Pomological Watercolor Collection, National Agricultural Library Special Collections, Beltsville, MD (POM00005237) / **51ac** Deborah Griscom Passmore, *Malus domestica: Black Sweet*, 1894. USDA Pomological Watercolor Collection, National Agricultural Library Special Collections, Beltsville, MD (POM00002594) / **51ar** J. Marion Shull, *Ananas comosus: Pineapple*, 1919. USDA Pomological Watercolor Collection, National Agricultural Library Special Collections, Beltsville, MD (POM00007369) / **51cl** Elsie E. Lower, *Diospyros*, 1908. USDA Pomological Watercolor Collection, National Agricultural Library Special Collections, Beltsville, MD (POM00001076) / **51c** Deborah Griscom Passmore, *Fragaria: Nick Ohmer*, 1903. USDA Pomological Watercolor Collection, National Agricultural Library Special Collections, Beltsville, MD (POM00004359) / **51cr** Amanda Newton, *Persea: Ultimate*, 1916. USDA Pomological Watercolor Collection, National Agricultural Library Special Collections, Beltsville, MD (POM000046070) / **51bl** Amanda Newton, *Pyrus communis: Easter*, 1911. USDA Pomological Watercolor Collection, National Agricultural Library Special Collections, Beltsville, MD (POM00006974) / **51bc** Elsie E. Lower, *Citrus nobilis: King*, 1911. USDA Pomological Watercolor Collection, National Agricultural Library Special Collections, Beltsville, MD (POM00006526) / **51br** Amanda Newton, *Punica granatum*. USDA Pomological Watercolor Collection, National Agricultural Library Special Collections, Beltsville, MD (POM00007510) / **52** Walter Crane, 'Francesca' wallpaper design, colour woodblock print, 1902 (detail). Photo The Stapleton Collection/Bridgeman Images / **53** George Brookshaw's *Pomona Britannica*, London, T. Bensley, 1812. Rare Book Division, The New York Public Library (b13640519) / **54** Ebenezer Emmons's *Natural History of New York*, Albany, Appleton & co. and Wiley & Putnam, 1842–94 / **55** Paul Cézanne, *Un coin de table (A Table Corner)*, *c.* 1895. Oil on canvas, 47 × 56.5 (18 ½ × 22 ¼). The Barnes Foundation, Philadelphia (BF711) / **56** Ernestine Eberhardt Zaumseil, *Branches and Vines Quilt*, *c.* 1875. Cotton, silk and wool, 223.5 × 218.4 (88 × 86). Gift of George E. Schoellkopf, 2013. The Metropolitan Museum of Art, New York (2013.958) / **57a** Kubo Shunman, *Rice Plant and Butter-Burs*, 19th century. Woodblock print, ink and colour on paper, 21 × 18.7 (8 ¼ × 7 ⅜). H. O. Havemeyer Collection, Bequest of Mrs. H. O. Havemeyer, 1929. The Metropolitan Museum of Art, New York (JP2166) / **57b** Charles K. Wilkinson, *Nebamun Supervising Estate Activities, Tomb of Nebamun* (detail), 1928 CE; original *c.* 1400–1352 BCE. Tempera on paper, facsimile, 99 × 130 (39 × 51 3/16). Rogers Fund, 1930. The Metropolitan Museum of Art, New York (30.4.57) / **58–59** René Burri, *Former Summer Palace. Dead Lotus Flowers on Kunming Lake*, Beijing, 1964. © René Burri/Magnum Photos / **60** Anna Atkins's *Photographs of British Algae. Cyanotype Impressions*, 1843–53. Spencer Collection, The New York Public Library (b11861683) / **61** Piet Mondrian, *Red Amaryllis with Blue Background*, *c.* 1907. Watercolour on paper, 46.5 × 33 (18 ⅜ × 13). The Sidney and Harriet Janis Collection. Photo The Museum of Modern Art, New York/Scala, Florence / **62a** *The Orchid World*, monthly illustrated journal, ed. Gurney Wilson, Sussex, Haywards Heath, 1911 / **62b** *Omoto Nayoro*, edited by Mizuno Tadataka and illustrated by Sekine Untei, 1832. National Diet Library, Japan / **63** Nowruz still life, Iran, Qajar, 19th century. Oil on canvas, 91 × 77 (35 ⅞ × 30 ⅜). Private Collection / **64** *Erbario*, Italy, *c.* 1400. Lawrence J. Schoenberg Collection, University of Pennsylvania, Philadelphia (LJS 419) / **65** Robert John Thornton, 'The Night-Blowing Cereus', 1807. Hand-coloured mezzotint, 57 × 48 (22 ½ × 19). The Miriam and Ira D. Wallach Division of Art, Prints and Photographs: Print Collection, The New York Public Library (b13998666) / **66a** Georgia O'Keeffe, *Oriental Poppies*, 1927. Oil on canvas, 76.2 × 101.6 (30 × 40). Weisman Art Museum, University of Minnesota, Minneapolis (1937.1). © 2026 Georgia O'Keeffe Museum/DACS / **66b** Johann Jakob Walther, *Poppies*, 1654. Watercolour. Victoria and Albert Museum, London (9175.27) / **67** Rachel Ruysch, *Flowers in a Glass Vase*, 1704. Oil on canvas, 83.8 × 67 (33 × 26 ⅜). Detroit Institute of Arts (1995.67) / **68** Gareth McConnell, *Dream Meadow LI*, 2024. Photograph, 61 × 48.9 (24 × 19 ¼). Courtesy Gareth McConnell and Sorika / **69** Gareth McConnell, *Dream Meadow XXIV*, 2021. Photograph, 61 × 48.9 (24 × 19 ¼). Courtesy Gareth McConnell and Sorika / **70** Carleton E. Watkins, *Section of the Grizzly Giant with Galen Clark*, Mariposa Grove, Yosemite, 1865–66. Albumen silver print from glass negative, 52.2 × 40.6 (20 9/16 × 16). Elisha Whittelsey Collection, 1972. The Metropolitan Museum of Art, New York (1972.643.3) / **71al** Karl Blossfeldt, *Acanthus mollis*, 1928. Gelatin silver print, 25.8 × 19.9 (10 3/16 × 7 13/16). J. Paul Getty Museum, Los Angeles (84.XM.142.2) / **71ar** Karl Blossfeldt, *Scabiosa columbaria*, 1928. Gelatin silver print, 31.2 × 24.2 (12 ⅜ × 9 ⅝). Rijksmuseum, Amsterdam (RP-F-2008-51-2170-91) / **71bl** Karl Blossfeldt, *Papaver orientalis*, 1928–32. Gelatin silver print, 25.9 × 19.4 (10 3/16 × 7 5/8). J. Paul Getty Museum, Los Angeles (84.XM.142.6) / **71br** Karl Blossfeldt, *Adiantum pedatum (Maidenhair Fern)*, 1928. Gelatin silver print, 31.2 × 24.2 (12 ⅜ × 9 ⅝). Rijksmuseum, Amsterdam (RP-F-2008-51-2170-55) / **72a** Odilon Redon, *Nasturtiums*, 1905. Oil on canvas, 50.2 × 73 (19 ¾ × 28 ¾). Bequest of Edith Malvina K. Wetmore. Yale University Art Gallery, New Haven, CT (1966.79.10) / **72b** Garden Frescoes (detail), Villa of Livia, Rome, 38–20 BCE. Palazzo Massimo, Museo Nazionale Romano, Rome / **73** Paul Klee, *Sky Blossoms above the Yellow House (The Chosen House)*, 1917. Watercolour and gouache on primed linen on paper, bordered with watercolour, on cardboard, 23 × 15 (9 ⅛ × 6). Photo bpk/National-galerie, SMB, Museum Berggruen/Jens Ziehe / **74** Margaretha de Heer, *A Red Cabbage, a Snail, a Butterfly, a Dragonfly, a Bee and a Wood Louse, in a Landscape*, 17th century. Oil on panel, 39.2 × 28.5 (15 ½ × 11 ¼). Private Collection / **75al** Barbara Regina Dietzsch, *A Buttercup with a Butterfly and an Insect*, 18th century. Bodycolour and gum arabic on vellum, gold framing lines, 28.8 × 20.7 (11 ⅜ × 8 ⅛). Private Collection / **75ar** Barbara Regina Dietzsch (attribution uncertain), *Flowering 'Queen of the Night' (Cactus grandiflorus)*, 18th century. Watercolour, gouache, gum arabic varnish (partially) on parchment, 28.6 × 21.1 (11 ⅜ × 8 ⅜). Staatliche Museen zu Berlin, Kupferstichkabinett/Dietmar Katz / **75bl** Barbara Regina Dietzsch (after a design by Johann Jakob Dietzsch), *Cabbage Head with Pearl-bordered Fritillary Butterfly and Caterpillar*, before 1762. Watercolour and gouache on black primed parchment, gold border line, 28.7 × 20.6 (11 ⅜ × 8 ⅛). Staatliche Museen zu Berlin, Kupferstichkabinett/Reinhard Wittich / **75br** Barbara Regina Dietzsch, *A Thistle Branch with a Butterfly, Dragonfly, Ladybird and Spider*, 18th century. Gouache on paper, 28 × 20.8 (11 × 8.2). Private Collection / **76** Jethro Buck, *These Great Trees are Prayers*, 2020. 19 carat gold on hemp paper, 27.3 × 24.1 (10 ¾ × 9 ½). Courtesy Jethro Buck / **77a, c** Junpo's *Picture Book of Chrysanthemums*, Hikone, Yao Yasutoyo et al., 1691. Library of Congress, Washington, DC (2021666416) / **77b** Samuel Palmer, *Oak Trees, Lullingstone Park*, 1828. Pen, brush and brown ink with graphite, watercolour and gouache, heightened with gum arabic and scraping out, on grey wove paper, 29.5 × 46.8 (11 ⅝ × 18 ½). Purchased 1937. National Gallery of Canada (4357) / **78a** Winslow Homer, *Palm Trees, Florida*, 1904. Watercolour over graphite on paper, 50.2 × 35.2 (19 ¾ × 13 ⅞). Bequest of John T. Spaulding. Museum of Fine Arts, Boston (48.731) / **78b** Pierre Puvis de Chavannes, *Sous-bois* ('Undergrowth'), *c.* 1870–90. Watercolour, pencil on paper, 16 × 25 (6 ⅜ × 9 ⅞). MAN Museo d'Arte Provincia di Nuoro / **79** Albrecht Dürer, *Great Piece of Turf*, 1503. Watercolour and bodycolour on vellum, 40.3 × 31.1 (15 ⅞ × 12 ¼). The Albertina Museum, Vienna / **80** Pierrot Men, *Allée des Baobabs, Morondava, Madagascar*, digital photograph, 2012. © Pierrot Men / **81** Hengki Koentjoro, *Mangrove*, digital photograph, 2017. © Hengki Koentjoro / **83** Emily Carr, *Scorned as Timber, Beloved of the Sky*, 1935. Oil on canvas, 111.8 × 68.4 (44 ⅛ × 27). Emily Carr Trust. Vancouver Art Gallery (VAG 42.3.15) / **84** Cedric Morris, *Crisis*, 1938. Oil on canvas, 92.5 × 122.5 (36 ½ × 48 ¼). Private Collection. Photo Lyon & Turnbull/Bridgeman Images. © Estate of Cedric Morris. All rights reserved 2026/Bridgeman Images / **86** After Hashime Murayama, *Two Portuguese man o' war (Physalia arethusa) drifting half-submerged in the sea*, 20th century. Colour line block print, 30.4 × 19 (12 × 7 ½). Wellcome Collection, London (42029i) / **88a** Carved intaglio gemstone, 1st–3rd century CE. Carnelian, 1 × 0.8 (⅜ × ⁵⁄₁₆). Gift of Ambasador and Mrs. William L. Eagleton, Jr., B. A. 1948. Yale University Art Gallery, New Haven, CT (1986.100.6) / **88b** *The Aberdeen Bestiary*, England, *c.* 1200. Aberdeen University Library (MS 24, fol. 63r) / **89a** Incense burner in form of a praying mantis, Japan, 19th century. Iron, gold, lining of silver, 8.9 × 15.2 (3 ⅓ × 6). Howard Mansfield Collection, Gift of Howard Mansfield, 1936. The Metropolitan Museum of Art, New York (36.120.669a) / **89b** Jean Corbechon's *Livre des Propriétés des Choses*, 15th century. Bibliothèque Louis Aragon, Amiens (Ms. 399) / **90a** Peter Birkhäuser, *Moth*, 1944–5. Watercolour and crayon, 45 × 46 (17 ¾ × 18 ⅛). Foundation Birkhäuser-Oeri, www.birkhaeuser-oeri.ch / **90b** William Gouw Ferguson, *Six Butterflies and a Moth on a Rose Branch*, *c.* 1690. Oil on panel, 20.5 × 26 (8 ⅛ × 10 ¼). Purchased with a contribution from the Patrons of the National Galleries of Scotland, 2013. National Galleries of Scotland, Edinburgh (NG 2867) / **91** Ambrosius Bosschaert the Elder, *Flower Still Life*, 1614 (detail). Oil on copper, 30.5 × 38.9 (12 × 15 ⁵⁄₁₆). J. Paul Getty Museum, Los Angeles (83.PC.386) / **92** Joan Miró, *Flight of the Dragonfly in Front of the Sun*, 1968. Oil on canvas, 173.9 × 243.8 (68 ⁷⁄₁₆ × 96). Collection of Mr. and Mrs. Paul Mellon. National Gallery of Art, Washington, DC (1983.1.23). © 2026 Successió Miró/ADAGP, Paris and DACS London / **93, 96** Paul A. Robert's *Les Papillons dans la Nature*, Neuchâtel, Éditions Delachaux and Niestlé, 1934 / **95** Levon Biss, *Phasmid (Stick Insect) Eggs*, 2022. Photography, 152.4 × 152.4 (60 × 60). © Levon Biss / **96** Fabre's *Book of Insects*, New York, Dodd, Mead and Company, 1921 / **97** Martin Johnson Heade, *Blue Morpho Butterfly*, *c.* 1863–64. Oil on canvas, 31.1 × 25.4 (12 ¼ × 10). Crystal Bridges Museum of American Art, Bentonville (2006.92) / **98** Albrecht Dürer, *Stag Beetle*, 1505. Watercolour and gouache, 14.1 × 11.4 (5 ⁹⁄₁₆ × 4 ½). J. Paul Getty Museum, Los Angeles (83.GC.214) / **99a** Vincent van Gogh, *Sketch of Three Cicadas*, Saint-Rémy-de-Provence, July 1889. Pen and ink on paper, 20.1 × 17.8 (8 × 7 ⅛). Van Gogh Museum, Amsterdam (Vincent van Gogh Foundation) (F1445) / **99b** Qi Baishi, *Grasshopper*, *Album of Plants and Insects Painting*, 1924. Ink and colour on paper, 12.9 × 18.5 (5 ⅛ × 7 ⅜). National Art Museum of China, Beijing / **100–101** Kobayashi Kiyochika, *Fireflies at Ochanomizu*, *c.* 1880. Colour woodblock print, 21.2 × 33.1 (8 ⅜ × 13 ⅛). Gift of Carl Holmes. Los Angeles County Museum of Art (M.71.100.82) / **102** Tsukioka Yoshitoshi, *New Publication of Selected Insects*, Japan, 19th century. Colour woodblock print, 33.5 × 22.3 (13 ¼ × 8 ⅞). Gift of Hideo Tsunoda. Los Angeles County Museum of Art (M.80.131) / **103** Émile-Allain Séguy's *Insectes*, Paris, Éditions Duchartre et Van Buggenhoudt, 1929 / **104** Louise Bourgeois, *Spider*, 1995. Drypoint on vellum, 40.5 × 30 (16 × 11 ⅞). Private Collection. © 2026 The Easton Foundation/VAGA at ARS, NY and DACS, London / **105** J. O. Westwood's *The Cabinet of Oriental Entomology*, London, William Smith, 1848 / **106** Vincent van Gogh, *Giant Peacock Moth*, Saint-Rémy-de-Provence, May–June 1889. Oil on canvas, 33.5 × 24.5 (13 ¼ × 9 ¾). Van Gogh Museum, Amsterdam (Vincent van Gogh Foundation) (F0610) / **107a** Jan Augustin van der Goes, *Spider*, 1690–1700. Gouache and watercolour on parchment, 4.1 × 5.1 (1 ⅝ × 2 ⅛). Purchased with the support of the Vereniging Rembrandt, Rijksmuseum, Amsterdam (RP-T-1884-A-330F) / **107b** Unknown artist, *Ants*, *Northumberland Bestiary*, 1250–60. Pen-and-ink drawings tinted with bodycolour and translucent washes, leaf 21 × 15.7 (8 ⅜ × 6 ¼). J. Paul Getty Museum, Los Angeles (Ms. 100 (2007.16)) / **108** Ilse Bing, German, *Untitled (Spiderweb)*, 1951. Gelatin silver print, 50.8 × 40 (20 × 15 ¾). Bequest of Ilse Bing Wolff. The Nelson-Atkins Museum of Art, Kansas City (2001.9.30). Image courtesy of Nelson-Atkins Digital Production & Preservation. Courtesy Edwynn Houk Gallery, NY © Estate of Ilse Bing / **109a** Sir John Lubbock's *Ants, Bees and Wasps: A Record of Observations on the Habits of the Social Hymenoptera*, New York, Appleton, 1897 / **109b** Saitō Shōshū, 'Snail and Poem by Buson' from the series *Postcards of Haikai Poetry* (*Haikai ehagaki*). Colour lithograph, embossing, ink and metallic pigment on card stock, 8.8 × 13.8 (3 ½ × 5 ½). Leonard A. Lauder Collection of Japanese Postcards. Museum of Fine Arts, Boston (2002.1092) / **110** Fabre's *Book of Insects*, New York, Dodd, Mead and Company, 1921 / **111a** Ju Chao, *Silkworm*, China, Qing Dynasty, 1859. Folding fan mounted as an album leaf, ink and colour on alum paper, 18.4 × 53.3 (7 ¼ × 21). Gift of Robert Hatfield Ellsworth, in memory of La Ferne Hatfield Ellsworth, 1986. The Metropolitan Museum of Art, New York (1986.267.41) / **111b** Al-Qazwini's *Ajāʾib al-Makhlūqāt va-Gharāʾib al-Mawjūdāt* ('The Wonders of Creation'), *c.* 1540–45. Spencer Collection, The New York Public Library (Persian MS. 49) / **112** Mōri Baien's *Baien Kaifu*, Japan, 19th century. National Diet Library, Japan / **113** William Carmichael McIntosh's *A Monograph of the British Marine Annelids*, London, The Ray Society, 1873–1923 / **114** *Cynthia microcosmus*, 19th century. Print, 25 × 20 (9 ⅞ × 7 ⅞). Iconographia Zoologica Print Collection. Allard Pierson, University of Amsterdam (UBA274) / **115** George Frederick Kunz and Charles Hugh Stevenson's *The Book of the Pearl*, New York, The Century Co., 1908 / **116** Anne Vallayer-Coster, *Panaches de mer, lithophytes et coquilles* ('Seaweed, Lithophytes and Seashells'), 1769. Oil on canvas, 130 × 97 (51 ¼ × 38 ¼). Musée du Louvre, Paris (RF 1992 410) / **117, 118** 19th-century prints. Iconographia Zoologica Print Collection. Allard Pierson, University of Amsterdam (UBA274) / **119** Alexandre-Isidore Leroy de Barde, *Choix de Coquillages* ('Selection of Shells'), *c.* 1803–10. Gouache and watercolour on paper laid down on canvas, 125 × 90 (49 ¼ × 35 ½). Musée du Louvre, Paris (INV 23689) / **120** Ernst Haeckel's *Kunstformen der Natur*, Leipzig, Bibliographisches Institut, 1899–1904 / **121** Richard Collins, *A jellyfish of the Aequoreidae family feeding on a ctenophore*, digital photograph, 2023. © Richard Collins / **122** Helen Walne, *Two Stars*, Cape Town, South Africa, 2023. Courtesy Helen Walne / **123** William Saville-Kent's *The Great Barrier Reef of Australia*, London, W. H. Allen Co., 1893 / **124a** A. E. d'Audebert de Férussac and Alcide d'Orbigny's *Histoire Naturelle Générale et Particulière des Céphalopodes Acétabulifères*, Paris, A. Lacour, 1835 / **124b** Albertus Seba's *Locupletissimi Rerum Naturalium Thesauri Accurata Descriptio*, Amsterdam, J. Wetsten, G. Smith & Jansson-Waesberg, 1734–65 / **125** Yashima Gakutei, *Crabs Near the Water's Edge*, *c.* 1830. Woodblock print, ink and colour on paper, 20.3 × 18.6 (8 × 7 ⅜). Rogers Fund, 1919. The Metropolitan Museum of Art, New York (JP1138) / **126** Helen Walne, *Losing Time*, Knysna, South Africa, 2021. Courtesy Helen Walne / **128** François-Nicolas Martinet and Edme Billardon-Sauvigny's *Histoire naturelle des dorades de la Chine*, Paris, Louis Jorry, 1780 / **129** Mrs Charles Meredith's *Tasmanian Friends and Foes*, London, M. Ward, 1881. Wellcome Collection, London / **130a** *Oki-no-Kuni Sanbutsu Ezu Chūsho* ('Oki National Products Illustration Notes'), Japan, 1735. National Diet Library, Japan / **130b** Utagawa Hiroshige, *Trout*, *c.* 1832–33. Woodblock print, ink and colour on paper, 25.5 × 37.1 (10 ⅛ × 14 ⅝). Bequest of Richard P. Gale. Minneapolis Institute of Art (74.1.258) / **131** Étienne-Jules Marey, *Undulations of the Fins of a Skate Viewed from the Side*, 1891–92. Watercolour and ink after a chronophotographic film, 65.5 × 48 (25 ⅞ × 19). Collection La Cinémathèque française, Paris (No. ME 165) / **132–3** Henri Matisse, *Polynésie, la mer*, 1946. Gouache, cut-and-pasted papers mounted on canvas, 196 × 314 (77 ¼ × 123 ⅝). Photo Centre Pompidou, MNAM-CCI, Paris/Dist. GrandPalaisRmn/Bertrand Prévost. © 2026 Succession H. Matisse/DACS / **134** Al-Qazwini's *Ajāʾib al-Makhlūqāt va-Gharāʾib al-Mawjūdāt* ('The Wonders of Creation'), 17th century. University of St Andrews Library (ms32(o)) / **135** Japetus Steenstrup and Christian Frederik Lütken's *Spolia Atlantica. Bidrag til Kundskab om Klump*, Copenhagen, Bianco Lunos (F. Dreyer), 1898 / **136** Oskar Laske, *Fischpredigt* ('Fish Sermon'), 1914. Colour lithograph, 47.4 × 57.7 (18 ¾ × 22 ¾). Wien Museum (227809) / **137** Louis Renard's *Poissons, Ecrevisses et Crabes, de Diverses Couleurs et Figures Extraordinaires*, Amsterdam, L. Renard, 1719 / **138–9** Paul Klee, *Fish Magic*, 1925. Oil and watercolour on canvas on panel, 77.2 × 98.4 (30 ⅜ × 38 ¾). The Louise and Walter Arensberg Collection, 1950. Philadelphia Museum of Art (1950-134-112) / **140** Jonathan Couch's *A History of the Fishes of the British Islands*, London, Groombridge, 1868–69 / **141** Albarrán Cabrera, *The Mouth of Krishna #615*, 2016. Photography, pigments, gampi paper and gold leaf, 48 × 32 (19 × 12 ⅝). © Albarrán Cabrera / **142a** Bowl, Persian, 1275–1350 CE. Stonepaste, painted in black under turquoise glaze. The Hossein Afshar Collection. Photo Museum of Fine Arts, Houston/Bridgeman Images / **142b** Francis de Castelnau's *Expédition dans les parties centrales de l'Amérique du Sud*, Paris, P. Bertrand, 1850–59 / **143a** Utagawa Hiroshige, *Kochi and nasu* ('two flatheads and flowering eggplant'), from the *Uozukushi* ('Every Variety of Fish') series, *c.* 1832–33. Woodblock print, 25.1 × 36.2 (10 × 14 ⅜). Private Collection / **143b** Al-Qazwini's *Ajāʾib al-Makhlūqāt va-Gharāʾib al-Mawjūdāt* ('The Wonders of Creation'), *c.* 1540–45. Spencer Collection, The New York Public Library (Persian MS. 49) / **144** Charles Frederick Holder's *Living Lights. A Popular Account of Phosphorescent Animals and Vegetables*, London, Sampson Low, Marston, 1887 / **145** Robert Longo, *Untitled (Shark 7)*, 2008. Charcoal drawing on paper, mounted on aluminium board. 228.5 × 152 (90 × 59 ⅞). Private Collection. © 2026 ARS, NY and DACS, London / **146a** Mark Catesby's *The Natural History of Carolina, Florida and the Bahama Islands*, London, 1731–43. Wellcome Collection, London (EPB/F/963.v2) / **146b** Hugo de Fouilloy's *De avibus* ('The Aviary'), Flanders, *c.* 1270. J. Paul Getty Museum, Los Angeles (MS. Ludwig XV 3) / **147** Edward Donovan's *The Natural History of British Fishes*, London, Donovan and F. & C. Rivington, 1802. Rare Book Division, The New York Public Library (b11399861) / **148** Carp in waves gift cover (*fukusa*), Japan, Meiji period, 1868–1912. Silk, embroidered, metallic thread, 87.3 × 71.8 (34 ⅜ × 28 ¼). Gift of Mr. and Mrs. J. H. Wade. The Cleveland Museum of Art (1916.1324) / **149a** Marcus Elieser Bloch's *Ichtyologie, ou, Histoire Naturelle, Générale et Particulière des Poissons*, Berlin, the Author and François de la Garde, 1787. Rare Book Division, The New York Public Library (b13637086) / **149b** *Keramic Studio* magazine, Keramic Studio Pub. Co. Syracuse, NY, April 1913. Smithsonian Libraries and Archives, Washington DC / **150** Michael Andrews, *School IV: Barracuda under Skipjack Tuna*, 1978. Acrylic on canvas, 175.5 × 175.5 (69 ⅛ × 69 ⅛). Private

Collection. © The estate of Michael Andrews/Tate Images / **151** Louis Renard's *Poissons, Ecrevisses et Crabes, de Diverses Couleurs et Figures Extraordinaires*, Amsterdam, L. Renard, 1719 / **152** Jean Painlevé, *Buste d'Hippocampe* ('Seahorse Bust'), silver gelatin print, 1933. © 2026 ADAGP, Paris and DACS, London / **153a** Photo Krzysztof Wiktor/Adobe Stock / **153b** UME, *Aquarium*, USA, 2008. © UME/Magnum Photos / **155** Helen Walne, *Golden Curl*, Cape Town, South Africa, 2024. Courtesy Helen Walne / **156** Mosaic of marine life, House of the Geometric Mosaics, Pompeii, *c.* 100 BCE. 88 × 88 (34 ¾ × 34 ¾). Museo Archeologico Nazionale di Napoli (MANN 120177). Photo courtesy Carole Raddato / **158** Charles Francis Annesley Voysey (design), *Snake Amongst Flowers*, embroidered by unknown, *c.* 1895. Silk and linen, 104.1 × 69.9 (41 × 27 ½). Purchased with funds from MaryLou Boone, Heather and Paul Haaga, and Anne and Jim Rothenberg in honour of John Murdoch's service to the Huntington upon the occasion of his retirement. The Huntington, San Marino, CA (2012.25) / **160** Mark Catesby, *The Land Frog, the Water Frog and the Green Tree Frog*, *c.* 1722–26. Watercolour and bodycolour heightened with gum arabic, over pen and ink and graphite, 37.4 × 26.2 (14 ¾ × 10 ⅜). Royal Collection Trust (RCIN 926023) / **161a** Hans Hoffmann, *Bullfrog (Rana esculenta L.)*, 1575–85. Brush, watercolour, white heightening on paper, 6 × 6.5 (2 ⅜ × 2 ⅝). Museum of Fine Arts, Budapest (181) / **161b** Winslow Homer, *Mink Pond*, 1891. Transparent and opaque watercolour and graphite on heavy white wove paper, 35.2 × 50.8 (13 ⅞ × 20). Bequest of Grenville L. Winthrop. Harvard Art Museums/Fogg Museum Cambridge, MA (1943.304) / **162a** Gerrit Schouten, *Rana paradoxa*, 1818. Watercolour, 19.3 × 20.9 (7 ⅝ × 8 ¼). Iconographia Zoologica Print Collection. Allard Pierson, University of Amsterdam (UBA274) / **162b** Matt Clancy, *Frill-necked Lizard (Chlamydosaurus kingii)*, Kakadu National Park, Australia, 2013. © Matt Clancy / **163** Taigadō Giryō, *Frog*, 1830–65. Ink on paper, 93 × 56.2 (36 ⅝ × 22 ⅛). Cotsen 1985 Trust, 2018. Collection of the Honolulu Museum of Art (2018-16-02) / **164** *Draco viridis*, 1700–1880. Print, 26.9 × 21.4 (10 ⅝ × 8 ½). Iconographia Zoologica Print Collection. Allard Pierson, University of Amsterdam (UBA274) / **165a** Alfred Edmund Brehm's *Brehms Tierleben. Allgemeine kunde des Tierreichs*, Leipzig, Bibliographisches Institut, 1890–93 / **165b** Richard Sternfeld, *Die Reptilien und Amphibien mitteleuropas*, Leipzig, Quelle & Meyer, 1912 / **166** Yamamoto Masao, #853, *Nakazora* series, 1999. Gelatin silver print with hand applied ink. © Yamamoto Masao / **167** Nakamura Hōchū, *Three Turtles*, from the *Kōrin Gafu* ('Kōrin Picture Album'), 1802. Woodblock print, ink and colour on paper, 27.1 × 39 (10 ¾ × 15 ⅜). Gift of Robert S. and Betsy G. Feinberg. Harvard Art Museums/Arthur M. Sackler Museum (2018.349.4) / **168** Ernst Haeckel's *Kunstformen der Natur*, Leipzig, Bibliographisches Institut, 1899–1904 / **169** Dick Nguleingulei Murrumurru, *Freshwater Turtle*, Liverpool River Escarpment, West Arnhem Land, Australia. Aboriginal bark painting, natural earth pigments, 67 × 36 (26 ½ × 14 ¼). Photo courtesy Tooveys Antique & Fine Art Auctioneers & Valuers. © 2026 Dick Ngulei Ngulei Murrmurra/Copyright Agency. Licensed by DACS / **170** Wilhelm Kuhnert, *A Tree Frog, Water Frog and Fire Salamander*, 19th century. Oil on board, 29.8 × 19.3 (11 ¾ × 7 ⅝). Photo Alamy / **171** William Beebe's *Galapagos: World's End*, New York, G. P. Putnam's Sons, 1924 / **172** Maria Sibylla Merian's *Dissertatio de generatione et metamorphosibus insectorum Surinamensium*, Amsterdam, Joannes Oosterwyk, 1705 / **173a** Photo Paul Starosta/Getty Images / **173b** Rachel Ruysch, *Fruit Still Life with Stag Beetle and Nest*, 1717 (detail). Oil on canvas, 65.2 × 54.5 (25 ¾ × 21 ½). Staatliche Kunsthalle Karlsruhe (377) / **174** *The Rochester Bestiary*, *c.* 1230. The British Library, London (Royal MS 12 F XIII) / **175a** Anastasia Samoylova, *Gator*, 2017, from the *FloodZone* series. © Anastasia Samoylova / **175b** Kaballa, *Untitled*, *c.* 1940s–50s. Oil on paper, 37.9 × 51.5 (15 × 20 ⅜). Anonymous Gift. Photo Franko Khoury. National Museum of African Art, Smithsonian Institution (2002-11-4). © The Artist's Estate / **176** Johanna Helena Herolt (attr.), *Lizards, Insects, and Plants, Including Clover and Forget-Me-Nots*, 17th century. Transparent and opaque watercolour on parchment, 32.3 × 25.3 (12 ¾ × 10). Purchased as the gift of Mrs. Vincent Astor. The Morgan Library & Museum, New York (1973.3) / **177** Mansur, *A Chameleon*, 1612. Brush and ink with green bodycolour on discoloured paper, 11 × 13.8 (4 ⅜ × 5 ½). The Royal Collection Trust, London (RCIN 912081) / **178–9** Carstian Luyckx (attr.), *Forest with Snakes and Frogs*, 17th century. Oil on canvas, 52 × 64.5 (20 ½ × 25 ½). Private Collection. Photo Courtesy Koller Auctions, Zurich / **180** Sebastião Salgado, *A Marine Iguana (Amblyrhynchus cristatus), Galápagos, Ecuador*. Gelatin silver print, 2004.© 2026 DACS / **181** Albert Renger-Patzsch, *Snake Head*. Gelatin silver print, 1927. © 2026 Albert Renger-Patzsch /Archiv Ann und Jürgen Wilde/DACS / **182** Martin Johnson Heade, *Cattleya Orchid and Three Hummingbirds*, 1871 (detail). Oil on wood, 34.8 × 45.6 (13 ¾ × 18). Gift of The Morris and Gwendolyn Cafritz Foundation. National Gallery of Art, Washington, DC (1982.73.1) / **184** Eugène Rambert and Paul Robert's *Les Oiseaux dans la Nature*, Paris, Baillière, Lebet, 1880 / **185** Mrs N. E. Jones and Howard Jones's *Illustrations of the nests and eggs of birds of Ohio*, Circleville, Ohio, USA, 1886 / **186a** John Atkinson Grimshaw, *Thrush's Nest, Primroses, Pear Blossom*, 1862. Oil on board, 30.5 × 40.5 (12 ⅛ × 16). Private Collection / **186b** Detroit Publishing Company postcards, 1898–1931. The Miriam and Ira D. Wallach Division of Art, Prints and Photographs. The New York Public Library (b12647398) / **187** Caspar David Friedrich, *Owl in a Gothic Window*, *c.* 1837. Sepia and graphite on paper, 37.8 × 25.6 (15 × 10 ⅛). State Hermitage Museum, St Petersburg (OR-43908). Photo Album/Alamy / **188** Herbert Ponting, *Adélie Penguin and Chicks, Antarctica*, *c.* 1911. Gelatin silver print, 38.2 × 29 (15 ⅛ × 11 ½). Private Collection / **189** Fiona Pardington, *Sweet Kiwi, Whanganui Museum*, 2008. Gelatin silver print. Courtesy of Fiona Pardington and Starkwhite / **190a** Walther Klemm, *Turkeys*, colour woodcut, 1907. *Deutsche Kunst und Dekoration. Illustr. Monatshefte für moderne Malerei, Plastik, Architektur, Wohnungskunst u. künstlerisches Frauen-Arbeiten*, 1907. Heidelberg University Library (20.1907) / **190b** Vulture amulet, Egypt, 400–30 BCE. Glass, 2.6 × 2.3 (1 ⅛ × 1). Gift of J. Pierpont Morgan, 1917. The Metropolitan Museum of Art, New York (17.194.2483) / **191** Habiballah of Sava, *The Concourse of the Birds*, *c.* 1600. Farīd al-Dīn ʿAṭṭār's *Mantiq al-Tayr* (*Language of the Birds*), Iran, Isfahan, *c.* 1600. Fletcher Fund, 1963. The Metropolitan Museum of Art, New York (63.210.11) / **192** John Constable, *The Skylark, Dedham*, *c.* 1830. Oil on board, 23.9 × 20.3 (9 ½ × 8). Private Collection / **193a** Claude Monet, *The Magpie*, 1868–69. Oil on canvas, 89 × 130 (35 ⅛ × 51 ¼). Musée d'Orsay (RF 1984 164) / **193b** Lois Dodd, *Chickens*, 1957–58. Oil on linen, 106.7 × 137.2 (42 × 54). © 2026 Lois Dodd/VAGA at ARS, NY and DACS, London / **194** Robert Fulton's *The Illustrated Book of Pigeons*, London, Paris & New York, Cassell, Petter & Galpin, 1876 / **195** Albrecht Dürer, *Three Studies of a Bullfinch*, watercolour and gouache on paper, 1543. Real Monasterio de San Lorenzo de El Escorial, Madrid. Photo Fine Art Images/Heritage Images via Getty Images / **196** Rosa Bonheur, *The Wounded Eagle*, *c.* 1870. Oil on canvas, 147.6 × 114.6 (58 ⅛ × 45 ⅛). Gift of Jane and Justin Dart. Los Angeles County Museum of Art (M.78.37) / **197a** Pair of earflares with condors, Moche, Peru, 6th–7th century. Silver, gold, gilded copper, shell, diam. 10 (4). The Michael C. Rockefeller Memorial Collection, Bequest of Nelson A. Rockefeller, 1979. The Metropolitan Museum of Art, New York (1979.206.1245, .1246) / **197b** Diego Rivera, *Vultures on Cactus*, 1930. Watercolour over black chalk on Japan paper, 42.9 × 32.4 (17 × 12 ⅞). Bequest of Robert H. Tannahill. Detroit Institute of the Arts (70.331). © 2026 Banco de México Diego Rivera Frida Kahlo Museums Trust, Mexico, DF /DACS / **198–9** Hiroshi Sugimoto, *Hyena-Jackal-Vulture*, 1976. Gelatin silver print, 119.4 × 149.2 (47 × 58 ¾). © Hiroshi Sugimoto. Courtesy the Artist and Lisson Gallery / **200** Unknown photographer, *Untitled (Eagle Facing Left)*, *c.* 1850. Daguerreotype, 8.3 × 7 (3 ¼ × 2 ¾). Gift of the Hall Family Foundation. The Nelson-Atkins Museum of Art, Kansas City (2010.72.2.). Image courtesy of Nelson-Atkins Digital Production & Preservation / **201** *Illustrations of the American Ornithology of Alexander Wilson and Charles Lucian Bonaparte*, Edinburgh, Frazer & Co., 1835 / **202al** John Gould's *Monograph of the Ramphastidae, or family of Toucans*, London, the author, 1834 / **202ar** W. T. Greene, *Parrots in Captivity*, London, George Bell and Sons, 1884–87 / **202b** Henri Rousseau, *The Flamingos*, 1910. Oil on canvas, 113.8 × 162 (44 ⅞ × 63 ⅞). Private Collection / **203** John James Audubon, *American Flamingo*, 1838. Hand-coloured engraving and aquatint on Whatman wove paper, 87.6 × 58.5 (34 ½ × 23 ⅛). Gift of Mrs. Walter B. James. National Gallery of Art, Washington, DC (1945.8.431) / **204** William Degouve de Nuncques, *The Black Swan*, 1895. Pastel on cardboard, 38 × 47 (15 × 18 ⅝). Kröller-Müller Museum, Otterlo (KM 107.446). Photo A. Burkatovski/Fine Art Images/Fine-Art-Images/SuperStock / **205** Thomas Skottowe and T. R. Browne's *Select Specimens from Nature of the Birds Animals &c &c of New South Wales*, Newcastle, New South Wales, 1813. State Library of New South Wales, Sydney (423725) / **206** Cherry Kearton, *Kittiwakes at Home*, negative 1896, print 1905. Hand-coloured gravure half-tone print, 23 × 19.5 (9 ⅛ × 7 ¾). J. Paul Getty Museum, Los Angeles (84.XM.500.1) / **207a** Louis Agassiz Fuertes's *Album of Abyssinian Birds and Mammals*, Chicago, Field Museum of Natural History, 1930 / **207b** Morris Graves, *Shore Birds*, 1939. Gouache and watercolour on paper, 65.4 × 78.1 (25 ¾ × 30 ¾). Private Collection. © Morris Graves Foundation / **208** Indian School (possibly by Mansur), *Kingfisher*, *c.* 1620. Opaque watercolour and gold on paper, 24.7 × 16.2 (9 ¾ × 6 ⅜). Photo 2026 Museum of Fine Arts, Boston. All rights reserved/Museum purchase with funds from the Francis Bartlett Donation of 1912 and by contribution/Bridgeman Images / **209** Tsuruzawa Tansaku Morihiro, *Crane and Waves*, Edo period, latter half of the 18th century. Hanging scroll, ink and colour on silk, 84 × 85.6 (33 ⅛ × 33 ¾). Photo 2026 Museum of Fine Arts, Boston. All rights reserved/Bridgeman Images / **210** Ohara Koson (Shoson), *Diving Mallard*, *c.* 1910. Woodblock colour print, 24.1 × 25.1 (9 ½ × 10). Gift of the Estate of Dr. Eleanor Z. Wallace. Brooklyn Museum (2007.32.84) / **211a** Nakamura Hōchū, *Five Plovers and Waves*, from the *Kōrin Gafu* ('Kōrin Picture Album'), 1826. Woodblock print, ink and colour on paper, 25.6 × 36.8 (10 ⅛ × 14 ½). Rijksmuseum, Amsterdam (RP-P-1951-932-23) / **211b** Yamamoto Masao, #1664, *Kawa=Flow* series, 2016. Gelatin silver print. © Yamamoto Masao / **212** Italian Baroque pietra dura panel, attributed to the Opificio delle Pietre Dure Florence, 17th century. Wood, stone, fabric, 28 × 23 (11 ⅛ × 9 ⅛). Private Collection / **213a** Al-Qazwini's *Ajāʾib al-Makhlūqāt va-Gharāʾib al-Mawjūdāt* ('The Wonders of Creation'), 17th century. University of St Andrews Library (ms32(o)) / **213b** Eagle relief, Toltec, 900–1200 CE. Andesite or dacite, Maya blue, stucco, red pigment, 62.2 × 77.5 × 6.4 (24 ½ × 30 ⅝ × 2 ⅝). Gift of Frederic E. Church, 1893. The Metropolitan Museum of Art, New York (93.27.2). / **214a** Ibn al-Muqaffa's *Kalila wa-Dimna*, 1301–1400. Bibliothèque nationale de France, Paris (Arabe 3465) / **214b** Mark Catesby's *The Natural History of Carolina, Florida and the Bahama Islands*, London, Charles Marsh, Thomas Wilcox, and Benjamin Stichall, 1754 / **215** William James Webbe, *The White Owl*, 1856. Oil on board, 45 × 26.3 (17 ¾ × 10 ⅜). Private Collection / **216** François Le Vaillant's *Histoire naturelle des oiseaux de paradis et des rolliers*, Paris, Denné le jeune and Perlet, 1806 / **217** Latifa Medjdoub, *Laughing Cassowary*, Valenciennes, France, 2024. Interlaced wool, cotton, rayon and acrylic on frame, 28 × 28 (11 ⅛ × 11 ⅛). Photo Sofia Downes. Courtesy Latifa Medjdoub / **218** Andrés Silva Vignoli, *Pájaros en la Cabeza* ('The Birds in my Head'), 2025. Watercolour and oil on linen, each panel 80 × 60 (31 ½ × 23 ⅝). © Andrés Silva Vignoli. Courtesy of the Artist. / **219** Edith A. Ibbs, *A Peacock in a Crowned Alchemical Flask*, 1900–1909. Watercolour, 41.8 × 19.6 (16 ½ × 7 ¾). Wellcome Collection, London (38825i) / **220a** Étienne-Jules Marey, *Vol de Héron* ('A Flight of Herons'), *c.* 1883. Albumen print, 11.7 × 17.5 (4 ⅝ × 7). Private Collection / **220c** Étienne-Jules Marey, *Heron in Flight*, 1887. Gelatin silver print, 17 × 36.2 (6 ¾ × 14 ⅜). Purchased with the support of SuitSupply. Rijksmuseum, Amsterdam (RP-F-2015-41) / **220b** Étienne-Jules Marey, *Pelican in Flight*, *c.* 1882. Gelatin silver print. Private Collection / **221** Harold Edgerton, *Rising Dove*, *c.* 1934. Harold E. Edgerton Collection. MIT Museum, Cambridge, MA (HEE-NC-34002) / **223** Terri Weifenbach, *Des Oiseaux*, 2018. © Terri Weifenbach / **224** Alexey Savrasov, *Evening, Migration of Birds*, 1874. Oil on canvas, 82.5 × 65.8 (32 ½ × 26). Odesa Fine Arts Museum / **226** Simen Johan, *Untitled #174*, 2013. Digital C-print, 147.3 × 177.8 (58 × 70). © Simen Johan / **228a** *The Rochester Bestiary*, *c.* 1230. The British Library, London (Royal MS 12 F XIII) / **228b** Anna Pugh, *Mole in a Hole*, 2016. Acrylic, 22.2 × 46.3 (8 ¾ × 18 ¼). © Anna Pugh. All rights reserved 2026/Bridgeman Images / **229** Abu'l Hasan, *Squirrels in a Plane Tree*, Mughal, 1605–8. Opaque watercolour, 36.2 × 22.5 (14 ⅜ × 8 ⅞). The British Library, London (Johnson.1,30) / **230** Unknown artist, *Bats and Sycamore Leaves*, Japan, *c.* 1900. Woodblock print, 31.8 × 24.1 (12 ⅝ × 9 ½). The Peggy Macdowell Thomas Japanese Print Collection. Taubman Museum of Art, Roanoke, VA (1996.089) / **231a** Takahashi Biho, *Bat in Moon*, *c.* 1905. Colour woodblock print, 23.5 × 24.3 (9 ⅜ × 9 ⅝). Gift of the Estate of Dr. Eleanor Z. Wallace, Brooklyn Museum (2007.32.4) / **231b** Bhawani Das (attr.), *Great Indian Fruit Bat*, *c.* 1777–82. Pencil, ink and opaque watercolour on paper, 59.7 × 83.2 (23 ⅝ × 32 ⅞). Purchase and gifts from multiple donors and bequests, 2008. The Metropolitan Museum of Art, New York (2008.312) / **232a** Unknown artist, probably Sawar school, *Dog with pups*, *c.* 1780, India, 18th century. Gum tempera and ink on paper, 22 × 16.5 (8 ¾ × 6 ½). Gift of Mr. and Mrs. William E. Ward in memory of her father, Charles Svec. The Cleveland Museum of Art (1969.77) / **232b** Henri de Toulouse-Lautrec, *Little Dog*. Oil on panel, 1888. Private Collection. Photo Alamy / **233** Franz Marc, *Four Foxes*, 1913. Watercolour and blue chalk on paper, 44.5 × 38.7 (17 ½ × 15 ¼). Jane and Roger Wolcott Memorial, Gift of Thomas T. Solley, Eskenazi Museum of Art, Indiana University (75.21). Photo Eskenazi Museum of Art/Kevin Montague / **234** Francisco de Goya, *Half-Submerged Dog*, 1820–23. Wall covering transferred to canvas, 131.5 × 79.2 (51 ⅛ × 31 ¼). Museo del Prado (P000767) / **235a** Bill Traylor, *Black Dog Running*, *c.* 1939–42. Poster, paint and graphite on cardboard, 35.6 × 38.1 (14 × 15). Private Collection. Photo Courtesy Just Folk / **235b** Frederic Remington, *Moonlight, Wolf*, *c.* 1904. Oil on canvas, 51 × 66 (20 ⅛ × 26). Gift of the members of the Phillips Academy Board of Trustees. Addison Gallery of American Art, Andover, MA (1956.2) / **236a** Jocelyn Lee, 'Deer at Night, Maine', from *The Physical World* series, 2002. Archival pigment print. © Jocelyn Lee / **236b** Carl Otto Czeschka, Wiener Werkstätte, *Waldidyll* ('Forest Idyll'), 1910–11. Textile fragment, linen plain weave, blockprinted, 21.5 × 38.1 (8 ½ × 15). Purchased in memory of Thomas H. Fox. Los Angeles County Museum of Art (M.2005.50). Photo Museum Associates/LACMA. © The Artist's Estate / **237** Unknown artist, He-Goats, *Livre de la Chasse* ('Book of the Hunt'), French, *c.* 1430–40. Tempera colours, gold paint, silver paint and gold leaf, leaf 26.4 × 18.4 (10 ⅜ × 7 ¼). J. Paul Getty Museum, Los Angeles (Ms. 27, fol. 12v) / **238a** Takeuchi Seiho, *Bear in Snow*, 1940. Colour woodblock print, 40.6 × 50.2 (16 × 19 ⅞). Private Collection / **238b** Carl Ederer, *A Polar Bear Seated on Snow and Another Polar Bear Walking in Background*, 1900–1930. Chromolithograph, 105.4 × 78.4 (41 ½ × 30 ⅞). Library of Congress Prints and Photographs Division Washington, DC (97519003) / **239a** *Howling Wolf*, Southern Siberia, *c.* 500–200 BCE. Wood with shell inlays, 11.1 × 3.3 × 5.9 (4 ⅜ × 1 ⅜ × 2 ⅜). John L. Severance Fund. The Cleveland Museum of Art (1986.26) / **239b** Henri and Alphonse Milne-Edwards's *Recherches pour Servir à l'Histoire Naturelle des Mammifères*, Paris, G. Masson, 1868–74 / **240** Gertrude Abercrombie, *Untitled (Lighthouse, Whale and Black Sea)*, 1948. Oil on canvas, 49.5 × 59.7 (19 ½ × 23 ½). Private Collection. © The Artist's Estate / **241a** James Stewart, *The Dugong*, *c.* 1837. Watercolour, graphite, pen and brown ink on cream wove paper, 12.5 × 18.3 (5 × 7 ¼). Paul Mellon Collection. Yale Center for British Art (B1981.25.2303) / **241b** Archeological Museum of Thebes / **242** Kyle McBurnie, *Seal in a Kelp Forest*, Cortes Bank, California, 2012. © Kyle McBurnie / **243a** *Transactions of the Zoological Society of London*, 1869. Iconographia Zoologica Print Collection. Allard Pierson, University of Amsterdam (UBA274) / **243b** Edward Roper, *A School of Porpoises in the Doldrums or Equatorial Calms*, 1870. Watercolour, 17.5 × 25 (7 × 9 ⅞). Private Collection / **244a** George Stubbs, *The Kongouro from New Holland (Kangaroo)*, 1772. Oil on canvas, 60.5 × 71.5 (23 ⅞ × 28 ¼). Acquired through purchase, gifts from multiple donors, and bequests. National Maritime Museum, Greenwich, London (ZBA5754) / **244b** John Gould's *The Mammals of Australia*, London, Taylor & Francis and the author, 1863 / **245** William Govett's *Notes and Sketches Taken During a Surveying Expedition in N. South Wales and Blue Mountains Road*, 1830–35. State Library NSW, Sydney (ZEN 0OzgzDvePm) / **246a** Joris Hoefnagel, *Two Types of Armadillos with a Marmoset, a Coconut Palm, and a Pepper Plant*, from *Animalia Qvadrvpedia et Reptilia (Terra)* series, *c.* 1575–99. Watercolour and gold paint on parchment, 14.3 × 18.4 (5 ¾ × 7 ¼). Gift of Mrs. Lessing J. Rosenwald. National Gallery of Art, Washington, DC (1987.20.6.42) / **246b** Hans Hoffmann, *A Hare in the Forest*, *c.* 1585. Oil on panel, 62.2 × 78.4 (24 ½ × 30 ⅞). J. Paul Getty Museum, Los Angeles (2001.12) / **247a** Takagi Shunzan's *Honzō Zusetsu*, Japan, Edo period, 1850s. Iwase Bunko Library, Nishio-City / **247b** Jacopo Ligozzi, *Jerboa*, 1580–1600. 26 × 34.1 (10 ¼ × 13 ½). Uffizi Galleries, Florence / **248a** Rosa Bonheur, *Wild Boars in the Snow*, *c.* 1872–77. Oil on wood panel, 21 × 30.8 (8 ¼ × 12 ⅛). Bequest of Noah L. Butkin. The Cleveland Museum of Art (1980.231) / **248b** *Piggy Bank*, Java, Majapahit Dynasty, 1300s–1400s. Terracotta, 24.2 (9 ½). John L. Severance Fund. The Cleveland Museum of Art (1980.16) / **249a** Hans Hoffmann, *Un Marcassin* ('A Wild Boar Piglet'), 1578. Watercolour and gouache on vellum, 29.7 × 45.1 (11 ¾ × 17 ⅞). Private Collection / **249b** Adriaen Verdoel, *Sleeping Pig*, 17th century. Oil on oak panel, 15 × 25.5 (6 × 10 ⅛). National Museum in Warsaw (M.Ob.350 MNW) / **250** Kevin Horan, 'Sydney', 'Jake', 'Briede', 'Sherlock', from *Chattel: A Portrait Study* series, 2012–14. Archival pigment prints.© Kevin Horan. All other rights reserved / **251a** *Miniature camelid effigy*, Inca, 1400–1535 CE. Silver, gold, copper alloy, 5.1 × 3.8 × 1.3 (2 × 1 ½ × ½). Gift and Bequest of Alice K. Bache, 1974, 1977. The Metropolitan Museum of Art, New York (1974.271.36) / **251b** Henry Moore, *Sheep in Snow Scene*, 1974. Etching and aquatint in black, 32.4 × 40.9 (12 ⅞ × 16 ⅛). Private Collection. © 2026 The Henry Moore Foundation. All Rights Reserved, DACS/www.henry-moore.org / **252a** George Catlin, *Buffalo Bulls Fighting in Running Season, Upper Missouri*, 1837–39. Oil on canvas, 60.9 × 73.7 (24 × 29). Gift of Mrs. Joseph Harrison, Jr. Smithsonian American Art Museum, Washington, DC (1985.66.424) / **252b** Samuel Palmer, *A Cow Lodge with a Mossy Roof*, *c.* 1829. Watercolour, gouache, pen and black ink on wove paper, 26.7 × 37.5 (10 ⅝ × 14 ⅞). Paul Mellon Collection. Yale Center for British Art (B1975.4.1876) / **253a** Kees Timmer, *Hyena*, 1960. Oil on canvas, 70 × 60 (27 ⅝ × 23 ⅝). Gift Joop Schafthuizen. Museum Boijmans Van Beuningen, Rotterdam. Photo Studio Tromp. © The Artist's Estate / **253b** C. F. A. Kolb's *Wilde Thiere aller Länder und Zonen*, Esslingen, J. F. Schreiber, 1877 / **254** George Stubbs, *Whistlejacket*, *c.* 1762. Oil on canvas, 292 × 246.4 (115 × 97 ⅛). The National Gallery, London (NG6569) / **255** Mochizuki Gyokusen, *Black Bull*, second half 19th century. Two-panel folding screen, ink, colour and gold on paper, 153 × 170 (60 ¼ × 67). The John R. Van Derlip Fund; purchase from the collection of Elizabeth and Willard Clark. Minneapolis Institute of Art (2013.31.48) / **256a** Shaykh Muhammad, *Camel and Keeper*, 1556–57. Ink, opaque watercolour and gold on paper, 8.8 × 11.2 (3 ½ × 4 ½). Purchase, Charles Lang Freer Endowment. Freer Gallery of Art Collection. National Museum of Asian Art, Smithsonian Institution, Washington, DC (F1937.21) / **256b** Franz Marc, *The Blue Horses (The Horses in the Pasture)*, 1910. Tempera on paper, 61.5 × 82 (24 ¼ × 32 ⅜). Musée des Beaux-Arts de Liège (10129237) / **257a** Paris Petridis, *Bethlehem*, 2012. © Paris Petridis / **257b** Hilma af Klint, *Untitled*, 15 February 1932. Watercolour on paper, 31.5 × 47.5 (12 ½ × 18 ¾). The Hilma af Klint Foundation, Stockholm / **258a** Engraving of a group of oryx and addax antelopes, ostriches, giraffes, dogs and characters, Uweinat massif, Sudan (Karkur-Talh or Karkur Talh), Prehistory, 8000–3000 BCE. Photo Jean Bernard. All rights reserved 2026/Bridgeman Images / **258b** Unknown artist, Kunwinjku people, *Wambiddyer* ('Anteater'), Western Arnhem Land, Northern Territory, Australia, *c.* 1912. Natural earth pigments on eucalyptus bark, 49.8 × 50 (19 ⅝ × 19 ¾). Acquired with the Founding Donors' Fund 2010. National Gallery of Australia, Canberra (202465) / **259a** *Small bowl with mice*, Nazca, 100–500 CE. Ceramic with white and black slip, h. 5 (2), diam. 12 (4 ¾). Museum purchase, Peter Jay Sharp, Class of 1952, Fund. Princeton University Art Museum (2019-252) / **259b** Photo someshver/Shutterstock / **260** Photo © Sandro Vannini/Laboratoriorosso / **261a** Unknown artist, *Four Lions*, Safavid, Persia, 17th century. Ink and pencil heightened with colour on paper. Private Collection / **261b** Rembrandt van Rijn, *Study of a Lion at Rest*, *c.* 1645–50. Pen and brown ink, brown wash, 13.8 × 20.7 (5 ½ × 8 ¼). Musée du Louvre, Paris (RF 4721) / **262–3** Candida Höfer, *Zoologischer Garten Washington DC IV 1992*. C-print, 47 × 64.7 (18 ⅝ × 25 ½). © 2026 Candida Höfer/VG Bild-Kunst, Bonn and DACS, London / **264a** Ishikawa Mōkō, after Albrecht Dürer, *Rhinoceros*, Japanese, late Edo period, end of 18th century. Hanging scroll, ink and colour on silk, 40 × 49.5 (15 ¾ × 19 ½). William Sturgis Bigelow Collection. Museum of Fine Arts, Boston (11.8118) / **264b** Asher Brown Durand (attr.), *Banknote Vignette Showing a Zebra*, *c.* 1824–37. Etching, 7.2 × 8.2 (2 ⅞ × 3 ¼). Gift of Joseph Veach Noble, 2002. The Metropolitan Museum of Art, New York (2002.333.8) / **265a** Artist unknown, *Hippopotamus*, New Kingdom, Egypt, *c.* 1479–1425 BCE. Limestone, paint, 10.8 × 12 (4 ¼ × 4 ¾). Rogers Fund, 1923. The Metropolitan Museum of Art, New York (23.3.6) / **265b** Rajasthan School, *An Elephant Running Amok*, India, *c.* 1800. Gouache on paper, 16.7 × 24.4 (6 ⅝ × 9 ⅝). The British Museum, London (1921.0719.0.3) / **266a** Antoine-Louis Barye, *Rolling Tiger*, *c.* 1840. Watercolour and opaque watercolour on wove paper, 23.7 × 28.7 (9 ⅜ × 11 ⅜). Purchased by Special Subscription. Brooklyn Museum (10.96) / **266b** Photo Courtesy Jean Clottes / **267** Leo, or Sinhm, the zodiac sign, shown as a tiger in a cave, Kangra, *c.* 1800. Gouache with gold on paper, 18.8 × 13.5 (7 ½ × 5 ⅜). Presented by Wayne and Nancy Hunnicutt, 2007. Ashmolean Museum, University of Oxford (EA2007.255) / **268** Utagawa Kunimaro, *A Leopard Drawn from Life*, Japan, Edo period, 1860. Ink and colour on paper, 35.8 × 24.6 (14 ⅛ × 9 ¾). The Anne van Biema Collection. National Museum of Asian Art, Smithsonian Institution, Washington, DC (S2004.3.204) / **269a** Tracey Emin, *Teacup and Pancake*, Blue Cross Christmas card, 2025. © 2026 Tracey Emin. All rights reserved, DACS / **269b** Duncan Grant, *The Cat, Opussyquinusque*, *c.* 1932. Oil on board, 23 × 31 (9 ⅛ × 12 ¼). Collection of the Charleston Trust, Lewes (CHA/P/1). © 2026 Estate of Duncan Grant. All rights reserved, DACS / **270a** Aloys Zötl, *The Gibbons*. Watercolour, 1833. Private Collection / **270b** Andy Warhol, *Orangutan*, from the *Endangered Species* series, 1983. Colour screenprint on Lenox Museum Board, 96.8 × 96.8 (38 ⅛ × 38 ⅛). Private Collection. © 2026 The Andy Warhol Foundation for the Visual Arts, Inc./ Licensed by DACS, London / **271a** Edward Lear, *Vitoe. Nyctipithecus Infulatus*, 1846. John Edward Gray's *Gleanings from the Menagerie and Aviary at Knowsley Hall*, Knowsley, 1846 / **271b** Henri Rousseau, *Tropical Forest with Monkeys*, 1910. Oil on canvas, 129.5 × 162.5 (51 × 64). John Hay Whitney Collection. National Gallery of Art, Washington, DC (1982.76.7) / **272** Francis Bacon, *Skull of a Gorilla*, 1957. Oil on canvas, 152.1 × 118.1 (60 × 46 ½). Fractional gift from Jan Shrem and Maria Manetti Shrem. San Francisco Museum of Modern Art. Photo Ben Blackwell. © 2026 The Estate of Francis Bacon. All rights reserved. DACS / **273** William Walls, *Baby Gorilla*, 1900. Oil on canvas, 28.4 × 22.8 (11 ¼ × 9). Courtesy Zoological Society of London / **275** Félix Vallotton, *L'Arc-en-ciel* ('Rainbow'), *c.* 1909. Oil on cardboard, 34.7 × 28 (13 ¾ × 11 ⅛). Private Collection / **276** George Frederic Watts, *After the Deluge*, 1885–86. Oil on canvas, 104 × 178 (41 × 70 ⅛). Watts Gallery, Guildford. Photo akg-images.

Index

Page numbers in **bold** refer to illustrations and their accompanying captions.

Acknowledgments

This book is dedicated to Jackie, and to the feral wildlife of London SE16 in all its unruly variety – pigeons, crows, waterfowl, Rallidae riff-raff, cable-chewing squirrels et al. I would never have found my way back to the living world without you.

One day, we will land on solid ground again.

I would like to thank Robert Macfarlane, Sarah Perry, Harriet Rix, Helen Scales and Stephen Moss for their wonderful, beautifully judged texts.

My thanks also go to Jane Laing, Florence Allard, Sadie Butler, Lise Seguin and everyone at Thames & Hudson involved in the realization of this book. I am incredibly grateful to all of you for your indispensable contributions, insight, support and patience over the past few years. Additionally, thank you to Blanche Craig and Harry Allard for their invaluable research and advice.

Finally, I would like to express my deep gratitude to all the artists, galleries, museums, institutions, collectors and estates who have so generously granted us permission to feature their work – *Ark* would not exist without them.

About the Author

Renowned image alchemist Stephen Ellcock is a London-based curator, writer, researcher and online collector of images who has spent the last decade creating an ever-expanding virtual museum of art that is open to all via social media. His ongoing attempt at creating the ultimate social media 'Cabinet of Curiosities' has so far attracted more than 650,000 followers worldwide.

He is also the author of *Elements*, *Underworlds*, *The Cosmic Dance*, *All Good Things*, *The Book of Change*, *England On Fire*, with text by Mat Osman, and *Jeux de Mains* and *Clin D'Oœil*, collaborations with Cécile Poimboeuf-Koizumi.

About the Contributors

Robert Macfarlane is Professor of Literature and the Environmental Humanities at the Faculty of English at the University of Cambridge, and a fellow of the Royal Society of Literature. His writing about landscape, nature, memory, language and travel has won many prizes internationally, including the E. M. Forster Award for Literature from the American Academy of Arts and Letters in 2017. His books have been widely adapted for TV, film and radio and he has collaborated with artists, film-makers, photographers and musicians, including Johnny Flynn and Jackie Morris.

Sarah Perry is the author of the bestselling novels *Melmoth*, *The Essex Serpent* and *After Me Comes the Flood*, and non-fiction titles, *Essex Girls* and *Death of an Ordinary Man*. *The Essex Serpent* was awarded Waterstones Book of the Year and the British Book Awards Book of the Year and, in 2022, was adapted for Apple TV+. She is Chancellor of the University of Essex and a fellow of the Royal Society of Literature.

Harriet Rix is a tree science consultant and author of *The Genius of Trees: How Trees Mastered the Elements and Shaped the World*. Harriet has worked at the Tree Council and as secretary of Hedgelink, the national advisory board on hedgerows. Her writing and photography have been published in the *Financial Times*, *London Review of Books* and *Times Literary Supplement*.

Helen Scales is a marine biologist, broadcaster and author of many books, including *The Shell Spotter's Guide*, *What the Wild Sea Can Be* and *Spirals in Time*, as well as several titles for children, such as *Shells... and What They Hide Inside* and *Oceanopedia*. She writes regularly for *The Guardian* and *National Geographic*, and is a storytelling ambassador for the Save Our Seas Foundation.

Stephen Moss is one of Britain's leading nature writers and broadcasters. He was the original producer of the BAFTA award-winning *Springwatch*. His bestselling books include a series of 'bird biographies' and *Ten Birds that Changed the World*. A lifelong naturalist, Stephen has travelled to all seven of the world's continents to watch wildlife.

FRONT COVER AND PAGE 4 Jethro Buck's *Genesis* (2015) depicts the many realms of life on earth. The original image is featured on page 4, and has been remodelled for use on the cover. Courtesy Jethro Buck.

SPINE, BACK COVER AND ENDPAPERS Naruto Strait's whirlpools are a major tourist attraction in Japan. Their dramatic vortex, which can stretch up to 30 metres (10 feet) in diameter, is captured in Utagawa Hiroshige's *Awa Province: Naruto Whirlpools* (1855) (detail). Colour woodblock print. Gift of the Rodbell Family Collection. Clark Art Institute (2014.16.6).

PAGE 2 After the deluge receeds, Noah and his animals emerge from the Ark to repopulate earth, as depicted in this painting of around 1628 by Roelandt Savery.

PAGE 20 For her Petri dish art, including this work *Orange and Blue* (2021), Dasha Plesen cultivates bacteria and mould, using manufactured additions, such as sprinkles or pigments, to intervene in their growth and appearance.

PAGE 86 Portuguese man o' wars (*Physalia arethusa*) are siphonophores, meaning that they are not a single animal, but colonial organisms comprised of several genetically identical units called 'zooids' that all contribute to a different aspect of the whole 'animal'.

PAGE 182 North American artist Martin Johnson Heade travelled to Central and South America several times between 1860 and 1870. The tropical flowers and birds he observed on these trips were the basis of many paintings, including *Cattleya Orchid and Three Hummingbirds* (1871) (detail).

PAGE 226 ◎ The Golden snub-nosed monkey (*Rhinopithecus roxellana*), here captured by Simen Johan in *Untitled #174* (2013), is endemic to a small area of mountainous forests in China. They can be found at altitudes as high as 3,000 metres (10,000 feet) above sea level, where temperatures reach as low as -8.3 degrees Celsius.

First published in the United Kingdom in 2026 by
Thames & Hudson Ltd, 6–24 Britannia Street, London WC1X 9JD

First published in the United States of America in 2026 by
Thames & Hudson Inc., 500 Fifth Avenue, New York, New York 10110

Cover designed by Daniel Streat, Visual Fields, and Daniele Roa
Interior layout designed by Daniel Streat, Visual Fields

EU Authorized Representative: Interart S.A.R.L.
19 rue Charles Auray, 93500 Pantin, Paris, France
productsafety@thameshudson.co.uk
interart.fr

A CIP catalogue record for this book is available from the British Library

Library of Congress Control Number 2026937554

ISBN 978-0-500-02964-0
01

Printed and bound in China by C&C Offset Printing Co. Ltd

MIX
Paper | Supporting responsible forestry
FSC www.fsc.org FSC® C008047

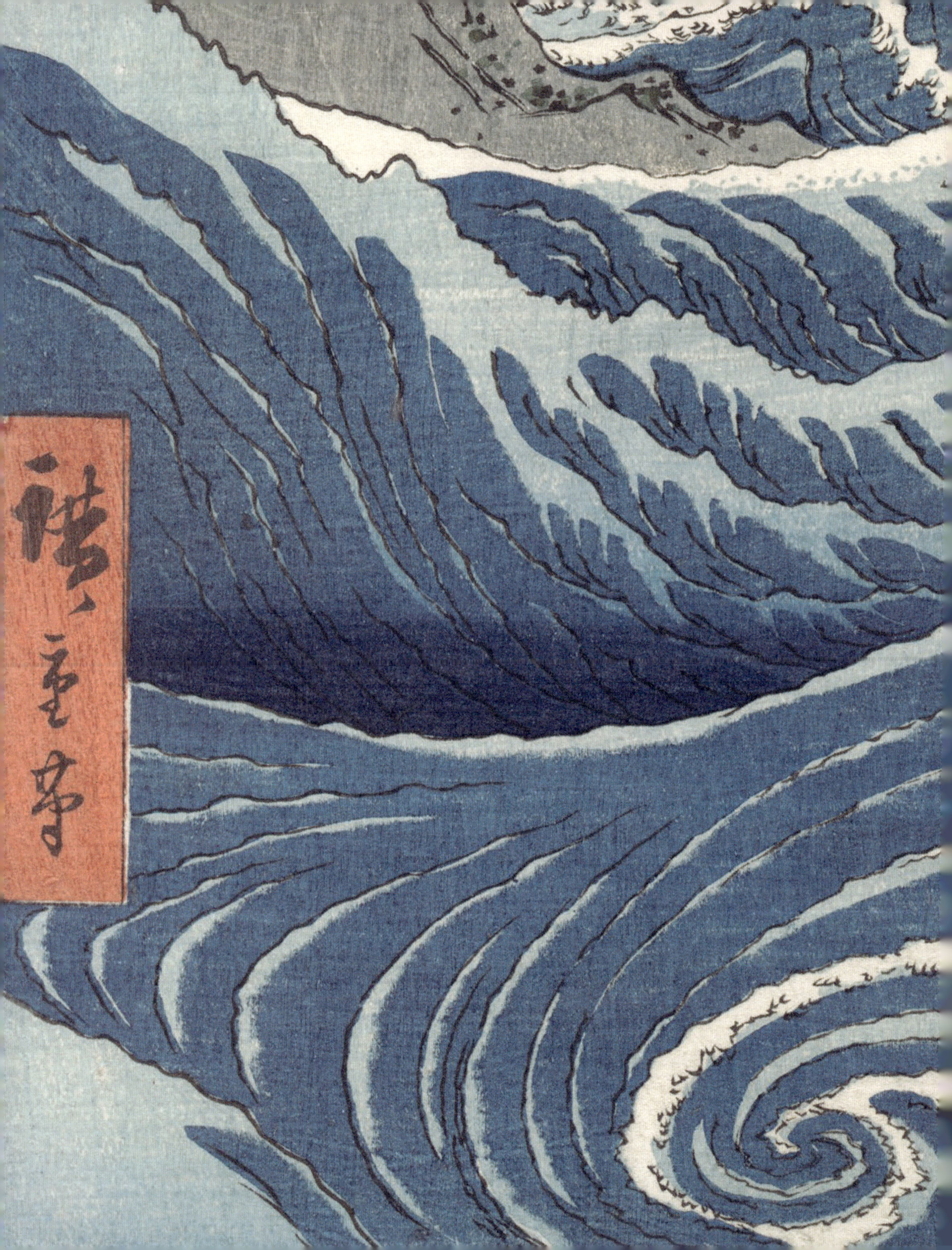
廣重筆